Physics for Engineering and Science

Physics for Engineering and Science

Fourth Edition

Michael E. Browne, PhD
Professor of Physics
University of Idaho

Schaum's Outline Series

New York   Chicago   San Francisco   Athens
London   Madrid   Mexico City   Milan
New Delhi   Singapore   Sydney   Toronto

MICHAEL E. BROWNE is a graduate of the University of California, Berkeley, where he received BS (engineering) and PhD (physics) degrees. He was a research scientist and manager at the Lockheed Research Laboratories, subsequently serving as professor of physics and chairman of the physics department at the University of Idaho. He has conducted research in solid-state physics and in science education, his most recent activity. His teaching experience ranges from introductory physics to graduate classes. He has published numerous research articles and coauthored two textbooks.

Copyright © 2020, 2013, 2010, 1999 by McGraw-Hill Education. All rights reserved. Printed in the United States of America. Except as permitted under the Copyright Act of 1976, no part of this publication may be reproduced or distributed in any form or by any means, or stored in a database or retrieval system, without the prior written permission of the publisher.

1 2 3 4 5 6 7 8 9   LOV   24 23 22 21 20 19

ISBN   978-1-260-45383-6
MHID   1-260-45383-9

e-ISBN   978-1-260-45384-3
e-MHID   1-260-45384-7

McGraw-Hill Education, the McGraw-Hill Education logo, Schaum's and related trade dress are trademarks or registered trademarks of McGraw-Hill Education and/or its affiliates in the United States and other countries and may not be used without written permission. All other trademarks are the property of their respective owners. McGraw-Hill Education is not associated with any product or vendor mentioned in this book.

This book is dedicated to Marguerite,
whose love, encouragement, and help
made so many things possible.

Preface

This book is intended for students who are taking, or who have already taken, calculus and who may be taking a formal physics class for the first time. A background in high school-level algebra, geometry, and trigonometry is assumed. The book will be a valuable supplemental text for students presently enrolled in a calculus-based physics course for engineering and science students. It is much more compact than traditional textbooks, and it stresses problem solving and provides many worked examples. Secondly, the book will be useful as a text for a person reviewing or strengthening his or her knowledge of physics, perhaps in preparation for a professional examination. The treatment is suitable for a person wishing to study physics on his or her own without enrolling in a formal course. The narrative discussion and explanations are sufficiently clear and complete so that the book may be used either as a text, or as a supplement to a longer text.

Life science students will find the approach here valuable, since the treatment is more rigorous than that in many liberal arts textbooks. Many examples relevant to medicine are presented. Although calculus is used, it is reviewed and explained when first encountered.

The crux of learning physics is to actually work through problems. In using this book, you should be an active learner. Try to work each of the sample problems and examples as you go along. Refer to the solutions only if you get stuck. As they say in training to run a marathon, "no pain, no gain." Only in the dictionary does success come before work. Learning about physics is a fascinating endeavor, and I hope you find it as much fun as I have.

I am deeply indebted to Luanne Semler for the outstanding work she did in typing the manuscript. Judy Breedlove made valuable contributions to the graphics work, and her efforts are much appreciated. My editors at McGraw-Hill were most helpful. Arthur Biderman helped launch the project, and Mary Loebig Giles guided it through to successful completion. The second edition of this book would not have reached fruition without the dedication, effort and talent of Hannah Turner, Editorial Manager at Techset Composition Ltd, in England. She is a wonder. I owe her.

MICHAEL E. BROWNE

Contents

CHAPTER 1 REVIEW OF MATHEMATICS   1

1.1 Symbols, Scientific Notation, and Significant Figures **1.2** Algebra **1.3** Geometry and Trigonometry **1.4** Vectors **1.5** Series and Approximations **1.6** Calculus

CHAPTER 2 MEASUREMENT AND PHYSICS   13

2.1 Units **2.2** Unit Conversion **2.3** Order-of-Magnitude Estimates Supplementary Problems

CHAPTER 3 MOTION IN ONE DIMENSION   20

3.1 Displacement and Velocity **3.2** Instantaneous Velocity and Acceleration **3.3** Constant Acceleration **3.4** Freely Falling Bodies **3.5** Summary of Key Equations Supplementary Problems

CHAPTER 4 MOTION IN A PLANE   35

4.1 Position, Velocity, and Acceleration **4.2** Constant Acceleration **4.3** Projectiles **4.4** Uniform Circular Motion **4.5** Relative Motion **4.6** Summary of Key Equations Supplementary Problems

CHAPTER 5 NEWTON'S LAWS OF MOTION   52

5.1 Newton's Third Law of Motion **5.2** Newton's First Law of Motion **5.3** Newton's Second Law of Motion **5.4** Applications of Newton's Laws **5.5** Summary of Key Equations Supplementary Problems

CHAPTER 6 CIRCULAR MOTION   71

6.1 Centripetal Force **6.2** Summary of Key Equations Supplementary Problems

CHAPTER 7 WORK AND ENERGY   80

7.1 Work **7.2** Kinetic Energy **7.3** Power **7.4** Summary of Key Equations Supplementary Problems

CHAPTER 8 POTENTIAL ENERGY AND CONSERVATION OF ENERGY   91

8.1 Potential Energy **8.2** Energy Conservation and Friction **8.3** Potential Energy of a Spring **8.4** Machines **8.5** Summary of Key Equations Supplementary Problems

CHAPTER 9   LINEAR MOMENTUM AND COLLISIONS   **104**

9.1 Linear Momentum **9.2** Impulse **9.3** Collisions in One Dimension **9.4** The Center of Mass **9.5** Rockets **9.6** Summary of Key Equations Supplementary Problems

CHAPTER 10   ROTATIONAL MOTION   **118**

10.1 Angular Variables **10.2** Rotational Kinetic Energy **10.3** Moment of Inertia Calculations **10.4** Torque **10.5** Rolling **10.6** Rotational Work and Power **10.7** Summary of Key Equations   Supplementary Problems

CHAPTER 11   ANGULAR MOMENTUM   **131**

11.1 Angular Momentum and Torque **11.2** Precession Supplementary Problems

CHAPTER 12   STATICS AND ELASTICITY   **140**

12.1 Rotational Equilibrium **12.2** Elasticity **12.3** Summary of Key Equations   Supplementary Problems

CHAPTER 13   OSCILLATIONS   **150**

13.1 Simple Harmonic Motion **13.2** Energy and SHM **13.3** SHM and Circular Motion **13.4** Pendulum **13.5** Damped Oscillations and Forced Oscillations **13.6** Summary of Key Equations   Supplementary Problems

CHAPTER 14   GRAVITY   **160**

14.1 The Law of Gravity **14.2** Gravitational Potential Energy **14.3** The Motion of Planets **14.4** Summary of Key Equations Supplementary Problems

CHAPTER 15   FLUIDS   **169**

15.1 Pressure in a Fluid **15.2** Buoyancy **15.3** Fluid Flow **15.4** Bernoulli's Equation **15.5** Summary of Key Equations   Supplementary Problems

CHAPTER 16   WAVES AND SOUNDS   **178**

16.1 Transverse Mechanical Waves **16.2** Speed and Energy Transfer for String Waves **16.3** Superposition of Waves **16.4** Standing Waves **16.5** Sound Waves **16.6** Standing Sound Waves **16.7** Beats **16.8** The Doppler Effect **16.9** Summary of Key Equations   Supplementary Problems

CHAPTER 17   TEMPERATURE, HEAT, AND HEAT TRANSFER   **190**

17.1 Temperature **17.2** Thermal Expansion **17.3** Heat and Thermal Energy **17.4** Heat Capacity and Latent Heat **17.5** Heat Transfer **17.6** Summary of Key Equations   Supplementary Problems

CHAPTER 18 THE KINETIC THEORY OF GASES **200**

18.1 The Ideal Gas Law 18.2 Molecular Basis of Pressure and Temperature 18.3 The Maxwell-Boltzmann Distribution 18.4 Molar Specific Heat and Adiabatic Processes 18.5 Summary of Key Equations Supplementary Problems

CHAPTER 19 THE FIRST AND SECOND LAWS OF THERMODYNAMICS **210**

19.1 The First Law of Thermodynamics 19.2 The Second Law of Thermodynamics 19.3 The Carnot Engine 19.4 The Gasoline Engine 19.5 Refrigerators and Heat Pumps 19.6 Entropy 19.7 Summary of Key Equations Supplementary Problems

CHAPTER 20 ELECTRIC FIELDS **223**

20.1 Properties of Electric Charge 20.2 The Electric Field 20.3 Motion of a Charged Particle in a Uniform Electric Field 20.4 Electric Field of a Continuous Charge Distribution 20.5 Summary of Key Equations Supplementary Problems

CHAPTER 21 GAUSS' LAW **234**

21.1 Electric Flux and Gauss' Law 21.2 Applications of Gauss' Law 21.3 Summary of Key Equations Supplementary Problems

CHAPTER 22 ELECTRIC POTENTIAL **244**

22.1 Electric Potential and Potential Energy 22.2 Electric Potential of a Point Charge 22.3 Finding the Field from the Potential 22.4 Potential of Continuous Charge Distributions 22.5 Potential of a Charged Conductor 22.6 Summary of Key Equations Supplementary Problems

CHAPTER 23 CAPACITANCE **253**

23.1 Calculation of Capacitance 23.2 Combinations of Capacitors 23.3 Energy Storage in Capacitors 23.4 Dielectrics 23.5 Summary of Key Equations Supplementary Problems

CHAPTER 24 CURRENT AND RESISTANCE **262**

24.1 Electric Current 24.2 Resistance, Resistivity, and Ohm's Law 24.3 Electric Power and Joule Heating 24.4 Summary of Key Equations Supplementary Problems

CHAPTER 25 DIRECT CURRENT CIRCUITS **268**

25.1 Resistors in Series and Parallel 25.2 Multiloop Circuits 25.3 RC Circuits 25.4 Summary of Key Equations Supplementary Problems

CHAPTER 26 MAGNETIC FIELDS **278**

26.1 The Magnetic Field 26.2 Motion of a Charged Particle in a Magnetic Field 26.3 Magnetic Force on a Current-Carrying Wire 26.4 Torque on a Current Loop 26.5 Summary of Key Equations Supplementary Problems

CHAPTER 27 SOURCES OF THE MAGNETIC FIELD **287**

27.1 Magnetic Fields due to Currents **27.2** Ampere's Law **27.3** Summary of Key Equations   Supplementary Problems

CHAPTER 28 ELECTROMAGNETIC INDUCTION AND INDUCTANCE **295**

28.1 Faraday's Law **28.2** Motional EMF **28.3** Inductance **28.4** Energy Storage in a Magnetic Field **28.5** Magnetic Materials **28.6** RLC Circuits **28.7** Summary of Key Equations   Supplementary Problems

CHAPTER 29 ALTERNATING CURRENT CIRCUITS **307**

29.1 Transformers **29.2** Single Elements in ac Circuits **29.3** The Series *RLC* Circuit and Phasors **29.4** Power in ac Circuits **29.5** Resonance in ac Circuits **29.6** Summary of Key Equations   Supplementary Problems

CHAPTER 30 ELECTROMAGNETIC WAVES **318**

30.1 Maxwell's Equations and the Wave Equation **30.2** Energy and Radiation Pressure **30.3** Polarization **30.4** Reflection and Refraction of Light **30.5** Total Internal Reflection **30.6** Summary of Key Equations   Supplementary Problems

CHAPTER 31 MIRRORS AND LENSES **329**

31.1 Plane Mirrors **31.2** Spherical Mirrors **31.3** Thin Lenses **31.4** Optical Instruments **31.5** Summary of Key Equations   Supplementary Problems

CHAPTER 32 INTERFERENCE **343**

32.1 Double Slit Interference **32.2** Multiple Slit Interference and Phasors **32.3** Interference in Thin Films **32.4** The Michelson Interferometer **32.5** Summary of Key Equations   Supplementary Problems

CHAPTER 33 DIFFRACTION **351**

33.1 Single Slit Diffraction **33.2** Resolution and Diffraction **33.3** The Diffraction Grating **33.4** Summary of Key Equations   Supplementary Problems

CHAPTER 34 SPECIAL RELATIVITY **360**

34.1 The Basic Postulates **34.2** Simultaneity **34.3** The Lorentz Transformation Equations **34.4** Time Dilation **34.5** Length Contraction **34.6** Relativistic Velocity Transformation **34.7** Relativistic Momentum and Force **34.8** Relativistic Energy **34.9** The Doppler Effect for Light **34.10** Summary of Key Equations   Supplementary Problems

CHAPTER 35 ATOMS AND PHOTONS **373**

35.1 Atoms and Photons **35.2** The Photoelectric Effect **35.3** The Compton Effect **35.4** Atomic Spectra and Bohr's Model of the Atom **35.5** Summary of Key Equations   Supplementary Problems

CHAPTER 36 QUANTUM MECHANICS   **382**

36.1 de Broglie Waves **36.2** Electron Diffraction **36.3** The Schrödinger Equation **36.4** A Particle in a Box **36.5** A Particle in a Finite Well and Tunneling **36.6** The Heisenberg Uncertainty Principle **36.7** Spin Angular Momentum **36.8** The Quantum Theory of Hydrogen **36.9** The Pauli Exclusion Principle **36.10** The Periodic Table **36.11** Summary of Key Equations Supplementary Problems

CHAPTER 37 NUCLEAR PHYSICS   **401**

37.1 Properties of the Nucleus **37.2** Nuclear Stability and Binding Energy **37.3** Radioactivity **37.4** Radioactive Decay Processes **37.5** Nuclear Reactions **37.6** Fission **37.7** Nuclear Fusion **37.8** Summary of Key Equations Supplementary Problems

APPENDIX   **416**

INDEX   **417**

*The laptop icon next to an exercise indicates that the exercise is also available as a video with step-by-step instructions. These videos are available on the Schaums.com website by following the instructions on the inside front cover.

CHAPTER 1

Review of Mathematics

Prior to studying this introductory physics course, a student should have completed high school courses in algebra, geometry, and trigonometry. Students should have, as a minimum, been studying calculus concurrently. By the second half of the course students should understand the basics of integral calculus. Most difficulties encountered in studying physics result from inadequate preparation in mathematics, so students should review this chapter if they have a weak background.

1.1 Symbols, Scientific Notation, and Significant Figures

It is important that you learn to use symbols, rather than numerical values, in doing calculations. Letters near the end of the alphabet, such as x, y, and z, are used for unknown variables. Letters such as a, b, and c are often used for given constant quantities. Greek letters are used for angular variables. Subscripts are used to provide added information. For example, the position of an object at time t_1 I label as x_1, and its position at time t_2 is x_2. Commonly encountered symbols are listed below:

$a = b$ means a is equal to b.
$a \neq b$ means a is not equal to b.
$a > b$ means a is greater than b, and $a \geq b$ means a is greater than or equal to b.
$a \gg b$ means a is much greater than b.
$a < b$ means a is less than b, and $a \leq b$ means a is less than or equal to b.
$a \ll b$ means a is much less than b.
$a \propto b$ means a is proportional to b.
$a \simeq b$ means a is approximately equal to b.
$a \sim b$ means a is of the order of magnitude of b; that is, a and b are equal to within a factor
     of 10 or so.
$n! = 1 \cdot 2 \cdot 3 \cdot 4 \cdots n$.
$y(x)$ means the quantity y depends on the value of x; that is, y is a function of x.
$\sum_i x_i = x_1 + x_2 + x_3 + \cdots + x_n$.

The Greek letter $\sum$ is used here because it corresponds to the letter S, which stands for "sum."

Very large and very small numbers are most conveniently expressed in **scientific notation**, as illustrated by the following examples:

$$10^3 = 10 \times 10 \times 10 = 1000 \qquad\qquad 10^5 = 100,000$$

$$10^{-2} = \frac{1}{100} = 0.01 \qquad\qquad 10^{-5} = \frac{1}{100,000} = 0.00001$$

$$275,000 = 2.75 \times 100,000 = 2.75 \times 10^5 \qquad 0.0032 = 3.2 \times 0.001 = 3.2 \times 10^{-3}$$

When powers of 10 are multiplied, the exponents are added, and when they are divided, the exponents are subtracted.

$$10^m \times 10^n = 10^{m+n} \quad \text{and} \quad \frac{10^m}{10^n} = 10^{m-n} \tag{1.1}$$

Thus

$$(2.75 \times 10^5)(2 \times 10^2) = (2.75 \times 2)(10^5 \times 10^2) = 5.50 \times 10^7$$

$$(4.02 \times 10^3)(3.00 \times 10^{-7}) = 12.06 \times 10^{-4}$$

$$(1.206 \times 10^1)(10^{-4}) = 1.206 \times 10^{-3}$$

Any number raised to the zero power is equal to one; that is, $x^0 = 1$.

When a number x is first raised to the nth power, and then the result is raised to the mth power, the result is $(x^n)^m = x^{nm}$. Thus, $(2^2)^3 = 4^3 = 64 = 2^6$.

When quantities are measured, there is usually some error involved. The number of digits whose values are known with certainty is the number of **significant figures**. For example, if the length of a room is measured to be 4.13 m, this means there can be uncertainty in the third decimal place; that is, the exact length may be 4.134 m, 4.133 m, 4.131 m, etc. Thus I say that 4.13 contains three significant figures. If a zero is given as the last digit to the right of the decimal point, it is significant. Thus 4.130 m contains four significant figures.

When no decimal point is given, some confusion can arise. Thus 400 contains only one significant figure. The number 4040 contains three significant figures. The number of significant figures is made clear by using scientific notation. Thus 4.00×10^2 contains three significant figures. Do not confuse significant figures with decimal places. For example, consider measurements yielding 2.46 seconds, 24.6 seconds, and 0.00246 second. These have two, one, and five decimal places, but all have three significant figures. If a number is written with no decimal point, assume infinite accuracy; for example, 12 means 12.000 . . . .

When two or more numbers are used in a calculation, the number of significant figures in the final answer is limited by the number of significant figures in the original data. For example, if the dimensions of a plot of ground are measured to be 40.2 m × 18.9 m, the area of the plot is, using your calculator, (40.2 m)(18.9 m) = 759.78 m². But since the dimensions are known to only three significant figures, the area cannot be known to more than three figures (not the five indicated by 759.78). Thus we must round the answer off to three significant figures, that is 760 m². **In general, when numbers are multiplied or divided, the number of significant figures in the answer equals the smallest number of significant figures in any of the original factors.**

Similar considerations apply when adding or subtracting numbers. The answer cannot be more accurate than any of the individual numbers added. Thus 12.0 + 1.665 + 2.0211 yields 15.6861. But 12.0 is known only to three significant figures, so the answer is not accurate to more than three significant figures, and we round it to 15.7. **When numbers are added or subtracted, the last significant figure in the answer is in the last column containing a number that results from a combination of numbers that are all significant.**

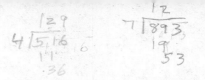

EXERCISES

Verify the following using your calculator and scientific notation:

1.1. $(2600)(0.00120)(5.11 \times 10^{-6}) = 1.59 \times 10^{-5}$

1.2. $\dfrac{1}{376} + \dfrac{4}{516} - \dfrac{7}{893} = 2.57 \times 10^{-3}$

1.3. $\left[16 + (246)^{1/2}\right]^{1/3} = 3.16$

1.4. $\dfrac{3.75 \times 10^{-5} + 0.00017}{0.0047 + (2 \times 10^{-2})^2} = 4.1 \times 10^{-2}$

1.5. $\dfrac{\left[(4)(36) - (9.3)^2\right]^{1/2}}{3.75 \times 10^{-5}} = 2.0 \times 10^{5}$

1.6. $\dfrac{(3.28 \times 10^{-2})(4.66 \times 10^{4})}{2.70 \times 10^3 + 1.60 \times 10^4} = 8.17 \times 10^{-2}$

1.7. $[16 + (246)^{1/2})]^{1/3} = 3.16$

1.2 Algebra

It will be necessary to solve equations in order to obtain an explicit expression for an unknown quantity we wish to know. The guiding principle is this: **Whatever is done on one side of an equation must be done on the other side as well.** Thus we may take the square root of both sides, or raise both sides to a power, or add or subtract the same quantity to each side, or multiply or divide each side by the same quantity. **WARNING:** Do not divide by zero. This process is not defined. For example, suppose we want to solve $x = v_0 t + \frac{1}{2} a t^2$ for a. First subtract $v_0 t$ from both sides. Then $\frac{1}{2} a t^2 = x - v_0 t$. Next multiply both sides by 2 and divide by t^2, yielding $a = 2x/t^2 - 2v_0/t$.

Manipulation of fractions frequently leads to errors. Review carefully the following:

Multiplication: $\left(\dfrac{a}{b}\right)\left(\dfrac{c}{d}\right) = \dfrac{ac}{bd}$ Example: $\left(\dfrac{2}{3}\right)\left(\dfrac{5}{7}\right) = \dfrac{10}{21}$

Division: $\dfrac{a/b}{c/d} = \dfrac{ad}{bc}$ Example: $\dfrac{2/3}{3/4} = \dfrac{(2)(4)}{(3)(3)} = \dfrac{8}{9}$

Addition: $\dfrac{a}{b} \pm \dfrac{c}{d} = \dfrac{ad \pm bc}{bd}$ Example: $\dfrac{1}{2} + \dfrac{5}{6} = \dfrac{(1)(6) + (2)(5)}{(2)(6)} = \dfrac{16}{12} = \dfrac{4}{3}$

EXERCISES

Solve for x:

1.8. $2x - 12 = 6$ Answer: $x = 9$

1.9. $3 = \dfrac{1}{1 + x}$ Answer: $x = -\dfrac{2}{3}$

1.10. $ax + 2 = bx - 6$ Answer: $x = \dfrac{8}{b - a}$

[handwritten annotations at top:] $\dfrac{3(2x+1)+2(x-2)}{(x-2)(2x+1)} = \dfrac{6x+3+4x-4}{2x^2+x-4x-2} = \dfrac{10x-1}{2x^2-3x-2}$

[handwritten:] $\dfrac{3}{x-2} = \dfrac{-2}{2x+1}$ $\dfrac{2x+1}{x-2} = -\dfrac{2}{3} \Rightarrow 2x+1 = -\dfrac{2}{3}x+\dfrac{4}{3}$

1.11. $\dfrac{3}{x-2} + \dfrac{2}{2x+1} = 0$ Answer: $x = \dfrac{1}{8}$

[handwritten:] $\dfrac{8}{3}x = \dfrac{1}{3}$ $8x = 1$ $x = \dfrac{1}{8}$

Frequently it is helpful to *factor* an equation, that is, to separate it into two or more parts that are multiplied together, as in the following examples:

Common factor: $ax + ay + az = a(x + y + z)$

Perfect square: $x^2 + 2xy + y^2 = (x + y)(x + y)$

Difference of squares: $x^2 - y^2 = (x - y)(x + y)$

Equations involving the square of the unknown variable are **quadratic equations**. Such an equation is of the form $ax^2 + bx + c = 0$. This equation can be factored into the form $(x - x_1)(x - x_2) = 0$. There are thus two solutions to a quadratic equation, $x = x_1$ and $x = x_2$, where

$$x_{1,2} = \frac{-b \pm \sqrt{b^2 - 4ac}}{2a} \tag{1.2}$$

x_1 results when the plus sign is used and x_2 results when the minus sign is used. Through physical reasoning, one can usually determine which solution is appropriate for a given problem.

EXERCISES

Solve the following quadratic equations:

[handwritten:] $(2x-5)(x-1) = x=1$, $x=\dfrac{5}{2}$

[handwritten:] $(x+4)(x-2)$, $x=-4$, $x=2$

[handwritten:] $(2x-3)(2x-3) \Rightarrow x=\dfrac{3}{2}$

1.12. $2x^2 - 7x + 5 = 0$ **1.13.** $x^2 + 2x - 8 = 0$ **1.14.** $4x^2 - 12x + 9 = 0$

Answer: $x_1 = \dfrac{5}{2}$ Answer: $x_1 = 2$ Answer: $x_1 = \dfrac{3}{2}$

$x_2 = 1$ $x_2 = -4$ $x_2 = \dfrac{3}{2}$

Frequently, more than one unknown variable is encountered. We require as many equations as we have unknowns in order to obtain a solution. Thus for two unknowns, say, x and y, we require two simultaneous equations in x and y. For three unknowns, three equations are required, etc. For example, suppose $2x + y = 1$ and $x - 2y = 8$. Solve one of these equations for one of the unknowns, say, x, and substitute this value in the second equation. If I start with $x - 2y = 8$ and add $2y$ to each side, $x = 8 + 2y$. Substitute this in the first equation. Then $2(8 + 2y) + y = 1$, or $16 + 4y + y = 1$, and $5y = -15$, so $y = -3$. Substitute this value back in one of the equations to obtain x. Thus $2x + y = 1$ yields $2x - 3 = 1$, $2x = 4$, $x = 2$.

Any number $y > 0$ can be expressed in the form $y = B^x$. Here x is the **logarithm** of the number y. B is the **base** of the logarithm. Two values of B are commonly used. For **common logarithms** (also called **base 10 logarithms**), $B = 10$ and $y = 10^x$, or $x = \log y$. The notation "log" is used to indicate base 10. We also encounter **natural logarithms**, for which $B = e = 2.718\ldots$ and $y = e^x$ or $x = \ln y$. The notation "ln" indicates the natural logarithm. Applying the rules for multiplying and dividing when using exponents yields the following relations:

$$\ln(AC) = \ln A + \ln C \quad \ln \frac{A}{C} = \ln A - \ln C \quad \ln A^n = n \ln A \tag{1.3}$$

The same rules apply for base 10 logarithms.

Calculators have keys to yield $\log y$ and $\ln y$, and inverse (or second function) keys to yield 10^x and e^x.

1.3 Geometry and Trigonometry

Some useful geometrical relations are given in Table 1-1.

TABLE **1-1**

Area of triangle of base b and height h	$\frac{1}{2}bh$
Circumference of circle of radius r	$C = 2\pi r$
Area of circle of radius r	$A = \pi r^2$
Volume of sphere of radius r	$V = 4\pi r^3/3$
Surface area of sphere of radius r	$A = 4\pi r^2$
Volume of cylinder of radius r and height h	$V = \pi r^2 h$

When two straight lines intersect, the opposite angles formed are equal (Fig. 1-1). Two angles are equal when the lines forming them are parallel (Fig. 1-2) or when the lines are mutually perpendicular (Fig. 1-3).

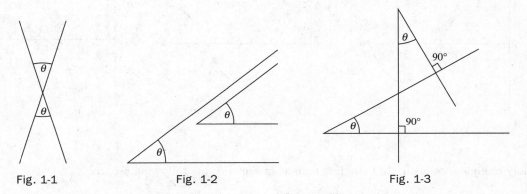

Fig. 1-1   Fig. 1-2   Fig. 1-3

The sum of the angles of a triangle is 180° (Fig. 1-4). A **right triangle** has one angle that is 90°. An **isosceles triangle** has two equal sides. An **equilateral triangle** has three equal sides, and each angle is 60°.

Two triangles are **similar** if two of their angles are equal. The sides of two similar triangles are proportional to each other; that is, $a_1/a_2 = b_1/b_2 = c_1/c_2$ (Fig. 1-5).

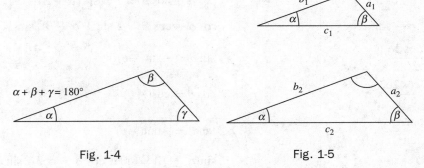

Fig. 1-4   Fig. 1-5

For a right triangle (Fig. 1-6) the following relations apply:

Pythagorean theorem $\qquad a^2 + b^2 = c^2 \qquad\qquad$ (1.4)

a = opposite side $\quad$ b = adjacent side $\quad$ c = hypotenuse

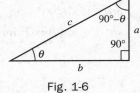

Fig. 1-6

The trigonometric functions sine, cosine, and tangent are defined as follows:

$$\sin \theta \equiv \frac{\text{side opposite } \theta}{\text{hypotenuse}} = \frac{a}{c} \quad \cos \theta \equiv \frac{\text{side adjacent to } \theta}{\text{hypotenuse}} = \frac{b}{c} \quad \tan \theta \equiv \frac{\text{side opposite } \theta}{\text{side adjacent to } \theta} = \frac{a}{b}$$

From the above definitions and the Pythagorean theorem, it follows that

$$\sin^2 \theta + \cos^2 \theta = 1 \quad \tan \theta = \frac{\sin \theta}{\cos \theta}$$

It is useful to measure angles in **radians** rather than degrees, where **2π radians** $= 360°$. Graphs of $\sin \theta$, $\cos \theta$, and $\tan \theta$ are shown in Fig. 1-7. Angles are shown in both degrees and radians.

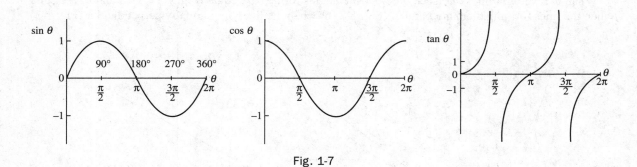

Fig. 1-7

For any triangle (not just right triangles) such as in Fig. 1-8, the following are true:

Law of sines　　　$\dfrac{\sin \alpha}{a} = \dfrac{\sin \beta}{b} = \dfrac{\sin \gamma}{c}$　　　(1.5)

Law of cosines　　$c^2 = a^2 + b^2 - 2ab \cos \gamma$　　(1.6)

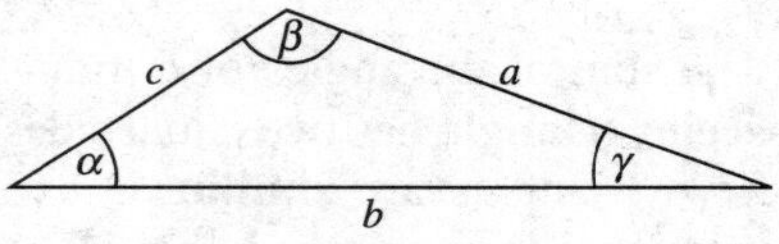

Fig. 1-8

Other useful relationships are:

$\sin 2\alpha = 2 \sin \alpha \cos \alpha$

$\cos 2\alpha = \cos^2 \alpha - \sin^2 \alpha$

$\sin^2 \alpha = \dfrac{1 - \cos 2\alpha}{2}$

$\cos^2 \alpha = \dfrac{1 + \cos 2\alpha}{2}$

$\tan 2\alpha = \dfrac{2 \tan \alpha}{1 - \tan^2 \alpha}$

$\sin \theta = \cos(90° - \theta)$

$\cos \theta = \sin(90° - \theta)$

$\sin \alpha \pm \sin \beta = 2 \sin\left[\tfrac{1}{2}(\alpha \pm \beta)\right] \cos\left[\tfrac{1}{2}(\alpha \mp \beta)\right]$

$\cos \alpha + \cos \beta = 2 \cos\left[\tfrac{1}{2}(\alpha + \beta)\right] \cos\left[\tfrac{1}{2}(\alpha - \beta)\right]$

$\cos \alpha - \cos \beta = 2 \sin\left[\tfrac{1}{2}(\alpha + \beta)\right] \sin\left[\tfrac{1}{2}(\alpha - \beta)\right]$

$\sin \theta = -\sin(-\theta)$

$\cos \theta = \cos(-\theta)$

$\tan \theta = -\tan(-\theta)$

$\sin(\alpha \pm \beta) = \sin \alpha \cos \beta \pm \cos \alpha \sin \beta$

$\cos(\alpha \pm \beta) = \cos \alpha \cos \beta \mp \sin \alpha \sin \beta$

1.4 Vectors

A **scalar quantity** is described by a single number, for example, the mass of a person or the density of water. A **vector quantity** is one that has both a **magnitude** and a **direction**. Examples of vectors are displacement, velocity, acceleration, force, and linear momentum. The magnitude of the displacement vector is called **distance**. The magnitude of the velocity vector is **speed**. For other vectors the magnitudes are not given special names. The magnitude of a vector is always a positive quantity.

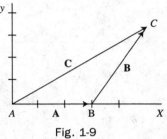

Fig. 1-9

The displacement vector can help us to understand how vectors are added. Suppose a person starts at point *A* and walks to point *B*, a distance of 3 m (Fig. 1-9). I represent this displacement by the vector **A**, an arrow three units long drawn from point *A* to point *B*. In books, vectors are indicated by boldface type. In your handwritten work, you should draw an arrow above a symbol to indicate a vector, like this: $\vec{A}$. Next the person walks 4 m from point *B* to point *C*, and this displacement is represented by the displacement vector **B**. The person could have accomplished the same net displacement by walking directly from *A* to *C*, a displacement represented by vector **C**. I say that $\mathbf{C} = \mathbf{A} + \mathbf{B}$. The magnitude of **C** (written as *C* with no boldface) can be determined by measuring the length of **C**. If this is done, one finds $C = 6\,\mathrm{m}$, not $3\,\mathrm{m} + 4\,\mathrm{m} = 7\,\mathrm{m}$, as would be the case with ordinary arithmetic. Thus when adding or subtracting vectors, one must be careful to take account of the directions of the vectors.

A vector can be drawn wherever you like on your paper. Thus we can slide vectors parallel to themselves in order to add them (Fig. 1-10). To add vectors **A** and **B**, slide them together so that they are aligned tail-head-tail-head. The vector $\mathbf{C} = \mathbf{A} + \mathbf{B}$ is then drawn from the starting point (the tail of **A**) to the final point (the head of **B**). This same procedure can be used for any vector, not just displacement vectors. The order in which the vectors are added does not matter: $\mathbf{A} + \mathbf{B} = \mathbf{B} + \mathbf{A}$. The sum of two or more vectors is called the *resultant vector* **R**. The addition of four vectors is shown in Fig. 1-11 where $\mathbf{R} = \mathbf{A} + \mathbf{B} + \mathbf{C} + \mathbf{D}$.

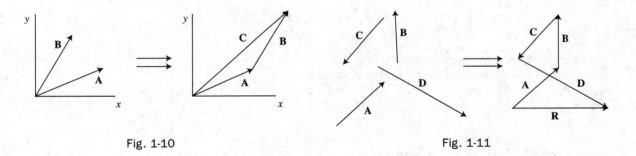

Fig. 1-10   Fig. 1-11

This graphical method of adding vectors is helpful conceptually, but to obtain accurate results, we must use trigonometry. We do this by resolving a vector into its **components**. Thus if A_x and A_y are the *x* and *y* components of **A**, $\mathbf{A} = A_x + A_y$. In Fig. 1-12, I have shown how to find the components of **A** by drawing a dashed construction line from the tip of **A** to the *x*- and *y*-axes, such that the construction line is perpendicular to the axis. In Fig. 1-13, I slid A_y to the right so that you can see $\mathbf{A} = A_x + A_y$ more clearly using the tail-head-tail-head graphical method of addition. From Fig. 1-12, I see

$$A_x = A\cos\theta, \qquad A_y = A\sin\theta, \qquad A = \sqrt{A_x^2 + A_y^2}, \qquad \tan\theta = \frac{A_y}{A_x} \qquad (1.7)$$

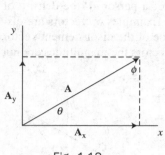

Fig. 1-12

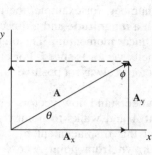

Fig. 1-13

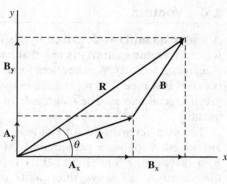

Fig. 1-14

We can use vector components to add vectors exactly (Fig. 1-14). Thus if $\mathbf{R} = \mathbf{A} + \mathbf{B}$, this requires that $R_x = A_x + B_x$ and $R_y = A_y + B_y$, where $R = \sqrt{R_x^2 + R_y^2}$ and $\tan \theta = R_y/R_x$, with θ the angle between $\mathbf{R}$ and the x-axis.

PROBLEM 1.1.   Find $\mathbf{R} = \mathbf{A} + \mathbf{B}$, given that $\mathbf{A}$ has magnitude 4 and is inclined at $30°$ above the positive x-axis and $\mathbf{B}$ has magnitude 2 and is inclined at $55°$ above the positive x-axis.

Solution   First find the x and y components of $\mathbf{A}$ and $\mathbf{B}$.

$$A_x = A \cos \theta_A = 4 \cos 30° = 3.46 \qquad A_y = A \sin \theta_A = 4 \sin 30° = 2.00$$
$$B_x = B \cos \theta_B = 2 \cos 55° = 1.15 \qquad B_y = B \sin \theta_B = 2 \sin 55° = 1.64$$
$$R_x = A_x + B_x = 4.61 \qquad\qquad R_y = A_y + B_y = 3.64$$

$$R = \sqrt{R_x^2 + R_y^2} = \sqrt{(4.61)^2 + (3.64)^2} = 5.87$$

$$\tan \theta = \frac{R_y}{R_x} = \frac{3.64}{4.61} = 0.790 \qquad \text{so } \theta = 38.3°$$

When using Cartesian coordinates, we use the unit vectors ($\mathbf{i}, \mathbf{j}, \mathbf{k}$) pointing along the ($x, y, z$) axes, respectively. Thus $\mathbf{A} = \mathbf{A_x} + \mathbf{A_y} + \mathbf{A_z} = A_x\mathbf{i} + A_y\mathbf{j} + A_z\mathbf{k}$, as shown in Fig. 1-15. If $\mathbf{R} = \mathbf{A} + \mathbf{B}$, then $\mathbf{R} = (A_x + B_x)\mathbf{i} + (A_y + B_y)\mathbf{j} + (A_z + B_z)\mathbf{k}$.

Two kinds of vector multiplication are defined. The **scalar product** (also called the **dot product**) is

$$\mathbf{A} \cdot \mathbf{B} = AB \cos \phi \qquad (1.8)$$

ϕ is the angle between $\mathbf{A}$ and $\mathbf{B}$, and it is between $0°$ and $180°$. A and B are the magnitudes of $\mathbf{A}$ and $\mathbf{B}$. The projection of $\mathbf{B}$ along $\mathbf{A}$ is $B \cos \phi$, so we can think of $\mathbf{A} \cdot \mathbf{B}$ as the product of the projection of $\mathbf{B}$ on $\mathbf{A}$ with the magnitude of A. We can also view $\mathbf{A} \cdot \mathbf{B}$ as the product (projection of $\mathbf{B}$ on $\mathbf{A}$) (magnitude of B), as illustrated in Fig. 1-16.

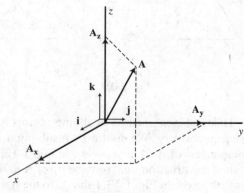

Fig. 1-15

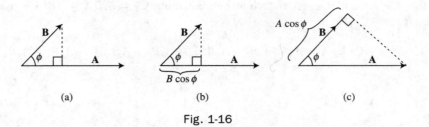

Fig. 1-16

Applying these ideas to the unit vectors (**i**, **j**, **k**) yields

$$\mathbf{i} \cdot \mathbf{i} = \mathbf{j} \cdot \mathbf{j} = \mathbf{k} \cdot \mathbf{k} = (1)(1)\cos 0° = 1$$

and

$$\mathbf{i} \cdot \mathbf{j} = \mathbf{i} \cdot \mathbf{k} = \mathbf{j} \cdot \mathbf{k} = (1)(1)\cos 90° = 0 \tag{1.9}$$

Thus, one obtains the following useful expression for $\mathbf{A} \cdot \mathbf{B}$:

$$\mathbf{A} \cdot \mathbf{B} = (A_x\mathbf{i} + A_y\mathbf{j} + A_z\mathbf{k}) \cdot (B_x\mathbf{i} + B_y\mathbf{j} + B_z\mathbf{k}) = A_xB_x + A_yB_y + A_zB_z \tag{1.10}$$

A second kind of vector multiplication is the **vector product** (also called the **cross product**), defined as

$$\mathbf{A} \times \mathbf{B} = \mathbf{C}, \quad \text{where the magnitude of } \mathbf{C} \text{ is } |\mathbf{C}| = AB\sin\phi \tag{1.11}$$

ϕ, the angle between $\mathbf{A}$ and $\mathbf{B}$, lies between $0°$ and $180°$. $\mathbf{C}$ is directed perpendicular to the plane of $\mathbf{A}$ and $\mathbf{B}$. To determine the direction of $\mathbf{C}$, use the following **right-hand rule** (see Fig. 1-17). Slide $\mathbf{A}$ and $\mathbf{B}$ together so their tails touch. Next, using your **right hand**, point your fingers along $\mathbf{A}$. Curl your fingers toward $\mathbf{B}$. Your right thumb will point up in the direction of $\mathbf{C} = \mathbf{A} \times \mathbf{B}$. $\mathbf{C}$ points in the direction a right-hand screw would advance when $\mathbf{A}$ is rotated toward $\mathbf{B}$. Note that $\mathbf{B} \times \mathbf{A} = -\mathbf{A} \times \mathbf{B}$. If $\mathbf{A}$ and $\mathbf{B}$ are parallel, $\phi = 0$ and $|\mathbf{A} \times \mathbf{B}| = 0$. If $\mathbf{A}$ and $\mathbf{B}$ are perpendicular, $\phi = 90°$, $\sin 90° = 1$, and $|\mathbf{A} \times \mathbf{B}| = AB$.

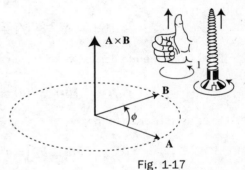

Fig. 1-17

For (**i**, **j**, **k**), we have $\mathbf{i} \times \mathbf{i} = \mathbf{j} \times \mathbf{j} = \mathbf{k} \times \mathbf{k} = 0$, $\mathbf{i} \times \mathbf{j} = -\mathbf{j} \times \mathbf{i} = \mathbf{k}$, $\mathbf{j} \times \mathbf{k} = -\mathbf{k} \times \mathbf{j} = \mathbf{i}$, and $\mathbf{k} \times \mathbf{i} = -\mathbf{i} \times \mathbf{k} = \mathbf{j}$. Using these relations, one can write $\mathbf{C} = \mathbf{A} \times \mathbf{B}$ in terms of the components of $\mathbf{A}$ and $\mathbf{B}$.

$$\mathbf{C} = \mathbf{A} \times \mathbf{B} = (A_yB_z - A_zB_y)\mathbf{i} + (A_zB_x - A_xB_z)\mathbf{j} + (A_xB_y - A_yB_x)\mathbf{k} \tag{1.12}$$

$\mathbf{A} \times \mathbf{B}$ can also be expressed as a determinant.

$$\mathbf{A} \times \mathbf{B} = \begin{vmatrix} \mathbf{i} & \mathbf{j} & \mathbf{k} \\ A_x & A_y & A_z \\ B_x & B_y & B_z \end{vmatrix} \tag{1.13}$$

To multiply this determinant, repeat the first two columns, as in Fig. 1-18. Imagine three downward sloping arrows, as shown. Multiply together the three elements on each arrow, and add them together to obtain $\mathbf{C}_1 = A_yB_z\mathbf{i} + A_zB_x\mathbf{j} + A_xB_y\mathbf{k}$. Next imagine three upward sloping arrows, as in Fig. 1-19. Again multiply together the three elements on each arrow, obtaining $C_2 = A_y B_x \mathbf{k} + A_z B_y \mathbf{i} + A_x B_z \mathbf{j}$. Subtract C_2 from C_1 to obtain

$$\mathbf{C} = \mathbf{C}_1 - \mathbf{C}_2 = \mathbf{A} \times \mathbf{B} = (A_y B_z - A_z B_y)\mathbf{i} + (A_z B_x - A_x B_z)\mathbf{j} + (A_x B_y - A_y B_x)\mathbf{k}$$

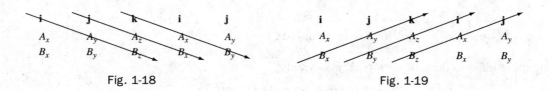

Fig. 1-18   Fig. 1-19

1.5   Series and Approximations

The **binomial expansion** is

$$(1 + x)^n = 1 + nx + \frac{n(n-1)}{2!}x^2 + \frac{n(n-1)(n-2)}{3!}x^3 + \cdots + \tag{1.14}$$

This yields the following useful approximations when $x \ll 1$:

$$(1 + x)^n \simeq 1 + nx \qquad (1 + x)^{-1} = \frac{1}{1 + x} \simeq 1 - x$$

$$(1 + x)^{1/2} \simeq \sqrt{1 + x} \simeq 1 + \frac{x}{2} \qquad (1 - x)^{-1} = \frac{1}{1 - x} \simeq 1 + x \tag{1.15}$$

If $f^{(n)}(a)$ is the nth derivative of $f(x)$ evaluated at $x = a$, then the Taylor's series is

$$f(x) = f(a) + f'(a)\frac{x - a}{1!} + f''(a)\frac{(x - a)^2}{2!} + f'''(a)\frac{(x - a)^3}{3!} + \cdots + \tag{1.16}$$

Other useful series follow:

$$(a + b)^n = a^n + \frac{n}{1!}a^{n-1}b + \frac{n(n-1)}{2!}a^{n-2}b^2 + \cdots + \tag{1.17}$$

$$e^x = 1 + x + \frac{x^2}{2!} + \frac{x^3}{3!} + \cdots + \tag{1.18}$$

$$\ln(1 \pm x) = \pm x - \frac{1}{2}x^2 \pm \frac{1}{3}x^3 - \cdots \tag{1.19}$$

In the series for $\sin x$, $\cos x$, and $\tan x$, x is in radians.

$$\sin x = x - \frac{x^3}{3!} + \frac{x^5}{5!} - \cdots - \tag{1.20}$$

$$\cos x = 1 - \frac{x^2}{2!} + \frac{x^4}{4!} - \cdots - \tag{1.21}$$

$$\tan x = x + \frac{x^3}{3} + \frac{2x^5}{15} + \cdots + \quad |x| < \frac{\pi}{2} \tag{1.22}$$

If $x \ll 1$,

$$\sin x \approx x \qquad e^x \approx 1 + x \qquad \cos x \approx 1 - \tfrac{1}{2}x^2 \tag{1.23}$$

$$\ln(1 \pm x) \approx \pm x \qquad \tan x \approx x \tag{1.24}$$

1.6 Calculus

The *derivative* of $y(x)$ with respect to x is

$$\frac{dy}{dx} = \lim_{\Delta x \to 0} \frac{\Delta y}{\Delta x} = \lim_{\Delta x \to 0} \frac{y(x + \Delta x) - y(x)}{\Delta x} \tag{1.25}$$

The derivative is the **slope** of a graph of y versus x.

Some commonly encountered derivatives follow. Here a is a constant, u and v are functions of x, and p is any number, positive or negative.

$$\frac{da}{dx} = 0 \qquad \frac{dx^p}{dx} = px^{p-1} \qquad \frac{d}{dx}\ln(x) = \frac{1}{x} \qquad \frac{d}{dx}\left(\frac{dy}{dx}\right) = \frac{d^2y}{dx^2}$$

$$\frac{d(ay)}{dx} = a\frac{dy}{dx} \qquad \frac{d}{dx}\left(\frac{1}{x}\right) = -\frac{1}{x^2} \qquad \frac{d}{dx}(\sin x) = \cos x$$

$$\frac{d}{dx}(u + v) = \frac{du}{dx} + \frac{dv}{dx} \qquad \frac{df}{dx} = \frac{df}{du}\frac{du}{dx} \qquad \frac{d}{dx}(\cos x) = -\sin x$$

$$\frac{d}{dx}(uv) = v\frac{du}{dx} + u\frac{dv}{dx} \qquad \frac{d}{dx}(e^{bx}) = be^{bx} \qquad \frac{d}{dx}(\tan x) = \frac{1}{\cos^2 x}$$

The integral of a function $y(x)$ between limits x_1 and x_2 is equal to the area under a graph of y versus x between the limits x_1 and x_2. Some useful integrals are given here.

$$\int x^n \, dx = \frac{x^{n+1}}{n+1} \text{ (provided } n \neq -1) \qquad \int e^{ax} \, dx = \frac{1}{a}e^{ax}$$

$$\int \frac{dx}{x} = \int x^{-1} \, dx = \ln x \qquad \int \ln ax \, dx = (x\ln ax) - x$$

$$\int \frac{dx}{a + bx} = \frac{1}{b}\ln(a + bx) \qquad \int xe^{ax} \, dx = \frac{e^{ax}}{a^2}(ax - 1)$$

$$\int \frac{dx}{(a + bx)^2} = -\frac{1}{b(a + bx)} \qquad \int \frac{dx}{a + be^{cx}} = \frac{x}{a} - \frac{1}{ac}\ln(a + be^{cx})$$

$$\int \frac{dx}{a^2 + x^2} = \frac{1}{a}\tan^{-1}\frac{x}{a} \qquad \int \sin ax \, dx = -\frac{1}{a}\cos ax$$

$$\int \frac{dx}{a^2 - x^2} = \frac{1}{2a}\ln\frac{a + x}{a - x}(a^2 - x^2 > 0) \qquad \int \cos ax \, dx = \frac{1}{a}\sin ax$$

$$\int \frac{dx}{x^2 - a^2} = \frac{1}{2a}\ln\frac{x - a}{x + a}(x^2 - a^2 > 0) \qquad \int \sin^2 ax \, dx = \frac{x}{2} - \frac{\sin 2ax}{4a}$$

$$\int \frac{x \, dx}{a^2 \pm x^2} = \pm\frac{1}{2}\ln(a^2 \pm x^2) \qquad \int \tan ax \, dx = -\frac{1}{a}\ln(\cos ax) = \frac{1}{a}\ln(\sec ax)$$

$$\int \frac{dx}{\sqrt{a^2 - x^2}} = \sin^{-1}\frac{x}{a}(a^2 - x^2 > 0) \qquad \int \cos^2 ax \, dx = \frac{x}{2} + \frac{\sin 2ax}{4a}$$

$$\int \frac{dx}{\sqrt{x^2 \pm a^2}} = \ln\left(x + \sqrt{x^2 \pm a^2}\right)$$

$$\int \frac{dx}{\sin^2 ax} = -\frac{1}{a}\cot ax$$

$$\int \frac{x\,dx}{\sqrt{a^2 - x^2}} = -\sqrt{a^2 - x^2}$$

$$\int \frac{dx}{\cos^2 ax} = \frac{1}{a}\tan ax$$

$$\int \frac{x\,dx}{\sqrt{x^2 \pm a^2}} = \sqrt{x^2 \pm a^2}$$

$$\int \tan^2 ax\,dx = \frac{1}{a}(\tan ax) - x$$

$$\int \frac{dx}{(a^2 + x^2)^{3/2}} = \frac{x}{a^2\sqrt{a^2 + x^2}}$$

$$\int \sin^{-1} ax\,dx = x(\sin^{-1} ax) + \frac{\sqrt{1 - a^2 x^2}}{a}$$

$$\int \frac{x\,dx}{(a^2 + x^2)^{3/2}} = -\frac{1}{\sqrt{a^2 + x^2}}$$

$$\int \cos^{-1} ax\,dx = x(\cos^{-1} ax) - \frac{\sqrt{1 - a^2 x^2}}{a}$$

$$\int x\sqrt{a^2 - x^2}\,dx = -\frac{1}{3}(a^2 - x^2)^{3/2}$$

$$\int \frac{dx}{(x^2 + a^2)^{3/2}} = \frac{x}{a^2\sqrt{x^2 + a^2}}$$

$$\int x\sqrt{x^2 \pm a^2}\,dx = \frac{1}{3}\left(x^2 \pm a^2\right)^{3/2}$$

$$\int \frac{x\,dx}{(x^2 + a^2)^{3/2}} = -\frac{1}{\sqrt{x^2 + a^2}}$$

$$\int \sqrt{a^2 - x^2}\,dx = \frac{1}{2}\left(x\sqrt{a^2 - x^2} + a^2 \sin^{-1}\frac{x}{a}\right)$$

$$\int \sqrt{x^2 \pm a^2}\,dx = \frac{1}{2}\left[x\sqrt{x^2 \pm a^2} \pm a^2 \ln\left(x + \sqrt{x^2 \pm a^2}\right)\right]$$

m and *n* are integers.

$$\int_0^\pi \sin(mx)\sin(nx)\,dx = \int_0^\pi \cos(mx)\cos(nx)\,dx = \begin{cases} \dfrac{\pi}{2} & \text{if } m = n \\ 0 & \text{if } m \neq n \end{cases}$$

$$\int_0^\pi \sin(mx)\cos(nx)\,dx = \begin{cases} \dfrac{2m}{m^2 - n^2} & \text{if } m - n \text{ is odd} \\ 0 & \text{if } m - n \text{ is even} \end{cases}$$

CHAPTER 2

Measurement and Physics

Physics is the study of all aspects of the universe. Physics is about understanding how everything works, from nuclear reactors to nerve cells to spaceships. By "understand" I mean that we can predict what will happen, given a certain set of conditions. Physics is based on experimental observations that enable us to formulate theories, which in turn enable us to predict future behaviors. It is through detection of systematic patterns in our measured observations that we are able to formulate theories. Particularly simple and widely applicable theoretical ideas are often called the **laws of nature**. These laws are not inviolate, but simply represent our best understanding of how nature behaves, given certain limiting assumptions. Typically theories are based on simplified **models** of real physical systems. Thus Newton's theory of mechanics is applicable when objects do not move very fast, but if high speeds are involved, Einstein's more accurate theory of relativity must be used. New theories in physics do not invalidate earlier ideas, but they instead extend their range of applicability. We will see that the idea of using idealized models of nature is very helpful. In this way we can avoid the overwhelming complexity that makes many physical systems intractable to mathematical analysis. For example, we will study the behavior of projectiles neglecting air effects. Although this is not realistic, it is a fairly good approximation in many cases, and it gives us good insight and understanding of what is going on.

At this time it is not completely clear just what the limits of physics are. Can we hope to understand the mind by means of physics and chemistry? How about emotions? Or religious ideas, like God and Heaven? No one knows for sure. This much we do know, however: We have come as far as we have by making use of experimental observations and measurements that form the basis for theories which are most effectively expressed using mathematics. It may even be that the physical world is simply a manifestation of mathematics. In reading a physics book, you may get the mistaken impression that everything in physics is already figured out. This is definitely not so. Unfortunately, books are written mostly about things that are known, not about the things we don't understand. There is still a lot to be learned, and there always will be.

2.1 Units

Length, mass, and time play a fundamental role in describing nature, and these are the three quantities on which we base our measurements. Length, mass, and time are called **dimensions**. We use the **International System** of units (also called the **metric system** or the **SI system**, for the French term **Système International**). In this system, the unit of **length** is the **meter** (m), the unit of **time** is the **second** (s), and the unit of **mass** is the **kilogram** (kg). Units for other quantities, such as force, energy, and power, are derived from these basic units. Mass is a measure of how much "stuff" we have. It is proportional to the weight of an object, but weight is a force and has a different meaning from mass. Do not confuse the two.

We frequently use prefixes to obtain units of a more convenient size. Here are examples of commonly encountered prefixes:

1 nanometer = 1 nm = 10^{-9} m (a little bigger than an atom diameter)
1 micrometer = 1 μm = 10^{-6} m (a human blood cell is about 7 μm)
1 millimeter = 1 mm = 10^{-3} m (a pencil lead is about 0.5 mm in diameter)
1 centimeter = 1 cm = 10^{-2} m (the diameter of a ballpoint pen)
1 kilometer = 1 km = 10^{3} m (about 0.6 mi)

1 microgram = 1 μg = 10^{-6} g = 10^{-9} kg (mass of a small dust particle)
1 milligram = 1 mg = 10^{-3} g = 10^{-6} kg (a raindrop is about 2 mg)
1 gram = 1 g = 10^{-3} kg (the mass of a penny is about 2.5 g)

1 nanosecond = 1 ns = 10^{-9} s (time for light to travel 30 cm)
1 microsecond = 1 μs = 10^{-6} s (time for a rifle bullet to travel about 1 mm)
1 millisecond = 1 ms = 10^{-3} s (about 14 ms between human heart beats)

In solving problems, be careful always to use units of meters, seconds, and kilograms, even if the problem is stated using different units (such as feet or hours or minutes).

2.2   Unit Conversion

We sometimes encounter data given in units other than those used in the SI system. In this case we must convert the units to the SI system using known conversion factors. Table 2.1 shows such factors. Suppose, for example, you wish to express a speed of 30 miles per hour in meters per second, the correct SI units. Using the conversion table, you see that

$$1 \text{ mile} = 1610 \text{ m} \quad \text{and} \quad 1 \text{ hour} = 60 \text{ min} = 3600 \text{ s}$$

Wherever "mile" appears, write 1610 m, and where "hour" appears, write 3600 s.

$$\text{Thus, } 30 \text{ mi/h} = 30\left(\frac{1610 \text{ m}}{3600 \text{ s}}\right) = 13.4 \text{ m/s}$$

PROBLEM 2.1. A paint sprayer can paint a surface at the rate of 6.00 gal/h. Express this rate in liters per minute (L/m).

Solution From Table 2.1,

$$7.48 \text{ gal} = 0.0283 \text{ m}^3 \quad \text{and} \quad 1 \text{ L} = 10^{-3} \text{ m}$$

so

$$1 \text{ gal} = \frac{0.0283}{7.48} \text{ m}^3 = \left(\frac{0.0283}{7.48}\right)(10^3 \text{ L}) \quad \text{and} \quad 1 \text{ h} = 60 \text{ min}$$

Thus,

$$6 \text{ gal/h} = 6\frac{(0.0283)}{7.48}(10^3 \text{ L})\left(\frac{1}{60 \text{ min}}\right) = 0.38 \text{ L/min}$$

PROBLEM 2.2. An acre is defined such that 640 acres = 1 mi^2. How many square meters are in 1 acre?

Solution From Table 2.1,

$$1 \text{ mi} = 1.609 \text{ km} = 1609 \text{ m}$$

$$1 \text{ acre} = \tfrac{1}{640} \text{ mi}^2 = \tfrac{1}{640}(1609 \text{ m})^2 = 4.05 \times 10^3 \text{ m}^2$$

TABLE 2-1. Conversion Factors

Length
1 inch (in) = 2.54 centimeters (cm)
1 foot (ft) = 0.3048 meter (m)
1 mile (mi) = 5280 ft = 1.609 kilometers (km)
1 m = 3.281 ft
1 km = 0.6214 mi
1 angstrom (Å) = 10^{-10} m
1 light year = 9.461×10^{15} m
1 astronomical unit (AU) = 1.496×10^{11} m
1 parsec (pc) = 3.09×10^{16} m

Mass
1 slug = 14.59 kilograms (kg)
1 kg = 1000 grams = 6.852×10^{-2} slug
1 atomic mass unit (amu) = 1.6605×10^{-27} kg
(1 kg has a weight of 2.205 lb where the
 acceleration due to gravity is 32.174 ft/s^2)

Time
1 day = 24 h = 1.44×10^3 min = 8.64×10^4 s
1 year (yr) = 365.24 days = 3.156×10^7 s

1 h = 60 min = 3600 s

Speed
1 mi/h = 1.609 km/h = 1.467 ft/s = 0.4470 m/s
1 km/h = 0.6214 mi/h = 0.2778 m/s = 0.9113 ft/s

Volume
1 liter (L) = 10^{-3} m^3 = 1000 cm^3 = 0.03531 ft^3
1 ft^3 = 0.02832 m^3 = 7.481 US gallons (gal)
1 U.S. gal = 3.785×10^{-3} m^3 = 0.1337 ft^3

Force
1 pound (lb) = 4.448 newtons (N)
1 N = 10^5 dynes (dyn) = 0.2248 lb

Work and energy
1 joule (J) = 0.7376 ft · lb = 10^7 ergs
1 kilogram-calorie (kcal) = 4186 J
1 Btu(60°F) = 1055 J
1 kilowatt-hour (kWh) = 3.600×10^6 J
1 electron volt (eV) = 1.602×10^{-19} J

Angle
1 radian (rad) = 57.30°
1° = 0.01745 rad

Pressure
1 pascal (Pa) = 1 N/m^2 = 1.450×10^{-4} lb/in^2
1 lb/in^2 = 6.895×10^3 Pa
1 atmosphere (atm) = 1.013×10^5 Pa = 1.013 bar

$$= 14.70 \text{ lb/in}^2 = 760 \text{ torr}$$

Power
1 horsepower (hp) = 550 ft · lb/s = 745.7 W
1 watt (W) = 0.7376 ft · lb/s

Note that the factor 1609 must be squared here. Failure to square the numerical factor is a common error. Be careful.

PROBLEM 2.3. A geologist finds that a rock sample has a volume of 2.40 in^3. Express this volume in cubic centimeters and in cubic meters.

Solution 1 in = 2.54 cm, so $V = 2.40 \text{ in}^3 = 2.40(2.54 \text{ cm})^3 = 39.3 \text{ cm}^3$. Also 1 cm = 10^{-2} m, so $V = 39.3 \text{ cm}^3 = (39.3)(10^{-2} \text{ m})^3 = 3.93 \times 10^{-5} \text{ m}^3$.

PROBLEM 2.4. A race car has an acceleration of 8 mi/h/s. Express this acceleration in meters per second per second.

Solution

$$a = 8 \frac{\text{mi/h}}{\text{s}} = 8 \frac{(1609 \text{ m})}{(\text{s})(3600 \text{ s})} = 3.58 \text{ m/s}^2$$

In every equation in physics the **dimensions must be consistent** in every term in the equation. For example, consider the equation

$$\text{distance} = \text{speed} \times \text{time}$$

The dimension of distance is length (L), the dimension of speed is length per time (L/T), and the dimension of time is time (T). Examining the dimensions of each side of the equation shows

$$L = \frac{L}{T} \times T$$

Time cancels out on the right side, so both sides of the equation have the dimension of length. This illustrates the fact that the units of each variable in an equation are to be included in the calculation. They are multiplied and divided as though they are algebraic quantities. It is helpful to check the units for each term in an equation, since any inconsistency indicates an error has been made, and you are thereby alerted to search for it.

PROBLEM 2.5.   Determine whether or not each of the following equations is dimensionally correct. M, L, and T indicate dimensions of mass, length, and time, respectively. The symbols used in the equations and their dimensions are as follows:

F = force	ML/T^2	(a)	$v = at$
x = distance	L	(b)	$t = \sqrt{2x/a}$
v = velocity	L/T	(c)	$v^3 = 2ax^2$
a = acceleration	L/T^2	(d)	$F = mvx$

Solution

(a)   This equation has dimensions $L/T = (L/T^2)(T) = L/T$ Consistent

(b) $T = \sqrt{L/LT^2} = \sqrt{T^2} = T$ Consistent

(c) $(L/T)^3 = (L/T^2)(L)^2$ or $L^3/T^3 = L^3/T^2$ Inconsistent

(d) $ML/T^2 = (M)(L/T)(L)$ or $ML/T^2 = ML^2/T$ Inconsistent

2.3   Order-of-Magnitude Estimates

Frequently, it is helpful to make a rough estimate of a desired result. Sometimes this is done to save time and to determine if a more careful calculation is merited. On other occasions inadequate data are available for a more exact calculation. Order-of-magnitude estimates can also help alert you to a possible error you have made in solving a problem. If you end up calculating the distance from New York to Los Angeles to be 5 million kilometers, you should readily recognize that you have made an error. When solving problems, ask yourself if your result makes sense. Do not mindlessly plug numbers into equations without a thought as to whether or not your results are reasonable.

In making rough estimates, you may be unsure of what values to use for some parameters. In many cases you will simply have to make an educated guess as to what numbers to use. These ideas are illustrated in the following examples.

PROBLEM 2.6.   It is a curious fact that most mammals seem to live for roughly the same number of heart beats. A small animal like a mouse or shrew has a rapid pulse rate and lives a relatively short time, whereas large mammals have slower pulse rates and live longer. This makes one wonder if we are given a certain number of heart beats to use up during our lifetimes. Probably not, I suspect, since exercise that increases your pulse rate is generally believed to improve your health and lengthen your life. How many times does a human heart beat in an average lifetime?

Solution   A typical pulse rate is about 72 beats per minute. If you didn't know this, measure your own pulse. An average lifetime is about 70 years. One year is 365 days, one day is 24 hours, and one hour is 60 minutes. Thus the number of times your heart beats is about

$$N = (72)(70)(365)(24)(60) = 2.6 \times 10^9 \text{ beats}$$

PROBLEM 2.7.   I sometimes fantasize that when I am walking in New York, a car full of suspicious-looking characters races by. They stop, throw a briefcase into a trash dumpster, and then speed off, pursued a few seconds later by police cars with sirens screaming. I pull the briefcase out of the dumpster and peek into it. Full of hundred-dollar bills. I haven't decided for sure just what happens next, but I'm curious. About how much money does such a briefcase hold?

Solution   My own briefcase measures about $5 \times 12 \times 18$ in. A hundred-dollar bill measures about 6×2.5 in. I'm not sure how thick a bill is, but a ream of typing paper (500 sheets) is about 2 in thick. I'll assume the bills fill up the briefcase with no empty space. This is probably unrealistic and might lead me to overestimate the total dollar amount by a factor of 2, but I won't worry about that. Thus I estimate the number of bills in the briefcase to be

$$N = \frac{\text{volume of briefcase}}{\text{volume of one bill}} = \frac{5 \times 12 \times 18}{6 \times 2.5 \times 2/500} = 18{,}000$$

Each bill is worth \$100, so the total cash value is $\$1.8 \times 10^6 = \$1.8$ million.

SUPPLEMENTARY PROBLEMS

2.8.   The mass density (also called just "density") of a substance is the ratio of its mass in kilograms to its volume in cubic meters. A piece of lead of mass 11.97 g has a volume of 1.05 cm^3. Determine the density of lead in SI units.

2.9.   The airline distance between New York and London is 3469 mi. Express this distance in kilometers.

2.10.   A person is 5 ft 8 in tall. Express this height in meters.

2.11.   On many interstate highways the speed limit is now 70 mi/h. Express this in kilometers per hour, meters per second, and feet per second.

2.12.   A room is 16 ft 6 in long by 12 ft 4 in wide and has a ceiling height of 8 ft 0 in. What is the volume of the room in cubic feet and in cubic meters?

2.13.   The radius of the Sun is 7.0×10^8 m and its mass is 2.0×10^{30} kg. What is the density of the Sun (*density = mass/volume*)? The density of water is 1000 kg/m^3. What is the ratio of the Sun's density to that of water?

2.14.   A cord is a volume of cut wood 8 ft long, 4 ft high, and 4 ft wide. Express this volume in cubic meters.

2.15.   Washington, D.C., is located at about 75°W longitude and 38°N latitude. San Francisco is near 122°W longitude and 38°N latitude. Estimate how many minutes later the Sun rises in San Francisco than in Washington, D.C. About how far apart would you estimate these cities to be? (Assume you know the length of a day and the approximate diameter of the Earth.)

2.16.   As a prize on a TV quiz show, you are given a big plastic trash bag and allowed to go into a bank vault and take away all of the one-dollar bills you can carry in one trip. Assuming the bag doesn't rip open and that you have enough time to stuff it full, how much money would you estimate you could haul off?

2.17. Determine if the dimensions in each of the following equations are consistent. (See Problem 2.5 for definitions of symbols.)

(a) $v^2 = at$

(b) $x = vt + \dfrac{at^2}{2}$

(c) $max + \dfrac{mv^2}{2} = Fx^2$

(d) $v^2 = 2ax + \dfrac{x}{t}$

(e) $v^2 = \dfrac{Fx}{m}$

SOLUTIONS TO SUPPLEMENTARY PROBLEMS

2.8. $\text{Density} = \dfrac{\text{mass}}{\text{volume}} = \dfrac{11.97\,\text{g}}{1.05\,\text{cm}^3} = 11.4\,\dfrac{\text{g}}{\text{cm}^3} = 11.4\,\dfrac{(10^{-3}\,\text{kg})}{(10^{-2}\,\text{m})^3} = 1.14 \times 10^4\,\dfrac{\text{kg}}{\text{m}^3}$

2.9. $d = 3469\,\text{mi} = 3469(1.609\,\text{km}) = 5582\,\text{km}$

2.10. $h = 5\,\text{ft}\ 8\,\text{in} = (5)(0.3048\,\text{m}) + (8)(0.0254\,\text{m}) = 1.73\,\text{m}$

2.11. $v = 70\,\dfrac{\text{mi}}{\text{h}} = 70\,\dfrac{1.609\,\text{km}}{\text{h}} = 113\,\dfrac{\text{km}}{\text{h}}$

   $v = 113\,\dfrac{1000\,\text{m}}{3600\,\text{s}} = 31.3\,\text{m/s} = 31.3\,\dfrac{3.281\,\text{ft}}{\text{s}} = 103\,\text{ft/s}$

2.12. $16\,\text{ft}\ 6\,\text{in} = 16.5\,\text{ft}$ $12\,\text{ft}\ 4\,\text{in} = 12.33\,\text{ft}$

   $v = (16.5\,\text{ft})(12.33\,\text{ft})(8\,\text{ft}) = 1630\,\text{ft}^3 = 1630(0.3048\,\text{m})^3 = 46.1\,\text{m}^3$

2.13. $\text{Volume of sphere} = \frac{4}{3}\pi r^3 = \frac{4}{3}\pi (7 \times 10^8\,\text{m})^3 = 1.44 \times 10^{27}\,\text{m}^3$

   $\text{Density} = \dfrac{\text{mass}}{\text{volume}} = \dfrac{2 \times 10^{30}\,\text{kg}}{1.44 \times 10^{27}\,\text{m}^3} = 1.39 \times 10^3\,\text{kg/m}^3$

   $\dfrac{\text{Density Sun}}{\text{Density water}} = \dfrac{1390}{1000} = 1.39$

2.14. $v = (8\,\text{ft})(4\,\text{ft})(4\,\text{ft}) = 128\,\text{ft}^3 = 128(0.3048\,\text{m})^3 = 3.62\,\text{m}^3$

2.15.   San Francisco is θ degrees west of Washington, where $\theta = 122° - 75° = 47°$. A full rotation of the Earth (360°) requires 24 h, so the time delay for sunrise between SF and DC is

$$t = \frac{47°}{360°} \times 24\,\text{h} = 3.13\,\text{h} = 3\,\text{h}\ 8\,\text{min}$$

R_E, the radius of the Earth, is about 4000 mi or 6400 km. The cities are at 38°N latitude, so the distance from each city to the axis of the Earth is $R_E \cos 38° = r$. Thus each city travels along a circle of radius r. The distance between the cities is an arc of length s that subtends an angle of 47° at the center of the circle, so

$$s = \frac{47°}{360°} 2\pi r = \left(\frac{47°}{360°}\right)(2\pi)(6400 \cos 38°) = 4140\,\text{km}$$

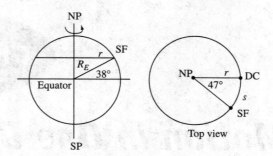

2.16. First estimate how much a dollar bill weighs. I believe a ream of paper (500 sheets of 8.5 in × 11 in) weighs about 6 lb. A dollar bill is about 6 in × 2.5 in, or 15 in². A sheet of typing paper is about 8 in × 11 in = 88 in², so one sheet of typing paper is equivalent to about 88/15 ≃ 6 dollar bills. Thus a dollar bill weighs about 6 lb/(6)(500) ≃ 0.002 lb. I estimate a person could carry about 60 lb, which amounts to 60/0.002 = $30,000. This might be off by a factor of 2, but it is definitely within a factor of 10 of an accurate calculation. Not too shabby.

2.17. (a) $\left(\dfrac{L}{T}\right)^2 = \left(\dfrac{L}{T^2}\right)(T) = \dfrac{L}{T}$ Inconsistent

    (b) $L = \left(\dfrac{L}{T}\right)T + \left(\dfrac{L}{T^2}\right)(T^2) = L + L$ Consistent

    (c) $M\left(\dfrac{L}{T^2}\right)(L) + M\left(\dfrac{L}{T}\right)^2 = M\left(\dfrac{L}{T^2}\right)L^2$ Inconsistent

    (d) $\left(\dfrac{L}{T}\right)^2 = \left(\dfrac{L}{T^2}\right)L + \dfrac{L}{T}$ Inconsistent

    (e) $\left(\dfrac{L}{T}\right)^2 = M\left(\dfrac{L}{T^2}\right)\dfrac{L}{M} = \left(\dfrac{L}{T}\right)^2$ Consistent

CHAPTER 3

Motion in One Dimension

The study of motion is called **kinematics**, and it is here that we begin our study of physics. We will follow a method that has proven very effective in science. We start by studying simple situations and then study gradually more complex physical problems. In this chapter we consider motion in one dimension, without regard to the forces that influence the motion. In the next chapter, we will extend the discussion to motion in two or three dimensions, but first we need a good understanding of the basic concepts involved, namely, displacement, velocity, and acceleration.

3.1   Displacement and Velocity

The position of an object moving along the x-axis is described by its x coordinate. The change in the object's position is its displacement Δx. If the object is at position x_1 at time t_1 and at x_2 at time t_2, then $\Delta x = x_2 - x_1$. Displacement is a vector. However, for motion in one dimension we can specify the displacement simply in terms of the x coordinate of the particle. If the particle is to the right of the origin, its coordinate is positive. If it is to the left of the origin, its coordinate is negative. We define the **average velocity** v as

$$\bar{v} \equiv \frac{x_2 - x_1}{t_2 - t_1} = \frac{\Delta x}{\Delta t} \tag{3.1}$$

If we choose our origin such that $x_1 = 0$ and $t_1 = 0$, then the position x at later time t is $x = vt$.

3.2   Instantaneous Velocity and Acceleration

If an object experiences a displacement Δx in a time Δt, its **instantaneous velocity** is

$$v \equiv \lim_{\Delta t \to 0} \frac{\Delta x}{\Delta t} = \frac{dx}{dt} \tag{3.2}$$

Velocity is a vector, but in one dimension we can indicate direction merely by giving the sign of the velocity. The magnitude of velocity is called **speed**. Speed is what a car's speedometer measures. Speed is always positive. Speed and velocity are measured in meters per second. **Velocity is the slope of a graph of x versus t**, as illustrated in Fig. 3-1. When the slope is positive, the object is moving to the right. When the slope is negative, the object is moving to the left. When the slope is zero, the object is stopped.

The rate at which velocity is changing is measured by **acceleration**. Thus if an object has velocity v_1 at time t_1 and velocity v_2 at time t_2, its **average acceleration** is

$$\overline{a} \equiv \frac{v_2 - v_1}{t_2 - t_1} = \frac{\Delta v}{\Delta t} \tag{3.3}$$

and its **instantaneous acceleration** is

$$a \equiv \lim_{\Delta t \to 0} \frac{\Delta v}{\Delta t} = \frac{dv}{dt} \tag{3.4}$$

Acceleration has units of velocity/time: $\mathrm{m/s/s} = \mathrm{m/s^2}$.

When thinking of the units of acceleration, never say to yourself, "meters per square second." Instead, always say, "meters per second (pause) per second." This makes clear the idea that acceleration is a measure of how much the velocity is changing each second. Drag racers describe a car's acceleration in units of "miles per hour per second." Thus if a car can go from 0 to 60 mi/h in 6 s, its acceleration is $10\,\mathrm{mi/h \cdot s}$. We always measure acceleration in units of meters per second per second, but the drag racer's mixed units convey the idea of acceleration more clearly.

It is best to avoid the use of the common word "deceleration." Describe acceleration simply as positive or negative. Note that *negative acceleration does not necessarily mean "slowing down."* When velocity and acceleration both have the same sign, the object speeds up. When velocity and acceleration have opposite signs, the object slows down.

Illustrative graphs of displacement, velocity, and acceleration for a moving object are shown in Fig. 3-2. Note that v can be deduced from the x versus t curve by remembering that v is the slope of x versus t. Similarly, a can be deduced from v versus t, since a is the slope of v versus t.

Acceleration is the **second derivative of displacement**. Thus

$$a = \frac{dv}{dt} = \frac{d}{dt}\left(\frac{dx}{dt}\right) = \frac{d^2x}{dt^2} \tag{3.5}$$

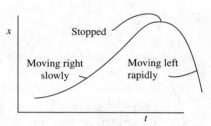

Fig. 3-1

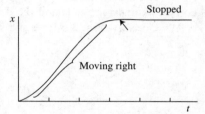

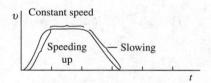

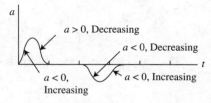

Fig. 3-2

3.3 Constant Acceleration

Many interesting phenomena involve motion with constant acceleration. In this case, it is easy to obtain expressions for velocity and displacement by integrating the acceleration. Thus if

$$a = \frac{dv}{dt} = \text{constant}$$

then

$$v = \int a\,dt = at + c_1$$

We can determine the constant by observing that if at time $t = 0$ the velocity has initial value v_0, then $v_0 = 0 + c_1$, so $c_1 = v_0$ and

$$v = at + v_0 \tag{3.6}$$

We can integrate the velocity to obtain the displacement x.

$$v = \frac{dx}{dt}$$

so

$$x = \int v\, dt = \int (at + v_0)\, dt$$
$$= \tfrac{1}{2}at^2 + v_0 t + c_2$$

If at time $t = 0$, the value of x is x_0 (the initial position), then $x_0 = c_2$ and

$$x = x_0 + v_0 t + \tfrac{1}{2}at^2 \tag{3.7}$$

In most problems, it is convenient to choose the origin at the position of the object at $t = 0$, that is, to set $x_0 = 0$. When this is done, Eq. 3.7 becomes

$$x = v_0 t + \tfrac{1}{2}at^2 \tag{3.8}$$

We can check that Eqs 3.6 and 3.8 are correct by differentiating them. Thus the derivative of x yields the correct expression for v, and the derivative of v yields the constant acceleration a.

We can solve Eq. 3.6 for t and substitute the result in Eq. 3.8. When this is done, we obtain

$$t = \frac{v - v_0}{a} \qquad x = v_0 \frac{(v - v_0)}{a} + \frac{1}{2}a\frac{(v - v_0)^2}{a^2}$$
$$2ax = 2v_0 v - 2v_0^2 + v^2 - 2v_0 v + v_0^2 \tag{3.9}$$
$$v^2 = v_0^2 + 2ax$$

The above equations are so important that it is worthwhile to place them all together and memorize them.

$$\boxed{\begin{aligned} &\text{If } a = constant, \text{ then} \\ &v = v_0 + at \\ &x = x_0 + v_0 t + \tfrac{1}{2}at^2 \\ &v^2 = v_0^2 + 2ax \end{aligned}} \tag{3.10}$$

The case of zero acceleration (constant velocity) is important and results in simple equations.

$$\boxed{\begin{aligned} &\text{Thus if } a = 0, \text{ then} \\ &v = v_0 \text{ (constant)} \\ &x = v_0 t \end{aligned}} \tag{3.11}$$

CAUTION: Do not use Eq. 3.11 if acceleration is not zero. Failure to heed this admonition is a common source of error.

PROBLEM 3.1.   A motorist drives for 2 h at 100 km/h and for 2 h at 80 km/h. What is the average speed of the motorist?

Solution $\bar{v} = \dfrac{\text{total distance}}{\text{total time}} = \dfrac{(100\,\text{km/h})(2\,\text{h}) + (80\,\text{km/h})(2\,\text{h})}{2\,\text{h} + 2\,\text{h}} = 90\,\text{km/h}$

PROBLEM 3.2. A motorist drives 120 km at 100 km/h and 120 km at 80 km/h. What is the average speed of the motorist for the trip?

Solution

$$\bar{v} = \frac{\text{total distance}}{\text{total time}} = \frac{120\,\text{km} + 120\,\text{km}}{120\,\text{km}/100\,\text{km/h} + 120\,\text{km}/80\,\text{km/h}}$$

$$= \frac{240\,\text{km}}{1.2\,\text{h} + 1.5\,\text{h}} = 88.9\,\text{km/h}$$

Observe that in Problem 3.1, the average speed was halfway between the high speed and the low speed, because the motorist drove equal *times* at each speed. Here, however, the motorist drove equal *distances* at each speed but drove for a longer time at the lower speed, so the average speed is closer to the lower speed and is *not* halfway in between. Remember, *average* means "time average."

PROBLEM 3.3. In good weather the drive from Seattle to Spokane, Washington, on Interstate 90 takes 3 h 51 min at an average speed of 105 km/h. In winter, however, it is not unusual to average only 80 km/h. How long would the trip take at this average speed?

Solution Express the time in hours. Thus $t_1 = 3\,\text{h}\ 51\,\text{min} = (3 + 51/60)\,\text{h} = 3.85\,\text{h}$. Since $x = v_1 t_1 = v_2 t_2$,

$$t_2 = \frac{v_1}{v_2} t_1 = \frac{105}{80}(3.85\,\text{h}) = 5.05\,\text{h}$$

$$= 5\,\text{h} + (0.05\,\text{h})(60\,\text{min/h}) = 5\,\text{h}\ 3\,\text{min}$$

PROBLEM 3.4. A cheetah is the fastest land mammal, and it can run at speeds of about 101 km/h for a period of perhaps 20 s. The next fastest land animal is an antelope, which can run at about 88 km/h for a much longer time. Suppose a cheetah is chasing an antelope, and both are running at top speed. (a) If the antelope has a 40-m head start, how long will it take the cheetah to catch him, and how far will the cheetah travel in this time? (b) What is the maximum head start the antelope can have if the cheetah is to catch him within 20 s (at which time the cheetah runs out of breath)?

Solution

(a) The speeds are constant, so Eq. 3.10, $x = vt$, applies. Both animals run for the same time, but the cheetah must run 40 m extra. Thus

$$x_C = v_C t = x_A + 40 \qquad\qquad\qquad\qquad \text{(i)}$$

and

$$x_A = v_A t \qquad\qquad\qquad\qquad \text{(ii)}$$

Substitute (ii) in (i) and solve for t:

$$v_C t = v_A t + 40 \qquad (v_C - v_A)t = 40 \qquad t = \frac{40}{v_C - v_A}$$

The speeds must be expressed in meters per second, not kilometers per hour.

$$v_C = 101\,\text{km/h} = 101\left(\frac{1000\,\text{m}}{3600\,\text{s}}\right) = 28.1\,\text{m/s}$$

$$v_A = 88\,\text{km/h} = 24.4\,\text{m/s} \qquad t = \frac{40\,\text{m}}{(28.1 - 24.4)\text{m/s}} = 10.8\,\text{s}$$

(b) Let h = head start distance and $t = 20\,$s for both animals. If the cheetah is to catch the antelope, then $x_C = x_A + h$.

So
$$x_C = v_C t \qquad x_A = v_A t$$

$$v_C t = v_A t + h \quad h = (v_C - v_A)t = (28.1 - 24.4)(20) = 74\,\text{m}$$

PROBLEM 3.5. A typical jet fighter plane launched from an aircraft carrier reaches a take-off speed of 175 mi/h in a launch distance of 310 ft. (a) Assuming constant acceleration, calculate the acceleration in meters per second per second. (b) How long does it take to launch the fighter?

Solution

(a) The plane starts from rest, so $v_0 = 0$. From Eq. 3.10, I choose the equation relating v, v_0, a, and x. I use this equation because I know v, v_0, and x and I want to find a. I do not use an equation involving t since I do not yet know t. If you were to start with one of the other equations, you would eventually reach the correct answer, but more algebra would be involved. With practice you will learn which equation to use for the easiest solution. Using Table 2.1 convert the data to SI units:

$$v = 175\,\text{mi/h} = (175)(0.447\,\text{m/s}) = 78.2\,\text{m/s}$$
$$x = 310\,\text{ft} = (310)(0.305\,\text{m}) = 94.6\,\text{m}$$
$$v^2 = v_0^2 + 2ax = 0 + 2ax$$
$$a = \frac{v^2}{2x}$$
$$= \frac{(78.2\,\text{m/s})^2}{2(94.6\,\text{m})} = 32.3\,\text{m/s}^2$$

(b) From Eq. 3.10, $v = v_0 + at = 0 + at$, so

$$t = \frac{v}{a} = \frac{78.2\,\text{m/s}}{32.3\,\text{m/s}^2} = 2.4\,\text{s}$$

PROBLEM 3.6. A motorist traveling 31 m/s (about 70 mi/h) passes a stationary motorcycle police officer. The police officer starts to move and accelerates in pursuit of the speeding motorist 2.5 s after the motorist passes. The motorcycle has a constant acceleration of 3.6 m/s². (a) How fast will the police officer be traveling when he overtakes the car? Draw curves of x versus t for both the motorcycle and the car, taking $t = 0$ at the moment the car passes the stationary police officer. (b) Suppose that for reasons of safety the policeman does not exceed a maximum speed of 45 m/s (about 100 mi/h). How long will it then take him to overtake the car, and how far will he have traveled?

Solution

(a) The car has a constant velocity and travels a distance x_c in time t:

$$x_c = v_c t$$

The motorcycle starts from rest ($v_0 = 0$) and moves a distance x_m in time $t - 2.5$ with constant acceleration:

$$x_m = \tfrac{1}{2}a(t - 2.5)^2$$

These curves are sketched here. When the motorcycle overtakes the car, both will have traveled the same distance. Thus

$$\tfrac{1}{2}a(t-2.5)^2 = v_c t$$

Substitute numerical values, and solve this quadratic equation for t, using Eqs 1.31 and 1.32.

$$\tfrac{1}{2}(3.6)(t-2.5)^2 = 31t, \qquad 1.8t^2 - 9t + 11.25 = 31t$$

$$1.8t^2 - 40t + 11.25 = 0$$

$$t = \frac{40 \pm \sqrt{(40)^2 - 4(1.8)(11.25)}}{(2)(1.8)}, \qquad t = 21.9\,\text{s or } 0.28\,\text{s}$$

The motorcycle did not start until $t = 2.5\,\text{s}$, so the solution we want is $t = 21.9\,\text{s}$.

$$v_m = v_0 + a(t-2.5) = 0 + 3.6\,\text{m/s}^2(21.9\,\text{s} - 2.5\,\text{s}) = 70\,\text{m/s} = 156\,\text{mi/h}$$

(b) Suppose the motorcycle accelerates for time t_1 over distance x_1 to a maximum speed $v = 45\,\text{m/s}$. It then continues at constant speed v for time t_2 and distance x_2 until it catches the car. The variables are then related as follows:

$$x_c = v_c(t_1 + t_2 + 2.5) \quad \text{(i)}, \qquad x_1 = \tfrac{1}{2}at_1^2 \quad \text{(ii)}, \qquad x_2 = v_m t_2 \quad \text{(iii)},$$

$$x_c = x_1 + x_2 \quad \text{(iv)}, \qquad v_m = at_1 \quad \text{(v)}$$

The preceding are five equations in five unknowns: x_1, x_2, t_1, t_2, and x_c. They can be solved simultaneously. The values of v_m and a are known, so Eq. (v) gives t_1 immediately. Substitute this value for t_1 in Eq. (ii) and x_1 is obtained. Now use Eqs (i), (iii), and (iv) to solve for the remaining three variables, x_2, t_2, and x_c. The results are $t_1 = 12.5\,\text{s}$, $x_1 = 281\,\text{m}$, $x_2 = 591\,\text{m}$, $t_2 = 13.1\,\text{s}$, $x_c = 872\,\text{m}$, and $t = t_1 + t_2 = 25.6\,\text{s}$.

PROBLEM 3.7. Suppose that motion studies of a runner show that the maximum speed he can maintain for a period of about 10 s is 12 m/s. If in a 100-m dash this runner accelerates with constant acceleration until he reaches this maximum speed and then maintains this speed for the rest of the race, what acceleration will he require if his total time is 11 s?

Solution Break the problem into two parts. While accelerating, the runner travels a distance x_1 in time t_1, and then runs the remaining distance x_2 in time t_2 at constant speed v.

$$x_1 + x_2 = 100 \quad \text{(i)}, \qquad t_1 + t_2 = 11 \quad \text{(ii)},$$

$$x_1 = \tfrac{1}{2}at_1^2 \quad \text{(iii)}, \qquad v = at_1 \quad \text{(iv)}, \qquad x_2 = vt_2 \quad \text{(v)}$$

These five equations can be solved for the five unknowns: a, x_1, x_2, t_1, and t_2. Substitute (iii) and (v) into (i): $\tfrac{1}{2}at_1^2 + vt_2 = 100$. Solve (ii) for t_2 and substitute into the above equation, first multiplying by 2:

$$at_1^2 + 2v(11 - t_1) = 200$$

Solve (iv) for t_1 and substitute it in the above equation:

$$a\left(\frac{v}{a}\right)^2 + 2v\left(11 - \frac{v}{a}\right) = 200$$

Solve for a:

$$a = \frac{v^2}{22v - 200}$$

Substitute $v = 12\,\text{m/s}$. The result is $a = 2.25\,\text{m/s}^2$.

In solving complicated problems like this, first be sure you are clear about exactly what is happening. Draw a little picture with a runner and the different parts of the race indicated. Label all the relevant quantities, using different symbols for the different unknown quantities. Here we had five unknowns, so we know we will need five independent equations to solve the problem. Write down the equations, and then solve them using algebra. Finally, substitute the numerical values.

3.4   Freely Falling Bodies

Consider an object moving upward or downward along a vertical axis. Let us neglect any air effects and consider only the influence of gravity on such an object. It has been found that all objects, large and small, experience the same acceleration due to the force of gravity. This acceleration varies slightly with altitude, but for objects near the surface of the Earth the acceleration is approximately constant. The acceleration is *always* directed downward, since it is caused by the downward force of gravity. We label the vertical axis the y-axis, with upward taken as the positive direction. We take $y = 0$ at some convenient point, such as sea level or floor level. We call the magnitude of the acceleration due to the force of gravity g. The value of g is approximately 9.80 m/s^2. The value of g is slightly smaller high in the mountains and slightly larger at low elevations, such as in Death Valley. Note that the acceleration of an object acted on only by the force of gravity is $-g$, since the acceleration is downward and hence negative. This is true whether the object is falling downward, moving upward, or momentarily stopped at its highest point. Equation 3.10 describes this situation. Taking y as our independent variable and setting $a = -g = $ constant, these equations become

$$
\begin{aligned}
a &= -g \\
v &= v_0 - gt \\
y &= y_0 + v_0 t - \tfrac{1}{2} g t^2
\end{aligned}
$$

(3.12)

In all of the following, neglect the influence of air. This is a fair approximation for objects that do not fall too far or too fast. Later, we will see how to incorporate air drag.

PROBLEM 3.8.   A rock is dropped from rest from the Golden Gate Bridge. How far will it have fallen after 1 s? After 2 s? After 3 s? How fast will it be moving at each time?

Solution   It is convenient to take the starting position of the rock as $y_0 = 0$. Thus, subsequent y values will be negative as the rock falls. The rock is initially at rest, so $v_0 = 0$. Thus Eq. 3.12 yields $y = -\tfrac{1}{2} g t^2$ and $v = -gt$. Substituting $t = 1, 2,$ and 3 s yields $y(1) = -4.9 \text{ m}$, $v(1) = -9.8 \text{ m/s}$, $y(2) = -19.6 \text{ m}$, $v(2) = -19.6 \text{ m/s}$, $y(3) = -44.1 \text{ m}$, and $v(3) = -29.4 \text{ m/s}$. Graphs of y versus t and v versus t are shown below. The graph of y is a parabola. Each succeeding second, the rock falls a greater distance as it gains speed. Note that the slope of the v versus t curve (the acceleration) is constant and negative.

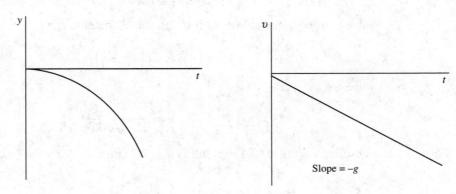

PROBLEM 3.9.   Using a slingshot, a kid shoots a rock straight up at 30 m/s from the top of the Rogun Dam (the world's highest dam) in Tajikistan. It finally strikes the water 325 m below its starting point. (Assume the face of the dam is vertical. Actually the dam slopes outward, but let's neglect this slight complication for now.) How high does the rock rise? How long is it in the air? How long would it have been in the air if it had been launched straight down? How long if dropped from rest? Sketch a graph of v versus t for each of these three cases.

Solution   Take $y = 0$ at the initial position. Let $y = h$ be the highest point reached. Let $y = -H$ be the surface of the water below. The initial velocity if thrown upward is $v_0 = 30\,\text{m/s}$. At the highest point, $v = 0$. Thus from Eq. 3.12,

$$v^2 = v_0^2 - 2gh = 0$$

so

$$h = \frac{v^2}{2g} = \frac{30^2\,\text{m}}{2(9.8)} = 45.9\,\text{m}$$

At the water $y = -H$, so Eq. 3.12 yields $-H = 0 + v_0 t - \frac{1}{2}gt^2$ where t is the time in the air. Solve this quadratic equation using Eq. 1.31.

$$4.9t^2 - 30t - 325 = 0 \quad t = \frac{30 \pm \sqrt{(-30)^2 - 4(4.9)(-325)}}{2(4.9)} \quad t = -5.64\,\text{s or } 11.8\,\text{s}$$

The physically meaningful solution is 11.8 s. The solution -5.64 s is allowed by the equations, but it does not correspond to what is happening here. If you imagined that the rock was projected upward from the water surface (where $y = -H$) at a time 5.64 s earlier than the actual launch, the equation is satisfied, but this is not what actually happened.

If the rock was initially launched straight down, then $v_0 = -30\,\text{m/s}$ and the equation for y becomes

$$-H = 0 + v_0 t - \tfrac{1}{2}gt^2 \quad \text{or} \quad -325 = -30t - 4.9t^2$$

Solving for t as before, $t = 5.64\,\text{s}$.

If dropped from rest, $v_0 = 0$, and $-H = 0 + 0 - \frac{1}{2}gt^2$, so $t = 8.14\,\text{s}$.

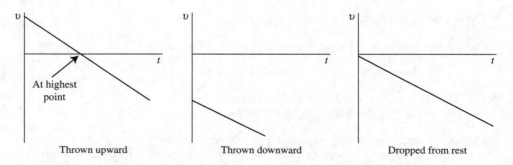

Thrown upward   Thrown downward   Dropped from rest

Note that when $v = 0$ for the rock thrown upward, it is at its highest point. Here the velocity is momentarily zero; a fraction of a second later it is moving downward and has negative velocity.

PROBLEM 3.10.   A ball is thrown straight up. Show that it spends as much time rising as it does falling back to its starting point.

Solution   At the peak of its flight $v = 0$. Thus $v = v_0 - gt_1 = 0$. Rise time is thus $t_1 = v_0/g$. Elevation is given by $y = v_0t - \frac{1}{2}gt^2$ assuming $y_0 = 0$.

When the ball returns to its starting point, $y = 0$. Thus $y = 0 = v_0t - \frac{1}{2}gt^2$ or $t = 2v_0/g = 2t_1$. The total time in the air is twice the rise time, so fall time = rise time.

3.5   Summary of Key Equations

Average velocity: $\bar{v} = \dfrac{x_2 - x_1}{t_2 - t_1}$

Instantaneous velocity: $v = \dfrac{dx}{dt}$

Average acceleration: $\bar{a} = \dfrac{v_2 - v_1}{t_2 - t_1}$

Instantaneous acceleration: $a = \dfrac{dv}{dt} = \dfrac{d^2x}{dt^2}$

If $a = 0$, then $v = v_0$ (constant) and $x = v_0t$

If $a = $ constant, then $v = v_0 + at$ and $x = x_0 + v_0t + \frac{1}{2}at^2$

For a freely falling object: An object dropped from rest will fall a distance
h in time t, where

$a = -g$
$v = v_0 - gt,$ $h = \frac{1}{2}gt^2$
$y = y_0 + v_0t - \frac{1}{2}gt^2$

SUPPLEMENTARY PROBLEMS

3.11. A graph of the displacement of a moving particle as a function of time is shown here. For this time interval, determine

(a) How many times the particle stopped?

(b) The total distance traveled.

(c) When was the particle moving fastest?

(d) How many times the particle returned to its starting point?

(e) The direction the particle was moving at $t = 6$ s.

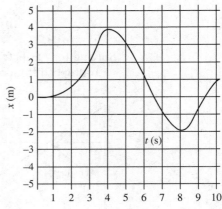

3.12. A Ferrari sports car can accelerate from rest to 96 km/h (about 60 mi/h) in 2.2 s. What is its average acceleration?

3.13. On a 40-km bike ride a cyclist rides the first 20 km at 20 km/h. What speed is required for the final 20 km if the average speed for the trip is to be (a) 10 km/h? (b) 30 km/h? (c) If the cyclist rides very fast for the final 20 km, what is the maximum value his average speed could approach?

3.14. The altitude of the space shuttle during the first 30 s of its ascent is described by the expression $y = bt^2$, where the constant b has the value 2.9 m/s^2. Using calculus, obtain an expression for the velocity and acceleration of the space shuttle during this period.

3.15. When a girder in a bridge undergoes small vibrations, its displacement can be described by $x = A \cos 2\pi f t$, where A is the amplitude of the vibration (that is, the maximum value of x) and f is the frequency of the vibration. Such a motion is called *simple harmonic motion*. Derive expressions for the velocity and acceleration for such motion.

3.16. The leaning tower of Pisa is 54.5 m tall. Supposedly Galileo investigated the behavior of falling objects by dropping them from the top of this tower and timing their descent. How long would it take an object to fall 54.5 m if air effects are negligible?

3.17. In 1939, a baseball nut named Joe Sprinz tried to set a world record for catching a baseball dropped from the greatest height. He tried to catch a ball dropped from a blimp 800 ft above him. On his fifth try he succeeded, but the ball slammed the glove into his face, breaking his jaw in 12 places, knocking out five teeth, and knocking him unconscious. As you might have guessed, the ball was really moving when it reached him. Air slows the ball appreciably, but you can get a pretty good idea of how fast the ball was moving even if you neglect air effects. How long would you estimate the ball took to fall 800 ft, and how fast was it going when it hit Joe?

3.18. A car is moving 60 km/h when the driver sees a signal light 40 m ahead turn red. The car can slow with acceleration $-0.5 g$ (where $g = 9.80$ m/s^2). What is her stopping distance assuming (a) zero reaction time? (b) A reaction time of 0.20 s between when she sees the red light and when she hits the brake?

3.19. A balloonist at an altitude of 800 m drops a package. One second later the balloonist drops a second package. (a) How far apart are the packages at the instant the second package is dropped? (b) How far apart are the packages when the first package hits the ground? (c) What is the time interval between when the two packages hit the ground?

3.20. A girl on top of a building drops a baseball from rest at the same moment a boy below throws a golf ball upward toward her with a speed of 20 m/s. The golf ball is thrown from a point 18 m below where the baseball is released. How far will the baseball have dropped when it passes the golf ball?

3.21. A typical jet liner lands at a speed of 100 m/s (about 224 mi/h). While braking, it has an acceleration of -5.2 m/s^2. (a) How long does it take to come to a stop? (b) What is the minimum length of the landing strip under these conditions?

3.22. A movie stunt man wishes to drop from a freeway overpass and land on the roof of a speeding truck passing beneath him. The distance he will fall from rest to the roof of the truck is 12 m, and the truck is moving 80 km/h. What horizontal distance away should the truck be when the stunt man jumps?

3.23. A ball is thrown upward with a speed of 12 m/s from the top of a building. How much time later must a second ball be dropped from the same starting point if it is to hit the ground at the same time as the first ball? The initial position is 24 m above the ground.

3.24. An object (perhaps a paratrooper) falls from an airplane and drops a vertical distance h. Upon striking snow-covered ground, the object stops with uniform acceleration a in a very small distance d. Determine the ratio a/g. A human has about a 50 percent chance of survival in such a fall if his or her acceleration does not exceed 50 g. Pilots have survived falls of 20,000 ft without a parachute, provided they landed in snow.

3.25. The stopping distance of a car depends on its speed in a way that is counterintuitive for many people. The stopping distance is *not* simply proportional to the speed; that is, if you double your speed, you do not merely double your stopping distance. On a dry road a car with good tires may be able to obtain a braking acceleration of -4.90 m/s^2. Calculate the stopping distance for a speed of 50 km/h and for 100 km/h.

3.26. In a movie, the FBI is investigating an assassination attempt on the life of the president. The setting is a parade in New York, and an amateur photographer has made a videotape of the passing motorcade. A careful examination of the tape shows in the background a falling object that turns out to be a pair of binoculars used by the would-be assassin. From the tape the FBI is able to determine that the binoculars fell the last 12 m before hitting the ground in 0.38 s. It is vital to them to know the height, and hence the building floor, from which the binoculars were dropped. Can this be determined from the given information? If so, from what height were the binoculars dropped?

3.27. The Earth travels in a nearly circular orbit about the Sun. The mean distance of the Earth from the Sun is 1.5×10^{11} m. What is the approximate speed of the Earth in its orbit around the Sun?

3.28. A driver traveling 100 km/h on a road in Montana sees a sheep in the road 32 m ahead. His reaction time is 0.70 s, and his braking acceleration is $-0.6\,g$ (where $g = 9.8$ m/s^2). Is he able to stop before he hits the animal? If so, what is his stopping distance? If not, at what speed does he hit the sheep?

3.29. Engineers at the Rand Corporation have designed a very high speed transit (VHST) vehicle that could radically reduce travel time between Los Angeles and New York. The 100-passenger car would be magnetically levitated and travel in an evacuated tube below the Earth's surface. On the 4800-km (3000-mi) trip from LA to New York, the car would accelerate for the first half of the trip and then coast to a stop in New York. On both legs of the trip the acceleration would have constant magnitude. An acceleration of about $0.4\,g$ (where $g = 9.8$ m/s^2) is the maximum a passenger can tolerate comfortably. Under these conditions, how long would the trip take? What maximum speed would be reached?

3.30. In an accident on a freeway, a sports car made skid marks 240 m long on the pavement. The police estimated the braking acceleration of the car to be $-0.9\,g$ under the road conditions prevailing. If this were true, what was the minimum speed of the sports car when the brakes were applied?

3.31. A driver traveling at night on a country road suddenly sees another car pull out into the roadway in front of him. He slams on his brakes and skids for a distance of 72.9 m until he hits the other car. If the driver slowed with constant acceleration of 4.8 m/s^2 for 4.5 s, with what speed would he hit the other car?

3.32. In order to take off, a 747 jetliner must reach a takeoff speed of approximately 95.0 m/s. If a 747 accelerates from rest with uniform acceleration of 2.20 m/s^2, what minimum length of runway is required for takeoff?

SOLUTIONS TO SUPPLEMENTARY PROBLEMS

3.11. (a) The particle is stopped when the slope is zero, that is, twice, at 4 s and 8 s.

    (b) The particle first went 4 m to the right, then returned to its starting point and continued on 2 m to the left. It then went back to the right a distance of 3 m, for a total distance moved of $4 + 4 + 2 + 3 = 13$ m.

    (c) The particle is moving fastest when the slope is greatest, which is near $t = 3$ s.

    (d) The particle returned twice to its starting point at $x = 0$.

    (e) At $t = 6$ s the slope is negative, so the particle is moving to the left.

3.12. $v_0 = 0 \quad v = 96$ km/h $= 96(1000$ m$/3600$ s$) = 26.7$ m/s

$$v = v_0 + at \quad a = \frac{v - v_0}{t} = \frac{26.7 - 0}{2.2} \text{ m/s}^2 = 12.1 \text{ m/s}^2$$

$$t_1 = \frac{20 \text{ km}}{20 \text{ km/h}} = 1 \text{ h}, \quad \text{so } t_2 = T - 1$$

3.13. (a) If $\bar{v} = 10$ km/h, then

$$10 \text{ km/h} = \frac{40 \text{ km}}{1 + t_2} \quad t_2 = 3 \text{ h} \quad \bar{v} = \frac{20 \text{ km}}{3 \text{ h}} = 6.7 \text{ km/h}$$

    (b) If $\bar{v} = 30$ km/h, then

$$30 \text{ km/h} = \frac{40 \text{ km}}{1 + t_2} \quad t_x = \frac{1}{3} \text{ h} \quad \bar{v} = \frac{20 \text{ km}}{1/3 \text{ h}} = 60 \text{ km/h}$$

(c) As $\bar{v}_2 \to \infty$, $t_2 \to 0$ and

$$\bar{v}_{max} = \frac{40\,km}{1\,h} = 40\,km/h$$

3.14. $\quad y = \frac{1}{2}bt^2 \quad v = \frac{dy}{dt} = 2bt \quad a = \frac{dv}{dt} = 2b$

3.15. $\quad x = A\cos 2\pi\,ft \quad v = \frac{dx}{dt} = -2\pi f\,A\sin 2\pi\,ft \quad a = \frac{dv}{dt} = -(2\pi f)^2 A\cos 2\pi\,ft$

3.16. $\quad y = y_0 + v_0 t - \frac{1}{2}gt^2$

Let $y_0 = 0$, $v_0 = 0$, and $y = -54.5$ m at ground. Then,

$$-54.5 = -\tfrac{1}{2}(9.8)t^2 \quad t = 3.33\,s$$

3.17. $\quad h = \frac{1}{2}gt^2 \quad t = \sqrt{\frac{2h}{g}} = \sqrt{\frac{2(800\,ft)(0.305\,m)}{9.8\,m/s^2}}\,m/ft = 7.06\,s$

$v = v_0 - gt = 0 - (9.8\,m/s^2)(7.06\,s) = 69.2\,m/s$ (about 155 mi/h)

3.18. $\quad v_0 = 60\,km/h = 60\,\dfrac{1000\,m}{3600\,s} = 16.7\,m/s$

(a) $v^2 = v_0^2 + 2ax = 0$ when stopped, so

$$x = -\frac{v_0^2}{2a}$$

$$= \frac{(16.7\,m/s)^2}{2(-0.5)(9.8\,m/s^2)} = 28.3\,m$$

(b) During the 0.20-s reaction time, the car travels a distance of

$$x_r = vt_r = (16.7\,m/s)(0.20\,s) = 3.33\,m$$

The total stopping distance is thus $28.3\,m + 3.3\,m = 31.6\,m$.

3.19. (a) In 1 s the first package falls a distance of $h = \frac{1}{2}gt^2 = \frac{1}{2}(9.8\,m/s^2)(1\,s)^2 = 4.9\,m$.

(b) The first package hits the ground after time t_1, where $800 = \frac{1}{2}gt_1^2$, so $t_1 = 12.8\,s$. The second package thus falls for 11.8 s, dropping a distance of

$$h_2 = \tfrac{1}{2}gt_2^2 = \tfrac{1}{2}(9.8\,m/s^2)(11.8\,s)^2 \quad h_2 = 680\,m$$

Thus the second package is $800\,m - 680\,m = 120\,m$ above ground when the first package hits.

(c) Since both packages take the same time to fall to the ground, the second package will reach the ground 1 second after the first.

3.20. The distance the baseball falls is $h = \frac{1}{2}gt^2$. The distance the golf ball rises is $y = v_0 t - \frac{1}{2}gt^2$.

$$h + y = 18, \quad so\ \tfrac{1}{2}gt^2 + v_0 t - \tfrac{1}{2}gt^2 = 18$$

$$v_0 = 20\,m/s, \quad so\ t = \frac{18}{v_0} = \frac{18\,m}{20\,m/s} = 0.90\,s$$

$$h = \tfrac{1}{2}gt^2 = \tfrac{1}{2}(4.9\,m/s^2)(0.90\,s)^2 = 3.97\,m$$

3.21. (a) $v = v_0 + at = 0$,

$$t = -\frac{v_0 a}{a} = -\frac{100 \text{ m/s}}{(2)(-5.2 \text{ m/s}^2)} = 19.2 \text{ s}$$

(b) $v^2 = v_0^2 + 2ax = 0$,

$$x = -\frac{v_0^2}{2a} = -\frac{(100 \text{ m/s})^2}{(-5.2 \text{ m/s}^2)} = 962 \text{ m}$$

3.22. To fall a distance $h = 12 \text{ m}$ from rest requires time t, where

$$h = \frac{1}{2}gt^2 \quad t^2 = \frac{2h}{g} = \frac{(2)(12 \text{ m})}{9.8 \text{ m/s}^2} \quad t = 1.56 \text{ s}$$

In 1.56 s, the truck moves a distance of $x = vt$:

$$x = \left(80 \frac{1000 \text{ m}}{3600 \text{ s}}\right)(1.56 \text{ s}) = 34.8 \text{ m}$$

The stunt man should drop when the truck is about 35 m away.

3.23. The time for the first ball to reach the ground is t_1, $y = y_0 + v_0 t - \frac{1}{2}gt^2$. Let $y = 0$ at starting point, so $y = -h = -24 \text{ m}$ at the ground.

$$v_0 = 12 \text{ m/s} \quad -24 = 0 + 12t - \frac{1}{2}(9.8)t^2 \quad 4.9t^2 - 12t - 24 = 0$$

$$t = \frac{12 \pm \sqrt{(12)^2 - 4(4.9)(-24)}}{(2)(4.9)} = 3.75 \text{ s or } -1.30 \text{ s}$$

The ball was thrown at $t = 0$, so it hits the ground at a later time, at $t = 3.75 \text{ s}$. The ball dropped from rest will require time t_2 to reach the ground, where

$$h = \frac{1}{2}gt_2^2 \quad t_2^2 = \frac{2h}{g} = \frac{2(24 \text{ m})}{9.8 \text{ m/s}^2} \quad t_2 = 2.21 \text{ s}$$

Thus the second ball should be dropped a time Δt later, where

$$\Delta t = t - t_2 = 3.75 \text{ s} - 2.21 \text{ s} = 1.54 \text{ s}$$

3.24. The object starts at rest, so $v_0 = 0$, $v_1^2 = v_0^2 - 2gy = 0 - 2g$, $y = $ distance dropped $= -h$, so $v_1^2 = 2gh$. For the stopping process, the initial velocity is v_1, so $v^2 = v_1^2 + 2a(-d)$. Note that the stopping force, and hence the stopping acceleration, is upward and positive. When stopped, $v^2 = 0 = v_1^2 - 2ad$ and $0 = 2gh - 2ad$. So $a/g = h/d$.

3.25. $v_1 = 50 \text{ km/h} = 13.9 \text{ m/s} \qquad v_2 = 100 \text{ km/h} = 27.8 \text{ m/s}$

$$v^2 = v_0^2 + 2ax = 0$$

when stopped. Thus

$$x = -\frac{v_0^2}{2a}$$

For $v_0 = v_1$,

$$x_1 = \frac{-(13.9 \text{ m/s})^2}{2(-4.9 \text{ m/s}^2)} = 19.7 \text{ m}$$

For $v_0 = v_2$,

$$x_2 = \frac{-(27.8\ \text{m/s})^2}{2(-4.9\ \text{m/s}^2)} = 78.7\ \text{m}$$

Doubling the speed increases the stopping distance by a factor of 4. Speed kills!

3.26. Suppose the binoculars were dropped from an altitude h. They would strike the ground after time t_1, where

$$h = \tfrac{1}{2}gt_1^2 \tag{i}$$

The time to reach a point 12 m above the ground is t_2, where

$$h - 12 = \tfrac{1}{2}gt_2^2 \tag{ii}$$

Also, we are told

$$t_1 - t_2 = 0.38\ \text{s} \tag{iii}$$

We thus have three simultaneous equations for the unknowns t_1, t_2, and h. Substitute (i) and (iii) into (ii):

$$\tfrac{1}{2}gt_1^2 - 12 = \tfrac{1}{2}g(t_1 - 0.38)^2$$

Solve

$$t_1 = 3.41\ \text{s}$$

$$h = \tfrac{1}{2}gt_1^2 = \tfrac{1}{2}(9.8\ \text{m/s}^2)(3.41\ \text{s})^2 = 57.1\ \text{m}$$

3.27. In 1 year the Earth travels a distance $2\pi r$, so

$$v = \frac{2\pi r}{T} = \frac{(2\pi)(1.5 \times 10^8\ \text{km})}{(365)(24\ \text{h})} = 108{,}000\ \text{km/h} = 30{,}000\ \text{m/s}$$

3.28. Stopping distance is

$$x_s = v_0 t_R + x_1 \quad v_0 = 100\ \text{km/h} = 27.8\ \text{m/s}$$

where

$$v^2 = v_0^2 + 2ax_1 = 0$$

so

$$x_s = v_0 t_R + \frac{v_0^2}{-2a} = (27.8\ \text{m/s})(0.70\ \text{s}) + \frac{1}{-2}\frac{(27.8\ \text{m/s})^2}{(-0.6)(9.8\ \text{m/s}^2)}$$

$$= 19.5\ \text{m} + 65.7\ \text{m} = 85.2\ \text{m}$$

The sheep is wiped out. The speed when $x = 32\ \text{m}$ (where the sheep is) is given by

$$v^2 = v_0^2 + 2ax \quad v^2 = (27.8\ \text{m/s})^2 + 2(-0.6)(9.8\ \text{m/s}^2)(32\ \text{m} - 19.5\ \text{m}) \quad v = 25.0\ \text{m/s}$$

Note that the driver travels 19.5 m before applying the brakes, so he slows only over a distance of $(32\ \text{m} - 19.5\ \text{m}) = 12.5\ \text{m}$.

3.29. Let $x = 2400\,\text{km} = $ one-half of the trip:

$$x_1 = v_0 t_1 + \tfrac{1}{2} a t_1^2 \quad v_0 = 0$$

so

$$t_1 \sqrt{\frac{2x_1}{a}} = \sqrt{\frac{(2)(2400 \times 10^3\,\text{m})}{(0.4)(9.9\,\text{m/s}^2)}}$$

$$= 1106\,\text{s}$$

Total trip time $= T = 2t_1 = 2212\,\text{s} \simeq 37\,\text{min}$

Speed $v = v_0 + at = 0 + (0.4)(9.8\,\text{m/s}^2)(1106\,\text{s}) = 4340\,\text{m/s} \simeq 9700\,\text{mi/h}$

3.30. $v^2 = v_0^2 + 2ax = 0$ if the sports car stopped at the end of the skid marks. Thus

$$v_0^2 = -2ax = -2(-0.9)(9.8\,\text{m/s}^2)(240\,\text{m}) \quad v_0 = 65\,\text{m/s} \simeq 146\,\text{mi/h}$$

3.31. $x = v_0 t - \tfrac{1}{2} a t^2$

$v = v_0 - at^2$

$v = \dfrac{x}{t} + \dfrac{1}{2} at - at = \dfrac{x}{t} - \dfrac{1}{2} at$

$v = \dfrac{72.9\,\text{m}}{4.5\,\text{s}} - \dfrac{(4.8\,\text{m/s}^2)(4.5\,\text{s})}{2} = 5.4\,\text{m/s}$

3.32. $v^2 = v_0^2 + 2ax$

$x = \dfrac{v^2 - v_0^2}{2a} = \dfrac{(95\,\text{m/s})^2 - 0}{2(2.20\,\text{m/s}^2)} = 2050\,\text{m}$

Motion in a Plane

Having described in the previous chapter how to define motion in one dimension, I will now extend these ideas to include motion in two or three dimensions. However, in many important problems (for example, projectiles, planetary orbits, oscillations of a pendulum), the motion is limited to a plane, so I will limit the discussion to this case. The basic features of three-dimensional motion readily follow from this treatment.

4.1 Position, Velocity, and Acceleration

For motion in one dimension, I was able to describe vector properties simply by assigning a plus or minus sign to them. Now, we must use more explicit notation to make clear the vector property. We specify the position of a particle by the **position vector r**. As the particle moves, **r** changes, as illustrated in Fig. 4-1. If at time t_1 the position vector is $\mathbf{r}_1$ and at time t_2 it is $\mathbf{r}_2$, the **displacement vector** for this time interval is defined as $\mathbf{r} = \mathbf{r}_2 - \mathbf{r}_1$.

The position, velocity, and acceleration vectors for a particle moving in the $x-y$ plane are:

$$\mathbf{r} = x\mathbf{i} + y\mathbf{j} \tag{4.1}$$

$$\mathbf{v} = \frac{d}{dt}\mathbf{r}(t) = \frac{d}{dt}(x\mathbf{i} + y\mathbf{j}) = \frac{dx}{dt}\mathbf{i} + \frac{dy}{dt}\mathbf{j} \tag{4.2}$$

$$= v_x\mathbf{i} + v_y\mathbf{j} = \mathbf{v}_x + \mathbf{v}_y$$

$$v_x = \frac{dx}{dt}, \qquad v_y = \frac{dy}{dt}, \qquad \mathbf{v}_x = \frac{dx}{dt}\mathbf{i}, \qquad \mathbf{v}_y = \frac{dy}{dt}\mathbf{j}$$

$$\mathbf{a} = \frac{dv_x}{dt}\mathbf{i} + \frac{dv_y}{dt}\mathbf{j} = a_x\mathbf{i} + a_y\mathbf{j} \tag{4.3}$$

$$a_x = \frac{dv_x}{dt} = \frac{d^2x}{dt^2}, \qquad a_y = \frac{dv_y}{dt} = \frac{d^2y}{dt^2}$$

The magnitudes of these vectors are:

$$r = |\mathbf{r}| = \sqrt{x^2 + y^2}, \qquad v = |\mathbf{v}| = \sqrt{v_x^2 + v_y^2}, \qquad a = |\mathbf{a}| = \sqrt{a_x^2 + a_y^2}$$

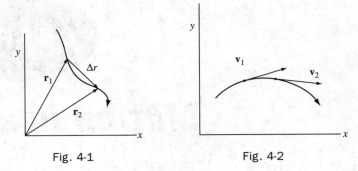

Fig. 4-1   Fig. 4-2

The velocity vector is directed tangent to the path of the particle (see Fig. 4-2). The acceleration vector can be oriented in any direction, depending on what is happening. The component of the acceleration vector that is parallel or antiparallel to **v** (that is, along the line tangent to the path) is called the **tangential acceleration**. A more descriptive name is "speeding up or slowing down acceleration." This is the kind of acceleration that measures changes in speed, as studied in the previous chapter.

4.2   Constant Acceleration

The equations obtained previously for motion in one dimension with constant acceleration apply here as well. *Caution: For problems involving motion in two dimensions, it is very important that you use subscripts to indicate if you are dealing with quantities associated with x or with y. Failure to do this is a common source of error.* If acceleration is constant, both a_x and a_y are constant.

$$
\begin{array}{ll}
x = x_0 + v_{0x}t + \frac{1}{2}a_x t^2 & y = y_0 + v_{0y}t + \frac{1}{2}a_y t^2 \\
v_x = v_{0x} + a_x t & v_y = v_{0y} + a_y t \\
v_x^2 = v_{0x}^2 + 2a_x x & v_y^2 = v_{0y}^2 + 2a_y y
\end{array}
\tag{4.4}
$$

These equations can be written in compact vector form.

$$
\mathbf{r} = \mathbf{r} + \mathbf{v}_0 t + \mathbf{a}t^2 \qquad \mathbf{v} = \mathbf{v}_0 + \mathbf{a}t \qquad \mathbf{a} = \text{constant}
\tag{4.5}
$$

In using these equations, take motion to the right as positive for x and motion upward as positive for y. The x-axis and the y-axis are normally horizontal and vertical, but any two perpendicular axes can be used, and equations of the above form will apply.

4.3   Projectiles

Consider an object that flies through the air subject to no force other than gravity and air resistance. The gravity force causes a constant downward acceleration of magnitude $g = 9.80\,\text{m/s}^2$. As a first approximation, I will neglect the effects of air and of variations in g. I assume the Earth is flat over the horizontal range of the projectiles. I neglect air effects mainly because they are complicated to include, not because they are insignificant in all cases. Despite these simplifying assumptions, we can still obtain a fairly good description of projectile motion. The path of a projectile is called its **trajectory**.

If air resistance is neglected, there is then no acceleration in the horizontal direction, and $a_x = 0$. The acceleration in the y direction is that due to gravity. It is constant and directed downward, so $a_y = -g$. It is convenient to choose $x_0 = 0$ and $y_0 = 0$ (that is, place the origin at the point where the projectile starts its motion). Further, we typically are concerned with the initial speed v_0 of the projectile. If the projectile is launched at an angle θ above horizontal, the initial velocity in the x direction and the initial velocity in the y direction can be expressed in terms of θ and v_0 using trigonometry.

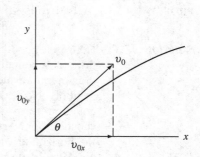

$$\boxed{v_{0x} = v_0 \cos \theta \quad \text{and} \quad v_{0y} = v_0 \sin \theta} \qquad (4.6)$$

Equation 4.3 thus becomes

$$\boxed{\begin{array}{ll} a_x = 0 & a_y = -g \\ v_x = v_0 \cos \theta = \text{constant} & v_y = v_0 \sin \theta - gt \\ x = (v_0 \cos \theta)t & y = (v_0 \sin \theta)t - \frac{1}{2}gt^2 \end{array}} \qquad (4.7)$$

From the equation for x, we can obtain $t = x/v_0 \cos \theta$. Substitute this in the equation for y and obtain

$$y = (\tan \theta)x - \left(\frac{g}{2v_0^2 \cos^2 \theta}\right)x^2 \qquad (4.8)$$

This is the equation of a parabola that passes through the origin.

A key feature of projectile motion is that **the horizontal motion is independent of the vertical motion**. Thus a projectile moves at a constant speed in the horizontal direction, independent of its vertical motion. This is illustrated in Fig. 4-3.

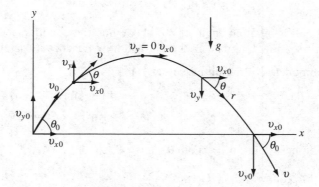

Fig. 4-3

We can gain insight into the meaning of Eq. 4.6 by viewing projectile motion in this way: First, if there were no gravity force and downward acceleration, in time t the projectile would move a distance $v_0 t$ in a straight inclined line. If now we imagine gravity "turned on," the effect would be to make the projectile fall away from the straight line path by a distance $\frac{1}{2}gt^2$. The super-position of these two effects results in the parabolic path observed (see Fig. 4-4).

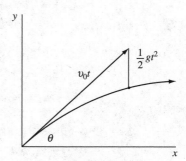

Fig. 4-4

PROBLEM 4.1.   A ball rolls off a table 80 cm high with a speed of 2.4 m/s. How far will it travel horizontally before striking the ground?

Solution   Here is a good problem-solving strategy to follow for any challenging problem. First, draw a schematic picture so that you are very clear on what is happening. In your drawing, indicate known quantities, and label unknown quantities with appropriate symbols. Next, decide what principles or laws you will apply, and write them down in equation form. Manipulate the equations to obtain the desired result, and finally, substitute numerical values. I solve this problem by reasoning as follows: I know that the horizontal velocity is constant (2.4 m/s), so I reason that if I knew the time in the air, I could find the horizontal distance since $x = v_x t$. Since falling is independent of moving sideways, I can find the time to fall down 80 cm, starting with zero vertical velocity. Thus I can find the time, and hence the horizontal distance. Notice that the problem statement did not ask for the time to fall. You have to realize on your own that the time must be found. This is typical of many "two-step" problems and is illustrative of many real-life problems such as those encountered in diagnosing a disease or designing a traffic control system. Finding the time of flight is involved in a majority of projectile problems. Here we recognize

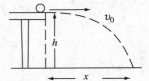

$$v_{0x} = 2.24 \text{ m/s}$$
$$v_{0y} = 0 \quad \text{(The ball rolls off horizontally.)}$$

The time to fall a distance h with zero initial vertical velocity is given by $h = \frac{1}{2} gt^2$, so $t = \sqrt{2h/g}$. The horizontal distance x is thus

$$v_{0x} t = v_{0x} 2\frac{h}{g} = (2.4 \text{ m/s})\sqrt{\frac{2(0.80 \text{ m})}{(9.8 \text{ m/s}^2)}}$$

$$= 0.97 \text{ m}$$

PROBLEM 4.2.   A golf ball is hit at an angle of 30° above horizontal with a speed of 44 m/s. How high does it rise, how long is it in the air, and how far does it travel horizontally?

Solution   The components of the initial velocity are:

$$v_{0x} = v_0 \cos \theta = 44 \cos 30 = 38.1 \text{ m/s}$$
$$v_{0y} = v_0 \sin \theta = 44 \sin 30 = 22 \text{ m/s}$$

At the highest point, $v_y = v_0 \sin \theta - gt = 0$, so

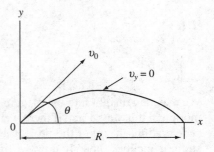

$$t = \frac{v_0 \sin \theta}{g} = \frac{(44 \text{ m/s})(\sin 30°)}{9.8 \text{ m/s}^2} = 2.24 \text{ s}$$

$$y = (v_0 \sin \theta)t - \frac{1}{2} gt^2$$

$$= (44 \text{ m/s})(\sin 30°)(2.24 \text{ s})$$

$$\quad - (0.5)(9.8 \text{ m/s}^2)(2.24 \text{ s})^2$$

$$= 24.7 \text{ m}$$

Another way of obtaining this answer is to use $v_y^2 = v_{0y}^2 - 2gy$. At the highest point $v = 0$, so $y = v_{0y}^2/2g = (22 \text{ m/s})^2/2(9.8 \text{ m/s}^2) = 24.7 \text{ m}$.

When the ball returns to ground level, $y = 0$, so T, the total time in the air, can be found from $y = (v_0 \sin \theta)T - \frac{1}{2}gT^2 = 0$. Thus, $T = 2(v_0 \sin \theta)/g = 2(44 \text{ m/s})(\sin 30°)/(9.8 \text{ m/s}^2) = 4.49 \text{ s}$. Thus the horizontal range ($x = R$ when $y = 0$) is

$$R = (v_0 \cos \theta)T = \frac{2v_0^2(\sin \theta \cos \theta)}{g} \tag{4.9}$$

Since $2 \sin \theta \cos \theta = \sin 2\theta$, we can write

$$R = \frac{(v_0^2 \sin 2\theta)}{g} \tag{4.10}$$

Thus

$$R = \frac{(44 \text{ m/s})^2(\sin 60°)}{9.8 \text{ m/s}^2} = 171 \text{ m}$$

From Eq. 4.9, we can see that the maximum range, for a given initial velocity, results when $\sin 2\theta$ is a maximum. Thus the range is a maximum when $2\theta = 90°$, or $\theta = 45°$. If air effects are taken into account, it turns out that the maximum range occurs for a slightly lower angle of elevation.

Inspection of Eq. 4.8 shows that two initial angles of elevation α and β will result in the same range provided $\alpha + \beta = 90°$. This is so because if α and β are complementary angles (they add to 90°), $\sin \alpha = \cos \beta$ and $\cos \alpha = \sin \beta$. Two such angles differ from 45° by the same amount, for example, 50° and 40°, 65° and 25°, and 71° and 19° (see Fig. 4-5).

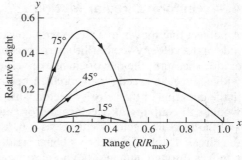

Fig. 4-5

PROBLEM 4.3. I've sometimes wondered if it is possible to throw a baseball high enough to hit the roof of a domed stadium like the King Dome in Seattle. I've seen a center fielder throw all the way from center field to home plate on the fly. Assume such a throw traveled 120 m horizontally and was thrown at an angle of elevation to maximize the range. How high would the ball go if thrown straight up?

Solution For the maximum range, $\theta = 45°$. From Eq. 4.9, I can determine v_0 given $R = 120 \text{ m}$ and $\theta = 45°$, Thus

$$120 \text{ m} = \frac{v_0^2 \sin 2(45°)}{9.8 \text{ m/s}^2} \qquad v_0 = 34.3 \text{ m/s}$$

If thrown straight up, $\theta = 90°$ and $v_{0y} = v_0 = 34.3 \text{ m/s}$. $v_y^2 = v_{0y}^2 - 2gy = 0$ at the highest point, so

$$y = \frac{v_0^2}{2g}$$

$$= \frac{(34.3 \text{ m/s})^2}{2(9.8 \text{ m/s}^2)} = 60 \text{ m}$$

Incidentally, notice that this result, maximum range $= 2 \times$ maximum height, is true generally.

PROBLEM 4.4. An archer standing on a cliff 48 m above the level field below shoots an arrow at an angle of 30° above horizontal with a speed of 80 m/s. How far from the base of the cliff will the arrow land?

Solution   I could use Eq. 4.7 to find x, since we are given $y = -48$ m, $\theta = 30°$, and $v_0 = 80$ m/s. This will require solving a quadratic equation. An alternate (but equivalent) approach is to find the time in the air and then determine the range from $x = (v_0 \cos \theta)t$. Thus

$$y = (v_0 \sin \theta)t - \tfrac{1}{2}gt^2 - 48$$

$$= (80 \sin 30°)t - (0.5)(9.8)t^2$$

$$4.9t^2 - 40t - 48 = 0$$

$$t = \frac{40 \pm \sqrt{(40)^2 - 4(4.9)(-48)}}{2(4.9)} = 9.23 \text{ s or} -1.06 \text{ s}$$

$$x = (v_0 \cos \theta)t = (80 \text{ m/s})(\cos 30°)(9.23 \text{ s}) = 639 \text{ m}$$

4.4   Uniform Circular Motion

An object that moves in a circle with constant speed is in **uniform circular motion**. Although the magnitude of the velocity vector (the speed) is constant, the direction of the velocity is changing. Recall that acceleration measures the rate of change of velocity. In the previous chapter, I discussed acceleration associated with changes in speed ("tangential" acceleration). Here we consider acceleration associated with a change in the direction of the velocity vector. This is what I would call "turning acceleration" but what other books would call **centripetal acceleration** or **radial acceleration**. Fig. 4-6 illustrates how the position vector **r** and the velocity vector **v** change as a particle moves around a circle. The velocity vector **v** is tangent to the circle. Think of **v** and **r** as being rigidly joined together, like the sides of a carpenter's square. When **r** moves through an angle θ, **v** moves through the same angle.

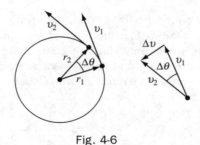

Fig. 4-6

Suppose the particle moves through a very small angle $\Delta\theta$. Slide the two velocity vectors together so that they form a small isosceles triangle with sides v_1, v_2, and Δv. The angle between the two sides of length v is $\Delta\theta$. Now consider the isosceles triangle formed by $\mathbf{r}_1$ and $\mathbf{r}_2$. If $\Delta\theta$ is very small, the short side of this triangle is approximately equal to the arc length Δs subtended by $\Delta\theta$. If $\Delta\theta$ is measured in radians, $\Delta s = r\Delta\theta$. These two isosceles triangles are similar triangles, so their sides are in the same ratio. Thus

$$\frac{\Delta v}{v} = \frac{\Delta s}{r} = \frac{v\Delta t}{r}$$

where

$$\Delta s = v\Delta t$$

so

$$\frac{\Delta v}{\Delta t} = \frac{v^2}{r}$$

As

$$\Delta t \to 0, \quad \text{then } a_c = \frac{\Delta v}{\Delta t} \quad \text{so} \quad \boxed{a_c = \frac{v^2}{r}} \tag{4.11}$$

This turning acceleration is called *centripetal acceleration* because this acceleration vector is directed "toward the center." One can see this, since $\mathbf{a} \simeq \Delta\mathbf{v}/t$ and $\Delta\mathbf{v}$ points inward toward the center of curvature. Note that the particle does not have to move in a full circle to experience centripetal acceleration. Any arc can be thought of as a small part of a circle. Remember, if an object turns left, it accelerates left. If it turns right, it accelerates right. Incidentally, you may have heard the expression "centrifugal acceleration." Forget you ever heard this term, and never, never use it. It will only confuse you. Long ago this term was used in connection with a confusing notion of fictitious forces.

An object can experience both centripetal (turning) acceleration and tangential (speeding up or slowing down) acceleration. In Fig. 4-7 some possible combinations for $\mathbf{v}$ and $\mathbf{a}$ for a moving car are shown. To understand the acceleration, resolve it into components parallel to $\mathbf{v}$ and perpendicular to $\mathbf{v}$. To tell if the car is turning right or left, imagine that you are the driver sitting with the velocity vector directed straight ahead in front of you. A forward component of acceleration means speeding up.

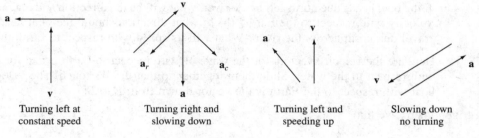

| Turning left at constant speed | Turning right and slowing down | Turning left and speeding up | Slowing down no turning |

Fig. 4-7

PROBLEM 4.5. A military jet fighter plane flying at 180 m/s pulls out of a vertical dive by turning upward along a circular path of radius 860 m. What is the acceleration of the plane? Express the acceleration as a multiple of g.

Solution
$$a = \frac{v^2}{r} = \frac{(180\,\text{m/s})^2}{860\,\text{m}} = 37.7\,\text{m/s}^2 = \frac{37.7}{9.8g} = 3.8g$$

4.5 Relative Motion

To describe motion, we must refer it to a **frame of reference**. Often we use a reference frame attached to the surface of the Earth or to the floor of the room. On a moving train, we might use the floor of the train car as the reference frame. If a person in a train moving at constant velocity drops a pencil, he will see it fall straight down. A person on the ground will see the pencil drop along a parabolic path. We frequently encounter such problems in connection with navigation. To solve such problems, label the tip of a velocity vector with a symbol representing the moving object. Label the tail of the velocity vector with a symbol representing the reference frame. To see how this works, consider the following example.

PROBLEM 4.6.   A person P walks at a speed of 1.5 m/s on a moving sidewalk SW in an airport terminal. The sidewalk moves at 0.8 m/s. How fast is the person moving with respect to the Earth E?

Solution   Draw to scale the velocity vector for the person with respect to the sidewalk, $v_{PSW} = 1.5$ m/s. Next, draw the velocity vector for the sidewalk with respect to the Earth, $v_{SWE} = 0.8$ m/s.

$$E \longrightarrow SW \qquad SW \longrightarrow P$$

Now slide these vectors parallel to themselves so that matching symbols are superimposed.

$$\begin{array}{c} SW \\ E \longrightarrow P \\ E \longrightarrow P \end{array}$$

The vector representing the velocity of the person P with respect to the Earth E is drawn from E to P, as shown. From the diagram we see $v_{PE} = 1.5$ m/s $+ 0.8$ m/s $= 2.3$ m/s. The usefulness of this technique is illustrated by the following, more complicated problem.

PROBLEM 4.7.   A river flows due east at 5 km/h. A motorboat can move through the water at 12 km/h. (a) If the boat heads due north across the river, what will be the direction and magnitude of its velocity with respect to the Earth? (b) In what direction should the boat head if it is to travel due north across the river? What will its speed with respect to Earth then be?

Solution   Here are the velocity vectors for the water W with respect to Earth E and for the boat B with respect to the water. Slide them together so that the Ws touch. The velocity of the boat with respect to the Earth is the vector drawn from E to B.

(a) Using trigonometry we find

$$v_{BE}^2 = (12 \text{ km/h})^2 + (5 \text{ km/h})^2$$

$$v_{BE} = 13 \text{ km/h}$$

$$\theta = \tan^{-1} \frac{5}{12} = 22.6° \text{ E of N}$$

The boat will travel 22.6° east of north.

(b)
$$v_{BE}^2 + v_{WE}^2 = v_{BW}^2$$

$$v_{BE}^2 = (12 \text{ km/h})^2 - (5 \text{ km/h})^2$$

$$v_{BE} = 10.9 \text{ km/h}$$

$$\theta = \sin^{-1} \frac{5}{12} = 24.6° \text{ W of N}$$

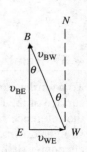

PROBLEM 4.8. A pilot with an airspeed (speed with respect to air) of 120 km/h wishes to fly due north. A 40-km/h wind is blowing from the northeast. In what direction should she head, and what will be her ground speed (speed with respect to the ground)?

Solution

$$(v_{PE} + v_{AE} \cos 45°)^2 + (v_{AE} \sin 45°)^2 = v_{PA}^2$$

$$(v_{PE} + 40 \cos 45°)^2 = (120)^2 - (40 \sin 45°)^2$$

$$v_{PE} = 88.3 \text{ km/h}$$

$$\theta = \sin^{-1} \frac{v_{AE} \sin 45°}{v_{PA}} = \sin^{-1} \frac{(40 \sin 45°)^2}{120} = 13.6° \text{ E of N}$$

The plane should head 13.6° east of north.

4.6   Summary of Key Equations

For constant acceleration in x and y directions,

$$x = x_0 + v_{0x}t + \tfrac{1}{2}a_x t^2 \qquad y = y_0 + v_{0y}t + \tfrac{1}{2}a_y t^2$$

$$v_x = v_{0x} + a_x t \qquad v_y = v_{0y} + a_y t$$

$$v_x^2 = v_{0x}^2 + 2a_x \qquad v_y^2 = v_{0y}^2 + 2a_y y$$

For projectiles,

$$v_{0x} = v_0 \sin \theta \qquad \text{and} \qquad v_{0y} = v_0 \cos \theta$$
$$a_x = 0 \qquad a_y = -g$$
$$v_x = v_0 \cos \theta = \text{constant} \qquad v_y = v_0 \sin \theta - gt$$
$$x = (v_0 \cos \theta)t \qquad y = (v_0 \sin \theta)t - \tfrac{1}{2}gt^2$$

The equation of the path is a parabola.

$$y = (\tan \theta)x - \left(\frac{g}{2v_0^2 \cos 2\theta} \right)x^2$$

The time in the air is

$$T = \frac{2v_0 \sin \theta}{g}$$

The horizontal range is a maximum for $\theta = 45°$, where

$$R = \frac{v_0^2 \sin 2\theta}{g}$$

The maximum height is

$$h = \frac{v_0^2 \sin^2 \theta}{2g}$$

For uniform circular motion the inward centripetal (radial) acceleration is

$$a_c = \frac{v^2}{r}$$

SUPPLEMENTARY PROBLEMS

4.9. An unidentified naval vessel is tracked by the Navistar Global Positioning System. With respect to a coordinate origin $(0, 0)$ fixed at a lighthouse beacon, the position of the vessel is found to be $x_1 = 2.0$ km west, $y_1 = 1.6$ km south at $t_1 = 0.30$ h and $x_2 = 6.4$ km west, $y_2 = 6.5$ km north at $t_2 = 0.60$ h. Using east–west as the x-axis and north–south as the y-axis, determine the average velocity in terms of its components. What are the direction and magnitude of the average velocity in kilometers per hour?

4.10. The track of a cosmic ray particle in a photographic emulsion is found empirically to be described by the expression $\mathbf{r} = (3t^3 - 6t)\mathbf{i} + (5 - 8t^4)\mathbf{j}$. Determine the velocity and acceleration.

4.11. A sandbag is dropped from rest from a hot air balloon at an altitude of 124 m. A horizontal wind is blowing, and the wind gives the sandbag a constant horizontal acceleration of 1.10 m/s^2. (a) Show that the path of the sandbag is a straight line. (b) How long does it take to hit the ground? (c) With what speed does it hit the ground?

4.12. A charged dust particle generated in an environmental study of smoke stack efficiency moves through a velocity selection device with constant acceleration $\mathbf{a} = 4\mathbf{j}$ m/s^2 with an initial velocity of $\mathbf{v} = 6\mathbf{i}$ m/s. Determine the speed and position of the particle when $t = 4$ s.

4.13. An artillery shell is fired so that its horizontal range is twice its maximum height. At what angle is it fired?

4.14. A motorcycle rider wants to jump a ditch 4 m wide. He leaves one side on a ramp that slopes up at $20°$ above horizontal. He lands at the same elevation at which he took off. His front wheel leaves the ground 1 m before the edge of the ditch and comes down 2 m past the far side of the ditch. What minimum take-off speed is required?

4.15. During World War I the Germans reportedly bombarded Paris from about 50 km away with a long-barreled cannon called the Big Bertha. Iraq was suspected of building a similar weapon to launch nuclear bombshells on Israel in 1992. Neglecting air resistance, estimate the muzzle velocity needed by the Big Bertha. Muzzle velocity is the initial speed at which the shell leaves the gun.

4.16. Migrating salmon are known to make prodigious leaps when swimming up rivers. The highest recorded jump by such a fish was 3.45 m upwards. Assuming the fish took off at an angle of $45°$ horizontal, with what speed did the fish leave the water?

4.17. A rifle bullet is fired with a speed of 280 m/s up a plane surface that is inclined at $30°$ above horizontal. The bullet is fired at an initial angle of elevation of $45°$ above horizontal (that is, at $15°$ above the plane surface). How far up the plane does it land? (Problems like this are discussed in I. R. Lapides, *Amer. J. Phys.,* **51** (1983), p. 806 and H. A. Buckmaster, *Amer. J. Phys.,* **53** (1985), p. 638.)

4.18. A girl throws a ball from a balcony. When the ball strikes the ground, its path makes an angle θ with the ground. What is the minimum value of θ?

4.19. A high-powered 7-mm Remington magnum rifle fires a bullet with a velocity of 900 m/s on a rifle range. Neglect air resistance. (a) Calculate the distance h such that a bullet will drop at a range of 200 m when fired horizontally. (b) To compensate for the drop of the bullet, when the telescope sight is pointed right at the target, the barrel of the gun is aligned to be slanted slightly upward, pointed a distance h above the target. The downward fall due to gravity then makes the bullet strike the target as desired. Suppose, however, such a rifle is fired uphill at a target 200 m distant. If the upward slope of the hill is $45°$, should you aim above or below the target, and by how much? What should you do when shooting on a *downhill* slope at $45°$ below horizontal?

4.20. The radius of the Earth is about 6370 km. Calculate the centripetal acceleration of a person at the equator.

4.21. An electric fan rotates at 800 revolutions per minute (rev/min). Consider a point on the blade a distance of 0.16 m from the axis. Calculate the speed of this point and its centripetal acceleration.

4.22. The fastest train in the United States is the Acela Express with a top speed of about 150 mph. Train passengers find the ride slightly uncomfortable if their acceleration exceeds 0.05g. (a) What is the smallest radius of curvature for a bend in the track that can be tolerated within this limit? (b) If the train had to go around a curve of radius 1.20 km, to what speed would the train have to be slowed in order not to exceed an acceleration of $0.05g$?

4.23. A race car driver increases her speed uniformly from 60 to 66 m/s in a period of 4.0 s while rounding a curve of radius 660 m. At the instant when her speed is 63 m/s, what is the magnitude of her tangential acceleration, her centripetal acceleration, and her total acceleration?

4.24. The pilot of a passenger jet with an airspeed of 700 km/h wishes to fly 1400 km due north. To move to the north, the pilot finds he must fly in a direction pointed 10° west of north. If the flight requires 1 h 54 min, what is the wind velocity?

4.25. A passenger in a car traveling 11 m/s (about 25 mi/h) notices that raindrops outside seem to be falling at an angle of about 60° with vertical. From this data, what would you estimate the speed of the falling raindrops to be? (Incidentally, because of air resistance, the rain is falling with constant velocity by the time it approaches the ground.)

4.26. A moving sidewalk in an airport terminal moves at 1.20 m/s. It is 80 m long. A man steps on the sidewalk and walks to the other end at a speed of 0.8 m/s with respect to the sidewalk. How long does it take him to reach the other end?

4.27. A river 86 m wide flows due west at 2.2 m/s. A man in a boat heads due south with respect to the water, moving at a speed of 4.8 m/s through the water. How long does it take him to cross the river? How far west of his starting point does he land?

4.28. Kate can swim 0.90 m/s. She tries to swim across a river that is flowing 1.80 m/s. She heads in a direction that will minimize her drift downstream, but she still lands 120 m downstream from the point directly across from where she started. In what direction did she swim, and how wide was the river?

4.29. The currents in the Strait of Juan de Fuca at the entrance to Puget Sound can be very swift. Travel there in a small fishing boat can be hazardous. Suppose the current is coming in from the open sea at a speed of 23 km/h, directed due east. A fisherman wants to travel north from Port Angeles to Victoria, British Columbia, a distance of about 48 km. He needs to make the trip in 2 h 15 min, but he isn't sure if his boat is fast enough. What minimum speed would he need? (A boat's speed is measured with respect to the water it moves through.)

4.30. For what launch angle is the range of a projectile equal to the maximum height it reaches?

4.31. A daredevil motorcycle rider wishes to jump a canyon of width w. He takes off horizontally with speed v. His landing spot on the opposite side of the canyon is a distance h below his take-off point. What minimum speed v is required for a safe landing on the other side? Express the answer in terms of g, w, and h.

SOLUTIONS TO SUPPLEMENTARY PROBLEMS

4.9.

$$\bar{v} = \frac{x_2 - x_1}{t_2 - t_1} = \frac{(-6.4\mathbf{i} + 6.5\mathbf{j}) - (-2.0\mathbf{i} - 1.6\mathbf{j})}{0.60 - 0.30}$$

$$\mathbf{v} = -8.4\mathbf{i} + 4.9\mathbf{j} \text{ km/h} \qquad v = \sqrt{(-8.4)^2 + (4.9)^2} = 9.72 \text{ km/h}$$

$$\theta = \tan^{-1}\frac{v_y}{v_x} = \frac{4.9}{-8.4} = -30.3°$$

that is,

30.3° N of W

4.10. $\mathbf{r} = (3t^2 - 6t)\mathbf{i} + (5 - 8t^4)\mathbf{j} \qquad \mathbf{v} = \dfrac{dr}{dt} = (6t - 6)\mathbf{i} - 32t^3\mathbf{j} \qquad \mathbf{a} = \dfrac{dv}{dr} = 6\mathbf{i} - 96t^2\mathbf{j}$

4.11. $x = v_{0x}t + \frac{1}{2}a_xt^2 = \frac{1}{2}a_xt^2$ since $v_{0x} = 0$ $y = v_{0y}t - \frac{1}{2}gt^2 = -\frac{1}{2}gt^2$ since $v_{0y} = 0$

(a) Thus,

$$t^2 = \frac{2x}{ax} = -\frac{2y}{g} \quad \text{or} \quad y = -\frac{g}{a_x} \times \text{equation of a straight line}$$

(b) At ground,

$$y = -124\,\text{m}, \quad \text{so } t^2 = -\frac{2(-124\,\text{m})}{9.8\,\text{m/s}^2} \quad t = 5.03\,\text{s}$$

(c) $$v_y = v_{0y} - gt = 0 - (9.8\,\text{m/s}^2)(5.03\,\text{s}) = -49.3\,\text{m/s}$$

$$v_x = v_{0x} + a_xt = 0 + (1.10\,\text{m/s}^2)(5.03\,\text{s}) = 5.53\,\text{m/s}$$

$$v = \sqrt{v_x^2 + v_y^2} = \sqrt{(5.53\,\text{m/s}^2 + (-49.3\,\text{m/s})^2} = 49.6\,\text{m/s}$$

4.12. $\mathbf{r} = \mathbf{v}_0t + \frac{1}{2}\mathbf{a}t^2 = 6t\mathbf{i} + \frac{1}{2}(4\mathbf{j})t^2$ at $t = 4\,\text{s}$, $\mathbf{r} = 24\mathbf{i} + 32\mathbf{j}$

4.13. $v_y^2 = v_{0y}^2 - sgy = 0$ at the highest point. $v_{0y} = v_0 \sin\theta$, so

$$y = \frac{v_0^2 \sin\theta}{2g}$$

The range from Eq. 4.8 is

$$R = \frac{2v_0^2}{g} \sin\theta\cos\theta$$

Here, $R = 2y$. Thus,

$$\frac{2v_0^2}{g}\sin\theta\cos\theta = 2\frac{v_0^2\sin^2\theta}{2g}, \quad \cos\theta = \frac{1}{2}\sin\theta \quad \tan\theta = 2, \quad \theta = 63.4°$$

4.14.   From Eq. 4.9, the range is

$$R = \frac{v_0^2}{g}\sin 2\theta \quad v_0^2 = \frac{Rg}{\sin 2\theta} \quad v_0^2 = \frac{(7\,\text{m})(9.8\,\text{m/s}^2)}{\sin 2(20°)} \cdot \quad v_0 = 10.3\,\text{m/s}$$

4.15.   From Eq. 4.9, the range is

$$R = \frac{v_0^2}{g}\sin^2 2\theta$$

θ would have been chosen for the maximum range, so $2\theta = 90°$.

$$v_0^2 = Rg = (50 \times 10^3\,\text{m})(9.8\,\text{m/s}^2) \quad v_0 = 700\,\text{m/s}$$

4.16.   At the highest point $v_y = 0$, so $v_y^2 = (v_0 \sin\theta)^2 - 2gh = 0$

$$(v_0 \sin\theta)^2 = 2gh \quad v_0^2 = \frac{2gh}{\sin\theta} = \frac{2(9.8\,\text{m/s}^2)(3.45\,\text{m})}{\sin 45°} \quad v_0 = 9.78\,\text{m/s}$$

4.17. The equation of the inclined plane is

$$y = \tan 30° \quad x = \frac{1}{\sqrt{3}}x$$

The equation of the parabolic path is given in Eq. 4.7:

$$y = (\tan \theta)x - \left(\frac{g}{2v_0^2 \cos^2 \theta}\right)x^2$$

The intersection of the parabola and the straight line occurs when

$$\frac{1}{\sqrt{3}}x = \tan(\tan \theta)x - \frac{g}{2v_0^2 \cos^2 \theta}x^2$$

Substitute $\tan 45° = 1, \cos 45° = 1/\sqrt{2}$ and simplify. Find $x = v_0^2/g[1 - (1/\sqrt{3})]$. From the $30°-60°-90°$ triangle we see that $x = s \cos 30° = s(\sqrt{3}/2)$. So

$$s = \left(\frac{2}{\sqrt{3}}\right)\left(1 - \frac{1}{\sqrt{3}}\right)\frac{v_0^2}{g} = \frac{2}{3}\left(\sqrt{3} - 1\right)\frac{v_0^2}{g} = \text{distance up the plane}$$

With $v_0 = 280 \, \text{m/s}$, $s = 3.90 \, \text{km}$. Another way of solving this problem is the following: If there were no gravity, the bullet would go straight to point P, a distance $v_0 t$ reached in time t. If gravity were "turned on," the bullet would also drop a distance $\frac{1}{2}gt^2$, to where it would hit the plane. By looking at the diagram, I see that

$$x = h + H$$
$$= \tfrac{1}{2}gt^2 + s \sin 30° \tag{i}$$

Also, $x = v_0 t \cos 45° = s \cos 30°$, so substitute

$$t = \frac{\cos}{v_0 \cos 45°}s \quad \text{and} \quad x = s \cos 30°$$

in Eq. (i), this yields

$$x = \frac{v_0^2}{g}\left(1 - \frac{1}{\sqrt{3}}\right)$$

as before, and

$$s = \frac{x}{\cos 30°} = \frac{2}{3}(\sqrt{3} - 1)\frac{v_0^2}{g}$$

4.18. At the ground, $\tan \theta = v_y/v_x$, where $v_x = v_{0x} = v_0 \cos \theta$.

$$v_y^2 = (v_0 \sin \theta)^2 + 2gh \quad \text{at } y = -h \text{ at ground}$$

Thus,

$$\tan \theta = \frac{\sqrt{(v_0 \sin \theta)^2 + 2gh}}{v_0 \cos \theta} = \left[\left(\frac{\sin \theta}{\cos \theta}\right)^2 - \frac{2gh}{(v_0 \cos \theta)^2}\right]^{1/2}$$

We need to minimize $\tan\theta$ with respect to variations in the launch angle θ. If $\tan\theta$ is a minimum, $\tan^2\theta$ is also a minimum. Minimizing $\tan^2\theta$ simplifies the math. Thus,

$$\frac{d(\tan^2\theta)}{d\theta} = \frac{2\sin\theta\cos\theta}{\cos^2\theta} - \frac{\sin^2\theta(-2\sin\theta)}{\cos^3\theta} - 2gh\frac{(-2)(-\sin\theta)}{\cos^3\theta} = 0$$

$$\frac{2\sin\theta}{\cos^3\theta}[\cos^2\theta + \sin^2\theta - 4gh] = 0 \quad \text{or} \quad \frac{\sin\theta}{\cos^3\theta}[1 - 4gh] = 0$$

This can only be true if $\sin\theta = 0$ or $\theta = 0°$. Substitute this back in $\tan\theta$:

$$\tan\theta = \frac{v_0\sin^2\theta + 2gh}{v_0\cos\theta}, \quad \sin 0° = 0, \quad \cos 0° = 1, \quad \text{so } \theta = \tan^{-1}\frac{\sqrt{2gh}}{v_0}$$

4.19. The time for the bullet to travel a range R is approximately

$$t = \frac{R}{v_x} \simeq \frac{R}{v_0} = \frac{200\,\text{m}}{900\,\text{m/s}} = 0.22\,\text{s}$$

In this time it drops a distance of $h = 1/2\,gt^2 = 1/2(9.9\,\text{m/s}^2)(0.22\,\text{s})^2$.

(a) $h = 0.242\,\text{m}$.

(b) The sight is aligned on the assumption that after 200 m the bullet will have moved 0.242 m *perpendicular to the path*. But when shooting up a 45° slope, the bullet falls back at 45° to the trajectory not perpendicular to it. Thus as shown in the drawing, the bullet will strike above the target by an amount H, where

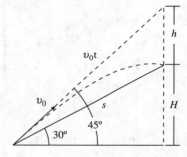

$$\frac{h}{h+H} = \sin 45° = \frac{1}{\sqrt{2}}, \text{ so } H = \left(\sqrt{2}-1\right)h$$

Thus when shooting uphill, you must aim low to hit the target. The same reasoning applies when shooting downhill. *You must aim low when shooting either uphill or downhill.* Once my friend won a $50 bet with a gun nut who couldn't believe this.

A way of reasoning this out qualitatively is to consider the limiting cases of shooting straight up or straight down. Then gravity doesn't deviate the bullet from the barrel axis at all, and you will definitely hit high.

4.20. The circumference of the Earth is $s = 2\pi R$. The Earth rotates once in time $T = 24\,\text{h}$, so the velocity of a point on the equator is

$$v = \frac{2\pi R}{T} = \frac{(2\pi)(6370 \times 10^3\,\text{m})}{(24)(3600\,\text{s})} = 463\,\text{m/s}$$

The centripetal (radial) acceleration is

$$a_c = \frac{v^2}{R} = \frac{(463\,\text{m/s})}{(6.37 \times 10^6\,\text{m})} = 0.034\,\text{m/s}^2 \simeq 0.003g$$

4.21.
$$v = \frac{2\pi r}{T}, \quad T = \text{time for 1 rev}, \quad T = \frac{1}{800}\,\text{min} = \frac{1}{800}(60\,\text{s}) = 0.075\,\text{s}$$

$$v = \frac{2\pi(0.16\,\text{m})}{0.075\,\text{s}} = 13.4\,\text{m/s}, \quad a_c = \frac{v^2}{r} = \frac{(13.4\,\text{m/s})^2}{0.16\,\text{m}} = 1120\,\text{m/s}^2$$

4.22.
$$a_c = \frac{v^2}{r} = 0.05g$$

(a) If $v = 70\,\text{m/s}$, then:

$$r = \frac{(70\,\text{m/s})^2}{(0.05)(9.8\,\text{m/s}^2)} = 10\,\text{km}$$

(b) If $r = 1.20\,\text{km}$, then:

$$v = \left[(1200\,\text{m})(0.05)(9.0\,\text{m/s}^2)\right]^{1/2} = 24\,\text{m/s}$$

4.23.
$$a_T \simeq \frac{\Delta V_T}{\Delta t} = \frac{66\,\text{m/s} - 60\,\text{m/s}}{4\,\text{s}} = 1.5\,\text{m/s}^2, \qquad a_c = \frac{v^2}{r} = \frac{(63\,\text{m/s})^2}{660\,\text{m}} = 6.01\,\text{m/s}^2$$
$$a = \sqrt{a_T^2 + a_c^2} = \sqrt{(1.5\,\text{m/s}^2)^2 + (6.0)\,\text{m/s}^2} = 6.20\,\text{m/s}^2$$

4.24. Ground speed is

$$v_{\text{PE}} = \frac{1400\,\text{km}}{1.9\,\text{h}} = 737\,\text{km/h}$$

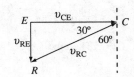

The velocity of the plane with respect to air is 700 km/h directed 10° west of north, so the velocity vectors can be drawn. Using the law of cosines,

$$v_{\text{AE}}^2 = v_{\text{PA}}^2 + v_{\text{PE}}^2 - 2v_{\text{PA}}v_{\text{PE}}\cos 10°$$
$$= (700)^2 + (737)^2 - 2(700)(737)\cos 10°$$
$$v_{\text{AE}} = 131\,\text{km/h}$$

4.25. Label car reference frame as C.

$$v_{\text{CE}} = 11\,\text{m/s}$$
$$v_{\text{RE}} = v_{\text{CE}}\tan 30° = 11\tan 30° = 6.4\,\text{m/s}$$
$$v_{\text{ME}} = v_{\text{MSW}} + v_{\text{SWE}} = 0.8\,\text{m/s} + 1.20\,\text{m/s} = 2.0\,\text{m/s}$$

4.26.
$$t = \frac{x}{v_{\text{ME}}} = \frac{80\,\text{m}}{2\,\text{m/s}} = 40\,\text{s}$$

4.27.
$$v_{\text{WE}} = 2.2\,\text{m/s}, \quad v_{\text{BW}} = 4.8\,\text{m/s}$$
$$v_{\text{BE}} = \sqrt{v_{\text{WE}}^2 + v_{\text{BW}}^2} = 5.28\,\text{m/s}$$

The speed perpendicular to the river is

$$v_{\text{BW}}, \text{ so } t = \frac{d}{v_{\text{BW}}} = \frac{86\,\text{m}}{4.8\,\text{m/s}} = 17.9\,\text{s}$$

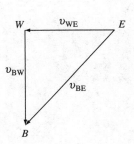

In this time he drifts downstream a distance of

$$L = v_{\text{WE}}t = (2.2\,\text{m/s})(17.9\,\text{s}) = 39.4\,\text{m}$$

4.28. Suppose she heads upstream at an angle θ from straight across. Her velocity component headed across the river is $v_{KW} \cos \theta$. Thus the time to cross is $t = W/v_{KW} \cos \theta$. In this time she drifts down-stream a distance $(v_{KE}/v_{KW} \cos \theta)W$.

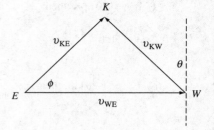

$$W = \text{river width}$$
$$v_{KW} = 0.90 \,\text{m/s}$$
$$v_{WE} = 1.80 \,\text{m/s}$$

But $v_{KE} \cos \theta = v_{WE} - v_{KW} \sin \theta$, so

$$x = \left(\frac{v_{WE} - v_{KW} \sin \theta}{v_{KW} \cos \theta} \right) W$$

since

$$v_{WE} = 2v_{KW}, \quad x = \left(\frac{2 - \sin \theta}{\cos \theta} \right) W \tag{i}$$

Minimize x with respect to θ:

$$\frac{dx}{d\theta} = \left[\frac{-\cos \theta}{\cos \theta} + \frac{(2 - \sin \theta)(\sin \theta)}{\cos^2 \theta} \right] W = 0$$

$$-1 + \frac{2 \sin \theta - \sin^2 \theta}{\cos^2 \theta} = 0, \quad -\cos^2 \theta + 2 \sin \theta - \sin^2 \theta = 0, \quad 2 \sin \theta = 1, \quad \theta = 30°$$

From the drawing, I see that $v_{KE} \sin \theta = v_{KW} \cos \theta$ and $v_{KE} \cos \theta = v_{WE} - v_{KW} \sin \theta$. Divide

$$\frac{v_{KE} \sin \theta}{v_{KE} \cos \theta} = \tan \phi = \frac{v_{KW} \cos \theta}{v_{WE} - v_{KW} \sin \theta} = \frac{\sqrt{3}/2}{2 - 1/2} = \sqrt{3}$$

so $\phi = 60°$. From Eq. (i),

$$x = \frac{2 - \sin \theta}{\cos \theta} \, W \quad \text{so} \quad 120 = \frac{2 - \sin 30°}{\cos 30°} W$$

Kate heads 30° upstream.

4.29. The boat's speed with respect to Earth is

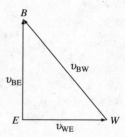

$$v_{BE} = \frac{48 \,\text{km}}{2.25 \,\text{h}} = 21.3 \,\text{km/h}$$

The water speed with respect to the Earth is $v_{WE} = 23 \,\text{km/h}$. Thus $v_{BW}^2 = (23 \,\text{km/h})^2 + (21.3 \,\text{km/h})^2$, and $v_{BW} = 31.3 \,\text{km/h}$.

4.30.
$$h = \frac{v_0^2 \sin^2 \theta}{2g}$$

$$R = \frac{v_0^2 \sin(2\theta)}{g} = \frac{2v_0^2 \sin \theta \cos \theta}{g}$$

Using $\sin(2\theta) = 2 \sin \theta \cos \theta$

$$h = R, \text{ so } \frac{v_0^2 \sin^2 \theta}{2g} = \frac{2v_0^2 \sin \theta \cos \theta}{g}$$

$$\sin \theta = 4 \cos \theta$$

$$\tan \theta = \frac{\sin \theta}{\cos \theta} = 4$$

$$\theta = 76°$$

4.31. The time to fall a distance h is given by

$$h = \tfrac{1}{2}gt^2$$

In this time t the motorcycle travels a horizontal distance w, where $w = vt$ thus, $v = \dfrac{w}{t} = \dfrac{w}{\sqrt{2h/g}} = w\sqrt{\dfrac{g}{2h}}$

Newton's Laws of Motion

I turn now to the question of dynamics, that is, the study of how forces influence motion. Dynamics is involved in understanding a wide range of phenomena. Begin by observing that forces result from interactions between two or more objects, so if A interacts with B, then B also interacts with A. Thus forces occur in pairs. The basic ideas needed are stated as Newton's laws of motion. It is helpful to consider the third law first.

5.1 Newton's Third Law of Motion

If body A pulls or pushes on body B, then body B also pulls or pushes on body A. The force on each body has the same magnitude, but the forces are oppositely directed.

The idea makes sense. Suppose Bill and Mary pull on opposite ends of a rope, as in a tug-of-war. Imagine that there is a spring scale in the rope, of the kind used in a market to weigh fish or vegetables. If Bill wants to know how hard he is pulling, he looks down and reads the scale. If Mary wants to know how hard she is pulling, she looks down and reads the same scale. Both persons will always see the same reading, and this force is called the **tension** in the rope. They pull with forces of equal magnitudes. The same is true if they are pushing. Suppose they are pushing on opposite sides of a bathroom scale. Each will read the same scale reading to determine how hard his or her push is. The readings must be the same for both people, since they read the same scale. This principle is true in *all* cases, no matter if the people are moving or if one is stronger or heavier. In SI units, force is measured in newtons, and $4.45\,\text{N} = 1\,\text{lb}$.

5.2 Newton's First Law of Motion

Motion must be referred to a reference frame in order to be described. Reference frames in which Newton's laws are valid are called **inertial frames**. More specifically, an inertial reference frame is one in which Newton's first law of motion is valid.

Consider a body acted on by no net force. If it is at rest, it will remain at rest. If it is moving, it will continue to move with constant velocity.

By a "net force" I mean the vector sum of all external forces acting on the object. For example, if you push on a book from the left with a horizontal force of 10 N and from the right with a force of 8 N, the net force acting is 2 N.

Any reference frame that moves with constant velocity with respect to an inertial reference frame is also an inertial reference frame. A reference frame that is accelerating with respect to an inertial reference frame will *not* be an inertial reference frame.

We will usually use a reference frame attached to the Earth's surface (the "laboratory frame"), but this is not in fact an inertial frame, since the Earth is rotating, and rotating objects are accelerating. However, for many purposes the rotation of the Earth is sufficiently slow so that it will have a negligible effect on our calculations. However, for certain phenomena the effects of the Earth's rotation are noticeable and must be included. Examples of the latter include motion of large air masses or of ocean currents, or the motion of an intercontinental ballistic missile or the trajectory of a long-range artillery shell.

5.3   Newton's Second Law of Motion

*If a net force **F** acts on an object of mass* m, *the object will have acceleration **a**, where*

$$\boxed{\mathbf{F} = m\mathbf{a}} \tag{5.1}$$

F is measured in newtons, **a** is measured in meters per second per second, and m is measured in kilograms.

Do not confuse mass with weight. Mass is proportional to weight at a given point, but as you move far away from the Earth, where the force of gravity on an object, and hence its weight, decreases, the mass does not change. Mass is a measure of the amount of "stuff" in an object.

The force of gravity acting on an object is called the *weight* of the object. This force can be written as

$$\boxed{W = mg} \tag{5.2}$$

where W is the weight in newtons of an object of mass m (in kilograms), and g is a quantity that depends on the mass of the Earth, the radius of the Earth, and on a universal gravity constant. For motion of objects near the surface of the Earth, g is almost constant, and I will take its value to be $g = 9.80\,\text{m/s}^2$. The quantity g is often called "the acceleration due to gravity." The reason for this term is the following: Suppose you release an object of mass m and allow it to fall freely under the influence of the gravity force. The net force acting on the object will be the gravity force, $-mg$. The force is negative because it is directed downward. Newton's second law becomes

$$ma = -mg, \quad \text{so } a = -g$$

Thus, g is indeed the magnitude of the acceleration of a freely falling object. However, even when the object is not falling (perhaps it is at rest on a table), it is still acted on by the gravity force mg. Then no acceleration occurs, and it is somewhat misleading to refer to g as the "acceleration due to gravity." It is best just to call g by the name "gee." Whatever you do, don't call g "gravity."

It is common practice to "weigh" objects in grams or kilograms. This is incorrect, since weight is measured in newtons, not kilograms. However, since mass and weight are proportional, no great harm is done if you merely want to compare two things. If you double the mass, you will double the weight. However, in your calculations in physics be careful to distinguish between these two distinct concepts. Do not use mass m in kilograms where you should be using weight W in newtons.

We will encounter two classes of problems. When the **net force is not zero, acceleration will result**. Using our previous kinematic equations, we can then determine the motion. When the **net force acting is zero, no acceleration occurs**. This situation is called **equilibrium**. If an object, or collection of objects, remains at rest, it is obviously in equilibrium, and we can then deduce what forces are acting, if some of the forces are known. I consider these two classes of applications in the following section. In what follows, I will make some simplifying approximations, unless I indicate otherwise. I will neglect friction. I will neglect the variation in g with altitude. I will neglect the weight of ropes, assuming they are light compared to the other objects. I will treat objects as point masses. In so doing, I will not have to worry about rotations of objects of finite size. Later, rotation will be taken into account in the study of the motion of extended rigid bodies. Finally, for the present I will consider problems where the forces all lie in a plane. This will not be a severe restriction, and it is not difficult to extend our treatment to forces in three dimensions. However, this is the situation for many interesting and practical problems, and the reduction in writing required helps make the concepts more clear.

5.4   Applications of Newton's Laws

In solving *all* problems involving forces, follow these procedures:

1. Draw a little picture, including sketches of people, cars, and so on, so that you are clear as to what is happening. Write down given information, and identify what is to be found.
2. Identify the forces that act *on* the object or system. These are called *external forces*. Do not include internal forces, for example, the forces between the atoms in the object. In our study of mechanics the only forces we will encounter are gravity, friction, normal forces exerted by surfaces, and tension (pulling forces, usually due to ropes). Show the forces in a **force diagram** (also called a **free-body diagram**). Slide all of the force vectors so that their tails are all at the one point that represents the object. Draw the force diagram with a straight-edge ruler, and do not make the drawing too small. It will usually occupy one-fourth page or more. Resolve the forces into components.
3. Determine the net force acting along each axis. Usually we use horizontal and vertical axes, but other perpendicular axes can be used if they are more convenient. For equilibrium problems, the net force will be zero. If the net force is not zero, use it to find the resulting acceleration, which can then be used to find velocity and displacement.
4. Use algebra to solve the equations obtained.
5. Substitute numerical values for the parameters in order to obtain a final answer.

5.4.1   Equilibrium Problems

If the net external force acting on an object of a system is zero, the object is in **equilibrium**. Thus if $\mathbf{F} = m\mathbf{a} = 0$, then $\mathbf{a} = 0$. This means the object is at rest or else moving with constant velocity. Most of the problems we will encounter concern objects that are at rest, for example, a person standing still or a building or other structure. If several external forces act, say, $\mathbf{F}_1, \mathbf{F}_2, \ldots, \mathbf{F}_x$, then $\mathbf{F} = \mathbf{F}_1 + \mathbf{F}_2 + \cdots + \mathbf{F}_x = \sum \mathbf{F}_{ix}$. This means $\sum F_{ix} = 0$ and $\sum F_{iy} = 0$. An easy way to apply these equations without having to worry about getting confused about the signs of the components F_{ix} and F_{iy} is simply to write

$$\boxed{F_{\text{left}} = F_{\text{right}} \quad \text{and} \quad F_{\text{up}} = F_{\text{down}}} \quad \text{in equilibrium} \tag{5.3}$$

where $F =$ total external force.

PROBLEM 5.1.   A book of mass 0.50 kg rests on a table. Draw the force diagram, and determine the upward force exerted by the table. Note: The force exerted by a surface perpendicular to the surface is called a **normal force**. In this context the word normal means "perpendicular." I will label such forces by the letter N.

Solution   The forces acting on the book are the force of gravity mg downward and the upward normal force exerted by the table. Since the book is in equilibrium, $F_{\text{up}} = F_{\text{down}}$ or $N = mg$. Thus $N = (0.50\,\text{kg})(9.8\,\text{m/s}^2) = 4.9\,\text{N}$. (Note: Remember that $1\,\text{kg} \cdot \text{m/s}^2 = 1\,\text{N}$.) Caution: Do not confuse weight mg in newtons with mass m in kilograms.

PROBLEM 5.2.   Once while elk hunting with a couple of mountain men in Idaho, our pickup truck got stuck in the mud. My compatriots got it out by using the following trick. They tied a steel cable tautly between the truck and a nearby tree in front of the truck. Then they pulled sideways with force F on the midpoint of the cable. Sure enough, the truck popped out of the mudhole. For such an arrangement, if the force F is 400 N (about 90 lb), what force does the cable exert on the truck if the angle θ in the drawing was 10°?

Solution   Let $T =$ tension in the cable = force on the truck. In a cable the tension is the same everywhere. Just before the truck moves, it is in equilibrium, so in the force diagram here,

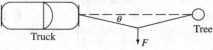

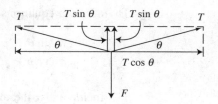

$$F_L = F_R \quad \text{or} \quad T\cos\theta = T\cos\theta$$

and

$$F_U = F_D \quad \text{or} \quad \sin\theta + T\sin\theta = F$$

So

$$T = \frac{F}{2\sin\theta} = 1150\,\text{N}$$

Thus pulling sideways has multiplied the force on the truck by a factor of almost 3 compared to what could have been obtained by pulling straight ahead. You might wonder where the "extra" force came from. The answer is that the tree is pulling on the cable, and this pull is also exerted on the truck. For this method to work, you need to use a steel cable that won't stretch, and it must be taut in order to minimize the angle θ.

Friction forces are encountered in many mechanical problems. When two surfaces are in contact, each can exert a force (called *friction*) on the other that is parallel to the surface. This force depends on the roughness of the surfaces. Friction forces are a little tricky in that they are *reactive forces*. By this I mean that they push back in response to another applied force. For example, if a book rests on a level table and you don't touch it, the friction force acting on it is zero. If you push lightly on the book with a force of 0.1 N, the friction force will push back with a force of 0.1 N, and the book won't move. If you increase your force to 0.2 N, the friction force also increases to 0.2 N, and still the book doesn't move. If you keep increasing your force, you will finally reach a point where the book does begin to move. The maximum friction force available has then been surpassed. The maximum friction force that can be exerted on the book depends on the nature of the book surface and the table surface and also on how strongly the two are pressed together, that is, on the normal force exerted by the table on the book. Thus we write the maximum friction force as

$$\boxed{F_f = \mu N} \tag{5.4}$$

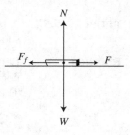

where μ = coefficient of friction, a dimensionless number.

The coefficient of friction depends on the nature of the two surfaces. Typically, μ is greater when the objects are stationary than when one object is sliding over the other, because sliding tends to break off the little sharp points (on a microscopic scale) sticking out of any surface. The coefficient of static friction is greater than the coefficient of kinetic (or sliding) friction.

PROBLEM 5.3. The coefficient of friction between a load of sand and the bed of the dump truck in which it is carried is 1.10. At what angle to horizontal does the truck bed have to be tilted before the sand starts to slide out?

Solution The force diagram is drawn here. It is easiest here to resolve the forces along axes parallel and perpendicular to the surface of the bed because the normal force and the friction force are already along these directions. Thus we have to resolve only the weight mg. It is not incorrect to use the x and y-axes, but so doing requires more algebra.

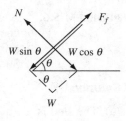

Just before the sand slides, it is in equilibrium, so

$$F_{\text{into the bed}} = F_{\text{out of the bed}} \quad \text{and} \quad F_{\text{down the slope}} = F_{\text{up the slope}}$$

Thus,

$$W\cos\theta = N \qquad W\sin\theta = F_f = \mu N$$

Divide these equations:

$$\frac{W\sin\theta}{W\cos\theta} = \frac{\mu N}{N} \quad \text{so } \tan\theta = \mu = 1.1 \quad \theta = 48°$$

Incidentally, the maximum angle for which no sliding occurs is called the "angle of repose." Read the fascinating novel by this name by Wallace Stegner. It is a great book, and Stegner gets all of his physics metaphors exactly right.

PROBLEM 5.4.   A mechanic tries to remove an engine from a car by attaching a chain to it from a point directly overhead and then pulling sideways with a horizontal force F. If the engine has mass 180 kg, what is the tension in the chain when it makes an angle of 15° with vertical? What is the force F?

Solution   Resolve forces, with $F_{up} = F_{down}$ and $F_L = F_R$:

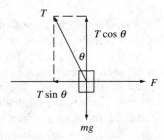

$$T\cos\theta = mg \qquad T\sin\theta = F$$

Divide

$$\frac{\sin\theta}{\cos\theta} = \frac{F}{mg} \qquad F = mg\tan\theta$$

$$F = (180\,\text{kg})(9.8\,\text{m/s}^2)\tan 15° = 473\,\text{N}$$

$$T = \frac{F}{\sin\theta} = \frac{473}{\sin 15°} = 1830\,\text{N}$$

PROBLEM 5.5.   A block and tackle is a simple machine used to lift heavy weights. For the arrangement shown here, what force F must be exerted to lift a load of weight W?

Solution   Let the "system" be the load and the lower pulley (of negligible weight). The force diagram is as shown. The tension in the rope is everywhere $T = F$.

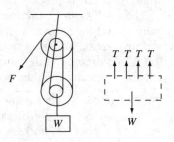

$$F_{up} = F_{down} \quad \text{so } 4T = W \text{ or } T = F = \frac{W}{4}$$

The ratio of the weight lifted to the force applied is called the **mechanical advantage** of the machine. Here, the MA = 4.

PROBLEM 5.6.   A girl moves her brother on a sled at a constant velocity by exerting a force F. The coefficient of friction between the sled and the ground is 0.05. The sled and rider have a mass of 20 kg. What force is required if (a) she pushes on the sled at an angle of 30° below horizontal and (b) she pulls the sled at an angle of 30° above horizontal?

Solution

(a) The sled is in equilibrium since the velocity is constant. Thus $F_{up} = F_{down}$ and $F_L = F_R$.

$$N = mg + F\sin 30°$$
$$F_f = \mu N = F\cos 30°$$

so

$$\mu(mg + F\sin 30°) = F\cos 30°$$

$$F = \frac{\mu mg}{\cos 30° - \mu \sin 30°}$$

$$F = \frac{(0.05)(20\,\text{kg})(9.8\,\text{m/s}^2)}{\cos 30° - 0.05\sin 30°} = 11.7\,\text{N}$$

(b) $N + F\sin 30° = mg$

$$F_f = \mu N = F\cos 30° \qquad \mu(mg - F\sin 30°) = F\cos 30°$$

$$F = \frac{\mu mg}{\cos 30° - \mu \sin 30°} = \frac{(0.05)(20\,\text{kg})(9.8\,\text{m/s}^2)}{\cos 30° + 0.05\sin 30°} = 11.0\,\text{N}$$

The force required in (b) is less than that for (a) because in (b) the force F angles up and supports some of the weight. This reduces N and hence F_f.

5.4.2 Nonequilibrium Problems

When a net force acts on a system, the system will have an acceleration given by $\mathbf{F} = m\mathbf{a}$. If we know $\mathbf{F}$, we can find the acceleration $\mathbf{a}$, and knowing $\mathbf{a}$ and the initial conditions, we can use the kinematic equations, Eq. 3.10, to find displacement and velocity.

PROBLEM 5.7.   A woman is wearing her seat belt while driving 60 km/h. She finds it necessary to slam on her brakes, and she slows uniformly to a stop in 1.60 s. What is the average force exerted on her by the seat belt (neglecting friction with the seat)? Express the result as a multiple of the woman's weight.

Solution   If she slows from v_1 to v_2 in time t, her average acceleration is

$$a = \frac{v_2 - v_1}{t} = \frac{0 - 60}{1.60\,\text{s}} = -10.4\,\text{m/s}^2$$

Thus, the average force exerted on her is $F = ma$. Her weight is $W = mg$, so

$$m = \frac{W}{g} \quad \text{and} \quad F = \frac{Wa}{g} = \frac{10.4}{9.8}\,W = 1.06\,W$$

PROBLEM 5.8.   A locomotive pulls 20 boxcars, each with a mass of 56,000 kg. The train accelerates forward with acceleration 0.05 m/s². (a) What is the force exerted by the coupling between the locomotive and the first car? (b) What is the force exerted by the coupling between the last car and the next to last car?

Solution   (a) View the 20 cars as the system. Then $F_1 = 20\,ma$, where $m = 56,000$ kg. Thus $F_1 = (20)(56,000\,\text{kg})$ $(0.05\,\text{m/s}^2) = 56,000\,\text{N}$. (b) View the last car as the system $F_L = ma = (56,000\,\text{kg})(0.05\,\text{m/s}^2) = 2300\,\text{N}$.

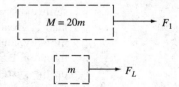

PROBLEM 5.9.   A woman (mass 50 kg) and her son (mass 25 kg) face each other on ice skates. Placing the palms of their hands together, they push each other apart, with the mother exerting an average force of 40 N on her son. What will be the acceleration of each during this process?

Solution   According to Newton's third law, each person will experience a force of the same magnitude, 40 N. Thus the accelerations will be

$$a_m = \frac{F}{m_m} = \frac{40\,\text{N}}{50\,\text{kg}} = 0.80\,\text{m/s}^2 \qquad a_s = \frac{F}{m_s} = \frac{40\,\text{N}}{25\,\text{kg}} = 1.60\,\text{m/s}^2$$

PROBLEM 5.10.   Once during the Great Depression of 1933 my father found temporary work driving a big flat bed truck loaded with steel. While going down a hill in Los Angeles, the brakes went out. This situation was a little stressful. If he hit something, either he would be injured in the crash, or else the steel would slide forward and wipe out the cab and driver. Fortunately an alert motorcycle cop saved the day by clearing traffic for a run-out. Suppose a load of steel is held in place only by friction, with a coefficient of friction of 0.4. What is the shortest stopping distance on level ground when moving 20 m/s (about 45 mi/h) if the load is not to slide forward into the cab?

Solution   The force diagram for the load is as shown here. The only horizontal force is the force of friction, and this is the net force acting on the load. Thus,

$$F = ma = -F_f$$
$$N = mg \quad \text{and} \quad F_f = \mu N$$

So

$$a = -\frac{\mu mg}{m} = -\mu g$$

From Eq. 3.10, $v^2 = v_0^2 + 2ax = 0$ when stopped. Thus,

$$x = \frac{v_0^2}{2a} = \frac{v_0^2}{2\mu g}$$

PROBLEM 5.11.   Here is a famous classic problem that will make you think. A rope is passed over a pulley suspended from a tree branch, and a stalk of bananas is tied to one end. A monkey hangs from the other end of the rope, and the bananas and the monkey are balanced. Now the monkey starts climbing up the rope. What will happen to the bananas? Will they stay in the same place, or will they move up away from the ground, or will they move down toward the ground?

Solution   Look at the force diagram for the monkey. His weight mg acts downward, and the rope tension T acts upward. If the monkey is to start moving up from rest, he must accelerate upward, which means there must be a net upward force acting on him. The net upward force on the monkey is $T - mg$, where T is the tension in the rope. But the tension is the same everywhere in a rope, so the tension at the end of the rope attached to the bananas is also T, greater than mg. Thus the bananas experience the same upward force as does the monkey, and so the bananas will move up with the same acceleration and velocity as the monkey. Both will move higher from the ground at the same rate. The net upward force on the system made up of the bananas plus the monkey is provided by the pulley.

PROBLEM 5.12.   Once my Boy Scout troop tried to improvise a scheme to pull an old ore car up out of a sloping mine shaft. The idea is illustrated here. We were going to divert a stream so that water ran into a bucket attached to the ore car. When enough water filled the car, it was supposed to move up the track. (a) If the ore car had a mass of 80 kg, and the track was inclined at 15° above horizontal, what mass of water would be required to start the car moving if friction was negligible? (b) If the friction coefficient between the car and the track was 0.20, what mass of water would be needed to start the car moving? (c) With friction present, suppose water is added until the car is just about to move. Now an additional 4 kg of water is added to the bucket with the wheels of the car locked. When the wheels are unlocked, how long will it take the car to move 34 m up the track?

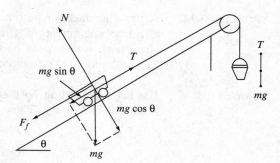

Solution

(a) Draw the force diagram. Resolve the forces into components parallel to the track and perpendicular to the track. In equilibrium, for the car, $mg \sin \theta = T$, and for the bucket, $T = Mg$. Thus, $Mg = mg \sin \theta$ and $M = m \sin \theta = 80 \sin 15° = 20.7$ kg.

(b) With friction present, in equilibrium, $T = mg \sin \theta + F_f = mg \sin \theta + \mu N$. $\mu N = mg \cos \theta$, so $T = mg \sin \theta + \mu mg \cos \theta$. For the bucket, $T = Mg$, so $Mg = mg(\sin \theta + \mu \cos \theta)$. $M = m$ ($\sin \theta + \mu \cos \theta = (80$ kg$)(\sin 15° + 0.20 \cos 15°) = 36.2$ kg.

(c) If $\Delta m = 4$ kg is added to the bucket, the net force will then be $F = \Delta mg$. Thus $F = \Delta mg = (m + M + \Delta m)a$.

$$a = \frac{\Delta m}{m + M + \Delta m} g = \frac{4}{80 + 36.2 + 4} (9.8 \text{ m/s}^2) = 0.33 \text{ m/s}^2$$

From Eq. 3.10, $x = v_0 t + \frac{1}{2} at^2$, where $v_0 = 0$, so

$$t = \sqrt{\frac{2x}{a}} = \sqrt{\frac{(2)134 \text{ m}}{0.33 \text{ m/s}^2}} = 14.4 \text{ s}$$

PROBLEM 5.13.   A person whose weight is 600 N stands on a bathroom scale in an elevator. What will the scale read when the elevator is (a) moving up or down at constant speed, (b) accelerating up with acceleration 0.5 g, (c) accelerating downward with acceleration 0.5 g, and (d) accelerating downward with acceleration g?

Solution   The force diagram includes two forces: the gravity force mg downward and the normal force N exerted upward by the surface of the scale. This normal force is the scale reading.

(a) Constant velocity means $a = 0$, so equilibrium and $N = W = 600 N$.
(b) $N - W = ma = +0.5mg = 0.5 W$, so $N = W + 0.5 W = 1.5 W$.
(c) $N - W = ma = -0.5mg = -0.5 W$, so $N = 0.5 W$.
(d) $N - W = ma = -mg = -W$, so $N = 0$.

The last case, when the scale reading is zero, represents what is called "effective weightlessness." The elevator is falling with acceleration $-g$, as is the person. Thus the person does not press down on the elevator. He is seemingly "weightless." This is the situation with the astronauts in an orbiting space vehicle. The vehicle and everything in it are falling freely, and hence they all seem weightless. You have probably seen pictures where the astronauts, their pencils, and their sandwiches and other loose equipment float weightlessly around the spaceship. Everything is falling toward the Earth with acceleration g. Because they are also moving sideways, they do not actually get closer to the Earth as they fall.

PROBLEM 5.14. A tire manufacturer performs road tests that show that a new design of tire has an effective coefficient of friction of 0.83 on a dry asphalt roadway. Under these conditions, what would be the stopping distance for a car traveling 50 km/h (about 31 mi/h) and 100 km/h (about 62 mi/h)?

Solution The net force acting on the car is the force of friction $F = \mu N$. Vertical forces are balanced, so $N = mg$. Thus

$$a = \frac{-F}{m} = -\frac{\mu mg}{m} = -\mu g \qquad v^2 = v_0^2 + 2ax = 0$$

when stopped, so

$$v_0^2 - 2\mu gx = 0 \quad x = \frac{v_0^2}{2\mu g}$$

At 50 km/h,

$$x = \frac{[(50)(1000 \, \text{m}/3600 \, \text{s})]^2}{(2)(0.83)(9.8 \, \text{m/s}^2)} = 11.9 \, \text{m}$$

At 100 km/h,

$$x = 47.4 \, \text{m}$$

Notice that the stopping distance varies as the *square* of the speed. Thus, doubling your speed increases your stopping distance by a factor of 4. SPEED KILLS!

PROBLEM 5.15. Consider a mass with initial velocity v_0. It can be launched as a projectile with its initial velocity elevated at angle θ above horizontal, or it can be launched up a frictionless plane inclined at angle θ above horizontal. In which case will the mass reach the greatest elevation, or will the maximum elevation be the same in each case? To answer this, calculate the maximum elevation reached in each case.

Solution For the projectile, $v_y^2 = v_{0y}^2 - 2gy_p$ from Eq. 3.10, $v_{0y} = v_0 \sin \theta$ and $v_y = 0$ at the highest point. Thus

$$y_p = \frac{v_0^2 \sin^2 \theta}{2g}$$

For the particle sliding up the plane, draw the force diagram.

$$F_{\text{into plane}} = F_{\text{out of plane}} \qquad mg \cos \theta = N$$

Thus the net vertical force is

$$F_y = -mg + N \cos \theta = -mg + mg \cos^2 \theta$$

Since $F_y = ma_y$,

$$a_y = \frac{F_y}{m} = \frac{-mg + mg \cos^2 \theta}{m} = -g + g \cos^2 \theta$$

$$v_y^2 = v_{0y}^2 + 2a_y y = 0$$

at the highest point, so

$$y = \frac{(v_0 \sin \theta)^2}{(2)(-g + g \cos^2 \theta)} = \frac{v_0^2 \sin^2 \theta}{2g(1 - \cos^2 \theta)}$$

Thus $y > y_p$, and the sliding mass rises higher.

PROBLEM 5.16. In an interesting lecture demonstration, I sometimes show the property of inertia as follows: I hold a lead brick (mass about 9 kg) in my hand and give it a mighty blow with a big hammer. I can hardly feel the force exerted on my hand. Were the hammer to strike my hand directly, it would surely be broken to smithereens. In the construction trades this trick of "backing" something with a massive object to reduce the force exerted on the "backer" is often used. To understand what is happening, approximate your hand as an isolated mass m in contact with a larger mass M (the brick). A force F is applied to the brick. Calculate the force f transmitted to your hand. What is f if $F = 100\,\text{N}$, $M = 9\,\text{kg}$, $m = 1\,\text{kg}$?

Solution First consider the block plus the hand as the "system" subject to force F. Then $F = (M + m)a$ and $a = F/(M + m)$. Now consider the hand alone, subject to force $f: f = ma$ where $a = F/(M + m)$, since hand and block move together. Thus

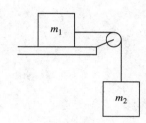

$$f = \frac{m}{M + m}F, \qquad f = \frac{1}{10 + 1}(100) = 9\,\text{N}$$

PROBLEM 5.17. Blocks of masses m_1 and m_2 are connected by a light string. The coefficient of friction between m_1 and the table surface is μ. Determine the acceleration of the blocks and the tension in the string if $m_1 = 4\,\text{kg}$, $m_2 = 3\,\text{kg}$, and $\mu = 0.30$.

Solution Draw the force diagram for each block. For m_1,

$$N = m_1 g \qquad F_f = \mu N = \mu m_1 g$$
$$F_{\text{net}} = T - F_f = T - \mu m_1 g = m_1 a$$

For m_2,

$$m_2 g - T = m_2 a$$

Solve

$$T = m_2 a - m_2 g$$

so

$$m_2 g - m_2 a - \mu m_1 g = m_1 a$$
$$a = \frac{m_2 - \mu m_1}{m_1 + m_2}g \qquad T = \frac{m_1 m_2}{m_1 + m_2}(1 + \mu)g$$

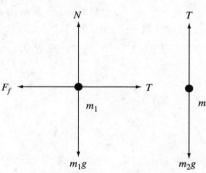

Substitute numbers: $T = 21.8\,\text{N}$, $a = 2.52\,\text{m/s}^2$.

Observe that we could also solve this problem by first considering both blocks together to form the system. In this case we would have

$$m_2 g - F = (m_1 + m_2)a$$

where $F = \mu N = \mu m_1 g$. This yields the acceleration immediately. We can then obtain the tension from $m_2 g - T = m_2 a$.

PROBLEM 5.18. A small block of mass m is placed on a wedge of angle θ and mass M. Friction is negligible. What horizontal force must be applied to the wedge so that the small block does not slide up or down the wedge surface?

Solution   First consider both blocks together: $F = (m + M)a$. Now look at the small block alone. The surface exerts a normal force on it. It does not move vertically, so $N \cos \theta = mg$. Horizontally, $N \sin \theta = ma$. Thus

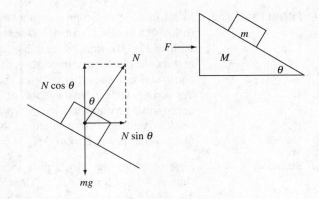

$$\left(\frac{mg}{\cos \theta} \right) \sin \theta = m \left(\frac{F}{m + M} \right)$$

$$F = (m + M)g \tan \theta$$

PROBLEM 5.19.   A packing crate of mass m is pulled across the floor at constant velocity by means of a cable attached to the front of the crate. The cable makes an angle θ with the floor. The coefficient of friction between the floor and the crate is μ. What value of θ will make the tension a minimum?

Solution   In equilibrium, $F_{\text{up}} = F_{\text{down}}$ and $F_L = F_R$.

$$N + T \sin \theta = mg$$
$$F_f = \mu N = T \cos \theta$$

Solve for T:

$$N = mg - T \sin \theta$$

$$\mu(mg - T \sin \theta) = T \cos \theta \qquad T = \frac{\mu mg}{\cos \theta + \mu \sin \theta}$$

Minimize T with respect to variations in θ by requiring that

$$\frac{dT}{d\theta} = 0, \qquad \text{so} \quad \frac{dT}{d\theta} = \mu mg \left[-\left(\frac{1}{\cos \theta + \mu \sin \theta} \right)^2 (-\sin \theta + \mu \sin \theta) \right] = 0$$

Thus

$$\frac{dT}{d\theta} = 0 \quad \text{if} \ -\sin \theta + \mu \cos \theta = 0, \quad \text{or} \ \mu = \tan \theta$$

PROBLEM 5.20.   Two masses are connected as shown here. Friction is negligible. What is the acceleration of each mass? What is the tension in the string?

Solution   Study the drawing carefully and you will see that when m_1 moves 2 cm, m_2 drops only 1 cm. Thus, $a_1 = 2a_2$. For m_1, $T = m_1 a_1$. For m_2, $m_2 a_2 = m_2 g - 2T$. Solve $m_2 a_2 = m_2 g - m_1 a_1 = m_2 g - 2m_1 a_2$:

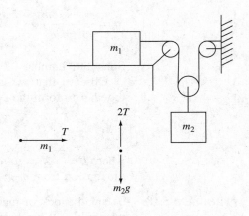

$$a_2 = \frac{m_2}{4m_1 + m_2} g \qquad a_1 = \frac{2m_2}{4m_1 + m_2} g$$

$$T = \frac{2m_1 m_2}{4m_1 + m_2} g$$

PROBLEM 5.21. On a level road the stopping distance for a certain car traveling at 80 km/h is 32 m. What would be the stopping distance for this car when going downhill on a 1 : 10 grade? (A grade of 1 : 10 means the elevation drops 1 m for a forward travel of 10 m along the roadway.)

Solution
$$v_0 = 80 \text{ km/h} = 80\left(\frac{1000 \text{ m}}{3600 \text{ s}}\right) = 22.2 \text{ m/s}, \quad \theta = \sin^{-1}\frac{1}{10} = 5.74°$$

On level ground, $-F_f = -\mu N = -\mu mg = ma$. So $a = -\mu g$ and

$$v_2 = v_0^2 + 2ax = 0, \quad x = \frac{-v_0^2}{-2\mu g},$$

$$\mu = \frac{v_0^2}{2gx} = \frac{(22.2 \text{ m/s})^2}{(2)(9.8 \text{ m/s}^2)(32 \text{ m})} = 0.79$$

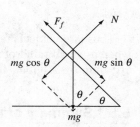

Let $a_s =$ the acceleration along the slope: $F_s = mg \sin \theta$ $-F_f = ma_s$. Thus $ma_s = mg \sin \theta - \mu mg \cos \theta$. The stopping distance on the slope is given by $v^2 = v_0^2 + 2a_s s = 0$ when stopped, where $a_s < 0$. Thus

$$s = \frac{v_0^2}{2a_s} = \frac{v_0^2}{2(g \sin \theta - \mu g \cos \theta)} = -\frac{(22.2 \text{ m/s})^2}{2(9.8 \text{ m/s}^2)[(\sin 5.7° (0.79) \cos 5.7°)]}$$

$$s = 36.8 \text{ m} \qquad \text{stopping distance on slope}$$

PROBLEM 5.22. An object falling through the air at high speed experiences a drag force which can be expressed approximately as

$$F_D = \frac{1}{2}\rho_A A C_D v^2$$

Here ρ_A is the density of air, C_D is a drag coefficient that depends on the shape and texture of the falling object, and A is the projected area of the object as seen looking up from the ground. C_D is a dimensionless number between 0 and 1. (a) Determine the maximum speed (called the **terminal velocity**) a falling object reaches in the presence of this drag force. (b) How does the terminal velocity of an object depend on its size? To answer this, calculate the ratio of the terminal velocities for two spherical hailstones, one of radius r_1 and a larger one of radius r_2.

Solution

(a) As the object falls, v gets larger and larger, until finally $F = \frac{1}{2}\rho_A A C_D v^2 - mg = ma = 0$. When $a = 0$, $v =$ constant, where $\frac{1}{2}\rho_A A C_D v_T^2 - mg = 0$. Thus,

$$v_T = \sqrt{\frac{2mg}{\rho_A A C_D}}$$

(b) For a sphere of ice, $m = (\text{density})(\text{volume}) = \frac{4}{3}\pi R^3 \rho$. The projected area seen from below is $A = \pi R^2$. Thus

$$v_T = \sqrt{\frac{2mg}{\rho_A A C_D}} = \sqrt{\frac{(2)(\rho)(4\pi R^3)g}{\rho_A \pi R^2 C_D}} = \sqrt{\frac{8\rho g}{\rho_A C_D}}\sqrt{R}$$

$$v_T = \kappa\sqrt{R}$$

where

$$\kappa = \sqrt{\frac{8\rho g}{\rho_A C_D}}$$

For two ice spheres,

$$\frac{v_{T_2}}{v_{T_1}} = \frac{\kappa\sqrt{r_2}}{\kappa\sqrt{r_1}} = \sqrt{\frac{r_2}{r_1}}$$

Thus large objects fall faster than small ones of the same shape. Big hailstones can flatten a wheat field, whereas small raindrops don't hurt it. Typical terminal velocities are as follows: 14.5 mi/h for a raindrop, 105 mi/h for a bullet, 140 mi/h for a person, and 145 mi/h for a 1000-lb bomb.

5.5　Summary of Key Equations

Newton's third law:　　　If A exerts force $\mathbf{F}_{AB}$ on B and B exerts force $\mathbf{F}_{BA}$ on A, then $\mathbf{F}_{AB} = -\mathbf{F}_{BA}$.

Newton's second law:　　$\mathbf{F} = m\mathbf{a}$.

Newton's first law:　　　If the net external force $\mathbf{F} = 0$, then $\mathbf{a} = 0$ and $\mathbf{v} =$ constant

Equilibrium:　　　　　　If $\mathbf{F} = 0$, then $\mathbf{a} = 0$, and $\mathbf{F}_{up} = \mathbf{F}_{down}$, and $\mathbf{F}_{left} = \mathbf{F}_{right}$.

SUPPLEMENTARY PROBLEMS

5.23. In an attempt to keep a packing box of mass m from sliding down a ramp inclined at angle 30° above horizontal, a woman exerts a horizontal force F. Assuming friction is negligible, what minimum force must she exert?

5.24. In a daredevil rescue attempt, a marine holds on to the landing gear of a hovering helicopter with one hand while with his other hand he reaches down to lift a buddy below him. If the upper marine has a weight of 450 N and the lower marine weighs 350 N, what force is exerted by the upper marine's upper arm and by his lower arm?

5.25. A crane lifts a mass of 200 kg with the arrangement shown here. Determine the force exerted by the boom and the tension in the cable.

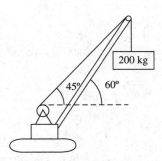

5.26. A uniform heavy cable of mass m is attached to two eyebolts on a ceiling. The line tangent to the cable makes an angle of 30° with the ceiling at each end of the cable. Determine the force exerted on each eyebolt and the tension at the midpoint of the cable.

5.27. What force F must be exerted on the block and tackle system shown here if the weight W is stationary? Determine the tensions T_1, T_2, T_3, T_4, and T_5. The pulleys have negligible weight.

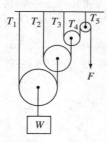

5.28. A steel ball bearing of mass 0.020 kg rests in a 90° groove in a track. What force does it exert on the track at point A and at point B?

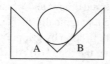

5.29. Find the tension in each cord for the 20-kg mass shown suspended here.

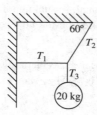

5.30. A painter who weighs 600 N stands on a platform, as shown here. The platform, paint, brushes, and so on weigh 400 N. What is the tension in the rope the painter is holding when the platform is motionless?

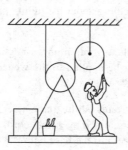

5.31. A painter of mass 80 kg sits in a bosun's chair of mass 10 kg. He pulls on the rope he is holding in order to accelerate himself up. In so doing, he presses down on the seat with a force of 392 N. (a) What is his acceleration? (b) What is the tension in the rope supporting the pulley?

5.32. You can make a simple accelerometer with which to measure approximately the acceleration of your Ferrari. Tie a small weight to the end of a string and let it hang vertically as a pendulum. When holding this instrument in your car while it is accelerating, the string will deviate from vertical by an angle θ. Derive an expression for the acceleration of the car as a function of θ. (You will have to make a protractor with which to measure the angle θ.)

5.33. Mass m_1 slides without friction on a plane inclined at $40°$ above horizontal. It is attached to a second mass m_2 by a light string. If $m_1 = 5\,\text{kg}$ and $m_2 = 4\,\text{kg}$, determine the acceleration of each block and the tension in the string.

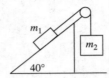

5.34. I always worry about being trapped on a high floor of a hotel when it catches fire. How will I escape? I have a plan worked out. I'll tie together my sheets and drapes and whatever else I can find and make a rope, which I will then slide down. Unfortunately, I've found that typically such a rope will support only about three-fourths of my weight. However, if I slow my descent with friction by gripping the rope, I might make it. (a) At what maximum acceleration can I descend without breaking the rope? (b) At what speed will I hit the ground 20 m below in my slowest descent? (c) At what speed will I hit the ground if I just jump for it with no rope?

5.35. A flea is a remarkable animal. It can leap to a height of about 32 cm (about 200 times its body length) when taking off at an angle of $60°$ above horizontal. Assuming a flea mass of $5.0 \times 10^{-7}\,\text{kg}$ and a push-off time of $10^{-3}\,\text{s}$, calculate (a) the average force exerted on the floor, expressed as a multiple of the flea's weight, and (b) the average acceleration of the flea during lift-off.

5.36. Some amateur hot air balloonists find themselves accelerating downward with acceleration a at a moment when the mass of the balloon plus the balloonists is M. They want to accelerate upward at this same rate, so they throw out ballast of mass m. Determine m.

5.37. When a small object such as a blood cell or a macromolecule or a silt particle falls through a viscous fluid under the influence of gravity, it experiences a drag force of the form bv, as long as v is small. Under these circumstances the particle achieves a steady velocity called the *sedimentation velocity*. Measurement of the sedimentation velocity enables us to learn something about the falling particle. Calculate the sedimentation velocity in terms of b, the particle mass, and g.

5.38. A small block of mass m can slide without friction on a wedge of mass M inclined at angle θ. What horizontal force F must be applied to the wedge if the small block is not to move with respect to the wedge?

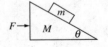

5.39. A principle used in certain interlock systems is illustrated here. Mass m slides on a vertical track attached to a base unit of mass M. The two masses are joined by a light cord, as shown. Mass m is released from rest. Determine how long it takes to fall a distance h to the base below.

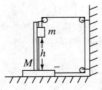

5.40. One type of cam arrangement operates on the principle illustrated here. A follower rod of mass 0.300 kg slides vertically in a lubricated bushing. At its end is a bearing that rests on a movable wedge of angle $15°$ and mass 0.150 kg. The wedge can move horizontally. Friction is negligible throughout the device. What horizontal force must be applied to the wedge to impart an upward acceleration of $2.0\,g$ to the follower rod?

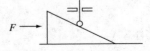

5.41. A 10.0 kg block rests on a horizontal frictionless table. A block of mass 4.0 kg rests on top of this block. The coefficient of friction between the two blocks is 0.40. If a sufficiently large horizontal force is applied to one of the blocks, they will begin to slide with respect to each other. Determine this maximum force if the force is applied to (a) the lower block and (b) the upper block.

5.42. A locomotive of mass M pulls a string of 10 boxcars, each of mass m. When the locomotive accelerates forward it exerts a force F on the coupling between the locomotive and the first boxcar. What force is then exerted on the tenth boxcar?

5.43. A conveyer belt moves at a constant speed of 1.20 m/s. A crate is dropped onto the belt. The coefficient of friction between the crate and the box is 0.70. (a) How long will it take for the crate to stop sliding on the belt? (b) How far from its starting point will the crate have moved before it comes to rest on the belt?

SOLUTIONS TO SUPPLEMENTARY PROBLEMS

5.23. $F\cos\theta = mg\sin\theta$, so $F = mg\tan\theta$.

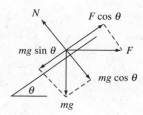

5.24. Lower arm:

$$T_L = 350\,\text{N}$$

Upper arm:

$$T_u = T_L + W$$
$$= 350\,\text{N} + 450\,\text{N}$$
$$= 800\,\text{N}$$

5.25. $F_{up} = F_{down}$ $F\sin 60° = mg + T\sin 45°$
$F_1 = F_1$ $F\cos 60° = T\cos 45°$

Divide:

$$\tan 60° = \frac{mg + T\sin 45°}{T\cos 45°}$$

$$T = \frac{mg}{\tan 60°\cos 45° - \sin 45°} = 3790\,\text{N}$$

$$F = \frac{\cos 45°}{\cos 60°}T = 5350\,\text{N}$$

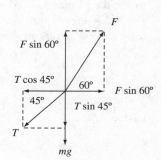

5.26. Draw the force diagram for one-half of the cable.

$$F_{\text{up}} = F_{\text{down}}, \qquad F_L = F_R$$
$$F \sin 30° = \tfrac{1}{2}mg, \qquad 0.5F = 0.5mg, \qquad F = mg$$
$$F \cos 30° = T \qquad T = 0.87mg$$

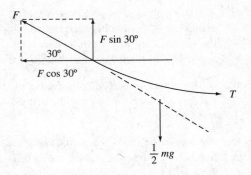

5.27. Draw the force diagram for each pulley alone. Thus,

$$2T_1 = W \qquad T_1 = 0.5\,W$$
$$2T_2 = T_1 \qquad T_2 = 0.5T_1 = 0.25\,W$$
$$T_3 = T_4 = F \qquad 2T_3 = T_2 \qquad T_3 = 0.125\,W = T_4 = F$$

5.28. By symmetry $N_A = N_B$ and $F_{\text{up}} = F_{\text{down}}$:

$$2N_a \sin 45° = mg \qquad N_A = N_B = \frac{1}{\sqrt{2}}mg$$

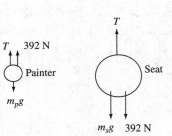

5.29.
$$T_3 = mg \qquad T_2 \sin 60° = mg \qquad T_2 \cos 60° = T_1$$
$$50T_1 = mg \tan 60° \qquad T = \frac{mg}{\sin 60°}T_{mg}$$
$$m = 20\,\text{kg, so } T_1 = 339\,\text{N} \qquad T_2 = 226\,\text{N} \qquad T_3 = 196\,\text{N}$$

5.30. Let the painter platform and lower pulley be the system. Three segments of the rope held pull up on this system, so $3T = 400\,\text{N} + 600\,\text{N}$ and $T = 333\,\text{N}$.

5.31. The seat pushes up on the painter with force $F = 392\,\text{N}$, so for the painter $T + 392 - m_p g = m_p a$. For the seat, $T - 392 - m_s g = m_s a$. Subtract these equations and solve for a: $a = 1.09\,\text{m/s}^2$ and $T = 479\,\text{N}$.

5.32. Divide:

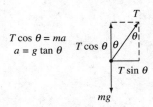

$$T \cos \theta = ma$$
$$a = g \tan \theta$$

5.33. Assume m_2 accelerates down. If this is wrong, a will turn out to be negative. For m_2, $m_2g - T = m_2a$. For m_1, $T - mg\sin 40° = m_1a$. Solve one equation for T and substitute in the other. Find $a = 0.86\,\text{m/s}^2$ and $T = 35.8\,\text{N}$. So m_2 goes down.

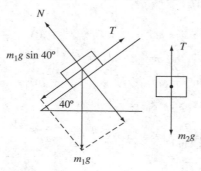

5.34. (a) $0.75\,mg - mg = ma$, so $a = -0.25\,g = -0.25(9.8\,\text{m/s}^2) = -2.45\,\text{m/s}^2$.

(b) $v^2 = v_0^2 + 2ay = 0 + 2(-2.45\,\text{m/s}^2)(-20\,\text{m})$, and $v = 9.9\,\text{m/s}$.

(c) Jump and $a = -g$, $v^2 = 0 + 2(-9.8\,\text{m/s}^2)(20\,\text{m})$, and $v = 19\,\text{m/s}$.

5.35. (a) $v_y^2 = v_{0y}^2 - 2gh = 0$ at the highest point. $v_{0y}^2 = (v_0\sin 60°)^2 = 2gh$, and $h = 0.32\,\text{m}$. Find $v_0 = 2.9\,\text{m/s}$:

$$v_0 = at = \frac{F}{m}t, \quad \text{so } F = \frac{mv_0}{t} = \left(\frac{v_0}{gt}\right)mg = 295\,mg!!!$$

(b) $a = \frac{v_0}{t} = 2.9 \times 10^3\,\text{m/s}^2 = 295g$. Amazing!

5.36. Going down, $-Mg + F = -Ma$ and $F = $ lift force. Going up, $F - (M-m)g = (M-m)a$. Solve simultaneously for m, and find

$$m = \frac{2a}{a+g}M$$

5.37. $bv - mg = ma$. v increases until $bv - mg = 0$. Thus $a = 0$ and $v = $ constant when $v = mg/b$.

5.38. Considering both masses as one object, $F = (m+M)a$. For the small block alone, $f_x = N\sin\theta = ma$ and $N\cos\theta = mg$.

$$\frac{N\sin\theta}{N\cos\theta} = \tan\theta = \frac{a}{g} \quad a = g\tan\theta$$

$$F = (m+M)\,a = (m+M)g\tan\theta$$

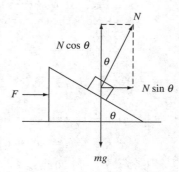

5.39. For both masses considered as the system, $T = (m+M)a_x$. Observe that when the base moves 1 cm to the right, mass m drops 2 cm, so $-a_y = 2a_x$ (down is negative). For mass m, $T - mg = -ma_x = -2ma_x$. Solve for a_x:

$$a_x = \left(\frac{2m}{5m+M}\right)g = -\frac{1}{2}a_y$$

$$-h = \frac{1}{2}a_yt^2 \quad t^2 = -\frac{2h}{a_y} = \frac{(5m+M)h}{2mg} \quad t = \sqrt{\frac{(5m+M)h}{2mg}}$$

5.40. Observe that when the wedge moves a distance x to the right, the rod moves up a distance y, when $\tan\theta = y/x$. Thus the vertical acceleration of the rod a_y is related to the horizontal acceleration of the wedge by $a_y = a_x\tan\theta$. The wedge exerts a normal force on the rod of mass m, so

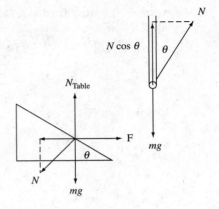

$$N\sin\theta - mg = ma_y \qquad (i)$$

The rod exerts a force N down on the wedge, so

$$F - N\sin\theta = Ma_x \qquad (ii)$$

Solve Eqs. (i) and (ii) for F, given $a_y = 2g = a_x\tan\theta$:

$$F = \left(3m\tan\theta + \frac{2M}{\tan\theta}\right)g = 13.3\,\text{N}$$

5.41. Let $M = 10\,\text{kg}$ and $m = 4\,\text{kg}$

(a) $$F = (M+m)a$$

$$\mu mg = mg = m\frac{F}{m+M}$$

$$F = \mu(M+m)g$$

(b) $F - \mu mg = ma$ for the upper block alone
$\mu mg = Ma$ for the lower block

So, $$F = \mu mg + m\left(\frac{\mu mg}{M}\right)$$

$$F = \mu\left(1+\frac{m}{M}\right)mg$$

One could also look at the two blocks together as one system, in which case $F = (M+m)a$.

This leads to the same result.

5.42. Let
F = force exerted by the locomotive
f = force applied to the tenth car
$f = ma$
$F = (10m)a$
So, $f = m\dfrac{F}{10m} = \dfrac{1}{10}F$

5.43. (a) $v = v_0 + at$, $v_0 = 0$ so $t = \dfrac{v}{a}$
$F = \mu mg = ma$, so $a = \mu g$, $t = \dfrac{v}{\mu g}$

$$t = \frac{1.20\,\text{m/s}}{(0.7)(9.8\,\text{m/s}^2)} = 0.18\,\text{s}$$

(b) $v^2 = v_0^2 + 2ax$

$$x = \frac{v^2}{2a} = \frac{v^2}{2\mu g} = \frac{(1.20\,\text{m/s})^2}{2(0.70)(9.8\,\text{m/s}^2)}$$

$$x = 0.105\,\text{m}$$

CHAPTER 6

Circular Motion

6.1 Centripetal Force

Whenever a moving object turns, its velocity changes direction. Since acceleration is a measure of the rate of change of velocity, an object that turns is accelerating. This kind of acceleration, called *centripetal* or *radial acceleration*, is related to v, the speed, and r, the radius of the curvature of the turn, by $a_c = v^2/r$. This relationship was obtained in Section 4.4. There we saw that there were two kinds of acceleration. Tangential acceleration (I called this "speeding up or slowing down" acceleration) measures the rate of change of speed. Radial, or centripetal, acceleration ("turning" acceleration) measures the rate of change of velocity associated with changing direction. Radial acceleration is directed perpendicular to the velocity vector and points toward the center of the arc on which the object is moving. If an object turns left, it accelerates left. If it turns right, it accelerates right.

From Newton's second law of motion we saw that in order for an object to accelerate, it must be subject to a net force. This is illustrated in Fig. 6-1. The amount of force needed to cause an object with speed v to curve along an arc of radius r is thus

$$F_c = ma_c = \frac{mv^2}{r}$$

(6.1)

Observe that F_c is not a *kind* of force. Many different kinds of forces can be used to make an object turn. For example, gravity causes the Moon to curve and travel a circular path around the Earth. The tension in a rope causes a tether ball to travel in a circle. The normal force exerted by a banked curve on a highway causes a car to travel a circular path. The friction force between a car's tire and the roadway causes the car to turn. When you draw a force diagram, *do not draw in a force labeled F_c.* Centripetal force is a way of using a force, not a kind of force.

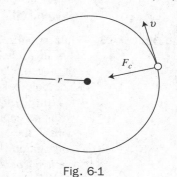

Fig. 6-1

An object does not have to travel in a complete circle to experience centripetal acceleration. However, frequently objects do travel around and around, as is the case with a spinning wheel or compact disk. Suppose that an object makes f rev/s. The **frequency** of revolution is f. One revolution per second is called **1 hertz** (a dumb way of labeling something, but we're stuck with it). Later we will extend this idea of frequency to anything that varies periodically, whether or not it moves in a circle. Thus the electricity in your house varies at 60 times per second, or 60 Hz. The AM radio station in my town broadcasts radio waves at a frequency of 1400 kHz. (The announcer always says, "KRPL at 1400 on your AM dial." He means 1,400,000 variations per second in the radio wave electric field.)

In one revolution an object travels a distance $2\pi r$. If it makes f rev/s, the distance traveled in 1 s is $2\pi rf$. The distance traveled per second is the speed v, so

$$v = 2\pi rf$$

(6.2)

One revolution is 2 radians (rad). The number of radians swept out per second is called the **angular frequency** or **angular velocity**, in **radians per second** (both terms are used). This symbol ω that looks like a small w is a lowercase Greek omega. It is measured in radians per second. Thus,

$$\omega = 2\pi f \qquad\qquad (6.3)$$

and

$$v = r\omega \qquad\qquad (6.4)$$

In terms of ω the centripetal force can be written

$$F_c = \frac{mv^2}{r} = mr\omega^2 \qquad\qquad (6.5)$$

When dealing with objects rotating at constant frequency, it is easiest to use Eq. 6.5. When an object is simply turning, such as a jet plane pulling out of a dive, use Eq. 6.1.

PROBLEM 6.1. A ball of mass 0.15 kg slides with negligible friction on a horizontal plane. The ball is attached to a pivot by means of a string 0.60 m long. The ball moves around a circle at 10 rev/s. What is the tension in the string?

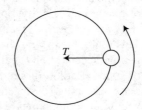

Solution The ball stays in the horizontal plane, so the upward normal force exerted by the table balances the downward force of gravity (the weight of the ball). In the horizontal plane only one force acts on the ball, the tension of the string. Thus a net force is acting on the ball, and it is not in equilibrium. It is accelerating with acceleration a_c. The force required to cause this acceleration is

$$F_c = T = mr\omega^2 = (0.15\,\text{kg})(0.60\,\text{m})\left(\frac{2\pi \times 0.1}{\text{s}}\right)^2 = 14\,\text{N}$$

Note that the units for ω and f are s^{-1}. Radians and revolutions are not "units" as such.

PROBLEM 6.2. In a popular carnival ride a person sits in a chair attached by means of a cable to a tall central post. The pole is spun, causing the rider to travel in a horizontal circle, with the cable making an angle θ with the vertical pole. A contraption like this is called a *conical pendulum*. Suppose the rider and chair have mass of 150 kg. If the cable length is 8 m, at what frequency should the chair rotate if the cable is to make an angle of 60° with vertical? What is the tension in the cable?

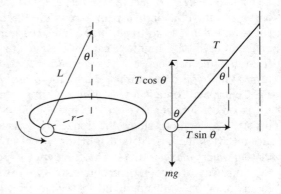

Solution Always begin by drawing the force diagram, as shown here. The rider is not moving up or down, so the vertical forces are in balance.

$$F_{\text{up}} = F_{\text{down}} \qquad T\cos\theta = mg$$

The component of the cable tension T directed horizontally, that is, toward the center of rotation, is a net force causing a centripetal acceleration toward the center. The magnitude of this required net force is given by Eq. 6.5: $T\sin\theta = mr\omega^2$, where $r = L\sin\theta$. Thus

$T \sin \theta = mL \sin \theta \omega^2$, $T = mL\omega^2$, and

$$T = \frac{mg}{\cos \theta} = \frac{(150 \, \text{kg})(9.8 \, \text{m/s}^2)}{\cos 60°} = 2950 \, \text{N}$$

$$\omega = 2\pi f = \sqrt{\frac{T}{mL}}$$

so

$$f = \frac{1}{2\pi}\left[\frac{2940 \, \text{N}}{(150 \, \text{kg})(8 \, \text{m})}\right]^{1/2} = 0.25 \, \text{Hz}$$

PROBLEM 6.3. On a level roadway the coefficient of friction between the tires of a car and the asphalt is 0.80. What is the maximum speed at which a car can round a turn of radius 25 m if the car is not to slip?

Solution The force of friction provides the needed turning force F_c. Thus

$$F_f = \mu N = \mu mg = \frac{mv^2}{r}$$

$$v^2 = \mu rg = (0.08)(25 \, \text{m})(9.8 \, \text{m/s}^2) \qquad v = 14 \, \text{m/s} = 31 \, \text{mi/h}$$

PROBLEM 6.4. A car traveling on a freeway goes around a curve of radius r at speed v. The roadway is banked to provide the necessary inward centripetal force in order for the car to stay in its lane. At what angle should the roadway be banked if the car is not to utilize friction to make the turn?

Solution Draw the force diagram. The car is not accelerating in the vertical direction, so $F_{\text{up}} = F_{\text{down}}$, and $N \cos \theta = mg$. The horizontal component of the normal force N provides the needed centripetal force:

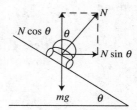

$$N \sin \theta = \frac{mv^2}{r}$$

Divide:

$$\frac{N \sin \theta}{N \cos \theta} = \frac{mv^2}{mgr} \qquad \tan \theta = \frac{v^2}{rg}$$

Caution: You might be tempted to resolve the weight and the normal force into components parallel and perpendicular to the road surface. You might then imagine that the components perpendicular to the surface are in balance. This is wrong because the road surface is not truly directed perpendicular to the plane of the paper. Our drawing is somewhat misleading in this respect. The normal force is actually larger than the car's weight since it must support the weight and also provide an inward force to make the car turn.

PROBLEM 6.5. In the "Human Fly" carnival ride a bunch of people stand with their backs to the wall of a cylindrical room. Once everyone is in place, the room begins to spin. The inward normal force exerted by the wall on the back of each person provides the needed centripetal force to ensure that each person travels in a circle of diameter equal to the diameter of the room. Once the room is spinning rapidly, the floor drops out from beneath the people. Friction between the wall and each person's back "glues" each one to the wall, although with some effort they can squirm and move about (like human flies on a wall). Personally, this is not my cup of tea. I get motion sickness, but kids love it. What minimum coefficient of friction is needed if the people are not to slip downward, assuming the room diameter is 8.0 m and the room spins at 18 rev/min?

Solution   For no slipping down, $F_f = mg$
and $N = mr\omega^2$.

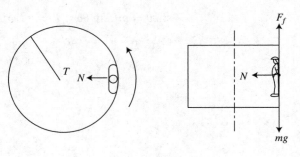

$$\omega = 2\pi f \qquad F_f = \mu N$$

Thus $\mu mr\omega^2 = mg$:

$$\mu = \frac{g}{r\omega^2} = \frac{g}{r(2\pi f)^2}$$

$$= \frac{9.8\,\text{m/s}^2}{(4\,\text{m})(2\pi)^2[(18/60)\text{s}^{-1}]^2}$$

$$\mu = 0.69$$

PROBLEM 6.6.   An F14 jet fighter traveling 260 m/s (about 580 mi/h) pulls out of a vertical dive by turning upward along a circu-lar arc of radius 2.4 km. What acceleration does the pilot experience? Express the result as a multiple of *g*. If the pilot's weight is 560 *N*, what force does the seat exert on him? (This is his "apparent weight.") Even with a pressurized suit, the maximum acceleration a person can experience without suffering brain hemorrhaging resulting in a blackout is about 11*g*. Thus a pilot must avoid turning too sharply so that he or she won't black out and the wings won't break off the airplane.

Solution $$a_c = \frac{v^2}{r} = \frac{(260\,\text{m/s})^2}{2400\,\text{m}} = 28.2\,\text{m/s}^2 = \frac{28.2}{9.8}g = 2.87g$$

Here *g* is the acceleration due to gravity. At the lowest point in the dive, gravity acts downward on the pilot with force *W*, and the seat pushes up with a normal force *N*, so

$$N - W = ma_c = \frac{mv^2}{r} = \frac{mg}{g}\frac{v^2}{r} = W\frac{a}{g}$$

so

$$N = W + W\left(\frac{a}{g}\right) = W\left(1 + \frac{a}{g}\right) = 560(1 + 2.87)\,N$$

$$= 3.87\,W = 2090\,\text{N}$$

PROBLEM 6.7.   A woman stands a distance of 2.40 m from the axis of a rotating merry-go-round platform. The coefficient of friction between her shoes and the platform surface is 0.60. What is the maximum number of revolutions per minute the merry-go-round can make if she is not to start slipping outward?

Solution   Friction provides the needed inward centripetal force, so $F_c = mr\omega^2 = F_f$. $F_f = \mu mg$, so

$$\omega^2 = (2\pi f)^2 = \frac{\mu mg}{mr} \qquad f = \frac{1}{2\pi}\sqrt{\frac{(0.60)(9.8\,\text{m/s}^2)}{2.40\,\text{m}}}$$

$$f = 0.25/\,\text{s} = (60)(0.25)\,\text{rev/min} = 15\,\text{rev/min}$$

PROBLEM 6.8.   Suppose you are driving at speed v_0 and find yourself heading straight for a brick wall that intersects the line of your path at 90°. Assuming that the coefficients of friction for stopping and for turning are the same, are your chances of avoiding a crash better if you continue straight ahead while braking or if you simply turn along a circular path at a constant speed?

Solution

The braking force is $F_f = \mu mg$, so the braking acceleration is $F_f/m = -\mu g$. From Eq. 3.9, $v^2 = v_0^2 + 2ax = 0$ when stopped. Thus the stopping distance is

$$x = -\frac{v_0^2}{2a} = \frac{v_0^2}{2\mu g}$$

For turning, friction provides the centripetal force, so

$$\frac{mv_0^2}{r} = \mu mg \qquad r = \frac{v_0^2}{\mu g}$$

Thus $x < r$, and it is better to brake than to turn.

PROBLEM 6.9.

The first space habitats built by humans will most likely be cylindrical in shape. It is envisioned that such a space habitat will be rotated about the axis of the cylinder in order to simulate the effect of gravity here on Earth. Thus an inhabitant will feel the floor pushing on his or her feet with a force of $N = mr\omega^2$ in order to cause him or her to move along a circular path as the space cylinder rotates. He or she will then experience a sort of "artificial gravity" pulling downward to counter the upward force of the floor. (Downward will be radially out.) Preliminary NASA designs have been developed for a cylinder about 6.4 km in diameter and 32 km in length. Later much larger structures could be built. At what rate would such a structure have to rotate in order to simulate the same acceleration due to gravity found here on Earth?

Solution

We require $mg = mr\omega^2$ so

$$\omega = 2\pi f = \sqrt{\frac{g}{r}}$$

$$f = \frac{1}{2\pi}\sqrt{\frac{9.8 \text{ m/s}^2}{6400 \text{ m}}} = 0.006/\text{s} = 0.37 \text{ rev/min}$$

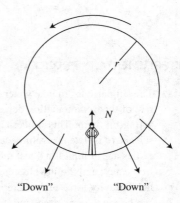

Note that your apparent weight $mr\omega^2$ decreases as r becomes smaller, that is, as you approach the axis of the cylinder. This could have important applications. For example, it is very difficult to hospitalize severely burned patients since they must lie on open wounds. Near the center of the space habitat, a person would feel "weightless" and could simply float above a bed with very little support.

PROBLEM 6.10.

The inward centripetal force required to cause any object to move in a circle can be very large when rigid objects are rotated at high frequency, as is the case in most machines. Spinning gears and wheels can be subject to huge forces that can cause them to fracture and cause serious damage. My father, a machinist, lost an eye when a grinding wheel he was using fractured and sent fragments in all directions. Some of the pieces struck his face. Experimental cars have been designed that are propelled by the energy stored in spinning flywheels (as an alternative to using internal combustion engines), but a limiting factor in the use of such machines is their ability not to fracture when rotated at high speed.

To gain an understanding of the forces involved when objects rotate, consider the following simple model. A very light rod of length L is rotated in a horizontal plane with one end fixed. At one end is attached a mass m_1, and at the center of the rod is attached a mass m_2. Determine the tension T_1 in the portion of the rod between m_1 and m_2 and the tension in the rod between m_2 and the axis of rotation.

Solution   The forces acting on m_1 and m_2 are drawn here. Note that the tension T_1 in the outer portion of the rod pulls *inward* on m_1 and *outward* on m_2. There must be a net inward centripetal force on each mass, so $T_2 > T_1$. Applying $F_c = ma_c$ to each mass, $T_1 = m_1 r_1 (2\pi f)^2$.

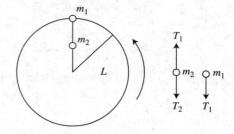

$T_2 - T_1 = m_2 r_2 (2\pi f)^2$, where $r_1 = L$, and $r_2 = \frac{1}{2}L$. Thus,

$$T_1 = m_1 L (2\pi f)^2$$

and

$$T_2 - m_1 L (2\pi f)^2 = m_2 \frac{L}{2} (2\pi f)^2, \text{ so } T_2 = \left(\frac{2m_1 + m_2}{2}\right) L (2\pi f)^2$$

If $m_1 = m_2$, we see that $T_1 = \frac{2}{3} T_2$. The tension is greater farther out, and it increases as the square of the frequency of rotation. Always wear eye protection when working around equipment or machinery with rotating parts.

6.2   Summary of Key Equations

$$\omega = 2\pi f \qquad s = r\theta \qquad a_c = \frac{v^2}{r} = r\omega^2 \qquad F_c = m\frac{v_2}{r} = mr\omega^2$$

SUPPLEMENTARY PROBLEMS

6.11.   The silt particles in lake water gradually settle to the bottom of the lake due to the force of gravity on them. The rate of this sedimentation depends on the size and shape of the particles and on the strength of the gravity force acting on them. This process can be very slow. A centrifuge is a laboratory device widely used in biological science to isolate macromolecules like nucleic acids (for example, DNA) or proteins by sedimentation. The molecules are in liquid in a test tube that is placed in an ultracentrifuge. This apparatus rotates the test tube at high frequency, and the inward centripetal force exerted by the bottom of the test tube, $mr(2\pi f)^2$, produces an artificial gravity force mg' where $g' = r(2\pi f)^2$. By rotating the sample at very high frequency the "effective g," g', can be made very large, and the resulting sedimentation time can be made short enough to separate out various macromolecules in reasonable times (instead of the thousands of years that would be required using normal gravity). What is the effective g' obtained when a sample is rotated at 70,000 rev/min at a distance of 5 cm from the axis?

6.12.   In an enduro motorcycle race a rider goes over the top of a small hill that is approximately spherical with a radius of curvature of 12 m. What is her maximum speed if she is not to become airborne?

6.13.   We saw in Problem 6.8 that when approaching a wall head on in your car at speed v_0, it is better to brake than to turn in order to avoid colliding with the wall. Suppose, however, that your line of motion intersects the plane of the wall at an angle θ. For what value of θ are your chances of barely avoiding a collision equal whether you brake or turn?

6.14.   Suppose the Earth is a sphere of radius 6370 km. If a person stood on a scale at the north pole and observed the scale reading (her weight) to be mg, what would the scale read if she stood on it at a point on the equator?

6.15. A small block of mass *m* slides with negligible friction in a horizontal circle on the inside of a conical surface. The axis of the cone is vertical, and the half angle of the cone is 60°. The block rotates at 1.20 rev/s. How high above the apex of the cone does the block slide?

6.16. In a ball mill used to polish stones, a cylindrical can of radius *r* is oriented with its axis horizontal and rotated at frequency *f*. The stones to be polished are immersed in a slurry containing grit and placed in the can. The rate of rotation of the can is chosen so that the stones fall away from the wall at an optimum position (determined experimentally). This same principle is used in a clothes dryer that tumbles wet clothes in order to dry them. Suppose it is desired to design a ball mill in which a rock will fall away from the wall at a given angle θ, where θ is the angle between vertical and the radius line from the axis of the cylinder to the rock. At what frequency should the can be rotated to achieve this result?

6.17. A very light rod of length 3*L* is pivoted at one end so that it can rotate about a vertical axis. A mass *M* is attached to the rod a distance *L* from the pivot and a second mass 2*M* is attached to the opposite end from the pivoted end. What is the ratio of the tension at a distance of *L*/2 from the pivot to the tension in the rod at a distance of 3*L*/2 from the pivot?

6.18. A centrifuge has many useful applications, e.g., as a means of separating uranium isotopes of nearly equal masses in order to obtain the relatively rare form needed for use as a fuel in a nuclear reactor. The centrifuge is also widely used to separate various entities important in biology, such as blood cells or macromolecules, such as proteins and nucleic acids. The rate (the sedimentation velocity) at which a macromolecule settles to the bottom of a solution depends on *g*, the acceleration due to gravity. The "effective *g*" in a rotating centrifuge can be made much larger than the normal value of $9.8 \, \text{m/s}^2$ by using a high rate of rotation. Very large forces can be experienced by the components of such an ultracentrifuge, so careful safety precautions must be taken when using this instrument because fracture of a component could send fragments flying and cause serious injury.

Consider a centrifuge in which the material of interest is positioned 8.0 cm from the axis of rotation. At how many revolutions per second must the centrifuge rotate in order to create an "effective *g*" of 100,000 *g*?

SOLUTIONS TO SUPPLEMENTARY PROBLEMS

6.11.
$$g' = a_c = r(2\pi f)^2 = (0.05 \, \text{m})(2\pi)^2 \left(\frac{70{,}000}{60 \, \text{s}}\right)^2$$

$$= 2.69 \times 10^6 \, \text{m/s}^2 = \frac{2.69 \times 10^6}{9.8} g = 2.74 \times 10^5 g$$

6.12. At the maximum allowable speed, the ground is not pushing up at all on the motorcycle, so the gravity force *mg* provides the needed centripetal force to keep the motorcycle moving along the circular arc of the hill's surface. Thus,

$$mg = m\frac{v^2}{r} \qquad \text{so } v^2 = rg = (12 \, \text{m})(9.8 \, \text{m/s}^2) \qquad v = 10.8 \, \text{m/s} \quad (\text{about } 24 \, \text{mi/h})$$

6.13. The two possible paths, braking or turning, are shown here. The braking distance is *s*, given by

$$v^2 = v_0^2 - 2as = 0, \quad s = \frac{v_0^2}{2a}$$

$$a = \frac{F_f}{m} = \frac{\mu mg}{m} = \mu g, \quad \text{so } s = \frac{v_0^2}{2\mu g}$$

For turning,

$$\frac{mv_0^2}{r} = F_f = \mu mg, \quad \text{so } r = \frac{v_0^2}{\mu g}$$

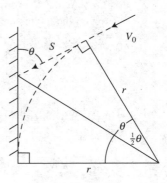

From the drawing I see that $s/r = \tan \theta/2$, so

$$\frac{v_0^2/2\mu g}{v_0^2/\mu g} = \frac{1}{2} = \tan \frac{\theta}{2} \qquad \frac{\theta}{2} = 26.6° \quad \theta = 53.1°$$

6.14. The gravity force mg acts on the person in a direction toward the center of the Earth, and the normal force N exerted by the scale acts radially outward. The net force inward must have a magnitude of $mr\omega^2$. Thus,

$$mr\omega^2 = mg - N \qquad N = mg - mr\omega^2 = mg - mr(2\pi f)^2 = mg - mr\frac{2\pi^2}{T}$$

where

$$T = \frac{1}{f} = 24\,\text{h} = mg - mg\frac{r}{g}\left(\frac{2\pi}{T}\right)^2 = mg\left[1 - \frac{r}{g}\left(\frac{2\pi}{T}\right)^2\right] = mg\left[1 - \frac{6.37 \times 10^6}{9.8}\left(\frac{2\pi}{(24)(3600)}\right)^2\right]$$

$$N = 0.997\,mg$$

Thus a person's apparent weight at the equator would be slightly less than his or her true weight.

6.15. The inward component of the normal force exerted by the conical surface provides the needed centripetal force, and the vertical component of the normal force balances the downward force of gravity.

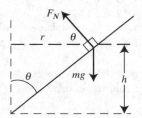

$$N\sin\theta = mg \qquad N\cos\theta = mr(2\pi f)^2$$

Divide:

$$\frac{\sin\theta}{\cos\theta} = \frac{mg}{mr(2\pi f)^2}$$

$$r = \frac{g}{(\tan\theta)(2\pi f)^2} \qquad h = \frac{r}{\tan\theta} = \frac{g}{(2\pi f)^2}\frac{1}{\tan^2\theta}$$

$$g = 9.8\,\text{m/s}^2 \qquad f = 1.20/\text{s} \qquad \theta = 60°$$

$$h = 0.057\,\text{m}$$

6.16. At the moment the rock falls away from the wall, the inward normal force exerted on the rock by the wall drops to zero. Thus, at this instant the inward centripetal force acting on the rock is just the inward component of the gravity force $mg\cos\theta$. Thus,

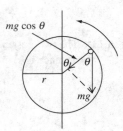

$$mr(2\pi f)^2 = mg\cos\theta, \quad \text{so } f = \frac{1}{2\pi}\sqrt{\frac{g\cos\theta}{r}}$$

6.17. The rod rotates with angular frequency ω.

Let $T_1 = $ tension in inner portion of the rod

$T_2 = $ tension in outer portion of the rod

For mass M, $T_1 - T_2 = ML\omega^2$

For mass $2M$, $T_2 = (2M)(3L)\omega^2 = 6ML\omega$

so $T_1 = T_2 + ML\omega^2 = 6ML\omega^2 + ML\omega^2 = 7ML\omega^2$

$$\frac{T_1}{T_2} = \frac{7}{6}$$

6.18. Require $mg_{\text{eff}} = mr\omega^2$, $g_{\text{eff}} = 10^5 g$

$$\omega = 2\pi f, \text{ so } f = \frac{1}{2\pi}\left(\frac{g_{\text{eff}}}{r}\right)^{1/2} = \frac{1}{2\pi}\left[\frac{(10^5)(9.8\,\text{m/s}^2)}{0.08\,\text{m}}\right]^{1/2}$$

$$f = 557\,\text{rev/s}$$

CHAPTER 7

Work and Energy

We have seen how to use displacement, velocity, acceleration, and force to describe the behavior of some simple mechanical systems. With this background we now develop an alternate approach based on Newton's laws that is simpler and very widely applicable. The concepts of work and energy will enable us to extend our previous analysis to complicated systems like the human body, the ecosystem in which we live, and chemical and nuclear reactions.

7.1 Work

You probably have a pretty good idea of what is meant by the term work. When you push on a heavy packing crate and slide it across the floor, you do work on it. You get tired because you have used energy to do the work. Suppose a constant force F_x acts in the x direction and causes an object to move a distance of Δx. We define the work done by the force as

$$W = F_x \Delta x \qquad (7.1)$$

Since work is the product of force and distance, it is measured in units of newton-meters. Work is such an important concept that its unit is given a special name, the **joule**. 1 newton-meter (N · m) = 1 joule (J). There are two important points worth recognizing here. First, in order for a force to do work on an object, the object must move. If you stand holding motionless a heavy concrete block in your hands, you may get tired, but you are not doing work on the block. Second, only forces directed along the line of motion of an object do work on the object. When you push a crate across the floor, the force of gravity does no work on the crate because the gravity force is directed downward, perpendicular to the direction of motion of the crate.

Consider a constant force F that acts on an object that is moved. If the force makes an angle θ with the direction of motion (which I take to be the x axis), the work done by F is

$$W = F_x \Delta x = F \cos \theta \, \Delta x \qquad (7.2)$$

Observe that when $90° < \theta < 180°$, $\cos \theta < 0$ and the work done by F is negative. What this means is that instead of the force doing work on the moving object, the object is doing work on whatever generates the force. Some books use this idea of negative work, but it is not necessary to do so, and I find the idea needlessly confusing. It is sort of like saying that when you draw money out of the bank, you make a negative deposit. I like to keep things simple.

PROBLEM 7.1. A logger drags a heavy log across level ground by attaching a cable from the log to a bulldozer. The cable is inclined upward from horizontal at an angle of 20°. The cable exerts a constant force of 2000 N while pulling the log 16 m. How much work is done in dragging the log?

Solution

$$W = F \cos \theta \, \Delta x = (2000\,N)(\cos 20°)(16\,m) = 3.0 \times 10^4\,J$$

If the displacement Δx is written as a vector, we can write Eq. 7.2 in the following useful form:

$$W = F \cos \theta \, \Delta x = \mathbf{F} \cdot \Delta \mathbf{x} \qquad (7.3)$$

This is an important relation. **MEMORIZE IT**.

Frequently the forces we encounter are not constant. They change in strength and direction as the object on which they act moves. For example, suppose a force $F(x)$ that is a function of position acts on an object that moves from position x_1 to position x_2. We can imagine this displacement to consist of many small steps Δx_i. The total work done is the sum of the work done to make each little step.

$$W = \sum F(x) \Delta x_i$$

In the limit that $\Delta x \to 0$, this may be written as an integral.

$$W = \int_{x_1}^{x_2} F(x)\,dx \qquad (7.4)$$

Equation 7.4 has a simple (and useful) interpretation in terms of a graph of force versus displacement, as illustrated in Fig. 7-1. The work done in each little step dx is just $F_x\,dx$. But $F_x\,dx$ is the area of the small shaded rectangle in the drawing. The total work done in going from x_1 to x_2 is thus seen to be equal to the area under the force versus displacement curve.

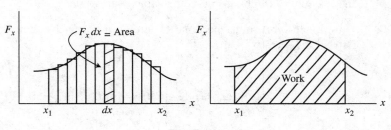

Fig. 7-1

An important example of a nonconstant force is the force exerted by a spring or a rubber band. Suppose a mass is attached to one end of a spring and placed on a frictionless horizontal surface. The other end of the spring is attached to a fixed point. Take the position of the mass when the spring is unstretched to be $x = 0$. If the mass is then displaced an amount x from its equilibrium position, the spring exerts a force F on it, where

$$F = -kx \qquad (7.5)$$

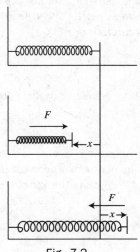

This kind of force is called a **Hooke's law force**, and k is the **spring constant**. The minus sign indicates that this spring force is a **restoring force**. That means that the spring force always tries to make the attached mass move back toward its equilibrium position. When x is positive, F is negative. When x is negative, F is positive. This is illustrated in Fig. 7-2.

This Hooke's law force is a big deal because it is a first approximation to the forces we encounter in many situations, for example, the forces on a girder in a bridge or between ions in a crystal. We will study it in greater detail later.

Fig. 7-2

If you want to stretch the spring from its equilibrium position (at $x = 0$) to a point where it is stretched an amount x, you must exert a force opposite to the force exerted by the spring. Of course, to actually get the mass to move from rest, you would have to exert a force a teensy bit bigger than the spring force, but if you don't mind taking forever to move the mass, a force equal in strength to the spring force will suffice. Hence the work done in stretching a spring a distance x is

$$W = \int_0^x F \, dx = \int_0^x kx \, dx$$

$$\boxed{W = \tfrac{1}{2}kx^2} \tag{7.6}$$

PROBLEM 7.2. A force of 120 N will stretch a spring 2 cm. What is the spring constant of the spring? If the spring were cut in half, what would then be the spring constant?

Solution $F = -kx \cdot k$ is always positive since F and x always have opposite signs. Thus, the magnitude of k is

$$k = \frac{120 \, \text{N}}{0.02 \, \text{m}} = 6000 \, \text{N/m}$$

If the entire spring stretched by 2 cm, half of its length stretched by 1 cm, whereas the force acting on it is still 120 N, so the spring constant for this smaller spring (half the original one) is 12,000 N · m. Thus the small spring is stiffer than the longer one. You can test this for yourself by joining rubber bands together. The longer the chain of rubber bands, the "softer" the spring they make. Note that the tension in a spring, like the tension in a light string, is the same everywhere.

PROBLEM 7.3. How much work must be done to stretch a spring by 2 cm if the spring constant is 640 N · m?

Solution

$$W = \tfrac{1}{2}kx^2 = \tfrac{1}{2}(640 \, \text{N} \cdot \text{m})(0.02 \, \text{m})^2 = 0.13 \, \text{J}$$

7.2 Kinetic Energy

Suppose that a single constant force F acts on a particle in its direction of motion and causes it to accelerate, increasing the speed from an initial value v_0 up to a final value v. Recall that for an object with constant acceleration, $v^2 = v_0^2 + 2ax$. Substitute $a = F/m$. Then,

$$v^2 = v_0^2 + \frac{2Fx}{m}$$

The work done by the force is $W = Fx$, so

$$\boxed{W = \tfrac{1}{2}mv^2 - \tfrac{1}{2}mv_0^2} \tag{7.7}$$

This is a very important equation. **MEMORIZE IT**.

We define the **kinetic energy** of an object of mass m with speed v as

$$\boxed{\text{KE} = \tfrac{1}{2}mv^2} \tag{7.8}$$

Energy, like work, is measured in joules: Eq 7.7 is the **work-energy theorem**. It says that **the work done on a particle is equal to the increase in the kinetic energy of the particle**. The work-energy theorem is valid even if the force is varying. Thus,

$$W = \int_{x_0}^{x} F_x\, dx = \int_{x_0}^{x} ma\, dx = \int_{x_0}^{x} m\frac{dv}{dt}\, dx = \int_{v_0}^{v} m\frac{dx}{dt}\, dv = \int_{v_0}^{v} mv\, dv$$

$$= \tfrac{1}{2}mv^2 - \tfrac{1}{2}mv_0^2$$

In the discussion above, I imagined an external force acting on a particle to speed it up and increase its kinetic energy. I could equally well consider a case in which a particle experiences a retarding force that slows it down. In this case the kinetic energy would decrease, and the particle would do work on an external entity. Using the analysis above, we could deduce that the loss in kinetic energy would be equal to the work done by the particle. This is what happens in your car engine when a rapidly moving gas molecule strikes the piston and pushes on it. The gas molecule slows down and in so doing, does work on the piston, and this work is then transferred into making the car move forward. This observation leads us to a pretty good working definition of the meaning of *energy*.

DEFINITION: The energy of a system is a measure of its ability to do work.

Energy can also be used to change the state of matter, for example, to melt a solid into a liquid. Energy takes many forms. Kinetic energy ("motion energy") is just one form of energy. Other forms of energy include light, thermal energy, chemical energy, mass energy (sometimes called "nuclear energy"), electric energy, magnetic energy, gravitational energy, and sound. These are only rough descriptors, and some of the categories overlap. Some types of energy are grouped under the broad title "potential energy," a term used to describe energy when it is more or less "stored" for future use. Thermal energy includes the kinetic energy of moving atoms as well as the stored potential energy associated with stretched electric bonds between atoms.

PROBLEM 7.4.   Calculate the kinetic energy of each of the following:

(a) The Earth orbiting the Sun $m = 5.98 \times 10^{24}\,\text{kg}$ $v = 2.98 \times 10^4\,\text{m/s}$
(b) Car-driving 60 mi/h $m = 1500\,\text{kg}$ $v = 27\,\text{m/s}$
(c) World-class sprinter $m = 80\,\text{kg}$ $v = 10\,\text{m/s}$
(d) Rifle bullet $m = 0.01\,\text{kg}$ $v = 1000\,\text{m/s}$
(e) Nitrogen molecule in air $m = 4.6 \times 10^{-26}\,\text{kg}$ $v = 500\,\text{m/s}$

Solution   Using $\text{KE} = \tfrac{1}{2}mv^2$ yields the following interesting results:

(a) $2.66 \times 10^{33}\,\text{J}$ (b) $5.47 \times 10^5\,\text{J}$ (c) $4 \times 10^3\,\text{J}$ (d) $5 \times 10^3\,\text{J}$ (e) $5.8 \times 10^{-21}\,\text{J}$

7.3 Power

DEFINITION: Power is the rate of doing work or of transferring energy.

If work dW is done in time dt, the instantaneous power used is

$$\boxed{P = \frac{dw}{dt}}$$

(7.9)

If a hot object radiates away energy dE in time dt, the power it radiates is

$$\boxed{P = \frac{dE}{dt}}$$

(7.10)

Power is measured in units of joules per second (J/s). The concept of power is so important that the unit of power is given its own name, the watt. $1 \text{ W} = 1 \text{ J/s}$. Other commonly encountered power units are the microwatt $(1 \mu\text{W} = 10^{-6} \text{ W})$, the milliwatt $(1 \text{ mW} = 10^{-3} \text{ W})$, the kilowatt $(1 \text{ kW} = 10^3 \text{ W})$, and the megawatt $(1 \text{ MW} = 10^6 \text{ W})$. The British system of units uses the horsepower unit: $1 \text{ hp} = 746 \text{ W}$.

If the power is constant, we may write $W = Pt$ or $E = Pt$. When utility companies sell electric energy, they measure the energy sold in a unit called the kWh. They do this because a joule is a very small and inconvenient unit for their purposes.

$$1 \text{ kWh} = (1000 \text{ J/s})(3600 \text{ s}) = 3.6 \times 10^6 \text{ J}$$

Note that a watt is a unit of power that does not depend on the kind of work being done or on the kind of energy transferred. You may associate the term "watt" with electricity (as in a 100-W light bulb), but this unit applies to all kinds of work and energy. Historically, people were slightly confused and used different units for different kinds of energy. For example, in chemistry and nutrition we encounter energy measured in calories or kilocalories $(1 \text{ cal} = 4.186 \text{ J})$. Architects use British thermal units (Btus) per hour to characterize building heating systems $(1 \text{ Btu} = 1054 \text{ J})$. Electronic engineers measure energy levels in a crystal using the electronvolt $(1 \text{ eV} = 1.6 \times 10^{-19} \text{ J})$.

If force $\mathbf{F}$ causes a particle to undergo a displacement $\mathbf{ds}$, the work done is $dW = \mathbf{F} \cdot \mathbf{ds}$. Since $\mathbf{ds}/dt = \mathbf{v}$, the power provided by the force is

$$P = \frac{dW}{dt} = \frac{\mathbf{F} \cdot \mathbf{ds}}{dt} \quad \text{or} \quad \boxed{P = \mathbf{F} \cdot \mathbf{v}} \tag{7.11}$$

PROBLEM 7.5. Consider a car traveling at a steady speed of 60 km/h (16.7 m/s). It encounters a frictional force (rolling and air drag) of 520 N. At what power level does the engine deliver energy to the wheels?

Solution
$$P = Fv = (520 \text{ N})(16.7 \text{ m/s}) = 8.68 \text{ kW}$$

$$= (8.68 \text{ kW})\left(\frac{1}{0.746} \frac{\text{hp}}{\text{kW}}\right) = 11.6 \text{ hp}$$

Note: It is not uncommon to see car ads that tout an engine rated at something like 150 hp. However, only about 20 to 25 percent of this power output is delivered to the wheels. The rest is lost as heat. Further, this high-power output is obtained only when the engine is running at full throttle. A usable power level of 10 to 15 hp is about what is required to propel a car traveling a level road at moderate speed.

PROBLEM 7.6. Combustion of 1 gal of gasoline yields $1.3 \times 10^8 \text{ J}$. Consider a car that can travel 28 mi/gal at a speed of 90 km/h. Of the energy obtained from burning the gasoline, 25 percent goes into driving the car, with the rest dissipated as heat. What is the average frictional force acting on the car?

Solution   The time for the car to travel 28 mi is

$$t = \frac{(28 \text{ mi})(1.6 \text{ km/mi})}{90 \text{ km/h}} = 0.5 \text{ h}$$

$$= 1800 \text{ s}$$

The gasoline energy used in this time is that from 1 gal, $1.3 \times 10^8 \text{ J}$. Of this energy, 25 percent powers the wheels, so

$$P = (0.25)\frac{(1.3 \times 10^8 \text{ J})}{1800 \text{ s}} = 18 \text{ kW}$$

$$v = 90 \text{ km/h} = 90\frac{10^3 \text{ m}}{3600 \text{ s}} = 25 \text{ m/s}, \qquad F = \frac{P}{v} = \frac{18,000 \text{ W}}{25 \text{ m/s}} = 720 \text{ N}$$

PROBLEM 7.7. The first human-powered airplane to cross the English Channel was the Gossamer Albatross. A person turned the propeller by means of a bicycle pedaling mechanism. In order to keep the plane flying, he had to deliver 0.3 hp to the drive mechanism throughout the duration of the 2 h, 49 min flight. Human muscles have an efficiency of about 20 percent; that is, 20 percent of the energy released in your body goes into doing mechanical work. (a) What total energy did the pilot use during this flight? (b) A Big Mac hamburger provides about 500 kcal. How many such hamburgers would the pilot have to eat to obtain enough energy for the flight across the channel?

Solution

(a) $t = 2\,\text{h}\,49\,\text{min} = (2)(3600\,\text{s}) + (49)(60\,\text{s}) = 10{,}140\,\text{s}$

$$W = Pt = (0.3\,\text{hp})(746\,W/\text{hp})(10{,}140\,\text{s}) = 2.27 \times 10^6\,\text{J}$$

If the pilot used energy E, then 20 percent of this energy went into the work of pedaling. Thus,

$$0.2E = 2.27 \times 10^6\,\text{J} \quad \text{and} \quad E = 1.13 \times 10^7\,\text{J}$$

(b) $1\,\text{kcal} = 4.2\,\text{kJ}$ so the number of burgers required is

$$n = \frac{1.13 \times 10^7\,\text{J}}{(5 \times 10^5\,\text{cal})(4.2\,\text{J/cal})} = 5.4$$

PROBLEM 7.8. A girl riding her bicycle on a level road is traveling at 20 km/h while sitting upright. She finds that when she leans low over the handlebars, while still doing work at the same rate, that her speed increases to 22 km/h. By what factor did she reduce the drag force acting on her by leaning low over the handlebars?

Solution She maintained power constantly, so $P = F_1 v_1 = F_2 v_2$. Thus,

$$F_2 = \frac{v_1}{v_2} \qquad F_1 = \frac{20}{22} \qquad F_1 = 0.91 F_1$$

PROBLEM 7.9. In terms of energy consumption, walking and bicycling are much more efficient means of transportation than traveling by car or airplane. A hiker can cover about 30 mi with an additional food intake of 2000 kcal (above what is needed for sitting in one place). A bicyclist can travel about 100 mi with this same amount of supplementary food intake. One gallon of gasoline releases about $1.3 \times 10^8\,\text{J}$ of energy. Calculate about how many "miles per gallon" (mi/gal) a hiker and a bicyclist would get if they used gasoline as a source of energy. An average car gets about 24.7 mi/gal.

Solution 2000 kcal is equivalent to G gal of gas, where

$$G = \frac{(2000\,\text{cal})(4.2\,\text{J/cal})}{1.3 \times 10^8\,\text{J}} = 0.06\,\text{gal}$$

For a hiker,

$$\text{mi/gal} = \frac{30\,\text{mi}}{0.06\,\text{gal}} = 470\,\text{mi/gal}$$

For a bicyclist,

$$\text{mi/gal} = \frac{100\,\text{mi}}{0.06\,\text{gal}} = 1560\,\text{mi/gal}$$

PROBLEM 7.10.   A car of mass 1600 kg with its gears in neutral is observed to reach a constant terminal speed of 110 km/h after coasting a long way down a 10 percent grade. (This is a slope that drops 1 m for every 10 m traveled along the roadway.) The car experiences a frictional drag force bv^2. What power would be required to drive this car at 90 km/h on a level highway?

Solution   When coasting at constant speed downhill, the downhill component of the gravity force $mg \sin \theta$ just balances the friction force F.

$$v = 110 \, \text{km/h} = 30.6 \, \text{m/s}$$
$$v_2 = 90 \, \text{km/h} = 25 \, \text{m/s}$$
$$F = bv_1^2 = mg \sin \theta$$
$$b = \frac{mg}{v_1^2} \sin \theta$$

On level ground,

$$P = Fv_2 = (bv_2^2)(v_2) = bv_2^3$$
$$= \left(\frac{mg \sin \theta}{v_1^2}\right)(v_2^3) = \frac{(1600 \, \text{kg})(9.8 \, \text{m/s}^2)(0.1)(25 \, \text{m/s})^3}{(30.6 \, \text{m/s})^2} = 2.67 \times 10^4 \, \text{W}$$
$$= (2.67 \times 10^4 \, \text{W})\left(\frac{1 \, \text{hp}}{746 \, \text{W}}\right) = 35.8 \, \text{hp}$$

7.4   Summary of Key Equations

Work: $W = F \cos \theta \qquad W = \int_{x_1}^{x_2} F(x) \, dx$

Hooke's law: $F = -kx$

Work to stretch a spring: $W = \frac{1}{2} kx^2$

Kinetic energy: $KE = \frac{1}{2} mv^2$

Work-energy theorem: $W = \frac{1}{2} mv_2^2 - \frac{1}{2} mv_1^2$

Power: $P = \dfrac{W}{t} \quad \text{or} \quad P = \dfrac{E}{t}$

In general: $P = \dfrac{dW}{dt} \quad \text{or} \quad P = \dfrac{dE}{dt}$

SUPPLEMENTARY PROBLEMS

7.11.   A man pushes at constant speed a 50-kg refrigerator a distance of 14 m across a level floor where the coefficient of friction is 0.40. How much work does he do?

7.12.   A toy dart gun utilizes a spring with a spring constant of 60 N · m. How much work must be done to compress this spring a distance of 3.2 cm?

7.13.   People have survived falls from great heights, provided that they landed in snow or foliage or some other such material to cushion the impact. A Russian pilot survived a fall from 22,000 ft when he landed on a sloping snow bank. The chances of survival depend on how you are oriented on impact. If you land head first, your chances of survival are greatly reduced. A typical adult of mass 80 kg landing flat on his back has about a 50 percent chance of survival if the impact force does not exceed 1.20×10^5 *N*. The terminal speed with which a person falling from a great height will strike the ground is about 140 mi/h, or 63.0 m/s. At this impact speed, what depth of snow would be required if the average stopping force is not to exceed that corresponding to 50 percent survival chance?

7.14. The engine in a car traveling on a level road must overcome air resistance and must do work in deforming the tires as they roll (road resistance). At 70 km/h, the effective drag forces due to these effects are approximately equal. The road resistance is essentially independent of speed, however, whereas the air resistance varies approximately as the square of the speed. (a) By what factor will the power delivered to the wheels increase if the speed is doubled? (b) By what factor will the number of miles per gallon be reduced if the speed is doubled?

7.15. A rock of mass *m* is dropped from rest from a point a distance *h* above the ground. Use the work-energy theorem to calculate the speed of the rock when it hits the ground.

7.16. A Jaguar car of mass 2*M* is racing a small Austin Healy sports car of mass *M*. Initially the Jaguar has half the kinetic energy of the Austin Healy, but when the Jaguar then speeds up by 10 m/s, the two vehicles have the same kinetic energy. What were the initial speeds of the two cars?

7.17. Suppose that energy *Q* is required to accelerate a car from rest to speed *v*, neglecting friction. How much added energy would be required to increase the speed from *v* to 2*v*?

7.18. In 1997 the Hale–Bopp comet provided a brilliant spectacle in the night sky. It passed harmlessly by us and will not return for another 2400 years, but its presence raised the specter of another comet colliding with Earth and causing extensive damage. It is speculated that a comet collision some 70 million years ago may have kicked up such a huge cloud of dust that it obliterated sunlight for several years and led to the extinction of the dinosaurs. It is estimated that the Hale–Bopp comet had a mass of about 2.7×10^{14} kg, and at its nearest approach was traveling about 63 km/s (about 140,000 mi/h). (a) What is the kinetic energy of such a comet? Express the answer in joules and in "megatons of TNT." The detonation of 1 million tons of TNT releases 4.2×10^{15} J of energy. (b) The detonation of 1 megaton of TNT will produce a crater of about 1 km diameter. The diameter of the crater is proportional to the one-third power of the energy released. What size crater would you expect Hale–Bopp to produce?

7.19. An engineer is asked to design a crash barrier for runaway trucks that get out of control descending a steep grade near Lewiston, Idaho. The specifications call for stopping a truck of mass 25,000 kg moving at 24 m/s with a stopping acceleration not to exceed 5.0*g* (where $g = 9.8 \, \text{m/s}^2$). (a) What spring constant is required? (b) How much will the spring have to compress? Does the design sound feasible to you?

7.20. An airplane experiences a drag force av^2 due to air passing over its surface. There is an additional *induced drag force* b/v^2 that results from the fact that the wings cause air to be pushed downward and slightly forward. From Newton's third law we see that the air will thus push back, exerting an upward lift force and a backward "induced drag" force. Thus, the total drag force can be expressed as $F = av^2 + b/v^2$. At constant speed, the engine must provide a forward force that balances this drag force. For a small single-engine airplane typical values of *a* and *b* might be $a = 0.12 \, \text{N} \cdot \text{s}^2/\text{m}^2$ and $b = 2.9 \times 10^7 \, \text{N} \cdot \text{m}^2/\text{s}^2$. Calculate the speeds, in terms of *a* and *b*, at which such a plane will have (a) the maximum horizontal range and (b) the maximum time in the air (maximum *endurance*).

7.21. A spring loaded barrier can be used to stop a runaway vehicle. What spring constant would be required if a vehicle of mass 2200 kg moving at 3.0 m/s is to be stopped in a distance of 80 cms?

7.22. In a bicycle race, a rider such as Lance Armstrong (mass = 70 kg) uses energy at a rate of about 6.0 W/kg of body mass. If he travels at an average speed of 11.0 m/s in a 120 km race, how much energy does he expend? Express the answer both in joules and in kilocalories. Note that a kilocalorie is called a "calorie" in nutrition. This is the calorie referred to in the nutrition label on a loaf of bread.

SOLUTIONS TO SUPPLEMENTARY PROBLEMS

7.11. $F_f = \mu mg \quad W = F_f x = \mu mgx \quad W = (0.40)(50\,\text{kg})(9.8\,\text{m/s}^2)(14\,\text{m}) = 2740\,\text{J}$

7.12. $W = \frac{1}{2}kx^2 = (0.5)(60\,\text{N} \cdot \text{m})(0.032\,\text{m})^2 = 0.03\,\text{J}$

7.13. $W = Fd = \frac{1}{2}mv^2 \quad d = \dfrac{mv^2}{2F} = \dfrac{(80\,\text{kg})(63.0\,\text{m/s})^2}{2(1.20 \times 10^5\,\text{N})} = 1.32\,\text{m}$

7.14. Let F_R = drag force due to road resistance and $F_A = bv^2$ = drag force due to air resistance.

(a) $P_1 = (F_R + F_A)v_1 = (F_R + bv_1^2)v_1 \quad P_2 = (F_R + bv_2^2)v_2 = F_R + b(2v_1)^2(2v_1)$

$\dfrac{P_2}{P_1} = \dfrac{2v_1(F_R + 4bv_1^2)}{(F_R + bv_1^2)v_1} \quad$ Given $F_R = bv_1^2 \quad \dfrac{P_2}{P_1} = \dfrac{2(F_R + 4F_R)}{F_R + F_R} = 5$

(b) Mileage varies as mi/gal = distance traveled/energy used = vt/Pt. Thus,

$$(\text{mi/gal})_1 = \frac{v_1}{P_1} \qquad (\text{mi/gal})_2 = \frac{v_2}{P_2} \qquad \frac{(\text{mi/gal})_2}{(\text{mi/gal})_1} = \frac{v_2}{v_1} \cdot \frac{P_1}{P_2} = \frac{2v_1}{v_1} \cdot \frac{1}{5} = \frac{2}{5}$$

Thus, mileage decreases to 40 percent of the value at 70 km/h when the speed is doubled.

7.15. The work done by the force of gravity is equal to the gain in kinetic energy of the rock. Thus, $W = Fh = mgh = \frac{1}{2}mv^2$, and $v = \sqrt{2gh}$.

7.16. Initially $\text{KE}_J = \frac{1}{2}\text{KE}_{AH}$, and $\frac{1}{2}(2M)v_J^2 = \frac{1}{2}(\frac{1}{2}Mv_{AH}^2)$. After speeding up $\frac{1}{2}(2M)(v_J + 10)^2 = \frac{1}{2}Mv_{AH}^2$. Divide these equations:

$$\frac{(v_J + 10)^2}{v_J^2} = \frac{2v_{AH}^2}{v_{AH}^2} \quad (v_J + 10)^2 = 2v_J^2$$

$$v_J + 10 = \sqrt{2}v_J \quad v_J = 24\,\text{m/s} \quad v_{AH} = 48\,\text{m/s}$$

7.17. Energy to go from 0 to v is $E_1 = Q = \frac{1}{2}mv^2$. Energy to go from 0 to $2v$ is $E_2 = \frac{1}{2}(2v)^2 = 4(\frac{1}{2}mv^2) = 4Q$. So added energy to go from 0 to $2v$ is $E_2 - E_1 = 3Q$.

7.18. (a) $\text{KE} = \frac{1}{2}mv^2 = (0.5)(2.7 \times 10^{14}\,\text{kg})(63 \times 10^3\,\text{m/s})^2 = 5.36 \times 10^{23}\,\text{J}$

$$= (5.36 \times 10^{23}\,\text{J})\left(\frac{1}{4.2 \times 10^{15}} \frac{\text{megaton}}{\text{J}}\right) = 1.28 \times 10^8\,\text{megaton TNT}$$

(b) $\dfrac{d_2}{d_1} = \left(\dfrac{E_2}{E_1}\right)^{1/3} = \left(\dfrac{1.28 \times 10^8\,\text{megaton}}{1\,\text{megaton}}\right)^{1/3} = 503\,\text{km}$

Thus, Hale–Bopp would probably blast a crater about 500 km in diameter.

7.19. The kinetic energy of the truck will do the work necessary to compress the spring.

$$W = \frac{1}{2}kx^2 = \frac{1}{2}mv^2 \tag{i}$$

The acceleration is not to exceed 5g, so

$$F = kx = ma = 5mg \tag{ii}$$

Substitute Eq. (ii) in Eq. (i):

$$\frac{1}{2}(5mg)x = \frac{1}{2}mv^2 \quad x = \frac{v^2}{5g} = \frac{(24 \text{ m/s})^2}{5(9.8 \text{ m/s}^2)} = 11.8 \text{ m}$$

$$k = \frac{5mg}{x} = \frac{(5)(25{,}000 \text{ kg})(9.8 \text{ m/s}^2)}{11.8 \text{ m}} = 1.04 \times 10^5 \text{ N} \cdot \text{m}$$

This doesn't sound like a great idea to me. It would require a mighty big spring, and one would have to have a latch to keep the compressed spring from shooting the truck back up the hill. But then, what do I know? I never thought Xerox photocopiers and CD players would work either.

7.20. (a) $F = av^2 + \dfrac{b}{v^2}, \qquad P = Fv = av^3 + \dfrac{b}{v}$

Fuel energy used in time t is

$$E = Pt = \left(av^3 + \frac{b}{v}\right)t$$

Range in time t is $x = vt$, so

$$E = \left(av^3 + \frac{b}{v}\right)\left(\frac{x}{v}\right) \quad \text{and} \quad x = \frac{Ev^2}{av^4 + b}$$

Maximize x by varying v, with E constant (total fuel):

$$\frac{dx}{dv} = 0 = \frac{2Ev}{av^4 + b} - \frac{Ev^2(4av^3)}{(av^4 + b)^2}$$

So

$$2v(av^4 + b) - 4av^5 = 0 \quad v = \left(\frac{b}{a}\right)^{1/4} \text{ for maximum range.}$$

(b) From the above,

$$E = \left(av^3 + \frac{b}{v}\right)t$$

Vary v to maximize t for longest flight time,

$$t = \frac{Ev}{av^4 + b} \quad \text{and} \quad \frac{dt}{dv} = 0 = \frac{E}{av^4 + b} - \frac{4Eav^4}{(av^4 + b)^2}$$

$$av^4 + b - 4av^4 = 0 \quad v = \left(\frac{b}{3a}\right)^{1/4} \quad \text{for maximum time in air.}$$

7.21. The kinetic energy of the vehicle is converted into potential energy of the spring.

$$\tfrac{1}{2}kx^2 = \tfrac{1}{2}mv^2$$

$$k = \frac{mv^2}{x^2} = \frac{(2200 \text{ kg})(3.0 \text{ m/s})^2}{(0.08 \text{ m})^2} = 3.10 \times 10^4 \text{ N/m}$$

7.22. $x = vt$, $t = \dfrac{x}{v}$

$W = Pt = P\dfrac{x}{v}$

$W = (6\,\text{W/kg})(70\,\text{kg})\left(\dfrac{120 \times 10^3\,\text{m}}{11\,\text{m/s}}\right) = 4.58 \times 10^6\,\text{J}$

$1\,\text{kcal} = 4186\,\text{J}$

So,

$$W = \frac{4.58 \times 10^6\,\text{J}}{4186\,\text{J/kcal}} = 1090\,\text{kcal}$$

CHAPTER 8

Potential Energy and Conservation of Energy

8.1 Potential Energy

Consider a hockey puck sliding across an ice rink. Because of its motion, it has kinetic energy. As it slides, it does work against the force of friction and steadily slows to a stop. When at rest, the puck has no kinetic energy. The kinetic energy of the puck has been lost to heat in doing work against friction, and we cannot get it back. Friction is an example of a **nonconservative force**. This means that the mechanical energy of an object or of a system is not conserved when friction forces are present. In physics we use conserved to mean "constant" or "not changing."

On the other hand, some forces (the force exerted by a stretched spring and the gravity force are important examples) are what I call "spring-back" forces. Most books call this kind of force a **conservative force**. Suppose, for example, you were to throw a ball straight up. When the ball leaves your hand, it has kinetic energy, but as it rises, the kinetic energy decreases until it is zero at the highest point. This kinetic energy is not "lost," as was the case when work was done against friction. The gravity force can pull the ball back down, allowing it to gain as much kinetic energy as was lost on the way up. If the ball were thrown up to rest on a window ledge, we might think of the lost kinetic energy as being stored, because at a later time we could push the ball off, allowing it to gain speed as it fell and thereby recover its lost kinetic energy. The moving ball could strike something on the ground and do work on it. The stored energy is called *potential energy* because it has the "potential" to do work. It is useful to describe this state of affairs by introducing the concept of **potential energy (PE)** $U(x, y, z)$. This is a scalar function associated with a conservative force. U depends only on the position of the object. If $W(A, B)$ is the work done on an object in moving it from point A to point B, then the potential energy function is defined such that

$$W(A, B) = U(A) - U(B) \tag{8.1}$$

The work done by gravity when an object of mass m moves from elevation y_1 to elevation y_2 is

$$W(y_1, y_2) = mg(y_1 - y_2) = U(y_1) - U(y_2) \tag{8.2}$$

This means the gravitational potential energy function $U(y)$ is of the form

$$U(y) = mgy + U_0 \tag{8.3}$$

where U_0 is an arbitrary constant.

From Eq. 8.1 we see that only differences in potential energy are significant. Thus, we choose the constant U_0 to have a value that will simplify our calculations. For example, we often choose the "zero" of gravitational potential energy to be at the surface of the Earth and set $y = 0$ there. In many problems (for example, for those involving projectiles) it is convenient to set $U = 0$ at the initial position of the projectile. If $U = 0$ at $y = 0$, then

$$U(y) = mgy \qquad (8.4)$$

MEMORIZE this important equation, Eq. 8.4. Note the following important points: Potential energy is measured in joules. For a conservative force, the work done by the force in moving an object from point A to point B depends *only* on the locations of points A and B, not on the path followed between them or on how fast the object moved from A to B. Any force that is velocity dependent is thus nonconservative. The potential energy function is a scalar function, like temperature or pressure, associated with each point in space. There is no potential energy function for nonconservative forces such as friction. We can place the "zero" point for potential energy wherever we want since only differences in potential energy matter. Thus, potential energy, unlike kinetic energy, can be either positive or negative (that is, we can go below sea level).

The above idea can be extended to varying forces. For example, for a conservative one-dimensional force $F(x)$, the work done in moving from x_1 to x_2 is

$$W(x_1, x_2) = \int_{x_1}^{x_2} F(x)\, dx = U(x_1) - U(x_2) = -[U(x_2) - U(x_1)]$$

This requires that

$$F(x) = -\frac{dU}{dx} \qquad (8.5)$$

Thus, if the potential energy function is known, we can find the conservative force function that gives rise to it. For gravity, $U = mgy$, so

$$F = -\frac{d}{dy}(mgy) = -mg \quad \text{as expected}$$

Equation 8.5 can be integrated to yield

$$U(x) = -\int F(x)\, dx \qquad (8.6)$$

We can apply the work–energy theorem to the above ideas.

Work done = gain in kinetic energy (KE)

$$W(x_1, x_2) = \tfrac{1}{2}mv_2^2 - \tfrac{1}{2}mv_1^2$$

But from Eq. 8.1, $W(x_1, x_2) = U(x_1) - U(x_2)$, so

$$U(x_1) - U(x_2) = \tfrac{1}{2}mv_2^2 - \tfrac{1}{2}mv_1^2$$

or

$$U(x_1) + \tfrac{1}{2}mv_1^2 = U(x_2) + \tfrac{1}{2}mv_2^2 \qquad (8.7)$$

Equation 8.7 is a remarkable result called the **law of conservation of energy**. It shows that the quantity $U + \tfrac{1}{2}mv^2$ stays constant at all points along the trajectory of a particle acted on by a conservative force. This quantity is called the **total mechanical energy E** of the object.

$$E = U + \tfrac{1}{2}mv^2 = \text{constant} \qquad (8.8)$$

or

$$PE + KE = E = \text{constant}$$

MEMORIZE Eq. 8.7 and be able to apply it to problems involving conservative forces. It is very important.

PROBLEM 8.1. A rock of mass m is dropped from rest at a point a height h above the ground. Use the conservation of energy principle to determine the speed of the rock when it strikes the ground. Neglect friction.

Solution

$$PE_1 + KE_1 = PE_2 + KE_2$$

$$mgh + 0 = 0 + \tfrac{1}{2}mv^2, \quad v = \sqrt{2gh}$$

Here, I chose ground level as the zero point of potential energy. The rock was initially at rest, so $KE_1 = 0$. Observe that this approach, using conservation of energy, is much simpler than the methods developed earlier.

PROBLEM 8.2. An Atwood's machine consists of two masses m_1 and m_2 joined by a light cord which passes over a pulley. Initially, the heavier mass is positioned a distance h above the floor. The masses are released from rest. At what speed are the masses moving when the heavier mass strikes the floor? Here $m_1 = 4\,\text{kg}$, $m_2 = 6\,\text{kg}$, and $h = 3\,\text{m}$. The cord is long enough so that the lighter mass does not reach the pulley. Devices like this are used in the construction of elevators.

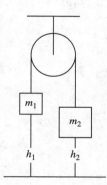

Solution

$$PE_1 + KE_1 = PE_2 + KE_2$$

$$m_1 gh_1 + m_2 gh + 0 = m_1 g(h_1 + h) + 0 + \tfrac{1}{2}(m_1 + m_2)v^2$$

$$v = \sqrt{2\left(\frac{m_2 - m_1}{m_2 + m_1}\right)gh}$$

$$= \sqrt{2\left(\frac{6 - 4}{6 + 4}\right)(9.8\,\text{m/s}^2)(3\,\text{m})} = 3.4\,\text{m/s}$$

PROBLEM 8.3. Water flows over Niagara Falls at a rate of about $6000\,\text{m}^3/\text{s}$, dropping a distance of 49 m. At what rate could electric power be generated if all of the potential energy loss of the water could be converted to electricity? One cubic meter of water has a mass of 1000 kg.

Solution

$$P = \frac{E}{t} = \frac{mgh}{t} = \left(\frac{m}{t}\right)gh$$

$$= (6000\,\text{m}^3/\text{s})(1000\,\text{kg/m}^3)(9.8\,\text{m/s}^2)(49\,\text{m}) = 2.88 \times 10^9\,\text{W} = 2.88\,\text{GW}$$

By comparison, note that a nuclear power plant might generate about 1000 MW of power.

PROBLEM 8.4. A skier passes over the crest of a small hill at a speed of 3.6 m/s. How fast will she be moving when she has dropped to a point 5.6 m lower than the crest of the hill? Neglect friction.

Solution

$$PE_1 + KE_1 = PE_2 + KE_2$$

$$mgh_1 + \tfrac{1}{2}mv_1^2 = mgh_2 + \tfrac{1}{2}mv_2^2$$

$$v_2 = \sqrt{v_1^2 + 2g(h_1 - h_2)}$$

$$= \sqrt{(3.6\,\text{m/s})^2 + 2(9.8\,\text{m/s}^2)(5.6\,\text{m})}$$

$$= 11.1\,\text{m/s}$$

PROBLEM 8.5.   In an amusement park roller coaster ride, a car starts from rest at point *A* and races through a loop-the-loop. What is the minimum height *h* from which the car can start if it is not to leave the track at point *B*? The loop has radius *R*.

Solution   If the car is just about to leave the track at point *B*, the normal force exerted on the car by the track at this point is zero. The only force acting on the car is then *mg*, and this must provide the needed centripetal force to keep the car moving along the circular track. Thus,

$$mg = \frac{mv^2}{R}$$

We can find the speed *v* as a function of the starting elevation *h* by applying the conservation of energy principle.

$$mgh + 0 = mgR + \tfrac{1}{2}mv^2$$

Thus,

$$v^2 = Rg \qquad gh = gR + \tfrac{1}{2}Rg \qquad h = \tfrac{3}{2}R$$

8.2   Energy Conservation and Friction

The law of conservation of energy can be applied to systems where nonconservative forces like friction act. If a system does work against friction, the mechanical energy of the system will decrease. Thus, if W_f is the work done against friction, then

Initial energy − energy lost to friction = final energy

$$E_1 - W_f = E_2$$

$$\boxed{U_1 + \tfrac{1}{2}mv_1^2 - W_f = U_2 + \tfrac{1}{2}mv_2^2} \tag{8.9}$$

PROBLEM 8.6.   Near Lewiston, Idaho, is a steep grade heavily traveled by logging trucks. Several serious accidents have occurred when trucks lost their brakes and careened down the hill at high speed. Runaway truck ramps have been built, which it is hoped will stop vehicles with no brakes. Suppose that a truck traveling 40 m/s encounters a ramp inclined up at 30° above horizontal. Loose gravel on the ramp provides a frictional force to help slow the truck as it moves up the ramp. The gravel has an effective coefficient of friction of 0.50. How far along the ramp would such a truck travel before coming to a stop?

Solution

$$N = mg \cos \theta$$

$$F_f = \mu N$$

$$= \mu mg \cos \theta$$

$$U_1 + \text{KE}_1 - W_f = U_2 + \text{KE}_2$$

$$0 + \tfrac{1}{2}mv^2 - F_f s = mgh + 0 \quad h = s \sin \theta$$

$$\tfrac{1}{2}mv^2 - (\mu mg \cos \theta)s = mgs \sin \theta$$

$$s = \frac{v^2}{2g(\sin \theta + \mu \cos \theta)} = \frac{(40 \, \text{m/s})^2}{2(9.8 \, \text{m/s}^2)(\sin 30° + 0.5 \cos 30°)}$$

$$= 87.5 \, \text{m}$$

PROBLEM 8.7.   A package of mass m is dropped onto a conveyor belt moving at speed v. The coefficient of friction between the package and the belt is μ. (a) How far does the package move before it stops sliding on the belt? (b) How much work is done by the belt (including work against friction) before the package stops sliding?

Solution   Initially, the package has no horizontal velocity and the belt slides under it. Friction accelerates the package for a time t, until it reaches velocity v. Then the slipping stops and no more work is done against the friction force. During the slipping process, the package moves a distance x and the belt moves a distance x_B.

(a) $F_f = \mu mg = ma \quad a = \mu g$

$$v^2 = v_0^2 + 2ax = 0 + 2ax \quad x = \frac{v^2}{2a} = \frac{v^2}{2\mu g}$$

$$v = v_0 + at = 0 + \mu g t \quad t = \frac{v}{\mu g}$$

The package slides a distance $\Delta x = x_B - x$ on the belt, where

$$x_B = vt = v\left(\frac{v}{\mu g}\right) = \frac{v^2}{\mu g} \qquad \Delta x = \frac{v^2}{\mu g} - \frac{v^2}{2\mu g} = \frac{v^2}{2\mu g}$$

(b)   The work done against friction is $W_f = F_f \Delta x$.

$$W_f = (\mu mg)\left(\frac{v^2}{2\mu g}\right) = \frac{1}{2}mv^2$$

The package gains kinetic energy $\text{KE} = \tfrac{1}{2}mv^2$. Thus, the total work done by the belt is

$$W = W_f + \text{KE} = \frac{1}{2}mv^2 + \frac{1}{2}mv^2 = mv^2$$

8.3   Potential Energy of a Spring

The force kx exerted by a spring is a conservative force. The work done in compressing or stretching a spring is stored as potential energy and can be used later to do work. We saw in Eq. 7.6 that the work done in stretching or compressing a spring of spring constant k by a distance x is $\frac{1}{2}kx^2$. Thus, the potential energy of a spring is

$$U = \tfrac{1}{2}kx^2 \tag{8.10}$$

Here, x is the displacement from the unstretched position of the end of the spring.

If a mass m is attached to the end of a spring and then allowed to oscillate back and forth, the energy of the system will remain constant.

$$\frac{1}{2}kx^2 + \frac{1}{2}mv^2 = E = \text{constant} \tag{8.11}$$

PROBLEM 8.8. A mass m resting on a frictionless horizontal table is attached to the end of a spring of spring constant k. The other end of the spring is fixed. The mass is displaced a distance A from its equilibrium position and released from rest. What is the maximum speed of the mass as it oscillates?

Solution The total energy of the system is constant, so the kinetic energy will be greatest when the potential energy is a minimum, and this occurs when $x = 0$. Thus,

$$E = U_1 + \text{KE}_1 = U_2 + \text{KE}_2$$

$$\frac{1}{2}kA^2 + 0 = 0 + \frac{1}{2}mv^2 \quad v = \sqrt{\frac{k}{m}}\,A$$

PROBLEM 8.9. An archery bow exerts a Hooke's law force kx on an arrow when the string is pulled back a distance x. Suppose that an archer exerts a force of 220 N in drawing back an arrow a distance of 64 cm. What is the spring constant of the bow? With what speed will an arrow of mass 24 g leave the bow?

Solution

$$k = \frac{F}{x} = \frac{220\,\text{N} \cdot \text{m}}{0.64\,\text{m}} = 344\,\text{N} \cdot \text{m} \quad \text{PE}_1 + \text{KE}_1 = \text{PE}_2 + \text{KE}_2$$

$$\frac{1}{2}kx^2 + 0 = 0 + \frac{1}{2}mv^2 \quad v = \sqrt{\frac{k}{m}}x$$

$$x = \sqrt{\frac{344\,\text{N} \cdot \text{m}}{0.024\,\text{kg}}}\,(0.64\,\text{m}) = 76.6\,\text{m/s}$$

PROBLEM 8.10. A crazy bungee cord jumper (there is no other kind) who weighs 800 N ties an elastic cord to his ankle and leaps off a high tower. The cord has an unstretched length of 30 m, and one end is attached to the point where the jumper starts. The effective spring constant of the elastic cord is $200\,\text{N} \cdot \text{m}$. How far will the jumper fall before the cord stops his descent?

Solution Let the lowest point in the jump be $h = 0$. The initial kinetic energy and the kinetic energy at the lowest point are both zero, so energy conservation yields $mgh = 0 + \frac{1}{2}kx^2$, where $x = h - 30$. Substitute $mg = 800\,\text{N}$ and $k = 200\,\text{N/m}$, and solve.

$$h^2 - 68h + 900 = 0 \quad h = 68 \pm \sqrt{(68)^2 - 4(900)} = 50\,\text{m or } 18\,\text{m}$$

The correct solution is $h = 50$ m. The solution $h = 18$ m corresponds to the jumper rebounding and compressing the bungee cord "spring," but a cord does not compress like a spring.

8.4 Machines

A simple machine is a device used to magnify a force or to change a small displacement into a large one. Common machines are a lever, an inclined plane, a block and tackle, a hydraulic jack, or a combination of gears. Typically, work is done on the machine (the input work W_1), and then the machine in turn does some output work W_2. The energy state of the machine does not change appreciably during this process, so if friction is negligible, $W_1 = W_2$, based on the idea of energy conservation. Very often the input and output forces are constant, in which case $W_1 = W_2$ yields

$$F_1 d_1 = F_2 d_2 \quad \text{or} \quad \boxed{F_2 = \frac{d_1}{d_2} F_1} \tag{8.12}$$

Here, F_1 acts over a distance d_1 and F_2 acts over a distance d_2. The *mechanical advantage* of the machine is defined as

$$\text{MA} = \frac{F_2}{F_1} \tag{8.13}$$

PROBLEM 8.11.   A pry bar is a device used to lift heavy objects (for example, a piano or large piece of machinery) a small distance, usually in order to place a wheeled dolly under the object. It consists of a long rod that rests on a fulcrum a short distance from the lifting end of the bar. Suppose the fulcrum of a pry bar is 3 cm from the load, and the point where you push down on the other end is 1.50 m from the fulcrum. What minimum force would you have to exert to lift a load of 2000 N? If you move the end of the bar down 4 cm, how much will you lift the load?

Solution   If the bar rotates through a small angle $\Delta\theta$, then

$$d_1 = L_1 \Delta\theta \quad \text{and} \quad d_2 = L_2 \Delta\theta$$
$$F_1 L_1 \Delta\theta = F_2 L_2 \Delta\theta$$
$$F_1 = \frac{L_2}{L_1} \quad F_2 = \left(\frac{0.03 \text{ m}}{1.50 \text{ m}}\right)(2000 \text{ N})$$
$$F_1 = 40 \text{ N}$$

For similar triangles,

$$\frac{d_1}{d_2} = \frac{L_1}{L_2} \quad d_2 = \frac{L_2}{L_1} d_1 = \frac{0.03 \text{ m}}{1.50 \text{ m}}(0.04 \text{ m}) = 0.008 \text{ m} = 8 \text{ mm}$$

Note that a small input force results in a large output force, but the price one pays is that a large input displacement produces only a small output displacement.

PROBLEM 8.12.   Sketched here is a differential hoist of the kind used in a shop to lift an engine out of a car. The pulleys have teeth that mesh with a continuous chain. The top pulleys are welded together, and there are 18 teeth on the outer pulley and 16 teeth on the inner pulley. Thus, when the pulley makes one revolution, 18 links of the chain are pulled up and 16 links are lowered, resulting in lifting the load. What is the mechanical advantage of this machine?

Solution

Consider what happens when the top pulley makes one revolution, that is, when the worker pulls 18 links of chain toward herself with force F_1. Let L = length of one link. The input work is thus $W_1 = 18\,LF_1$. The loop of chain that goes down to the load is thus shortened by 18 links and lengthened by 16 links, with a net shortening of $18L - 16L = 2L$. Shortening the loop by $2L$ lifts the load by L (try this with a piece of string to convince yourself of this tricky feature). Thus the output work is $W_2 = F_2L$. Neglecting friction,

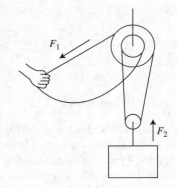

$$W_1 = W_2 \quad \text{or} \quad 18LF_1 = F_2L$$

The mechanical advantage of the hoist is thus $\mathrm{MA} = F_2/F_1 = 18$.

PROBLEM 8.13.   My sailboat trailer is equipped with a windlass that I use to pull my boat out of the water. It consists of a crank handle 30 cm long attached to the shaft of a small gear with 12 teeth. This small gear meshes with a larger gear with 36 teeth. Attached to this large gear is a drum of radius 2 cm on which is wound the line attached to the boat. (For you landlubbers, a line is a rope.) What tension can I apply to the line when I push on the crank with a force of 80 N?

Solution

Consider what happens when the crank makes one revolution. My hand moves a distance $d_1 = 2\pi R_1$. The large gear moves $\frac{12}{36} = \frac{1}{3}$ revolution. The line is thus pulled a distance $d_2 = 2\pi R_2/3$.

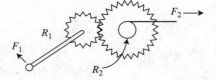

$$F_1 d_1 = F_2 d_2, \quad \text{so} \quad F_2 = \frac{d_1}{d_2}F_1 = \frac{2\pi R_1}{2\pi R_2/3}F_1 = 3\frac{R_1}{R_2}F_1$$

$$F_2 = 3\left(\frac{30\,\text{cm}}{2\,\text{cm}}\right)(80\,\text{N}) = 3600\,\text{N}$$

The mechanical advantage of the winch (neglecting friction) is 45. Amazing!

8.5   **Summary of Key Equations**

If U = potential energy: $F = -\frac{dU}{dx}$

Gravitational potential energy: $U = mgy$

Spring potential energy: $U = \frac{1}{2}kx^2$

Law of conservation of energy: $E = U + \frac{1}{2}mv^2 = \text{constant}$

With friction present: $U_1 + \frac{1}{2}mv_1^2 - W_f = U_2 + \frac{1}{2}mv_2^2$ where $W_f = F_f d$

Machines: $W_1 = W_2 \quad \text{or} \quad F_1 d_1 = F_2 d_2$

SUPPLEMENTARY PROBLEMS

8.14. Suppose that a rock of mass m thrown straight up will rise to a height h_1 in the absence of air drag. If a constant drag force of 0.1 mg acts, to what fraction of the height h_1 will the rock now rise? We have seen that in the absence of friction, the rise time is equal to the fall time. Is this still true if friction is present? Use energy conservation to reason this out.

8.15. A boy on a bridge throws a rock with speed v, and it lands in the water a distance h below. Calculate the speed with which the rock hits the water when it is thrown (a) horizontally, (b) at 45° above horizontal, and (c) straight down.

8.16. Downhill ski racers always push off at the starting gate, assuming that so doing will give them a better time for the race. Calculate the speed of a racer after he has dropped 4.0 m in elevation below his starting point for the case of starting from rest and for starting with an initial speed of 1.0 m/s.

8.17. Once many years ago I gave a lecture demonstration to illustrate the conservation of energy by means of the following setup. I tied a bowling ball to one end of a string, and I fastened the other end to the ceiling of the lecture hall. Holding the bowling ball while standing on a tall step ladder, I intended to show that when released from rest at the end of my nose, the ball would not swing higher and smash me in the face on the return swing. (Try this sometime if you want to experience a real game of "chicken." It's scary.) The demonstration made quite an impression on the class, but not for the reason I expected. Although the string was strong enough to hold the ball when it was motionless, when I let it go, the string broke at the bottom of the arc and the ball went bouncing around the room going "Boing, boing, boing" and scattering kids in every direction. Believe me, a bowling ball will really bounce on concrete. I'm sure they remember it to this day, long after they've forgotten about conservation of energy. Suppose the ball weighed 80 N and the string was 4.0 m long and had a breaking strength of 120 N. What is the maximum starting angle with vertical from which I could have released the ball without having the string break?

8.18. You probably heard about the guy who fell off the top of the Empire State Building. At the 93rd floor a lady standing by an open window heard him say, "So far, so good. So far, so good." At the 54th floor they heard him say, "So far, so good. So far, so good." Sort of like heading for the final exam in your physics class sometimes. Suppose someone had clocked this guy with a radar gun and found he was moving 12.0 m/s at one floor and 15.6 m/s one floor lower. Use energy principles to determine the distance between floors.

8.19. Masses m_1 and m_2 are connected as shown here. Mass m_1 slides on a surface where the coefficient of friction is μ. Determine its speed after m_2 has fallen a small distance h.

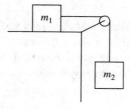

8.20. A small mass m is released from rest at the top of a frictionless spherical surface. At what angle with vertical will it leave contact with the sphere?

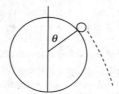

8.21. In a bottling plant, containers travel along various conveyor belts between different units, such as sterilizers, fluid dispensers, capping machines, and labelers. At one point a bottle of mass m starts at rest and slides a distance s down a ramp inclined at angle θ above horizontal. There it strikes a spring of spring constant k. By how much does it compress the spring?

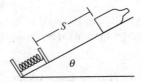

8.22. A student designs a toy dart gun to propel a 12-g dart with a speed of 12.0 m/s by means of a spring that is to be compressed 2.00 cm to launch the dart. What spring constant is required?

8.23. A 60-kg student finds she can run up a flight of stairs in a football stadium in 12 s. The flight has 120 steps, each 20 cm high. At what power level is she doing work? If her muscles are 20 percent efficient, how much energy does she use in this exercise? Express the result in kilocalories, remembering one Big Mac hamburger yields about 500 kcal. How much weight could you lose this way? To burn off 1 lb of fat, you must use about 3500 kcal.

8.24. A workman pushes a packing crate of weight W a distance s up a plane ramp inclined at angle θ above horizontal. How much work does he do? He pushes parallel to the plane. What force must he exert? What is the mechanical advantage of this simple machine?

8.25. A woman uses the block and tackle shown here to lift a heavy weight W. Use energy principles to determine the mechanical advantage of this simple machine.

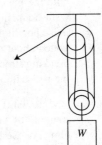

8.26. A rocket of mass m is moving straight up with constant acceleration. At one point its velocity is v, and at an elevation h higher its velocity is $1.4\,v$. How much work was done by the rocket engine during this period?

8.27. A Chinese windlass, of the kind used to lift water from wells in ancient times, is constructed with two cylinders of radii R_1 and R_2 mounted on a horizontal shaft. The shaft is turned by a crank handle of length L. Rope is wound on the cylinders as shown here, and the hanging loop of the rope lifts a bucket of weight W. What is the mechanical advantage of this machine?

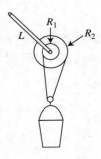

8.28. The starting point on a water slide is a height H above the surface of the water pool below. The bottom of the slide is a height h above the water. A girl starts from rest at the top of the slide and leaves the slide at the bottom in a horizontal direction. How far from the bottom of the slide in a horizontal direction will she hit the water? Neglect friction.

8.29. Near my house is a hill used by kids for sledding. At the bottom of the hill is a level field, on the far side of which is a deep ditch. If you start too far up on the hill, you will end up in the ditch before your sled comes to a halt. How far up the hill, measured along the surface, can a kid start without danger of ending up in the ditch a distance 200 m away, subject to the following conditions? The hill is a plane inclined at 30° above horizontal, and friction on it is negligible. On the flat field friction slows the sled, and the coefficient of friction there is 0.10.

8.30. A boy of weight 600 N ties a rope to a tree branch that overhangs a pond. He plans to swing down on the rope, letting go to fall into the water when he reaches his lowest point. The rope will support a weight of 1200 N, and the boy swings in an arc of radius 4 m on the way down. At what height above his release point can the boy start without the rope breaking?

SOLUTIONS TO SUPPLEMENTARY PROBLEMS

8.14. With no friction: $0 + \frac{1}{2}mv^2 = mgh_1 + 0,\quad h_1 = \sqrt{v^2/2g}$

With friction: $\frac{1}{2}mv^2 - (0.1\,mg)h_2 = mgh_2,\quad h_2 = 0.91h_1$

    In the case where friction is present, consider two points A and B on the trajectory, both at the same elevation. Point A is reached on the way up, and point B is reached on the way down. Since both points are at the same elevation, the potential energy of the rock is the same at both points. However, since work was done against friction in going from A to B, the kinetic energy at B must be less than at A, so the rock is moving more slowly at B than at A. This is true for any two points at the same elevation along the trajectory, and hence the time required to fall will be greater than the rise time when friction is present. A more difficult problem is that of determining how friction affects the *total* time in the air. I leave that for you. It makes my head hurt.

8.15. $0 + mgh = \frac{1}{2}mv^2 + 0,\quad$ so $v = \sqrt{2gh}$

    The angle at which the rock is thrown doesn't enter into the result, so the speed at the water is independent of the angle at which the rock is thrown. This is counterintuitive for many people since they imagine that throwing the rock straight down will somehow give it greater speed at the ground. Not so. Note, however, that the horizontal component of the velocity will differ in the three cases, as will the vertical component of velocity.

8.16. From rest:

$$0 + mgh = \frac{1}{2}mv_1^2 + 0 \quad v_1 = \sqrt{2gh} = \sqrt{2(9.8\,\text{m/s}^2)(4\,\text{m})} = 8.85\,\text{m/s}$$

With $v_0 = 1\,\text{m/s}$, $\quad \frac{1}{2}mv_0^2 + mgh = \frac{1}{2}mv_2^2 + 0$, so $v_2 = 8.91\,\text{m/s}$. Note that $v_2 \neq v_1 + 1$!!!

8.17. The string must provide enough upward force to balance the weight plus provide the radial force mv^2/r needed to make the ball curve upward. The tension in the string will thus be greatest at the lowest point in the arc, where the gravity force is directed straight down and the ball is moving fastest.

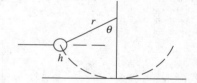

$$T - mg = \frac{mv^2}{r} \quad \text{and} \quad mgh = \frac{1}{2}mv^2$$

$$h = r - r\cos\theta \qquad T - mg = 2mg(1 - \cos\theta)$$

$$\cos\theta = 1 - \frac{T - mg}{2mg} = 1 - \frac{120 - 80}{(2)(80)},\quad \theta = 41.4°$$

8.18. $mgh_1 + \frac{1}{2}mv_1^2 = \frac{1}{2}mgh_2 + \frac{1}{2}mv_2^2,\quad h_1 - h_2 = \frac{1}{2}(v_2^2 - v_1^2) = 5.07\,\text{m}$

8.19. Take $h = 0$ as the initial position of m_2. Then $\text{KE}_1 + \text{PE}_1 - W_f = \text{KE}_2 + \text{PE}_2$.

$$0 + 0 - \mu m_1 gh = \frac{1}{2}(m_1 + m_2)v^2 + m_2 g(-h) \quad v = \sqrt{\frac{2(m_2 + \mu m_1)gh}{m_1 + m_2}}$$

8.20. $mg\cos\theta = \dfrac{mv^2}{r} \qquad mgr + 0 = mgh + \dfrac{1}{2}mv^2 \qquad h = r\cos\theta$

$$\cos\theta = \frac{v^2}{rg} = 2 - 2\cos\theta \qquad \cos\theta = \frac{2}{3} \quad \theta = 48.2°$$

8.21. Let $h = 0$ at the bottle's lowest point and $h = (s + x)\sin\theta$ at the bottle's initial position. Here $x = $ distance spring is compressed. Thus,

$$mgh + 0 = 0 + \frac{1}{2}kx^2 + 0 \quad x^2 - \frac{2mg}{k}(s + x)\sin\theta = 0$$

Let

$$A = \frac{2mg \sin \theta}{k}, \quad \text{then } x^2 - Ax - As = 0$$

$$x = \frac{-A \pm \sqrt{A^2 + 4As}}{2} = 0.486 \, \text{m or } -0.434 \, \text{m}$$

The spring is thus compressed by 0.486 m. The solution -0.434 m corresponds to the spring when it is stretched out.

8.22. $\frac{1}{2}kx^2 = \frac{1}{2}mv^2, \quad k = m\left(\frac{v}{x}\right)^2 = (0.012 \, \text{kg})\left(\frac{12 \, \text{m/s}}{0.02 \, \text{m}}\right)^2 = 4320 \, \text{N} \cdot \text{m}$

8.23. $P = \frac{W}{t} = \frac{mgh}{t} = \frac{(60 \, \text{kg})(9.8 \, \text{m/s}^2)(120)(0.20 \, \text{m})}{12 \, \text{s}} = 1176 \, W$

$$\text{Work} = Pt = (1176 \, W)(12 \, \text{s}) = 1.41 \times 10^4 \, \text{J}$$

$$\text{Work} = (1.41 \times 10^4 \, \text{J})\left(\frac{1 \, \text{kcal}}{4186 \, \text{J}}\right) = 3.37 \, \text{kcal}$$

If energy E is used, $0.2E = $ work or $E = 5$ work or 16.9 kcal.

$$\text{Weight loss} = \frac{16.9 \, \text{kcal}}{3500 \, \text{kcal/lb}} = 0.005 \, \text{lb}$$

It is tough to lose weight by exercising. It is better to eat less and reduce your calorie intake. Exercise is beneficial for other reasons.

8.24. The crate is lifted vertically a distance $h = s \sin \theta$, so

$$\text{Work} = Fs = Wh = Ws \sin \theta$$

and

$$F = W \sin \theta \qquad MA = \frac{W}{F} = \frac{1}{\sin \theta}$$

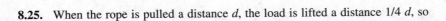

8.25. When the rope is pulled a distance d, the load is lifted a distance $1/4 \, d$, so

$$Fd = W\left(\frac{d}{4}\right) \qquad MA = \frac{W}{F} = 4$$

8.26. Work done by engine = gain in energy:

$$W = Fh = mgh + \frac{1}{2}m(1.4v)^2 - \frac{1}{2}mv^2 \qquad W = m(gh + 0.48v^2)$$

8.27. Input work for one revolution is $W_1 = 2\pi LF$. The rope is pulled up a distance $d_1 = 2\pi R_2$ and lowered a distance $2\pi R_1$. The loop of rope is shortened by $2\pi R_2 - 2\pi R_1$, so the load is lifted by half this amount.

$$d_2 = \frac{1}{2}(2\pi R_2 - 2\pi R_1) = \pi (R_2 - R_1) \qquad MA = \frac{d_1}{d_2} = \frac{2\pi R_2}{\pi(R_2 - R_1)} = \frac{2R_2}{R_2 - R_1}$$

8.28. Speed at bottom of the slide is v,

$$\tfrac{1}{2}mv^2 = mg(H - h)$$

Horizontal distance traveled after leaving the slide is

$$x = vt$$

t is the time to fall distance h,

$$h = \tfrac{1}{2}gt^2$$

Solve for x: $x = [2\,g(H - h)]^{1/2}\left[\dfrac{2h}{g}\right]^{1/2}$

8.29. Sliding down, energy is conserved, so

$$mgh = \tfrac{1}{2}mv^2, \quad \text{where } h = L\sin 30°$$

Sliding across the field, kinetic energy is lost to work done against friction,

$$\tfrac{1}{2}mv^2 = Fd \quad F = \mu mg$$

Thus, $\tfrac{1}{2}mv^2 = mgL\sin 30° = \mu mgd$

$$L = \frac{\mu d}{\sin 30°}$$

$$L = \frac{(0.10)(200\,\text{m})}{\sin 30°} = 40\,\text{m}$$

8.30. $mgh = \tfrac{1}{2}mv^2$

$\dfrac{mv^2}{L} = T - mg$

$\dfrac{2\,mgh}{L} = T - mg$

$h = \dfrac{L}{2\,mg}(T - mg)$

CHAPTER 9

Linear Momentum and Collisions

A linebacker tries to tackle a big running back and is knocked over. A hailstorm flattens a wheat field in northern Idaho. Ninety cars pile up on a fog-covered freeway in southern California. Energetic subatomic particles cause radiation damage that incapacitates an electronic component in a space probe. All of these events involve collisions, and it turns out that they can most readily be understood in terms of linear momentum. This important concept has far-reaching philosophical consequences as well, and its conservation is related to our ideas about the homogeneity of space throughout the universe. I won't go into this aspect further, but perhaps this introductory treatment will whet your appetite to delve deeper into physics.

9.1   Linear Momentum

The linear momentum **p** of a particle of mass m with velocity **v** is a vector quantity defined as

$$\mathbf{p} = m\mathbf{v} \tag{9.1}$$

Usually, I refer to linear momentum simply as "momentum," with the understanding that by this I mean "linear momentum." Also, when dealing with one-dimensional motion, I do not use vector notation and simply refer to momentum as being positive or negative, as is done with velocity.

The most general formulation of Newton's second law of motion is

$$\mathbf{F} = \frac{d\mathbf{p}}{dt} \tag{9.2}$$

For a system for which the mass is constant,

$$F = \frac{d\mathbf{p}}{dt} = \frac{d(m\mathbf{v})}{dt} = m\frac{d\mathbf{v}}{dt} = m\mathbf{a}$$

This is the result stated previously. For systems where mass is not constant (for example, a rocket ejecting exhaust gases), Eq. 9.2 must be used.

Kinetic energy (KE) can be expressed in terms of momentum. If the magnitude of the momentum is $p = mv$, then

$$\text{KE} = \frac{1}{2}mv^2 = \frac{p^2}{2m} \tag{9.3}$$

An interesting conclusion can be drawn concerning the behavior of two particles isolated from the outside world. This never happens in reality, but often it is a good approximation. Suppose the momenta of the particles are $\mathbf{p}_1$ and $\mathbf{p}_2$. If particle 1 exerts force $\mathbf{F}_{21}$ on particle 2 and particle 2 exerts force $\mathbf{F}_{12}$ on particle 1, then according to Newton's third law,

$$\mathbf{F}_{21} = -\mathbf{F}_{12} \quad \text{or} \quad \mathbf{F}_{12} + \mathbf{F}_{21} = 0$$

Since

$$\mathbf{F}_{12} = \frac{d\mathbf{p}_1}{dt} \quad \text{and} \quad \mathbf{F}_{21} = \frac{d\mathbf{p}_2}{dt}$$

then

$$\frac{d\mathbf{p}_1}{dt} + \frac{d\mathbf{p}_2}{dt} = \frac{d}{dt}(\mathbf{p}_1 + \mathbf{p}_2) = 0$$

The total momentum of the system is $\mathbf{p} = \mathbf{p}_1 + \mathbf{p}_2$, and the time rate of change of $\mathbf{p}$ is zero; thus

$$\boxed{\mathbf{p} = \mathbf{p}_1 + \mathbf{p}_2 = \text{constant}} \tag{9.4}$$

Equation 9.4 is the **law of conservation of linear momentum** applied to an isolated system of two particles. Further, one can reason that the same law will apply to an isolated system of many particles as well.

Another way of expressing Eq. 9.4 is to say that the momentum of the system at one time is equal to the momentum of the system at a later time.

$$(\mathbf{p}_1 + \mathbf{p}_2)_{\text{init}} = (\mathbf{p}_1 + \mathbf{p}_2)_{\text{final}} \tag{9.5}$$

If the vector momentum $\mathbf{p}$ is constant, then each of its x, y, and z components must also be constant.

PROBLEM 9.1.   A truck of mass 3000 kg traveling 5 m/s strikes a sedan stopped at a signal light. The two vehicles stick together. If the mass of the sedan is 2000 kg, at what speed does it move immediately after the collision?

Solution
$$P_{\text{before}} = P_{\text{after}} \qquad m_1 v_1 + m_2 v_2 = (m_1 + m_2)V$$
$$(3000\,\text{kg})(5\,\text{m/s}) + 0 = (3000\,\text{kg} + 2000\,\text{kg})V \quad V = 3.57\,\text{m/s}$$

PROBLEM 9.2.   A cannon of mass 1200 kg fires a 64-kg shell with a muzzle velocity of 62 m/s (this is the speed of the shell with respect to the cannon). Immediately after firing, what is the velocity V of the cannon and the velocity v of the shell with respect to the Earth?

Solution   The initial momentum of the system is zero. Just after firing, the cannon has velocity V (V is negative, since the cannon recoils to the left) and the shell moves to the right with velocity v with respect to the Earth. The muzzle velocity is $62\,\text{m/s} = v - V$:

$$0 = m_1 V + m_2 v \quad 0 = m_1 V + m_2 (V + 62)$$
$$0 = 1200\,V + 64(V + 62) \quad V = -3.14\,\text{m/s} \quad v = 58.9\,\text{m/s}$$

9.2   Impulse

Since $\mathbf{F} = d\mathbf{p}/dt$, $d\mathbf{p} = \int \mathbf{F}\,dt$ and the change in momentum $\Delta \mathbf{p}$ over a time interval t_1 to t_2 is

$$\Delta \mathbf{p} = \int_{t_1}^{t_2} \mathbf{F}\,dt = \bar{\mathbf{F}}\,\Delta t \tag{9.6}$$

Here, $\Delta \mathbf{p} = \mathbf{p}_2 - \mathbf{p}_2$ and $\Delta t = t_2 - t_1$ and $\bar{\mathbf{F}}$ is the average force that acts during the time interval Δt. $\Delta \mathbf{p}$ is called the *impulse* of the force $\mathbf{F}$.

Note the following important observation. A moving particle has momentum and kinetic energy, but it *does not* carry with it a force, contrary to what many people imagine. The force required to cause a particle to stop (that is, to reduce its momentum to zero) depends on how big the momentum change is and on *how quickly* the momentum change occurs. A long collision time results in a smaller force, and a short collision time results in a larger force. If your head hits the hard surface of a car dashboard and stops quickly, a large (and possibly fatal) force will be applied to your head. If your head hits a cushioning air bag and stops slowly, the force will be greatly reduced. There are many examples of the consequences of this. When a parachutist lands, he bends his knees to lengthen the collision time with the Earth, thereby reducing the force he experiences. A boxer rolls with the punch to minimize its effect. When you catch a hardball bare-handed, you draw your hand back to reduce the sting.

PROBLEM 9.3.   High-speed photography reveals that when a bat strikes a baseball, a typical collision time is about 2 ms. If a speed of 45 m/s is imparted to a ball of mass 0.145 kg, what average force is exerted by the bat?

Solution
$$\bar{F}\Delta t = \Delta p = m\Delta v, \quad \bar{F} = \frac{(0.145\,\text{kg})(45\,\text{m/s})}{0.002\,\text{s}} = 3260\,\text{N}$$

PROBLEM 9.4.   A ball of mass m and velocity v strikes a wall at an angle θ and bounces off at the same speed and at the same angle. If the collision time with the wall is Δt, what is the average force exerted on the wall?

Solution   The component of the momentum perpendicular to the wall is $mv \sin\theta$ just perpendicular to the wall is $mv \sin\theta$ just before the collision and $-mv \sin\theta$ just after the collision. This means the change in momentum (the impulse) is

$$\Delta p_x = mv \sin\theta - (-mv \sin\theta)$$
$$= 2mv \sin\theta$$

Thus,

$$\bar{F} = \frac{\Delta p}{\Delta t} = \frac{2mv \sin\theta}{\Delta t}$$

Note that when the ball bounces back (as opposed to simply sticking to the wall), the force is *increased*.

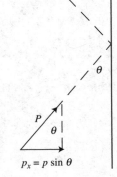

PROBLEM 9.5.   Firefighters sometimes use a high-pressure fire hose to knock down the door of a burning building. Suppose such a hose delivers 22 kg of water per second at a velocity of 16 m/s. Assuming the water hits and runs straight down to the ground (that is, it doesn't bounce back), what average force is exerted on the door?

Solution   Consider what happens in 1 s, that is,

$$\Delta t = 1\,\text{s} \qquad \bar{F} = \frac{\Delta p}{\Delta t} = \frac{m\Delta v}{\Delta t} = (22\,\text{kg/s})(16\,\text{m/s}) = 352\,\text{N}$$

9.3   Collisions in One Dimension

When two particles collide, the forces they exert on each other are much larger than any external forces acting. Thus, we may assume that external forces are negligible, with the consequence that the momentum of the system remains constant. This means that **a system's momentum just before a collision is the same as the momentum just after the collision**. In a collision some kinetic energy of the particles is converted to heat, sound, elastic distortion, and so on. Such collisions are called **inelastic collisions**. Sometimes the loss in kinetic energy is negligible (as when two billiard balls collide). Such collisions are called **elastic collisions**, and for them the kinetic energy of the system is conserved before and after the collision. Of course, the kinetic energy of an individual particle can change, but the combined kinetic energy of both of the particles remains the same.

When two objects stick together, the collision is perfectly inelastic. If the particles bounce apart, it is hard to say at a glance if the collision was elastic or inelastic. In the problems you will encounter here, you will have to be told if the collision was elastic or inelastic. If mass m_1 has velocity v_1 and mass m_2 has velocity v_2 just before a perfectly inelastic collision (the particles stick together), the velocity V just after the collision is determined by

$$m_1 v_1 + m_2 m_2 = (m_1 + m_2)V \quad \text{or} \quad V = \frac{m_1 v_1 + m_2 v_2}{m_1 + m_2} \tag{9.7}$$

If the collision is perfectly elastic, masses m_1 and m_2 can have different velocities V_1 and V_2, respectively after the collision. Since the kinetic energy remains constant,

$$m_1 v_1 + m_2 v_2 = m_1 V_1 + m_2 V_2 \tag{9.8}$$

and

$$\tfrac{1}{2} m_1 v_1^2 + \tfrac{1}{2} m_2 v_2^2 = \tfrac{1}{2} m_1 V_1^2 + \tfrac{1}{2} m_2 V_2^2 \tag{9.9}$$

These two equations can be solved for the two unknown final velocities, V_1 and V_2. Multiply Eq. 9.9 by 2 and rearrange:

$$m_1\left(v_1^2 - V_1^2\right) = m_2\left(v_2^2 - V_2^2\right)$$

Factor both sides of this equation:

$$m_1(v_1 - V_1)(v_1 + V_1) = m_2(v_2 - V_2)(v_2 + V_2) \tag{9.10}$$

Rearrange Eq. 9.8:

$$m_1(v_1 - V_1) = m_2(v_2 - V_2) \tag{9.11}$$

Divide Eq. 9.10 by Eq. 9.11:

$$v_1 + V_1 = v_2 + V_2 \quad \text{or} \quad v_1 - v_2 = -(V_1 - V_2) \tag{9.12}$$

$v_1 - v_2$ is the relative velocity of particle 1 with respect to particle 2 before the collison, and $V_1 - V_2$ is the same quantity after the collision. Thus Eq. 9.12 yields the interesting result that the relative speed of one particle with respect to the other does not change in a collision. The relative velocity of each particle changes direction in a collision, but the relative speed is constant.

We can solve Eq. 9.12 for V_2 and substitute it back into Eq. 9.8, thereby yielding one equation for V_1. The same thing can be done with V_1 to find V_2. After some algebraic labor, we find:

$$V_1 = \left(\frac{m_1 - m_2}{m_1 + m_2}\right)v_1 + \left(\frac{2m_2}{m_1 + m_2}\right)v_2 \tag{9.13}$$

$$V_2 = \left(\frac{2\,m_1}{m_1 + m_2}\right)v_1 + \left(\frac{m_2 - m_1}{m_1 + m_2}\right)v_2 \tag{9.14}$$

These equations yield interesting results for some simple special cases.

1. **$m_1 = m_2$**. Here $V_1 = v_2$ and $V_2 = v_1$. The particles exchange speeds. This is approximately what happens when pool balls collide.
2. **m_2 initially at rest**. Equations 9.13 and 9.14 become

$$V_1 = \left(\frac{m_1 - m_2}{m_1 + m_2}\right)v_1 \tag{9.15}$$

$$V_2 = \left(\frac{2m_1}{m_1 + m_2}\right)v_1 \tag{9.16}$$

If $m_2 \gg m_1$ (like a golf ball hitting a brick wall), we see that $V_1 \simeq -v_1$ and $V_2 \simeq v_2 = 0$. The big object remains at rest, and the small one bounces back with its speed unchanged.

If $m_1 \gg m_2$ (a locomotive hitting your parked motorcycle), then $V_1 = v_1$ and $V_2 = 2v_1$. The incident big particle continues with no change in speed, and the small stationary particle takes off with twice the speed of the incoming particle.

PROBLEM 9.6.   Here's an entertaining lecture demonstration. Place a small ball bearing of mass m_2 on top of a larger superball of mass $m_1 (m_1 \gg m_2)$. Hold the two together at shoulder height (call this height h above the floor), and drop the two simultaneously onto a concrete floor. The result is impressive. The steel ball bearing takes off like a bat out of hell. The first time I did this, I broke the overhead fluorescent lights. Calculate the height to which the ball bearing would rise (if it doesn't hit the lights on the ceiling) if the two are dropped from height h. Assume all collisions are elastic. (*Hint*: Imagine that first the superball collides elastically with the floor, and then when it rebounds, it meets the falling ball bearing that is right behind it.)

Solution   The superball hits the floor with speed v_1 where conservation of energy during the fall yields

$$0 + m_1 gh = \frac{1}{2}m_1 v_1^2 + 0 \qquad v_1 = \sqrt{\frac{2g}{h}}$$

The superball bounces up with speed v_1 and collides with the ball bearing, whose velocity is $v_2 = -v_1$. After the collision the velocity of the ball bearing is given by Eq. 9.16.

$$V_2 = \left(\frac{2\,m_1}{m_1 + m_2}\right)v_1 + \left(\frac{m_2 - m_1}{m_1 + m_2}\right)v_2 \quad \text{where } m_1 \gg m_2$$

$$V_2 \cong 2v_1 - v_2 = 2v_1 - (-v_1) = 3v_1 = 3\sqrt{\frac{2g}{h}}$$

Applying conservation of energy to the rising ball bearing yields the height h' to which the ball bearing rises.

$$\frac{1}{2}m_2V_2^2 + 0 = 0 + m_2gh' \quad \text{or} \quad h' = \frac{1}{2g}\left(3\sqrt{\frac{2g}{h}}\right)^2 = 9h!!!$$

Maybe we could use this technique to launch space vehicles or to send raw materials to manufacturing plants on the Moon. Or how about lifting water to elevated levees in the Himalayan mountains in Nepal? (Too late: People thought of this idea long ago. It's called a *hydraulic ram*.)

9.4 The Center of Mass

In trying to understand the behavior of a system of particles or of an extended object like a baseball bat, it is useful to introduce the concept of the *center of mass* (CM). The CM is what I would call the "balance point," that is, like the fulcrum of a teeter-totter. For two particles of equal mass, the center of mass lies midway between them on the line joining them. For an object like a brick, the center of mass is at the geometrical center.

If a system consists of particles of mass m_1 at position $\mathbf{r}_1$, m_2 at position $\mathbf{r}_2, \ldots,$ and m_N at $\mathbf{r}_N$, the position of the center of mass is defined to be

$$\mathbf{R} = \frac{m_1\mathbf{r}_1 + m_2\mathbf{r}_2 + \cdots + m_N\mathbf{r}_N}{m_1 + m_2 + \cdots + m_n} \tag{9.17}$$

or

$$\boxed{\mathbf{R} = \frac{m_1\mathbf{r}_1 + m_2\mathbf{r}_2 + \cdots + m_N\mathbf{r}_N}{M}} \tag{9.18}$$

Here M is the total mass of the system. The x, y, and z coordinates of the center of mass are

$$x = \frac{m_1x_1 + m_2x_2 + \cdots + m_Nx_N}{M} \tag{9.19}$$

$$y = \frac{m_1y_1 + m_2y_2 + \cdots + m_Ny_N}{M} \tag{9.20}$$

$$z = \frac{m_1z_1 + m_2z_2 + \cdots + m_Nz_N}{M} \tag{9.21}$$

Observe that the total momentum of a system subject to no external forces is constant.

$$\mathbf{P} = m_1\mathbf{v}_1 + m_2\mathbf{v}_2 + \cdots + m_N\mathbf{v}_N = \frac{d}{dt}(m_1\mathbf{r}_1 + m_2\mathbf{r}_2 + \cdots + m_N\mathbf{r}_N)$$

$$= \frac{d}{dt}(M\mathbf{R}) = \text{a constant}$$

Thus we can think of the behavior of the system like this: Imagine all of the mass of the system concentrated at the CM point. If no external forces act, this CM point will then move with constant velocity. It can be seen also that if external forces act, the CM moves as if it were a particle of mass M subject to those forces. For example, consider an artillery shell moving through the air along a parabolic path. If the shell explodes in midair, pieces will fly off in all directions, but the center of mass point will continue moving along the parabolic path as if nothing had happened.

PROBLEM 9.7.   Two masses are placed on the *x*-axis; 4 kg is at $x = 1$ m and 2 kg is at $x = 4$ m. What is the position of the CM?

Solution

$$x = \frac{(4\,\text{kg})(1\,\text{m}) + (2\,\text{kg})(4\,\text{m})}{4\,\text{kg} + 2\,\text{kg}} = 2\,\text{m}$$

We see that the CM is closer to the larger mass. The distance from each mass to the CM point is in the inverse ratio of the masses; that is, $d_1/d_2 = m_2/m_1$. Here $m_1 = 4$ kg, $m_2 = 2$ kg, $d_1 = 1$ m, and $d_2 = 2$ m.

We can find the CM of a continuous mass distribution by breaking the object into many little pieces, each of mass dm and volume dV. The mass density ρ of a material is defined as the mass per unit volume, $\rho = dm/dV$. If a small-volume element is located at (x, y, z), the coordinates of the center of mass become

$$x = \frac{1}{M}\int x\,dm = \frac{1}{M}\int \rho x\,dV \tag{9.22}$$

$$y = \frac{1}{M}\int y\,dm = \frac{1}{M}\int \rho y\,dV \tag{9.23}$$

$$z = \frac{1}{M}\int z\,dm = \frac{1}{M}\int \rho z\,dV \tag{9.24}$$

For symmetric objects of uniform density, the CM is at the geometric center. Note that the CM does not have to be within the object.

PROBLEM 9.8.   A uniform sheet of metal is cut into a right triangle. Its surface mass density (in kilograms per square meter) is σ. Determine the position of the CM.

Solution   To find the *x* coordinate of the CM, break the triangle into narrow strips parallel to the *y*-axis. The mass of a strip is $dm = \sigma y\,dx$, where $y\,dx$ is the area of the strip. Note that $y/x = b/a$, so $dm = \sigma bx\,dx/a$. Eq 9.23 thus yields

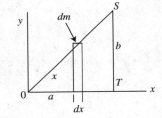

$$x = \frac{1}{M}\int_0^a \frac{\sigma bx^2}{a}\,da = \frac{\sigma ba^3}{3Ma} \qquad M = \frac{1}{2}ab\sigma, \quad \text{so } x = \frac{2}{3}a$$

A similar calculation can be carried out to find *Y*. However, we need not go to the work of doing this. Rather, notice that the *x* coordinate of the CM is two-thirds of the way from the vertex at *O*, so the *y* coordinate of the CM is two-thirds of the way from vertex *S*, at $Y = b/3$.

An ingenious way of solving this problem without using integrals is the following: Observe that the CM of each little strip must lie at its midpoint. Thus the CM of the triangle must lie somewhere on the line *OP*, where point *P* is the midpoint of side *ST*. By similar reasoning, if we break the triangle into strips parallel to the *x*-axis, we see that the CM of each strip must lie at its midpoint, and so the CM of the triangle must lie along line *QS*. Thus the CM of the triangle lies at the point where *OP* and *QS* intersect. The location of this point is found using geometry. Remember, THINK before you start calculating madly. Elegance is the essence of mathematics and physics!

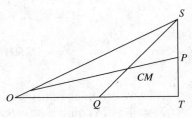

PROBLEM 9.9.   Maybe you've had an experience like this. One day while canoeing with my grandkids, I tried to climb out of the canoe onto the dock. As I stepped out of the canoe, it moved off in the opposite direction, dumping me in the lake. Great screams of laughter all around. Suppose I (of mass 90 kg) started 2 m from the midpoint of the canoe (mass 30 kg) and walked 4 m along the canoe toward the opposite end. How far through the water would the canoe move?

Solution   Assuming no net external force acts on the system, the CM will not move. The CM of a uniform canoe is at its center. Take $x = 0$ at the dock. The CM of the system is a distance X from the dock. Initially,

$$X = \frac{m_1 x_1 + m_2 x_2}{M}$$

After moving,

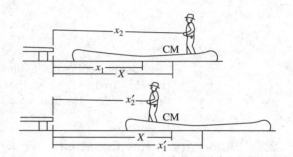

$$X' = \frac{m_1 x_1' + m_2 x_2'}{M}$$

$X' = X$, since the CM doesn't move,

$$m_1 x_1 + m_2 x_2 = m_1 x_1' + m_2 x_2'$$
$$x_2 = x_1 + 2 \qquad x_2' = x_1' - 2$$

The canoe moves a distance d of $x_1' - x_1$:

$$m_1 x_1 + m_2 (x_1 + 2) = m_1 x_1' + m_2 \left(x_1' - 2 \right)$$

Solve with $m_1 = 30\,\text{kg}$ and $m_2 = 90\,\text{kg}$. Find $d = x_1' - x_1 = 3\,\text{m}$.

Note that once you conclude that the CM is 0.5 m from the person, you could arrive at the answer by looking at the drawing. The distance between the person and the midpoint of the canoe is 2.0 m, so the CM must be 0.5 m from the person and 1.5 m from the canoe midpoint (a ratio of 1 : 3) since the masses 30 kg and 90 kg are in this ratio.

9.5   Rockets

Consider a rocket plus fuel that is initially at rest. When fuel is ejected out the back of the rocket, it acquires momentum, and so the rocket must move forward to acquire opposite momentum to cancel the fuel's momentum, since the total momentum of the system remains constant. Suppose that a rocket of mass M is moving at speed v with respect to the Earth. Now a mass of fuel Δm is ejected with speed v_e with respect to the rocket. This means the fuel is moving with velocity $v - v_e$ with respect to the Earth, and the rocket now moves forward with mass $(M - \Delta m)$ and velocity $v + \Delta v$.

Momentum conservation requires that $Mv = (M - \Delta m)(v + \Delta v) + \Delta m(v - v_e)$. Simplify $M\Delta v = \Delta m(v_e)$. The change in mass of the rocket is $M = -\Delta m$. In the limit $\Delta v \to dv$ and $\Delta m \to dm$, we obtain $M\,dv = -v_e\,dM$. Integrate:

$$\int_{v_i}^{v_f} dv = -v_e \int_{M_i}^{M_f} \frac{dM}{M}$$

$$v_f - v_i = v_e \, \ln\left(\frac{M_i}{M_f}\right) \tag{9.25}$$

The force propelling the rocket is called the *thrust*.

$$\text{Thrust} = F = M\frac{dv}{dt} = -v_e\frac{dM}{dt} \tag{9.26}$$

$\frac{dM}{dt} < 0$, so the thrust is positive (forward)

PROBLEM 9.10.   The Saturn V rocket had a mass of 2.45×10^6 kg, 65 percent of which was fuel. In the absence of gravity and starting at rest, what would be the maximum velocity attained (the "burnout velocity")? The fuel exhaust velocity was 3100 m/s.

Solution   From Eq. 9.25,

$$v - 0 = v_e \ln\frac{M_i}{M_f} = 3100 \ln\frac{M}{0.35M} = 3250\,\text{m/s}$$

In the preceding discussion of rockets, I have not included the effect of gravity. If a rocket is fired straight up, the thrust of the rocket will be reduced by the weight *mg*, and this will reduce the final velocity in Eq. 9.25 by *gt*, where *t* is the burn time for the fuel.

9.6   Summary of Key Equations

$$\mathbf{p} = m\mathbf{v}$$

$$\mathbf{F} = \frac{d\mathbf{p}}{dt}$$

If $\mathbf{F}_{ext} = 0$, then $\mathbf{p} = \mathbf{p}_1 + \mathbf{p}_2 + \mathbf{p}_3 + \cdots + = \text{constant}$

In *all* collisions, $\mathbf{p}_{before} = \mathbf{p}_{after}$

In *elastic collisions*, $KE_{before} = KE_{after}$

Impulse: $\Delta\mathbf{p} = \int \mathbf{F}\,dt = \bar{\mathbf{F}}\Delta t$

Elastic collision: If v_1 and V_1 are initial and final velocities of m_1 and v_2 and V_2 are initial and final velocities of m_2, then:

$$m_1,\ v_1,\ m_2,\ v_2, \qquad V_1 = \left(\frac{m_1 - m_2}{m_1 + m_2}\right)v_1 + \left(\frac{2\,m_2}{m_1 + m_2}\right)v_2$$

$$m_1,\ V_1,\ m_2,\ V_2, \qquad V_2 = \left(\frac{2m_1}{m_1 + m_2}\right)v_1 + \left(\frac{m_1 - m_2}{m_1 + m_2}\right)v_2$$

CM: $\mathbf{R} = \dfrac{m_1\mathbf{r}_1 + m_2\mathbf{r}_2 + \cdots + m_N\mathbf{r}_N}{m_1 + m_2 + \cdots + m_N}$

Rockets: $v_f - v_i = v\ \ln\dfrac{M_f}{M_i}$

Rocket thrust: $F = -v_e\dfrac{dM}{dt}$

SUPPLEMENTARY PROBLEMS

9.11. Calculate the KE and momentum of a particle of mass 0.020 kg and speed 65 m/s.

9.12. Once I tried to determine the speed of an arrow from my bow in the following way. I stuffed a big cardboard box full of newspapers (total mass 2.0 kg). I placed the box on a table where the coefficient of friction was 0.30. Then I fired my arrow horizontally into the box. The arrow mass was 0.030 kg. The arrow stuck in the box and caused the box to slide 24 cm. What was the speed of the arrow?

9.13. If a soldier shoots an enemy with the intent of killing him, he wants the bullet to deliver as much energy as possible. On the other hand, if you simply want to knock someone down (as in riot control), you want to deliver the maximum force. To see that these are not the same considerations, calculate the energy transferred to a very massive wooden block and the average force exerted on the block when a bullet of mass 0.008 kg and velocity 400 m/s makes a collision of duration 6.0 ms. Consider the case of a rubber bullet that bounces back from the block with no loss in speed (a perfectly elastic collision) and the case of an aluminum bullet that sticks in the block (a perfectly inelastic collision).

9.14. Marilyn Vos Savant writes a newspaper column in which she answers questions sent in by readers. Although her credentials list her as having recorded the highest IQ test score ever, she came up with the wrong answer to this stickler. Suppose you are driving your car at high speed and face two choices. You can hit a brick wall head on, or you can hit an oncoming car identical to yours and moving at the same speed as you. In both cases assume your car sticks to whatever it hits and that the collision times are the same in both cases. In truth, the collision time might be a little longer if you hit another car because modern cars are designed to be "crushable" in order to lengthen the collision time. Calculate the force experienced in each case. (Most people think it is better to hit the wall.)

9.15. A football running back of mass 90 kg moving 5 m/s is tackled head on by a linebacker of mass 120 kg running 4 m/s. They stick together. Who knocks whom back, and how fast are they moving just after the tackle?

9.16. A machine gun fires 4.8 bullets per second at a speed of 640 m/s. The mass of each bullet is 0.014 kg. What is the average recoil force experienced by the machine gun?

9.17. Two astronauts, Joe and Katie, each of mass $2M$, are floating motionless in space. Joe throws a compressed air cylinder of mass M with speed v toward Katie. She catches it and throws it back with speed v (with respect to herself). Joe again catches the cylinder. What will be his speed after so doing?

9.18. The cabin section of a spacecraft is separated from the engine section by detonating the explosive bolts that join them. The explosive charge provides an impulse of $400\,\text{N}\cdot\text{s}$. The cabin has a mass of 1000 kg, and the engine compartment has a mass of 1400 kg. Determine the speed with which the two parts move apart.

9.19. Water impinges on a fixed turbine blade with velocity v. The blade is curved so that it deflects the water by $180°$ and directs it back in its initial direction with no loss in speed. The mass of water striking the blade per unit time is μ. What force is exerted on the blade?

9.20. An object of mass m and velocity v collides elastically with a stationary object and continues in the same direction with speed $0.25\,v$. What is the mass of the object that was initially stationary?

9.21. A rocket in deep outer space turns on its engine and ejects 1 percent of its mass per second with an ejection velocity of 2200 m/s. What is the initial acceleration of the rocket?

9.22. A boy of mass 40 kg stands on a log of mass 60 kg. The boy walks along the log at 2 m/s. How fast does the log move with respect to the shore?

9.23. The objects shown here are constructed by bending a uniform wire. Determine the approximate position of the CM for each using symmetry and graphical methods (as opposed to using equations).

9.24. Suppose a big 18-wheeler truck crashes head-on into a small Toyota sedan. Most of us know intuitively that it is safer to be driving the big truck than the small car in such an accident. This is true. However, many people imagine that the truck is safer for the driver because the truck exerts a larger force in the collision than does the small car. This is not true. The force experienced by each vehicle is the same, as shown by Newton's Third Law of Motion. Why, then, is the truck driver less likely to suffer serious injury? To understand this, calculate the force experienced by the truck driver and by the car driver in a collision of duration time t in the following simplified example. Assume the vehicles stick together after the collision.

$$\text{Car mass} = M \qquad \text{Truck mass} = 10\,M \qquad \text{Initial car speed} = \text{initial truck speed} = v$$

$$\text{Truck driver mass} = \text{car driver mass} = m$$

9.25. Water droplets are released from a leaking pipe at uniform intervals of 0.10 s. The seventh drop is released just as the first drop hits the floor. How far below the release point is the location of the CM of the five drops in the air at the time the first drop hits the floor?

9.26. A uniform semicircular plate of thickness t and radius a is pivoted about an axis perpendicular to the plate and passing through its center. How far from the center is the CM located?

SOLUTIONS TO SUPPLEMENTARY PROBLEMS

9.11. $KE = \frac{1}{2}mv^2 = (0.5)(0.020\,\text{kg})(65\,\text{m/s})^2 = 42.25\,\text{J}$

$P = mv = (0.020\,\text{kg})(65\,\text{m/s}) = 1.3\,\text{kg}$

9.12. $mv = (m + M)V \qquad \frac{1}{2}(m + M)V^2 = F_{f^s} = \mu(m + M)gs$

$v = \left(\dfrac{m + M}{m}\right)\sqrt{2\mu g s} = 80\,\text{m/s}$

9.13. The rubber bullet has the same KE before and after the collision, so it transferred no energy to the block. The force it exerted on the block is

$$F_r \cong \frac{\Delta p}{\Delta t} = \frac{mv - (-mv)}{\Delta t} = \frac{2\,mv}{\Delta t} = \frac{(2)(0.008\,\text{kg})(400\,\text{m/s})}{s} = 1070\,\text{N}$$

For the aluminum bullet, $mv = (m + M)V$. Loss in momentum of the bullet is

$$\Delta p = mv - mV = mv - \left(\frac{m^2}{m + M}\right)v$$

$$= \frac{mM}{m + M}v \simeq mv \quad \text{since } M \gg m$$

$$F = \frac{\Delta p}{\Delta t} \simeq \frac{mv}{\Delta t} = 535\,N$$

Energy transferred to the block is

$$\Delta KE = \frac{1}{2}mv^2 - \frac{1}{2}mV^2 \simeq \frac{1}{2}mv^2 - \frac{1}{2}m\left(\frac{m}{m + M}\right)^2 v^2$$

$$\Delta KE \simeq \frac{1}{2}mv^2 \quad \text{if } M \gg m$$

Thus, the rubber bullet exerts a greater force and is more likely to knock the block over, and the aluminum bullet transfers more energy to the block.

9.14. If your car comes to a stop, $\Delta p = p - 0 = p = mv$. The force it experiences is

$$F \cong \frac{\Delta p}{\Delta t} = \frac{mv}{\Delta t}$$

Thus, the force is the same whether you hit the wall or the oncoming car. Of course, if the other car had more momentum (for example, was larger or moving faster), you would be knocked back and you would experience a larger force. The linebacker knocks back the running back.

9.15. $m_1 v_1 + m_2 v_2 = (m_1 + m_2)V \qquad V = \dfrac{m_1 v_1 + m_2 v_2}{m_1 + m_2}$

$$= \frac{(90\,\text{kg})(5\,\text{m/s}) + (120\,\text{kg})(-4\,\text{m/s})}{90\,\text{kg} + 120\,\text{kg}} = -0.14\,\text{m/s}$$

9.16. $\bar{F} = \dfrac{\Delta p}{\Delta t \Delta} = \dfrac{mv}{\Delta t} = (4.8/\text{s})(0.014\,\text{kg})(640\,\text{m/s}) = 43\,N$

9.17. Joe throws the cylinder: $0 = Mv_c + 2Mv_J$ and $v_c - v_J = v$, so

$$v_c = \frac{2v}{3}, \qquad v_J = -\frac{v}{3}$$

Katie catches the cylinder:

$$M\left(\frac{2v}{3}\right) + 0 = (M + 2M)v_{K_c}, \qquad v_{K_c} = \frac{2}{9}v$$

Katie throws the cylinder:

$$(M + 2M)\left(\frac{2v}{9}\right) = Mv'_c + 2Mv_K, \quad v_K - v'_c = v$$

so,

$$\frac{2}{3}v = v'_c + 2(v + v'_c), \quad v'_c = -\frac{4}{9}v$$

Joe catches the cylinder:

$$(2M)\left(-\frac{v}{3}\right) + M\left(-\frac{4}{9}v\right) = (2M + M)v'_J, \quad v'_J = -\frac{10}{27}v$$

The final speed of Joe and the cylinder is $0.37v$. Check: The total momentum of the system must remain zero. Thus,

$$(M + 2M)v'_J + 2Mv_K = 0$$

$$(3M)\left(-\frac{10}{27}v\right) + 2M\left(\frac{5}{9}v\right) = 0$$

using

$$v_K = v + v'_c = v - \frac{4}{9}v = \frac{5}{9}v, \quad 0 = 0 \quad \text{OK}$$

9.18. $\Delta p = \bar{F}\,\Delta t = mv$

Thus,

$$v = \frac{\Delta p}{m} \qquad v_1 = \frac{400\,\text{N} \cdot \text{s}}{1000\,\text{kg}} = 0.4\,\text{m/s}$$

$$v_2 = \frac{400\,\text{N} \cdot \text{s}}{1400\,\text{kg}} = 0.286\,\text{m/s} \qquad v = v_1 + v_2 = 0.686\,\text{m/s}$$

9.19. $F \cong \dfrac{\Delta p}{\Delta t} = \dfrac{mv - (-mv)}{\Delta t} = \dfrac{2mv}{\Delta t} = 2\mu v$

9.20. From Eq. 9.15,

$$V_1 = \frac{m_1 - m_2}{m_1 + m_2} v_1 \qquad \frac{v}{4} = \frac{m - m_2}{m + m_2} v_1 \qquad m_2 = 0.6\,\text{m}$$

9.21. From Eq. 9.26,

$$F = -v_c \frac{dM}{dt} = Ma \qquad a = -\frac{v_c}{M}\frac{dM}{dt} = -(2200\,\text{m/s})(-0.01) = 22\,\text{m/s}^2$$

9.22. The momentum of the system is zero: $m_B v_B + m_L v_L = 0$.

$$m_B(v - v_L) + m_L v_L = 0 \quad v_L = \frac{m_B}{m_B + m_L} v \quad v_L = \left(\frac{40}{40 + 60}\right)(2\,\text{m/s}) = 0.8\,\text{m/s}$$

9.23.

The CM for each straight section is at its midpoint. I treat each of them as a point mass. For the U shape, I found the CM for the two parallel sides and weighted this as worth $2M$. I then combined this with M for the bottom section.

9.24. Assume vehicles have velocity V after the collision.

$$10Mv - Mv = (10M + M)V, \text{ so } V = \frac{9}{11} v$$

Change in momentum of truck is $\Delta P_t = 10M\Delta v$

$$\Delta P_t = 10M\left(v - \frac{9}{11}v\right) = \frac{20}{11}Mv$$

Force on truck is

$$F_t = \frac{\Delta P_t}{t} = \frac{20}{11}\frac{Mv}{t}$$

Change in momentum of car is

$$\Delta P_c = M\left(v + \frac{9}{11}v\right) = \frac{20}{11}Mv$$

Force on car is

$$F_c = \frac{\Delta P_c}{t} = \frac{20}{11}\frac{Mv}{t}$$

As expected, $F_t = F_c$
For the truck driver,

$$f_T = \frac{\Delta P_T}{t} = \frac{1}{t}(mv - mV) = \frac{1}{t}\left(mv - \frac{9}{11}mv\right) = \frac{2}{11}\frac{mv}{t}$$

For the car driver,

$$f_C = \frac{\Delta P_C}{t} = \frac{1}{t}\left(mv + \frac{9}{11}mv\right) = \frac{20}{11}\frac{mv}{t}$$

Thus, $f_C = 10 f_T$

A heavier car is indeed safer in a collision!

9.25. Let $t = 0.10\,\text{s} =$ time between droplets

$y_n = \frac{1}{2}gt_n^2 =$ distance nth drop has fallen

$Y =$ position of CM below release point

$$Y = \frac{1}{2}g\left(\frac{0.1^2 + 0.2^2 + 0.3^2 + 0.4^2 + 0.5^2}{5}\right) = \frac{1}{2}g\left(\frac{0.55}{5}\right) = 0.055\,g$$

$Y = (0.055\,\text{s}^2)(9.8\,\text{m/s}^2) = 0.54\,\text{m}$

The distance to the floor is $h = \frac{1}{2}gT^2$

$h = \frac{1}{2}(9.8\,\text{m/s}^2)(0.6\,\text{s})^2 = 1.76\,\text{m}$

Note that the CM of the droplets in the air is above the midpoint of their fall.

9.26. The density of the plate is $\rho = \dfrac{\text{mass}}{\text{volume}} = \dfrac{M}{\frac{1}{2}\pi a^2 t}$

Divide the plate into semi-circular pieces, each of width dr. The CM is then a distance R from the center, where

$R = \dfrac{1}{M}\displaystyle\int_0^a r(\rho \pi r t)\,dr$. Here $\rho \pi r t\,dr = dm$

$$R = \frac{1}{M}\rho\pi t\left(\frac{1}{3}a^3\right) = \frac{1}{M}\left(\frac{2M}{\pi a^2 t}\right)\pi t\left(\frac{a^3}{3}\right)$$

$$R = \frac{2}{3}a$$

CHAPTER 10

Rotational Motion

Rotational motion plays an important role in nature, and here we investigate the behavior of rigid bodies when they rotate. A rigid body is one that does not deform as it moves. The equations involved here are similar to those that describe linear translational motion.

10.1   Angular Variables

Consider a planar object rotating about an axis perpendicular to its plane. We describe the position of a point on the object by the coordinates r and θ, where θ is measured with respect to the x-axis, as in Fig. 10-1. When the object turns through an angle θ, the point moves a distance s along the arc. We define the angle θ in **radians** as

$$\theta = \frac{s}{r} \quad \text{or} \quad \boxed{s = r\theta} \qquad (10.1)$$

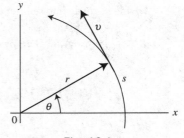

Fig. 10-1

You can see that if θ is doubled, the arc length s will also be doubled. Since θ is the ratio of two lengths, it is a dimensionless quantity. The circumference of a circle is $s = 2\pi r$ so θ for a full circle is 2π. Thus 2π rad $= 360°$. It is easy to convert radians to degrees or degrees to radians using a ratio.

$$\frac{\theta(\text{radians})}{\theta(\text{degrees})} = \frac{2\pi}{360°}$$

EXERCISE 10.1

Express $45°$, $60°$, $90°$, and $170°$ in radians. Express 1.0 rad, 0.6 rad, 7.25 rad, $\pi/2$ rad, and π rad in degrees.

Solution

$$45° = \frac{45}{360}(2\pi) = 0.78 \text{ rad} \qquad 60° = \frac{60}{360}(2\pi) = 1.05 \text{ rad}$$

$$90° = \frac{90}{360}(2\pi) = \frac{\pi}{2} \text{ rad} = 1.57 \text{ rad} \qquad 170° = \frac{170}{360}(2\pi) = 2.97 \text{ rad}$$

$$1.0 \text{ rad} = \frac{1}{2\pi}(360°) = 57.3° \qquad 0.6 \text{ rad} = \frac{0.6}{2\pi}(360°) = 34.4°$$

$$7.25 \text{ rad} = \frac{7.25}{2\pi}(360°) = 415° = 55° \text{ (subtract } 360°)$$

$$\frac{\pi}{2} \text{ rad} = \frac{0.5\pi}{2\pi}(360°) = 90° \qquad \pi \text{ rad} = \frac{\pi}{2\pi}(360°) = 180°$$

The linear velocity in meters per second of a point as it moves around a circle is called the *tangential velocity*:

$$v = \frac{ds}{dt} = r\frac{d\theta}{dt}$$

We define the *angular velocity* ω in radians per second as $\omega = d\theta/dt$. Thus

$$\boxed{v = r\omega} \tag{10.2}$$

If the point is accelerating along its path with tangential acceleration a, then

$$a = \frac{dv}{dt} = r\frac{d\omega}{dt}$$

We define the *angular acceleration* α in radians per second, as $\alpha = d\omega/dt = d^2\theta/dt^2$. Thus

$$\boxed{a = r\alpha} \tag{10.3}$$

We have seen previously how to describe linear motion with constant acceleration. If we simply divide the earlier equations by r, we obtain the equations that describe the rotational motion. For example, the equation $v = v_0 + at$ becomes

$$\frac{v}{r} = \frac{v_0}{r} + \frac{at}{r} \quad \text{or} \quad \omega = \omega_0 + \alpha t$$

Table 10-1 summarizes the parallels between linear motion and rotation. Previously, I used the variable x to represent displacement along a straight line. Now, I am using the letter s to remind you that the "linear" motion is along a curved arc. However, the motion is still one-dimensional.

TABLE 10-1

ROTATIONAL MOTION (α = CONSTANT)	LINEAR MOTION (α = CONSTANT)
$\omega = \omega_0 + \alpha t$	$v = v_0 + at$
$\theta = \frac{1}{2}(\omega_0 + \omega)t$	$s = \frac{1}{2}(v_0 + v)t$
$\theta = \omega_0 t + \frac{1}{2}\alpha t^2$	$s = v_0 t + \frac{1}{2}at^2$
$\omega^2 = \omega_0^2 + 2\alpha\theta$	$v_2 = v_0^2 + 2as$

Angular velocity ω and angular acceleration α are actually vector quantities, but as long as we keep the axis of rotation fixed, we do not need to worry about their vector nature. To keep things simple, I will just consider a counterclockwise rotation (as viewed from above) as positive.

It is common to describe rotating objects by specifying their **frequency** of revolution in revolutions per second. Since 1 rev is 2π rad,

$$\boxed{\omega = 2\pi f} \tag{10.4}$$

Equations 10.1 through 10.4 are important. **MEMORIZE** them. Sometimes one encounters rotation rates given in revolutions per minute. Be certain always to change to revolutions per second. Also, in using the above equations, be certain to use radians, not degrees. Be very careful about this or you will make errors.

PROBLEM 10.1. An electric drill rotates at 1800 rev/min. Through what angle does it turn in 2 ms? If it reaches this speed from rest in 0.64 s, what is its average angular acceleration?

Solution

$$\theta = \omega t = 2\pi f t = (2\pi)\left(\frac{1800}{60\,\text{s}}\right)(0.002\,\text{s}) = 0.37\,\text{rad} = 22°$$

$$\alpha = \frac{\omega}{t} = \frac{2\pi f}{t} = (2\pi)\left(\frac{1800}{60\,\text{s}}\right)\left(\frac{1}{0.64\,\text{s}}\right) = 295\,\text{rad/s}^2$$

PROBLEM 10.2.   A light chopper consists of a disk spinning at 40 rev/s in which is cut a hole of diameter 1.0 mm, a distance of 5.4 cm from the axis. A very thin laser beam is directed through the hole parallel to the axis of the disk. The light travels at 3×10^8 m/s. What length of light beam is produced by the chopper?

Solution   The hole is in front of the beam for a time t, where

$$d = vt = r\omega t, \quad \text{so } t = \frac{d}{r\omega}$$

In time t the laser beam travels a distance of L:

$$L = ct = \frac{cd}{r\omega} = \frac{cd}{2\pi f r} = \frac{(3 \times 10^8 \text{ m/s})(0.001 \text{ m})}{(2\pi)(40 \text{ rev/s})(0.054 \text{ m})} = 2.2 \times 10^4 \text{ m}$$

It is important to recognize that when a rigid body rotates, every point has the same angular velocity and the same angular acceleration. However, the linear speed and the tangential acceleration are not the same for all points. They increase for points farther from the axis. Note also that in addition to a possible tangential acceleration, each point has a centripetal acceleration directed inward. Centripetal acceleration and tangential acceleration are perpendicular vectors, and consequently the magnitude of the total acceleration is $a = \sqrt{a_c^2 + a_t^2}$.

10.2   Rotational Kinetic Energy

Imagine a rotating object to consist of lots of little pieces of mass. The piece m_i is at a distance r_i from the axis, and all rotate with the same angular velocity. The total kinetic energy associated with their rotational motion is

$$K_R = \sum K_i = \sum \frac{1}{2} m_i v_i^2 = \frac{1}{2} \sum m_i r_i^2 \omega^2$$
$$= \frac{1}{2} \left(\sum m_i r_i^2 \right) \omega^2$$

The quantity in parentheses is called the *moment of inertia*. A more descriptive name might be the "rotational mass," since it plays the same role for rotation that mass does for translational motion. $\frac{1}{2} I \omega^2$ is analogous to $\frac{1}{2} m v^2$.

$$\boxed{I = \sum m_i r_i^2} \tag{10.5}$$

In terms of the moment of inertia the rotational kinetic energy can be expressed as

$$\boxed{K_R = \frac{1}{2} I \omega^2} \tag{10.6}$$

A key feature of the moment of inertia is that it depends both on the amount of mass and on how far from the axis the mass is located. For example, when an ice skater spins with her arms in close to her body, she has a certain moment of inertia. If she then moves her arms outward, her moment of inertia increases appreciably, even though her mass does not change. This gives rise to some profound effects, as we shall see.

PROBLEM 10.3.   A molecule consists of identical atoms of mass m placed at each vertex of a regular hexagon of side a. Calculate the moment of inertia of the molecule about (a) the z-axis that is perpendicular to the plane of the hexagon and passing through its center, and (b) the y-axis that passes through two diametrically opposite atoms.

Solution   All atoms are the same distance a from the z-axis perpendicular to the plane, so

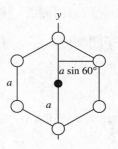

$$I_z = 6ma^2$$

To calculate I_y, note that two atoms are on the y-axis, so $r = 0$ for them. For the other four atoms,

$$r = a\sin 60° = \frac{\sqrt{3}}{2}a$$

Thus,

$$I_y = 4ma^2\left(\frac{3}{4}\right) = 3ma^2$$

PROBLEM 10.4.   As an alternative to the use of internal combustion engines, experimental cars have been designed that are propelled by the energy stored in a large spinning flywheel. Unfortunately, it turns out to be a challenging engineering problem to store enough energy in this way. The flywheels are huge and spinning so fast that they are hazardous. If you spin them too fast, the centripetal force required to hold the wheel together exceeds its breaking strength, and it can fracture and fly apart and go through your floorboard and blast you to kingdom come. A little golf-cart-type car might manage 20 mi/h using 5 hp. (a) Calculate the energy used to travel 20 mi. (b) Suppose you stored this much energy in a big steel flywheel with moment of inertia $18\,\text{kg} \cdot \text{m}^2$. (Such a wheel might have a mass of 100 kg and a radius of 60 cm.) At what frequency, in revolutions per minute, would it have to rotate?

Solution   To go 20 mi requires $1\,\text{h} = 3600\,\text{s}$.

(a) $E = Pt = (5\,\text{hp})(746\,\text{W/hp})(3600\,\text{s}) = 1.3 \times 10^7\,\text{J}$
(b) $\text{KE} = \frac{1}{2}I\omega^2$ $\omega = 2\pi f = (2\,\text{KE}/I)$ $f = 194/\text{s} = 11{,}700\,\text{rev/min}$.

10.3   Moment of Inertia Calculations

Consider a continuous object to be composed of many small pieces of mass dm. Then Eq. 10.5 becomes

$$\boxed{I = \int r^2\,dm} \tag{10.7}$$

If the mass is spread throughout the volume with density ρ ($\rho = \text{mass/volume}$), the mass in a volume dV is $dm = \rho\,dV$. For a surface mass density σ, $dm = \sigma\,dA$. For a line mass density λ, $dm = \lambda\,dx$. In these cases the moment of inertia can be written

$$\boxed{I = \int \rho r^2\,dV \quad \text{or} \quad I = \int \sigma r^2\,dA \quad \text{or} \quad I = \int \lambda x^2\,dx} \tag{10.8}$$

Table 10-2 lists moments of inertias for some common shapes.

TABLE 10-2 Moment of Inertia, I

Hoop or cylindrical shell about its axis	MR^2
Solid cylinder or disk	$\frac{1}{2}MR^2$
Rod about perpendicular axis through center	$\frac{1}{12}MR^2$
Rod about perpendicular axis through end	$\frac{1}{3}MR^2$
Rectangular plate $a \times b$ about perpendicular axis though center	$\frac{1}{12}M(a^2 + b^2)$
Solid sphere	$\frac{2}{5}MR^2$
Spherical shell	$\frac{2}{3}MR^2$

PROBLEM 10.5.   Calculate the moment of inertia of a hoop of mass M and radius R about its axis.

Solution   All of the mass elements dm are the same distance R from the axis, so Eq. 10.7 yields

$$I = \int r^2 dm = R^2 \int dm = MR^2$$

This result applies to a hollow cylinder (like a pipe) as well.

PROBLEM 10.6.   Calculate the moment of inertia of a uniform rod of mass M and length L rotated about an axis perpendicular to the rod and passing through one end.

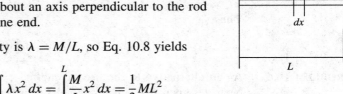

Solution   The linear mass density is $\lambda = M/L$, so Eq. 10.8 yields

$$I = \int \lambda x^2 \, dx = \int_0^L \frac{M}{L} x^2 \, dx = \frac{1}{3}ML^2$$

PROBLEM 10.7.   Calculate the moment of inertia of a rectangular sheet of metal of mass M and of sides a and b about the edge of length a.

Solution

$$\sigma = \frac{M}{b} \qquad I = \int_0^b \left(\frac{M}{ab}\right)(r^2)(a\,dr) = \frac{1}{3}Mb^2$$

Note that the result does not depend on the length a.

PROBLEM 10.8.   Calculate the moment of inertia of a solid cylinder of mass M and radius R about its axis.

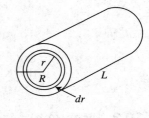

Solution

$$\rho = \frac{M}{\pi R^2 L} \qquad dV = 2\pi r \, dr L$$

$$I = \int_0^R \left(\frac{M}{\pi R^2 L}\right)(r^2)(2\pi r L \, dr) = \frac{1}{2}MR^2$$

Parallel-axis theorem (prove this for yourself): If the moment of inertia about an axis passing through the center of mass is I_{CM}, then the moment of inertia about a parallel axis displaced by a distance d from the center of mass axis is

$$\boxed{I = I_{CM} + Md^2} \tag{10.9}$$

PROBLEM 10.9.   The moment of inertia of a rod of mass M and length L about a perpendicular axis through its end is $\frac{1}{3}MR^2$. What is I about a parallel axis through the midpoint (the CM)?

Solution

$$I_{\text{end}} = I_{CM} + M\left(\frac{L}{2}\right)^2 = \frac{1}{3}ML^2, \quad \text{so } I_{CM} = \frac{1}{12}ML^2$$

PROBLEM 10.10.   The moment of inertia of a hoop about an axis through the center and perpendicular to the plane of the hoop is MR^2. What is I for a parallel axis through a point on the hoop?

Solution   Direct calculation is complicated, but using our theorem, it is easy to find I.

$$I = I_{CM} + MR^2 = MR^2 + MR^2 = 2MR^2$$

Perpendicular-axis theorem (prove this for yourself): The moment of inertia of a plane object about an axis perpendicular to the plane is equal to the sum of the moments of inertia about any two perpendicular axes in the plane. Thus if the x and y axes are in the plane,

$$I_z = I_x + I_y \tag{10.10}$$

PROBLEM 10.11. A square planar object has side a and mass M. What is its moment of inertia about (a) an axis through the center and perpendicular to the plane, and (b) an axis through diagonally opposite corners of the square? Use the previously obtained result (Problem 10.7) that the moment of inertia of a square about one edge is $\frac{1}{3}Ma^2$.

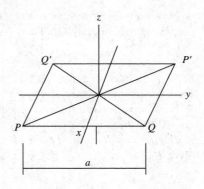

Solution Let the xy axes lie in the plane, passing through the center and parallel to the edges. From the parallel-axis theorem we find I_x.

$$I_{edge} = I_x + M\left(\frac{a}{2}\right)^2, \qquad \frac{Ma^2}{3} = I_x + \frac{Ma^2}{4}, \quad \text{so } I_x = \frac{Ma^2}{12}$$

By symmetry, $I_x = I_y$, so

$$I_z = I_x + I_y = \frac{Ma^2}{12} + \frac{Ma^2}{12} = \frac{Ma^2}{6}$$

Axes PP' and QQ' are perpendicular and symmetric, so $I_P = I_Q$. From the perpendicular-axis theorem, $I_z = I_p + I_Q = 2I_P$. So

$$I_p = \frac{1}{2}I_z = \frac{Ma^2}{12}$$

This approach is much easier than calculating I_P directly.

10.4 Torque

When a net force is applied to an object, it acquires a linear acceleration. The rotational quantity analogous to force is *torque*. For an object to acquire an angular acceleration, it must be subject to a net torque. *Torque* means "twist." The torque due a force F about a pivot P is τ, where the magnitude of the torque is

$$\boxed{\tau = Fr\sin\theta = Fr_\perp} \tag{10.11}$$

The distance from the pivot to the point of application of the force, as illustrated in Fig. 10-2, is r. The term $\tau_\perp = r\sin\theta$ is the *perpendicular lever arm* (also called the "lever arm" or "moment arm"). It is the shortest distance from the pivot to the line of action of the force. Torque is a vector, and its more complete definition is

$$\boxed{\tau = \mathbf{r} \times \mathbf{F}} \tag{10.12}$$

τ is perpendicular to the plane of $\mathbf{F}$ and $\mathbf{r}$. To find its direction, place the fingers of your right hand along $\mathbf{r}$. Curl them toward $\mathbf{F}$. Your thumb will point up along τ.

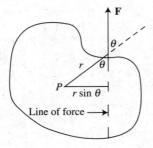 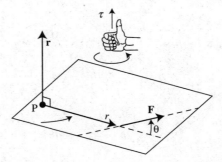

Fig. 10-2

Now consider a particle of mass m moving in a circle of radius r. Suppose a tangential force F acts on the particle and accelerates it. Then

$$F = ma \quad \text{or} \quad Fr = mra$$

$Fr = \tau$, the torque acting on the particle, and $a = r\alpha$, where α is the angular acceleration of the particle. Thus for a single particle, $\tau = mr^2\alpha$.

Now consider an extended object (like a disk) that can rotate. Imagine it consists of many small elements dm. All of them rotate with the same angular acceleration, so if the external force acting on each element is dF, $dF = a\,dm$, and $d\tau = r\,dF = ra\,dm = dmr^2\alpha$. Integrate to find the total external torque about the pivot P.

$$\tau_{\text{net}} = \int r^2\,dm\alpha = \alpha \int r^2\,dm = I\alpha$$

Thus

$$\boxed{\tau_{\text{net}} = I\alpha} \tag{10.13}$$

This equation is analogous to $F_{\text{net}} = ma$ for translational motion.

PROBLEM 10.12.   A string of negligible weight is wrapped around a pulley of mass M and radius R and tied to a mass m. The mass is released from rest, and it drops a distance h to the floor. Use energy principles to determine the speed of the mass when it hits the floor. Also, use Eq. 10.12 to determine this speed, as well as the tension in the string and the angular acceleration of the pulley.

Solution

$$KE_1 + PE_1 = KE_2 + PE_2$$

$$0 + mgh = \frac{1}{2}mv^2 + \frac{1}{2}I\omega^2 + 0$$

$$I = \frac{1}{2}MR^2 \quad \text{and} \quad v = R\omega$$

$$mgh = \frac{1}{2}mv^2 + \frac{1}{2}\left(\frac{1}{2}MR^2\right)\left(\frac{v}{R}\right)^2$$

$$v = \left(\frac{4gh}{2m + M}\right)^{1/2}$$

Applying Eq. 10.12 to the pulley yields $\tau = I\alpha$ so

$$TR = \frac{1}{2}MR^2\alpha \quad \alpha = \frac{2T}{MR}$$

Applying $F = ma$ to the mass m yields $T - mg = -ma$. Here a is the magnitude of the linear acceleration. I took up as positive (but acceleration is downward). But

$$a = \alpha R = \left(\frac{2T}{MR}\right)R = \frac{2T}{M}, \quad \text{so} \quad T - mg = -(m)\left(\frac{2T}{M}\right)$$

$$T = \left(\frac{M}{M + 2m}\right)mg \qquad a = \frac{2}{M + 2m}g \qquad \alpha = \frac{a}{R} = \frac{2g}{R(M + 2m)}$$

Since

$$v^2 = v_0^2 + 2ah = 0 + 2\left(\frac{2g}{M + 2m}\right)h, \quad \text{then} \quad v = \left(\frac{4gh}{M + 2m}\right)^{1/2}$$

10.5 Rolling

When a wheel of radius R rolls without slipping, a point on the circumference moves a distance ds when the wheel rotates through an angle $d\theta$, where $ds = R\,d\theta$. If this happens in time dt,

$$\frac{ds}{d\theta} = R\frac{d\theta}{dt} \quad \text{or} \quad v = R\omega \tag{10.14}$$

One can imagine the motion to consist of simple rotation about the point of contact with the ground. Viewed in this way, the kinetic energy is rotational kinetic energy, where

$$\text{KE} = \tfrac{1}{2}I\omega^2$$

By the parallel-axis theorem, Eq. 10.9, $I = I_{\text{CM}} + MR^2$. Thus

$$\text{KE} = \frac{1}{2}(I_{\text{CM}} + MR^2)\omega^2 = \frac{1}{2}I_{\text{CM}}\omega^2 + \frac{1}{2}MR^2\omega^2$$

Since $v = R\omega$,

$$\boxed{\text{KE} = \frac{1}{2}I_{\text{CM}}\omega^2 + \frac{1}{2}Mv^2} \tag{10.15}$$

This is an important result. It states that the kinetic energy of a rolling object is equal to the kinetic energy of translation of the center of mass CM (imagining all of the mass concentrated there) plus the kinetic energy of rotation about the CM.

PROBLEM 10.13. A hoop of radius R_H and mass m_H and a solid cylinder of radius R_C and mass m_C are released simultaneously at the top of a plane ramp of length L inclined at angle θ above horizontal. Which reaches the bottom first, and what is the speed of each there?

Solution The moment of inertia for each object is of the form $I_{\text{CM}} = kmR^2$, where $k = 1$ for a hoop and $k = \frac{1}{2}$ for a solid cylinder. Energy is conserved, so

$$mgh = \frac{1}{2}I_{\text{CM}}\omega^2 + \frac{1}{2}mv^2, \qquad mgh = \frac{1}{2}kmR^2\omega^2 + \frac{1}{2}mv^2$$

$$v = R\omega, \quad \text{so } v = \sqrt{\frac{2gh}{1+k}} \quad \text{where } h = L\sin\theta$$

Thus

$$v_H = \sqrt{\frac{2gh}{1+1}} = \sqrt{gh} \quad \text{and} \quad v_C = \sqrt{\frac{2gh}{1+1/2}} = \sqrt{\frac{4gh}{3}}$$

Thus we see that any solid cylinder will roll faster than a hoop of any size. Amazing.

10.6 Rotational Work and Power

Suppose that a tangential force pushes on an object and causes it to rotate through an angle $d\theta$. If the distance from the pivot to the point of application of the force is r, the work done by the force is $dW = F\,ds = Fr\,d\theta = \tau\,d\theta$. Thus

$$\boxed{W = \int \tau\,d\theta} \tag{10.16}$$

If this work is done in time dt, the power is

$$P = \frac{dW}{dt} = \tau\frac{d\theta}{dt} \quad \text{or} \quad \boxed{P = \tau\omega} \tag{10.17}$$

The work-energy theorem for rotational motion is

$$W = \int_{\theta_1}^{\theta_2} \tau \, d\theta = \frac{1}{2} I\omega_2^2 - \frac{1}{2} I\omega_1^2 \qquad (10.18)$$

PROBLEM 10.14.   A molecule in a microwave oven experiences a torque $\tau = \tau_0 \sin \theta$. How much work must be done to rotate the molecule from $\theta = 0°$ to $\theta = 180°$?

Solution

$$W = \int_0^\pi \tau \, d\theta = \int_0^\pi \tau_0 \sin \theta \, d\theta = -\tau_0 \omega s\theta \Big|_0^\pi = -\tau_0[-1 - (-1)] = 2\tau_0$$

10.7   Summary of Key Equations

Instantaneous angular velocity: $\omega = \dfrac{d\theta}{dt}$ $s = r\theta$

$$v = r\omega$$

Instantaneous angular acceleration: $\alpha = \dfrac{d\omega}{dt}$ $a = r\alpha$

Analogous Linear and Angular Quantities:

Linear impulse	$\bar{F}\,\Delta t$	$\longleftrightarrow$	Angular impulse	$\bar{\tau}\,\Delta t$
Linear displacement	s	$\longleftrightarrow$	Angular displacement	θ
Linear speed	v	$\longleftrightarrow$	Angular speed	ω
Linear acceleration	a	$\longleftrightarrow$	Angular acceleration	α
Mass (inertia)	m	$\longleftrightarrow$	Moment of inertia	I
Force	F	$\longleftrightarrow$	Torque	τ

If, in the equations for linear motion, we replace the linear quantities by the corresponding angular quantities, we get the corresponding equations for angular motion.

Linear: $F = ma$ $KE = \dfrac{1}{2}mv^2$ Work $= Fs$ Power $= Fv$

Angular: $\tau = I\alpha$ $KE = \dfrac{1}{2}I\omega^2$ Work $= \tau\theta$ Power $= \tau\omega$

In these equations, θ, ω, and α must be expressed in radians.

Motion with constant angular acceleration:

$$\theta = \theta_0 + \omega t + \frac{1}{2}\alpha t^2 \qquad \omega = \omega_0 + \alpha t \qquad \omega^2 = \omega_0^2 + 2\alpha(\theta - \theta_0)$$

Acceleration of particle on rotating body:

$$a_{\text{tan}} = R\alpha \qquad a_{\text{cent}} = R\omega^2$$

Moment of inertia:

$$I = \sum_{i=1}^n m_i R_i^2 \qquad I = \int \rho R^2 \, dV$$

Parallel-axis theorem:

$$I = I_{\text{CM}} + Md^2$$

Perpendicular-axis theorem (for a flat plate in the xy plane):

$$I_z = I_x + I_y$$

Kinetic energy of rotation:

$$K_R = \tfrac{1}{2}I\omega^2$$

Torque (direction is given by right-hand rule):

$$\tau = \mathbf{r} \times \mathbf{F} \quad \tau = rF\sin\theta = r_\perp F$$

Rolling kinetic energy:

$$\text{KE} = \frac{1}{2}I_{\text{CM}}\omega^2 + \frac{1}{2}Mv^2$$

SUPPLEMENTARY PROBLEMS

10.15. A disk initially at rest is given an angular acceleration of 12.0 rad/s². What is its angular velocity after 10 s? What is its frequency then in revolutions per minute (rev/min)? How many revolutions does it make during this time?

10.16. A car accelerates from rest to 20 m/s in 8 s. The wheels have radius 0.32 m. What is the average angular acceleration of the wheels?

10.17. A tether ball of mass 0.80 kg is attached to the top of a tall pole by a light cord 1.8 m long. What is the kinetic energy of the ball when the string makes an angle of 30° with vertical and the ball is rotating at 0.40 rev/s?

10.18. A circular disk of radius R has mass M. A hole of diameter R is cut in the disk, positioned as shown here. What is the moment of inertia for rotations about an axis perpendicular to the disk and passing through its center?

10.19. A playground merry-go-round is a metal disk of radius 2.5 m and mass 80 kg. Two kids, each of mass 40 kg, are riding on the outer rim of the disk. When they move halfway in toward the center, by what factor does the moment of inertia change?

10.20. What is the moment of inertia of a circular hoop of mass M and radius R about an axis that is tangent to the hoop and lies in its plane?

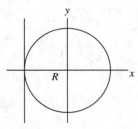

10.21. Two oppositely directed forces of equal magnitudes are applied perpendicularly to the ends of a rod of length L. The rod is pivoted a distance x from one end. What is the torque about the pivot P? Does the result depend on the value of x?

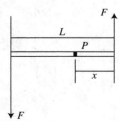

10.22. The head bolts on an engine must be tightened to the manufacturer's specification for proper operation. What is the minimum force you must apply to a 10-in wrench (25 cm) in order to exert a torque of 25 ft · lb (34 N · m)?

10.23. A bicycle chain passes over a front sprocket with 32 teeth and a rear sprocket of 16 teeth. The crank arm on which the bicyclist pushes is 15 cm long, and the bike wheels have a radius of 33 cm. (a) When the cyclist pushes with 80 N on the pedal, what force is applied to the ground by the rear tire? (b) When the pedals make 4 rev/s, how fast does the bicycle move?

10.24. A sphere of mass 0.036 kg and radius 1.2 cm rolls down an inclined plane. It is initially moving 0.48 m/s. How fast will it be moving after it has dropped 12 cm in elevation?

10.25. A streetcar is to be powered by the energy stored in a flywheel of mass 240 kg and radius 0.80 m. The flywheel is initially rotating at 4000 rev/min. How long could the flywheel provide power at a rate of 10 hp?

10.26. A string is wrapped around a uniform cylinder of mass M and radius R. A ball of mass m is attached to the string. The ball is released from rest and drops a distance h to the floor, causing the cylinder to spin. With what speed will the ball strike the floor?

10.27. Masses M and m are connected by a rod on length L and negligible mass. Diatomic molecules can be approximated by this model. (a) Find the distance of the center of mass from the mass M. (b) Show that the moment of inertia for rotations about an axis passing through the rod and perpendicular to it is a minimum if the axis passes through the center of mass. (c) Determine this minimum moment of inertia.

10.28. The moment of inertia of an object of irregular shape can be determined as follows. The object under study is attached to a light cylinder of negligible mass and radius R. The cylinder is mounted so that it can rotate about its vertical symmetry axis. A string is wrapped around the cylinder, and a constant tension T is applied to the cylinder. This is usually done by suspending a weight on the string and letting it fall to the floor. A length L of string is pulled off in time t. Determine an expression for the moment of inertia in terms of R, L, T, and t.

SOLUTIONS TO SUPPLEMENTARY PROBLEMS

10.15. $\omega = \omega_0 + \alpha t = 0 + (12\,\text{rad/s}^2)(10\,\text{s}) = 120\,\text{rad/s}$

$$\omega = 2\pi f \quad f = \frac{120\,\text{rad/s}}{2\pi} = 19.1\,\text{rev/s} = (19.1\,\text{rev/s})(60\,\text{s/min}) = 1150\,\text{rev/min}$$

$$\theta = \omega t = (120\,\text{rad/s})(10\,\text{s}) = 1200\,\text{rad} = 120\left(\frac{1\,\text{rev}}{2\pi\,\text{rad}}\right) = 19.1\,\text{rev}$$

10.16. $\omega = \dfrac{v}{r} = \dfrac{20\,\text{m/s}}{0.32\,\text{m}} = 62.5\,\text{rad/s} \quad \omega = \omega_0 + \alpha t = 0 + \alpha t \quad \alpha = \dfrac{\omega}{t} = \dfrac{62.5\,\text{rad/s}}{8\,\text{s}} = 7.8\,\text{rad/s}^2$

10.17. The ball moves in a circle of radius $L\sin 30°$: $\omega = 2\pi f$ and $I = mr^2$.

$$\text{KE} = \tfrac{1}{2}I\omega^2 = (0.5)(0.80)(1.8\sin 30°)^2(2\pi)^2(0.40)^2\,\text{J}$$
$$= 2.05\,\text{J}$$

10.18. Treat the hole as a small disk of negative mass superimposed on the larger solid disk (with no hole in it). Use the parallel-axis theorem to find I for the small disk. The mass of the small disk is

$$m = \frac{\pi(R/2)^2}{\pi R^2}M \qquad m = \frac{1}{4}M$$

$$I = I_{\text{big}} - I_{\text{small}} \quad \text{(since small disk has "negative" mass)}$$

$$= \frac{1}{2}MR^2 - \left[\frac{1}{2} + \left(\frac{M}{4}\right)\left(\frac{R}{2}\right)^2\right] = \frac{13}{32}MR^2$$

10.19. The moment of inertia of the disk is $I_D = \frac{1}{2}MR^2$. The moment of inertia of the two kids is $I_K = 2mr^2$. Thus

$$\frac{I_2}{I_1} = \frac{I_D + I_{K^2}}{I_D + I_{K^1}} = \frac{1/2\,MR^2 + 2m(R/2)^2}{1/2\,MR^2 + 2MR^2} = \frac{M+m}{M+4m} = \frac{1}{2}$$

10.20. The moment of inertia about the axis of the hoop (through its center) is MR^2. By the perpendicular-axis theorem, $I_z = I_x + I_y = MR^2$. By symmetry, $I_x = I_y$. Thus, $MR^2 = 2I_x$, $I_x = \frac{1}{2}MR^2$. By the parallel-axis theorem, $I = I_x + MR^2 = \frac{3}{2}MR^2$.

10.21. $\tau = F(L-x) + Fx = FL$ independent of x. A pair of forces like this is called a "couple." The net torque exerted is independent of where the pivot is located.

10.22.
$$\tau = Fr_\perp \quad F = \frac{\tau}{r_\perp} = \frac{34\,\text{N}\cdot\text{m}}{0.25\,\text{m}} = 136\,\text{N} = 31\,\text{lb}$$

10.23. (a) Suppose the pedal makes 1 rev. The work done is $W_1 = 2\pi rF_1$. The back sprocket then makes 0.5 rev, and the tire moves a distance of πR along the ground, exerting a force F_2. Thus

$$2\pi rF_1 = \pi RF_2 \quad F_2 = 2\frac{r}{R}F_1 = 2\left(\frac{15}{33}\right)(80\,\text{N}) = 72\,\text{N}$$

(b) In 1 s the pedals make 4 rev, so the back wheel makes $(16/32)(4) = 2$ rev, and the bike travels a distance of $s = (2)(2\pi R_2) = (4\pi)(0.33\,\text{m}) = 2.1\,\text{m}$, so $v = 2.1\,\text{m/s}$.

10.24. $\frac{1}{2}I\omega_1^2 + \frac{1}{2}mv_1^2 + mgh = \frac{1}{2}I\omega_2^2 + \frac{1}{2}mv_2^2 \quad I = \frac{2}{5}mv^2 \quad v = v\omega$

$$v_2 = \sqrt{v_1^2 + (10/7)gh} = \sqrt{(0.48\,\text{m/s})^2 + (10/7)(9.8\,\text{m/s}^2)(0.12\,\text{m})} = 1.38\,\text{m/s}$$

Note that mass and radius do not affect the answer. All spheres roll at the same rate.

10.25. $P = \frac{E}{t} \quad t = \frac{E}{P} \quad E = \frac{1}{2}I\omega^2 = \frac{1}{2}MR^2\omega^2$

$$t = \frac{MR^2\omega^2}{2P} = \frac{(240)(0.8)^2(2\pi)^2(4000/60)^2}{(2)(10)(746)}$$

$$= \frac{(240\,\text{kg})(0.8\,\text{m})^2(2\pi)^2(4000/60\,\text{s})^2}{(2)(10\,\text{hp})(746\,\text{W}/\text{h})} = 7.5\,\text{s}$$

10.26. Energy is conserved, so

$$mgh = \frac{1}{2}mv^2 + \frac{1}{2}I\omega^2$$

$$v = R\omega \quad I = \frac{1}{2}MR^2$$

$$mgh = \frac{1}{2}mv^2 + \frac{1}{2}\left(\frac{1}{2}MR^2\right)\left(\frac{v}{R}\right)^2$$

$$2mv^2 + Mv^2 = 4\,mgh$$

$$v = \left(\frac{4\,mgh}{2m+M}\right)^{1/2}$$

10.27. (a) $R = \dfrac{(M)(0) + (m)(L)}{M + m} = \dfrac{m}{M + m}L$

(b) Let the axis of rotation pass through a point a distance x from mass M.

$$I = Mx^2 + m(L - x)^2$$

Minimize I with respect to x

$$\frac{dI}{dx} = 0 = 2Mx + 2m(L - x)(-1)$$

$$0 = 2Mx - 2mL + 2mx$$

$$x = \frac{m}{M + m}L$$

Thus, we see that the moment of inertia is a minimum when the axis of rotation passes through the center of mass.

(c) $I = Mx^2 + m(L - x)^2$

$$I = M\left(\frac{mL}{M + m}\right)^2 + m\left(L - \frac{m}{M + m}L\right)^2$$

$$I = \frac{m^2 M}{(M + m)^2}L^2 + mL^2\left(\frac{M + m - m}{M + m}\right)^2 = \frac{m^2 ML^2}{(M + m)^2} + \frac{mM^2 L^2}{(M + m)^2}$$

$$I = \frac{mM}{M + m}L^2$$

10.28. $\theta = \dfrac{L}{R} = \dfrac{1}{2}\alpha t^2$

$\alpha = \dfrac{2\theta}{t^2} = \dfrac{2L}{Rt^2}$

$\tau = I\alpha = RT$

So, $I = \dfrac{RT}{\alpha} = \dfrac{RT}{2L/Rt^2} = \dfrac{R^2 T t^2}{2L}$

CHAPTER 11

Angular Momentum

Many aspects of rotational motion are analogous to translational motion. However, some rotational phenomena are bizarre (for example, gyroscope motion) and seem almost magical when first encountered. The theory of angular momentum has profound consequences in quantum mechanics and all of modern physics, and it has led to our understanding of atoms.

11.1 Angular Momentum and Torque

The angular momentum with respect to the origin of a particle with position $\mathbf{r}$ and momentum $\mathbf{p} = m\mathbf{v}$ is

$$\mathbf{L} = \mathbf{r} \times \mathbf{p} = \mathbf{r} \times m\mathbf{v} \qquad (11.1)$$

If the angle between $\mathbf{r}$ and $\mathbf{p}$ is θ, then the magnitude of L is

$$L = rp \sin \theta = mvr \sin \theta \qquad (11.2)$$

If $\mathbf{r}$ and $\mathbf{p}$ lie in the xy plane, $\mathbf{L}$ is along the z axis (Fig. 11-1). The time rate of change of the angular momentum is

$$\frac{d\mathbf{L}}{dt} = \frac{d}{dt}(\mathbf{r} \times \mathbf{p}) = \frac{d\mathbf{r}}{dt} \times \mathbf{p} + \mathbf{r} \times \frac{d\mathbf{p}}{dt} \qquad (11.3)$$

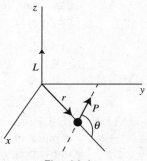

Fig. 11-1

The cross product of a vector with itself is always zero, so

$$\frac{d\mathbf{r}}{dt} \times \mathbf{p} = \frac{d\mathbf{r}}{dt} \times m\mathbf{v} = m(\mathbf{v} \times \mathbf{v}) = 0, \quad \text{since} \quad \frac{d\mathbf{r}}{dt} = \mathbf{v}$$

From Newton's second law, $\mathbf{F} = d\mathbf{p}/dt$. Thus

$$\boxed{\frac{d\mathbf{L}}{dt} = \mathbf{r} \times \mathbf{F}} \quad \text{or} \quad \boxed{\frac{d\mathbf{L}}{dt} = \vec{\tau}} \qquad (11.4)$$

If we think of a rigid body as a collection of particles of mass m_i, the z component of the angular momentum can be expressed in terms of the moment of inertia, $I = \sum m_i r_i^2$.

$$L_z = \sum m_i r_i v_i = \sum m_i r_i^2 \omega \quad \text{or} \quad \boxed{L_z = I\omega} \qquad (11.5)$$

This expression is analogous to $p = mv$, where I is like m and ω is like v. Apply Eq. 11.4 to a rigid body.

$$\frac{d\mathbf{L}}{dt} = \sum \mathbf{r}_i \times \mathbf{F}_i$$

All terms involving internal forces cancel, so

$$\frac{d\mathbf{L}}{dt} = \sum \mathbf{r}_i \times \mathbf{F}_{i,\text{ext}} \quad \text{or} \quad \boxed{\frac{d\mathbf{L}}{dt} = \vec{\tau}_{\text{ext}}} \tag{11.6}$$

Equation 11.6 is analogous to $\mathbf{F}_{\text{ext}} = d\mathbf{p}/dt$ for translational motion.

If no external torque acts on a system, the angular momentum of the system remains constant.

$$\boxed{\text{If } \tau = 0, \text{ then } \mathbf{L} = \text{constant.}} \tag{11.7}$$

This is the **law of conservation of angular momentum**.

PROBLEM 11.1.   In an interesting lecture demonstration, a student sits in a swivel chair. She has moment of inertia I_s about a vertical axis. She holds vertical the axis of a bicycle wheel of moment of inertia $I_B \ll I_s$ spinning with large angular velocity ω_1. The wheel is spinning counterclockwise, as viewed from above. Now she rotates the wheel axis by 180°. What happens?

Solution   No external torque acts (friction is negligible) so the angular momentum of the system remains constant. Initially $L_1 = I_w\omega_1$, directed upward. After the wheel axis is inverted, the angular momentum vector of the wheel will point downward, so the chair will rotate counterclockwise with angular velocity ω_2 so that the total angular momentum remains unchanged in magnitude and points upward.

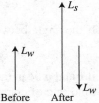

PROBLEM 11.2.   A disk is mounted with its axis vertical. It has radius R and mass M. It is initially at rest. A bullet of mass m and velocity v is fired horizontally and tangential to the disk. It lodges in the perimeter of the disk. What angular velocity will the disk acquire?

Solution   Angular momentum is conserved. The initial angular momentum is just the angular momentum of the bullet, $L_1 = mvR$. After the collision

$$L_2 = mR^2\omega + I\omega \quad \text{where } I = \frac{1}{2}MR^2 \text{ and } L_1 = L_2, \quad \text{so } \omega = \frac{mv}{(M + 2m)R}$$

PROBLEM 11.3.   The force of gravity on the Earth due to the Sun exerts negligible torque on the Earth (assuming the Earth to be spherical), since this force is directed along the line joining the centers of the two bodies. The Earth travels in a slightly elliptical orbit around the Sun. When it is nearest the Sun (the perihelion position), it is 1.47×10^8 km from the Sun and traveling 30.3 km/s. The Earth's farthest distance from the Sun (aphelion) is 1.52×10^8 km. How fast is the Earth moving at aphelion?

Solution   No torque acts, so the Earth's angular momentum is constant.

$$mr_1v_1 = mr_2v_2 \quad \text{or} \quad v_2 = \frac{r_1}{r_2}v_1 = \frac{1.47 \times 10^8}{1.52 \times 10^8}(30.3\,\text{km/s}) = 29.3\,\text{km/s}$$

11.2 Precession

When a torque acts on a system, the angular momentum **L** will change by an amount $\Delta\mathbf{L}$ in time Δt. From Eq. 11.4, $\mathbf{L} = \vec{\tau}\Delta t$, so the change in angular momentum is a vector directed in the direction of the torque. For a rigid body, where $L = I\omega$ and I is fixed, an increase in angular momentum (longer **L** vector) means the body speeds up and rotates more rapidly. A smaller **L** means the body slowed down. A change in direction of **L** means that the axis of rotation has changed direction. Three simple cases are shown in Fig. 11-2.

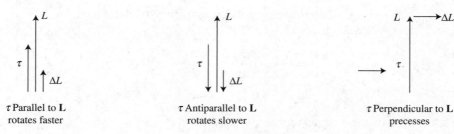

τ Parallel to **L**
rotates faster

τ Antiparallel to **L**
rotates slower

τ Perpendicular to **L**
precesses

Fig. 11-2

When the torque is directed perpendicular to **L**. $\Delta\mathbf{L}$ is also perpendicular to **L**. This means that **L** will rotate without changing magnitude. A spinning top exhibits this behavior. Gravity acts on the center of mass of the top and exerts a downward force (the weight) which produces the torque τ shown in Fig. 11-3. Think of the angular momentum vector **L** as "chasing" the torque vector. In Fig. 11-4 you can see that **L** rotates about the z-axis through a small angle $\Delta\phi$ in time Δt. This rotation about the z-axis is called *precession*. From Fig. 11-4 we can deduce an expression for the precessional frequency.

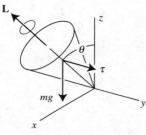

Fig. 11-3

For small $\Delta\phi$,

$$\Delta L = L \sin\theta \, \Delta\phi$$

Divide by Δt:

$$\frac{\Delta \mathbf{L}}{\Delta t} = L \sin\theta \frac{\Delta\phi}{\Delta t}$$

$$= \vec{\tau}$$

and $\omega_p = \Delta\phi/\Delta t =$ angular frequency of precession. Thus

$$\omega_p = \frac{\tau}{L \sin\theta} \qquad (11.8)$$

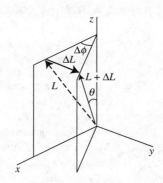

Fig. 11-4

In the above I assumed the **spin angular momentum**, $I\omega$, is large compared to the angular momentum associated with the precession. The faster the top spins, the more slowly it precesses. Even though the top is leaning to one side and would quickly fall were it not spinning, its spinning causes its axis to sweep out the surface of a cone as it precesses around the z-axis (Fig. 11-15). Amazing! In real tops the spin angular momentum is usually not large enough to justify the above approximation, and the motion consists of precession plus some complicated wobbling (called *nutation*). Thick books are written on the complex and fascinating motion of tops and gyroscopes.

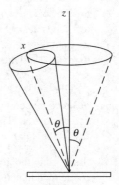

Fig. 11-5

PROBLEM 11.4.   A weighted bicycle wheel is adapted to have a long axle. The axle is approximately horizontal and is supported at a point $d = 18\,\text{cm}$ from the wheel. Essentially all of the mass of the wheel (13.2 kg) is at the rim a radius $r = 32.0\,\text{cm}$ from the axle. The wheel is spinning at 240 rev/min. What is the angular velocity of precession? Compare the spin angular momentum with the angular momentum associated with the precession.

Solution $$L \simeq L_s = I\omega, \qquad \tau = mgd$$

$I = mr^2$ for a hoop. The precession frequency is given by Eq. 11.8, with $\theta = 90°$.

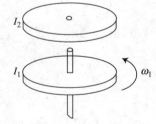

$$\omega_p = \frac{mgd}{mr^2\omega} = \frac{gd}{r^2\omega} = \frac{(9.8)(0.18)}{(0.32)^2(2\pi)^2(240/60)^2} = 0.685\,\text{rad/s}$$

$$= 0.182\,\text{rev/min}$$

The angular momentum due to precession is

$$L_p = md^2\omega_p = \frac{mgd^3}{r^2\omega}$$

$$\frac{L_p}{L_s} = \frac{mgd^3}{mr^4\omega^2} = \frac{gd^3}{r^4\omega^2} = \frac{(9.8)(0.18)^3}{(0.32)^4(2\pi)^2(240/60)^2} = 8.6 \times 10^{-3}$$

Thus

$$L_p \ll L_s$$

SUPPLEMENTARY PROBLEMS

11.5.   At an instant when a particle of mass m is at the position $(2, 3, -4)$, its velocity has components $(1, -1, 3)$. What is the angular momentum of the particle with respect to the origin?

11.6.   An arrangement encountered in disk brakes and certain types of clutches is shown here. The lower disk, of moment of inertia I_1, is rotating with angular velocity ω_1. The upper disk, with moment of inertia I_2, is lowered onto the bottom disk. Friction causes the two disks to adhere, and they finally rotate with the same angular velocity. Determine this final angular velocity if the initial angular velocity of the upper disk was (a) zero, (b) ω_2 in the same direction as ω_1, and (c) ω_2 in the opposite direction from ω_1.

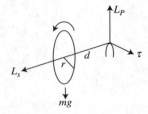

11.7.   A disk of radius R and moment of inertia I_1 rotates with angular velocity ω_0. The axis of a second disk, of radius r and moment of inertia I_2, is at rest. The axes of the two disks are parallel. The disks are moved together so that they touch. After some initial slipping the two disks rotate together. Find the final rate of rotation of the smaller disk.

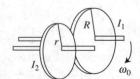

11.8.   A small mass m slides from rest down a smooth slope, dropping a distance h in elevation. It strikes the end of a rod of length d and mass M and sticks to it. The rod is initially at rest and fastened by a pivot P. How fast is the small mass moving just after the collision?

11.9.   Show that the precessional frequency of a top (given in Eq. 11.8) is independent of θ, the angle between vertical and the spin axis of the top.

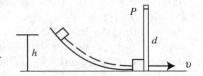

11.10. If the pilot of a small single-engine airplane is gliding with the engine off and pulls back on the control stick, the flaps will create a torque that makes the front of the plane rise. Suppose, however, that the engine is running and the propeller is turning clockwise, as viewed by the pilot. The propeller and engine now have significant angular momentum. Explain what will happen now if the pilot pulls back on the stick. What happens if he pushes forward on the stick, in an attempt to lose elevation?

11.11. If you played with Hula Hoops when you were a kid, you probably learned this trick. If you want to make the hoop turn to the right, run along behind it with a small stick in your hand. Give the top of the hoop a sharp blow directed to the right and parallel to the ground. This impulse will cause the axis of the hoop to rotate about a vertical axis, and the hoop will roll on at an angle θ to its original direction. Determine θ as a function of the average force F you apply and the duration Δt of the impulse. The hoop has mass M, radius R, and speed v. It does not slip on the ground.

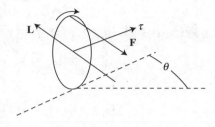

11.12. The evolution of a star depends on its size. If a star is sufficiently large, the gravity forces holding it together may be large enough to collapse it into a very dense object composed mostly of neutrons. The density of such a neutron star is about 10^{14} times that of the Earth. Suppose that a star initially had a radius about that of our Sun, 7×10^8 km, and that it rotated once every 26 days, as our Sun does. What would be the period of rotation (the time for 1 rev) if the star collapsed to a radius of 15 km?

11.13. A comet of mass 2×10^{14} kg moves in an elliptical orbit about the Sun. The Sun is at one focus of the elliptical orbit. At its point of closest approach to the Sun, approximately 10^6 km, the comet is moving at 5×10^6 m/s. What is the angular momentum of the planet with respect to an origin at the Sun?

11.14. A hockey stick of mass m and length d lies on the ice. It is struck perpendicularly a distance of $d/5$ from one end by a hockey puck. In the collision an impulse $F\Delta t$ is applied to the stick. Describe the subsequent motion of the stick. Assume the stick is a uniform rod.

11.15. A bowling ball of mass m and radius r is launched with speed v_0 (and with no spin) on an alley where the coefficient of friction is μ. How far will it travel before it rolls without slipping? What then will be its speed?

11.16. When I was in graduate school in Berkeley, we were as poor as church mice. My wife used to go to the UC Dental School in San Francisco (it was cheap), and while she was there, I would take our boys to the merry-go-round across the street in Golden Gate Park. It consisted of a metal disk (90-kg mass, 2.4-m radius) mounted on a vertical axis. A great game is played like this: Position eight kids (20 kg each) on the outer edge of the merry-go-round. Get it going as fast as you can (I could manage about 1 rev every 2 s). Then scream "Banzai!" or "Geronimo" (anything conveying bravery works), at which signal the kids all scramble toward the center. The merry-go-round takes off like a flying saucer. Any kids remaining on the perimeter are hurled off and go tumbling head over heels. Mothers drop like flies with heart attacks at the sight. Suppose for the parameters above, the kids all move in to a position 0.6 m from the axis. What then will be the frequency of revolution? Is the kinetic energy of the system conserved? Explain.

11.17. Most of us have seen ice skaters who can make themselves spin at a very high rate. They do this by first spinning slowly with arms outstretched. When their arms are pulled in, their moment of inertia decreases while their angular momentum remains constant. This results in an impressive increase in angular velocity. This effect is demonstrated in the classroom with the following arrangement. A student sits on a stool that can rotate freely. In each hand he holds a heavy dumbbell. He is given a push and starts to rotate slowly. Then he pulls his arms in. His angular momentum is conserved, and he begins to spin very fast. Calculate the effect for the following parameters: Each dumbbell has a mass of 3.2 kg. Each dumbbell is initially 0.90 m from the axis of rotation, and each is then pulled in to a distance of 0.25 m from the axis. The moment of inertia of the student plus the stool has a constant value of 2.50 kg-m^2. The student is initially rotating at 0.30 rev/s. (a) What is the rate of revolution after the dumbbells are pulled in? (b) Calculate the kinetic energy of the system before and after the dumbbells are pulled in. Explain any difference.

11.18. A simple merry-go-round consists of a uniform disk of radius R and mass M. A girl of mass m stands on it at a distance r from the axis of rotation. She walks at a constant distance r from the center with speed v with respect to the surface of the merry-go-round. Initially both the girl and the merry-go-round are stationary. At what angular velocity will the merry-go-round rotate?

SOLUTIONS TO SUPPLEMENTARY PROBLEMS

11.5. $\mathbf{L} = \mathbf{r} \times m\mathbf{v} = m(2\mathbf{i} + 3\mathbf{j} - 4\mathbf{k}) \times (\mathbf{i} - \mathbf{j} + 3\mathbf{k})$

$$= m \begin{vmatrix} \mathbf{i} & \mathbf{j} & \mathbf{k} \\ 2 & 3 & -4 \\ 1 & -1 & +3 \end{vmatrix} = m(9-4)\mathbf{i} + m(-4-6)\mathbf{j} + m(-2-3)\mathbf{k}$$

$$= 5m\mathbf{i} - 10m\,\mathbf{j} - 5m\,\mathbf{k}$$

11.6. (a) $L_1 = L_2 \qquad I_1\omega_1 = (I_1 + I_2)\omega \qquad \omega = \dfrac{I_1}{I_1 + I_2}\omega_1$

(b) $L_1 = I_1\omega_1 + I_2\omega_2 = (I_1 + I_2)\omega \qquad \omega = \dfrac{I_1\omega_1 + I_2\omega_2}{I_1 + I_2}$

(c) $L_1 = I_1\omega_1 - I_2\omega_2 = (I_1 + I_2)\omega \qquad \omega = \dfrac{I_1\omega_1 - I_2\omega_2}{I_1 + I_2}$

11.7. The contact force F has the same magnitude for each disk. Thus the torque equation becomes

$$Fr = I_2\alpha_2 \quad \text{and} \quad \omega_2 = \alpha_2 t = \frac{Frt}{I_2} \quad \text{so } Ft = \frac{I_2\omega_2}{r} \tag{i}$$

$$FR = I_1\alpha_1 \qquad \omega_1 = \omega_0 - \alpha_1 t = \omega_0 - \frac{FRt}{I_1} \tag{ii}$$

Since v at contact point is finally the same for both disks, $r\omega_2 = R\omega_1$. Substitute Eq. i in Eq. ii and solve.

$$\omega_2 = \frac{rRI_1}{r^2 I_1 + R^2 I_2}\omega_0$$

11.8. Energy conserved sliding down: $mgh = \frac{1}{2}mv^2$ and $v = \sqrt{2gh}$. Angular momentum conserved in collision: $mvd = mvd + I\omega$, where $V = \omega d$ and $I = \frac{1}{3}Md^2$ (rod).

$$m\sqrt{2gh}\,d = mVd + \tfrac{1}{3}Md^2\omega \quad V = \tfrac{3m\sqrt{2gh}}{3m+M}$$

11.9. From Fig. 11-3, $r = Wd \sin\theta$. From Eq. 11.8,

$$\omega_p = \frac{\tau}{L\sin\theta} = \frac{Wd\sin\theta}{L\sin\theta} = \frac{Wd}{L}$$

ω_p is thus independent of θ.

11.10. If you pull back on the stick, this creates a torque vector directed to your right. The angular momentum vector of the propeller is directed straight ahead. It tends to move toward the torque vector, so when you pull back on the stick, the plane goes up and to the right. When you push forward on the stick, the plane goes down and to the left. This is a noticeable effect in a small plane.

11.11. The force creates a torque vector directed straight ahead and parallel to the ground. The angular momentum vector of the rolling hoop is to the left. The hoop precesses about a vertical axis, with $\mathbf{L}$ moving toward the torque vector. In time Δt it precesses through an angle $\Delta \theta$. The frequency of precession is given by Eq. 11.8, with $\theta = 90°$.

Note that if the hoop doesn't slip, the friction force of the ground exerts a reaction force equal to the force applied at the top of the hoop. This friction force gives a torque about the center of mass in the same direction as the torque due to the applied force. Thus

$$\tau = FR + FR = 2FR$$

and

$$\Delta \theta = \omega_p \Delta t = \frac{\tau \Delta t}{L \sin \theta} = \frac{2FR\Delta t}{I\omega}, \quad I = MR^2, \quad \omega = \frac{v}{R}, \quad \text{so} \quad \Delta \theta = \frac{2F\Delta t}{Mv}$$

11.12. $L_1 = L_2$, so $I_1 \omega_1 = I_2 \omega_2$

$$I = \frac{2}{5}MR^2, \quad \omega = \frac{2\pi}{T}, \quad \text{so} \quad \frac{T_2}{T_1} = \frac{\omega_1}{\omega_2} = \frac{R_2^2}{R_1^2}$$

$$T_2 = \frac{R_2^2}{R_1^2}T_1 = \left(\frac{15}{7 \times 10^8}\right)^2 (26\,\text{days}) = 1.19 \times 10^{-14}\,\text{days} = 1.4\,\text{ms}$$

11.13. $L = mvd = (2 \times 10^{14}\,\text{kg})(5 \times 10^6\,\text{m/s})(10^9\,\text{m})$

$$= 10^{30}\,\text{kg} \cdot \text{m/s}$$

11.14. The center of mass of the stick acquires a velocity v, where

$$\Delta p = mv - 0 = F\Delta t \quad \text{or} \quad v = \frac{F\Delta t}{m}$$

The distance from the center of mass to the point of application of the force is

$$l = \frac{d}{2} - \frac{d}{5} = \frac{3d}{10}$$

Thus the torque applied about the CM is $\tau = 3/10 Fd$. The change in angular momentum is thus

$$\Delta L = \tau \Delta t$$
$$= \tfrac{3}{10}d\,F\Delta t$$
$$= I\Delta\omega = I(\omega - 0) = I\omega$$

From Table 9.1, $I = 1/12\,md^2$ for a rod rotating about its center. Thus

$$\Delta L = \frac{1}{12}md^2\omega = \frac{3}{10}dF\Delta t \quad \omega = \frac{18}{5}\frac{F\Delta t}{d}$$

Thus the CM of the stick slides with constant velocity v, and the stick rotates with constant angular velocity ω.

11.15. The friction force will exert a force that causes the velocity to *decrease* at a constant rate, that is,

$$v = v_0 - at = v_0 - \frac{F_f}{m}t, \quad F_f = \mu mg \quad \text{or} \quad v = v_0 - \mu gt$$

The friction force exerts a torque about the center of mass, $\tau = rF_f = r\mu mg$. This torque causes the angular momentum L (and also the angular velocity ω, where $L = I\omega$) to *increase* linearly. Thus

$$\omega = \omega_0 + \alpha t = 0 + \frac{\tau}{I}t = \frac{\mu mgr}{I}t.$$

For a sphere, $I = \frac{2}{5}mr^2$, so

$$\omega = \frac{5}{2}\frac{\mu g}{2r}t$$

When $v = r\omega$, the motion of the sphere will be pure rolling (until then it is also slipping). This will happen after time t, where

$$v_0 - \mu gt = r\left(\frac{5}{2}\frac{\mu g}{r}t\right) \quad \text{or} \quad t = \frac{3}{7}\frac{v_0}{\mu g}$$

The distance traveled in this time is

$$x = v_0 t - \frac{1}{2}\mu g t^2$$
$$= v_0\left(\frac{2}{7}\frac{v_0}{\mu g}\right) - \frac{1}{2}\mu g\left(\frac{2}{7}\frac{v_0}{\mu g}\right)^2 = \frac{12}{49}\frac{v_0^2}{\mu g}$$

At this time the speed will be

$$v = v_0 - \mu gt$$
$$= v_0 - \mu g\left(\frac{2}{7}\frac{v_0}{\mu g}\right) = \frac{5}{7}v_0$$

Once the ball's motion is pure rolling, its speed doesn't change. Note also that the final speed is independent of the value of the coefficient of friction, although the time required to reach this final speed does depend on μ.

11.16. No external torque acts, so the angular momentum of the system is conserved.

$$L_{\text{before}} = L_{\text{after}} \quad \text{or} \quad I_1\omega_1 = I_2\omega_2, \quad \omega = 2\pi f, \quad \text{so } f_2 = \frac{I_1}{I_2}f_1$$

Initially, $I_1 = 8mr^2 + \frac{1}{2}Mr^2$ ($I = \frac{1}{2}Mr^2$ for a disk). After moving in,

$$I_2 = 8m\left(\frac{r}{4}\right)^2 + \frac{1}{2}Mr^2$$

Thus

$$f_2 = \frac{8mr^2 + 0.5Mr^2}{8/16\,mr^2 + 1/2\,Mr^2}f_1 = \frac{16m + M}{m + M}f_1 \qquad f_2 = \frac{(16)(20) + 90}{20 + 90}(0.5/\text{s}) = 1.9/\text{s}$$

The merry-go-round speeds up by a factor of almost 4!

$$\text{KE}_1 = 8\left(\frac{1}{2}mr^2\omega_1^2\right) + \frac{1}{2}I\omega_1^2 \quad \left(I = \frac{1}{2}Mr^2\right) \qquad \text{KE}_2 = 8\left[\frac{1}{2}m\left(\frac{r}{4}\right)^2\omega_2^2\right] + \frac{1}{2}I\omega_2^2$$

$$\frac{\text{KE}_2}{\text{KE}_1} = \frac{1/4mr^2\omega_2^2 + 1/4Mr^2\omega_2^2}{4mr^2\omega_1^2 + 1/4Mr^2\omega_1^2} = \frac{m + M}{16m + M}\left(\frac{f_2}{f_1}\right)^2 = \frac{m + M}{16m + M}\left(\frac{16m + M}{m + M}\right)^2 = \frac{16m + M}{m + M} = 3.7$$

Thus kinetic energy is *not* conserved. The kids must pull hard and do work to move in to the center (try it yourself), and this work increases the kinetic energy of the system.

11.17. $L = I_1\omega_1 = I_2\omega_2$ or $I_1 f_1 = I_2 f_2$ since $\omega = 2\pi f$

$I_1 = I_0 + 2mR_1{}^2 = 2.5\,\text{kg}\cdot\text{m}^2 + 2(3.2\,\text{kg})(0.9\,\text{m})^2$

$I_1 = 7.68\,\text{kg}\cdot\text{m}^2$

$I_2 = I_0 + 2mR_2{}^2 = 2.5\,\text{kg}\cdot\text{m}^2 + 2(3.2\,\text{kg})(0.25\,\text{m})^2$

$I_2 = 2.90\,\text{kg}\cdot\text{m}^2$

$f_2 = \dfrac{I_1}{I_2}f_1 = \dfrac{7.68}{2.90}f_1 = 2.65 f_1$

(a) $f_2 = (2.65)(0.3\,\text{rev/s}) = 0.79\,\text{rev/s}$

(b) $\text{KE}_1 = \frac{1}{2}I_1\omega_1{}^2 = (0.5)(7.68\,\text{kg}\cdot\text{m}^2)(2\pi)^2(0.3\,\text{s})^2$

$\text{KE}_1 = 13.6\,\text{J}$

$\text{KE}_2 = \frac{1}{2}I_2\omega_2 = (0.5)(2.90\,\text{kg}\cdot\text{m}^2)(2\pi)^2(0.79\,\text{rev/s})^2$

$\text{KE}_2 = 3.57\,\text{J}$

The kinetic energy increases because the student does work when the dumbbells are pulled in.

11.18. Angular momentum is constant, so

$$I\omega - mv'r = 0$$

The speed of the girl with respect to the ground is

$$v' = v - r\omega$$

Also, $\qquad\qquad\qquad\qquad\qquad I = \frac{1}{2}MR^2$

Thus $\qquad\qquad\qquad \frac{1}{2}MR^2\omega - m(v - r\omega)r = 0$

$$\omega = \left(\frac{smvr}{mr^2 + MR^2}\right)$$

CHAPTER 12

Statics and Elasticity

We have seen that when an object is treated as a point mass, the condition that it be in equilibrium (that is, remain at rest or move in a straight line with constant speed) is that the net external force acting on it is zero. However, real objects are extended bodies that can rotate, and so we now continue our idea of equilibrium to encompass this possibility.

12.1 Rotational Equilibrium

If the angular velocity of a rigid body is not to change, no net external torque can act on the object, since

$$\sum \vec{\tau}_i = \frac{d\mathbf{L}}{dt} = I\frac{d\omega}{dt}$$

Consider an object that is at rest, that is, not translating and not rotating. Since calculation of torque requires identification of an origin about which to calculate the torque, we might imagine that the requirement of zero net torque would be influenced by the choice of an origin. This is not so. If an object is not rotating, it is not rotating about *every* point, even points outside the object. This will become more clear in the examples that follow. A judicious choice of the origin will simplify the algebra, but any choice of origin will work. (I call the origin about which the torque is calculated the "imaginary pivot" to help you visualize what is happening.) Note that a force that passes through a point exerts no torque about that point since its lever arm is then zero. Thus it is helpful to choose a pivot through which unknown forces pass, since then they drop out of the calculation and reduce the number of unknown variables. The complete conditions for the equilibrium of a rigid body are as follows:

The resultant external force is zero:

$$\boxed{\sum \mathbf{F} = 0} \tag{12.1}$$

The resultant external torque is zero about any origin:

$$\boxed{\sum \vec{\tau} = 0} \tag{12.2}$$

In most of the following I will discuss the case of **statics**, where the object is at rest and not translating or rotating. In such problems we usually want to find the forces acting on various parts of the system or structure. It is essential that a careful and accurate force diagram be drawn, showing the external forces acting on the system. All or part of a structure can be considered the "system." Recall that torque = (force)(perpendicular lever arm). Thus $\tau = Fr \sin \theta = Fr_\perp$. Most often errors occur in drawing the force diagram and in determining the lever arms. Study the following examples carefully and you will avoid these pitfalls.

When dealing with objects whose mass is distributed, treat all of the mass as if it is concentrated at the center of mass (CM). The gravity force exerts no torque about the CM point, which is why I think of the CM as the "balance point."

PROBLEM 12.1. Abe and Mary carry a uniform log of length 6 m and weight 150 N. Abe is 1 m from one end and Mary is 2 m from the other end. What weight does each person support?

Solution Draw the force diagram. If we imagine a pivot at Abe's position, and take a counterclockwise torque as positive, then Mary exerts a positive torque and the weight of the log exerts a negative torque. Using Eq. 12.2, $F_2(3\,\text{m}) - (150\,\text{N})(1\,\text{m}) = 0$ and $F_2 = 100$ N. From Eq. 12.1, $F + F_2 = 150$ N, so $F_1 = 50$ N. If instead we had chosen the pivot at Mary's position, the torque equation would yield $-F_1(3\,\text{m}) + (150\,\text{N})(1\,\text{m}) = 0$ and $F_1 = 50$ N as above.

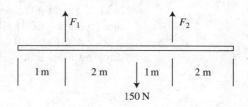

PROBLEM 12.2. A 5-kg beam 2 m long is used to support a 10-kg sign by means of a cable attached to a building. What is the tension in the cable and compressive force exerted by the beam?

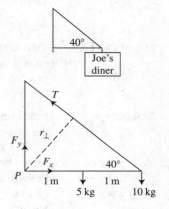

Solution Calculate the torque about a pivot at point P.

$$r_\perp = 2\sin 40° = 1.3\,\text{m}$$

$$T(1.3\,\text{m}) - 5\,\text{kg}(1\,\text{m}) - 10\,\text{kg}(2\,\text{m}) = 0$$

$$T = 19\,\text{kg} = (19)(9.8) = 188\,\text{N}$$

$$\sum F_x = 0 \qquad F_x - T\cos 40° = 0 \qquad F_x = 144\,\text{N}$$

PROBLEM 12.3. In kinesiology (the study of human motion), it is often useful to know the location of the CM of a person. This can be determined with the arrangement shown here. A plank of weight 40 N is placed on two scales separated by 2.0 m. A person lies on the plank and the left scale reads 314 N and the right scale reads 216 N. What is the distance from the left scale to the person's CM?

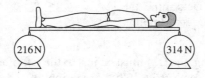

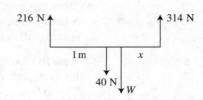

Solution Calculate the torque about the CM of the person: $(216\,\text{N})(2 - x) - (40\,\text{N})(1 - x) - (314\,\text{N})(x) = 0$ and $x = 0.80$ m. Note that the person's weight (which we could find using $\sum F_y = 0$) does not enter into the calculation because I chose the "pivot" at the CM.

PROBLEM 12.4.   A stepladder weighs 60 N and each half is 2.0 m long. A brace 1.0 m long (of negligible weight) is connected 0.50 m from the end of each half of the ladder. Assume no friction on the floor. What is the tension in the brace?

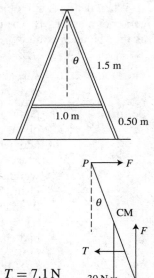

Solution   Draw the force diagram for one-half of the ladder. Calculate the torque about a pivot at the top of the ladder. F is the force due to the other half. Note that the weight of half the ladder is 30 N. From $\sum F_y = 0$, $F_N - 30 = 0$, and $F_N = 30$ N. From the drawing I see that

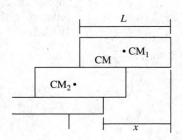

$$\sin \theta = \frac{0.5}{1.5}, \quad \theta = 19.5°$$

From $\sum \tau = 0$ about P,

$$F_N(2 \sin \theta) - (30)(1 \sin \theta) - T(1.5 \cos \theta) = 0$$

$$(30)(2)(\sin 19.5°) - (30)(\sin 19.5°) - 1.5\,T \cos 19.5° = 0 \quad T = 7.1 \text{ N}$$

PROBLEM 12.5.   Two identical bricks, each of length L, are stacked on a table so that the top brick extends as far as possible from the edge of the table. Determine this distance.

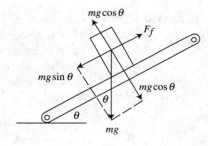

Solution   First consider the placement of only the top brick on the bottom brick. In order for the top brick not to tip, its CM (CM$_1$) must be just above the edge of the lower brick, that is, it must extend a distance $L/2$ beyond the lower brick edge. Now the CM of the combination of the two bricks must lie directly above the table edge. This CM is halfway between the CMs for the two identical bricks, that is, a distance $L/4$ from the outer end of the lower brick. Thus $x = 3L/4$.

PROBLEM 12.6.   A large number of packing crates, each with a base 0.60 m × 0.60 m and 1.20 m tall, is to be loaded by a conveyer belt moving at constant speed and inclined at angle θ above horizontal. The CM of a crate is at its geometrical center. The coefficient of friction between the belt and a crate is 0.40. As θ is increased, the crate will either tip over or begin to slip. Determine the critical angles for slipping and for tipping. Which will occur first?

Solution   Draw the force diagram. For tipping, the CM must be just to the left of the lower edge of the crate, so $\tan \theta_t = 0.60/1.20$, and $\theta_t = 26.6°$. For slipping, $F_f = mg \sin \theta_s$, and $F_f = \mu N = \mu mg \cos \theta_s$. So $mg \sin \theta_s = \mu mg \cos \theta_s, \tan \theta_s = \mu$, and $\theta_s = 21.8°$. Thus the crate will slip before it tips over as θ is increased.

PROBLEM 12.7.   A large piece of machinery is carried on a flatbed truck. Cleats keep the machinery from sliding, but only gravity keeps it from tipping. Its CM is 0.80 m above the truck bed and its base is 1.10 m from front to back. What is the shortest distance in which the truck can stop when traveling 10 m/s if the load is not to tip over?

Solution

When the load is just about to tip, it will do so about the front edge at point P. Thus the upward normal force of the bed is applied there, with $N = mg$. The backward force for decelerating the load is also applied at this point. If the load is not to tip, the torque about the CM must be zero. $v^2 = v_0^2 - 2ax = 0$ to stop, and $a = F/m$, so $x = mv_0^2/2F$. For $\tau = 0$, $0.55mg - 0.8F = 0$. Thus $F = 0.69mg$.

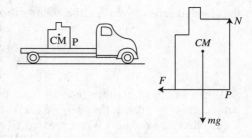

$$x = \frac{mv_0^2}{2F} = x = \frac{mv_0^2}{2(0.69mg)} = \frac{(10\,\text{m/s})^2}{2(0.69)(9.8\,\text{m/s}^2)} = 7.42\,\text{m}$$

12.2   Elasticity

Real materials are not perfectly rigid. When subjected to forces, they deform. For example, suppose one leg of a table is slightly longer than the others. The table will wobble, but if we put enough weight on the table, the longer leg will compress until the table is steady. However, using the methods we have examined so far, we cannot determine the exact force exerted by each leg. Such problems are indeterminate using our previous approach. A solution to finding the forces requires knowing something about the elastic properties of the leg material. If a substance deforms when subjected to a force, but returns to its initial shape when the force is removed, the substance is **elastic**.

Consider a cylinder of material of length L and cross-sectional area A. If a force F is applied along the axis of the cylinder, and this causes a change in length ΔL of the cylinder, then we define the **stress** and **strain** as

$$\text{Stress} \equiv \frac{F}{A} \tag{12.3}$$

$$\text{Strain} \equiv \frac{\Delta L}{L} \tag{12.4}$$

In a weak material, a small stress produces a large strain. For sufficiently small stresses, stress and strain are proportional. The constant of proportionality depends on the kind of material and on the nature of the deformation. The ratio of stress to strain is the **elastic modulus**.

$$\text{Elastic modulus} \equiv \frac{\text{stress}}{\text{strain}} \tag{12.5}$$

Suppose you pull or push on a cylinder of length L and cross-sectional area A with a force F directed along the axis. The material is subject to a **tensile stress**. **Young's modulus** is defined as

$$Y = \frac{\text{tensile strength}}{\text{tensile strain}} = \frac{F/A}{\Delta L/L} \tag{12.6}$$

If the stress exceeds the **elastic limit**, the material does not return to its original shape when the stress is removed.

The shear modulus measures a material's ability to resist changes in its shape. Suppose a piece of material in the form of a rectangular block (like a brick) has one face fixed and a force F applied to the opposite face, of area A. Imagine F applied parallel to the face, like the friction force (see Fig. 12-1). If the two faces are separated by distance h, and

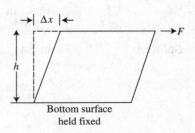

Fig. 12-1

the sheared face moves Δx, the **shear modulus** is

$$S = \frac{\text{shear stress}}{\text{sheer strain}} = \frac{F/A}{\Delta x/h} \tag{12.7}$$

When an object is subjected to a force from all sides, it is subject to a **pressure** P. If a force F acts perpendicular to an area A, the pressure exerted is $P = F/A$. This situation arises when an object is immersed in a fluid like air or water. When pressed from all sides, the volume V of an object will change by ΔV. The tendency for this is measured by the **bulk modulus** B defined by

$$B = \frac{\text{volume stress}}{\text{volume strain}} = -\frac{F/A}{\Delta V/V} = -\frac{P}{\Delta V/V} \tag{12.8}$$

The negative sign is inserted so that B is a positive number because ΔV is negative due to a positive pressure. In some tables of data the inverse of B, called the **compressibility**, is tabulated. A large bulk modulus means it is difficult to compress the material, whereas a large compressibility means it is easy to compress the material.

PROBLEM 12.8.   A steel beam used in the construction of a bridge is 10.2 m long with a cross-sectional area of $0.12\,\text{m}^2$. It is mounted between two concrete abutments with no room for expansion. When the temperature rises 10°C, such a beam will expand in length by 1.2 mm if it is free to do so. What force must be exerted by the concrete to keep this expansion from happening? Young's modulus for steel is $2.0 \times 10^{11}\,\text{N/m}^2$.

Solution   Imagine that the steel expands and that then the concrete exerts a compressional force to return it to its original length. From Eq. 12.6,

$$F = Y\left(\frac{\Delta L}{L}\right)A = (2 \times 10^{11}\,\text{N/m}^2)\left(\frac{1.2 \times 10^{-3}\,\text{m}}{10.2\,\text{m}}\right)(0.12\,m^2) = 2.8 \times 10^6\,\text{N}$$

This force could well crack the concrete. The forces involved in thermal expansion can be huge, which is why it is necessary to leave expansion space in joints in large structures like bridges and buildings.

PROBLEM 12.9.   A cube of Jell-O 6 cm on a side sits on your plate. You exert a horizontal force of 0.20 N on the top surface parallel to the surface and observe a sideways displacement of 5 mm. What is the shear modulus of the Jell-O?

Solution   From Eq. 12.7,

$$S = \frac{F/A}{\Delta x/h} = \frac{Fh}{A\Delta x} = \frac{Fh}{h^2\Delta x} = \frac{F}{h\Delta x} = \frac{0.20\,N}{(0.060\,\text{m})(0.005\,\text{m})} = 670\,\text{N/m}^2$$

PROBLEM 12.10.   When an object is submerged in the ocean to a depth of 3000 m, the pressure increases by about $3030\,\text{N/m}^2$. By how much does a piece of aluminum of volume $0.30\,\text{m}^3$ decrease in volume when lowered to this depth? The bulk modulus of aluminum is $7 \times 10^{10}\,\text{N/m}^2$.

Solution   From Eq. 12.8,

$$\Delta v = -\frac{PV}{B} = -\frac{(3030\,\text{N/m}^2)(0.30\,\text{m}^3)}{7 \times 10^{10}\,\text{N/m}^2} = -1.2 \times 10^{-8}\,\text{m}^3 = -12\,\text{mm}^3$$

12.3 Summary of Key Equations

Rotational equilibrium: $\quad \sum \mathbf{F} = 0 \quad \sum \vec{\tau} = 0$

Young's modulus: $\quad Y = \dfrac{\text{tensile strength}}{\text{tensile strain}} = \dfrac{F/A}{\Delta L/L}$

Shear modulus: $\quad S = \dfrac{\text{shear stress}}{\text{shear strain}} = \dfrac{F/A}{\Delta x/h}$

Bulk modulus: $\quad B = \dfrac{\text{volume stress}}{\text{volume strain}} = -\dfrac{F/A}{\Delta V/V} = -\dfrac{P}{\Delta V/V}$

$\text{Stress} \equiv \dfrac{F}{A} \qquad \text{Strain} \equiv \dfrac{\Delta L}{L}$

SUPPLEMENTARY PROBLEMS

12.11. In the crane here the boom is 3.2 m long and weighs 1200 N. The cable can support a tension of 10,000 N. The weight is attached 0.5 m from the end of the boom. What maximum weight can be lifted?

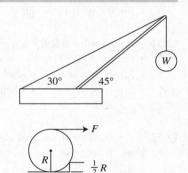

12.12. What horizontal force applied as shown here is required to pull a wheel of weight W and radius R over a curb of height $h = R/2$?

12.13. Two people carry a refrigerator of weight 800 N up a ramp inclined at 30° above horizontal. Each exerts a vertical force at a corner. The CM of the refrigerator is at its center. Its dimensions in the drawing are 0.72 m × 1.8 m. What force does each person exert?

12.14. A cylindrical shell of weight W and diameter $3R/2$ is placed upright on a horizontal surface. Two spheres, each of weight w and radius R, are placed in the cylinder. What are the contact forces exerted by the spheres on the cylinder? What is the maximum value of w for which the cylinder will not tip over?

12.15. A plank 7.2 m long of mass 20 kg extends 2.4 m beyond the edge of a cliff. How far beyond the edge of the cliff can a 15-kg child walk before the plank tips?

12.16. A string is wrapped many times around a cylinder, covering its surface. The cylinder is placed on a plane inclined at angle θ above horizontal. The end of the string is directed horizontally, where it is attached, thereby holding the cylinder in place. The coefficient of friction between the string-covered cylinder and the plane is μ. Determine the maximum value of θ for which the cylinder will remain in place and not start to move down the plane with the string unwinding.

12.17. To moor a ship, a sailor wraps a rope around a bollard (a cylindrical post). By pulling with a small force T_1, he can exert a much larger tension T_2 on the end of the rope attached to the ship because of the friction between the rope and the bollard. The coefficient of friction between the rope and the bollard is $\mu = 0.2$. If the sailor pulls with 400 N, how many turns are needed if he is to exert a force of 24,000 N on the ship?

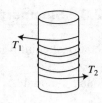

12.18. The separation between the front and back axles of a bicycle is 1.14 m, and the CM of the bike plus rider is 1.20 m above the ground. The coefficient of friction between the tires and the roadway is 0.60. Determine the braking deceleration when (a) both brakes are applied, (b) only the front brake is applied, and (c) only the back brake is applied.

12.19. A ladder 6 m long weighs 120 N. It leans against a smooth wall (negligible friction), making an angle of 50° with horizontal. The coefficient of friction with the floor is 0.5. How far up the ladder can an 800-N worker climb before the ladder starts to slip?

12.20. An oil drum of radius 28 cm and mass 20 kg is to be rolled up over a curb 15 cm high by applying a horizontal force directed through the symmetry axis of the drum. What minimum force is required?

12.21. A uniform cubical crate of mass 40 kg and side 1.20 m rests on a horizontal floor where the coefficient of friction is 0.80. A horizontal force is applied to the crate.
(a) If the force is applied to the top of the crate, what force is required for tipping?
(b) At what minimum height above the floor can the force be applied if the crate is to tip without starting to slide?

12.22. A mountain climber of mass 60 kg uses a nylon rope of diameter 6 mm and unstretched length 15 m. How much will the rope stretch when the climber is suspended from it? The Young's modulus for nylon is $0.38 \times 10^{10}\,\text{N/m}^2$.

SOLUTIONS TO SUPPLEMENTARY PROBLEMS

12.11. Calculate the torque about the base of the boom. The cable is one side of a triangle with angles 30°, 135°, and 15°.

$$(10{,}000\,\text{N})(3.2\sin 15°) - w(2.7\cos 45°) = 0 \quad w = 4340\,\text{N}$$

12.12. The torque due to F about the contact point must balance the torque due to gravity acting at the center of the sphere.

$$w\left(\frac{\sqrt{3}}{2}R\right) - F\left(\frac{3R}{4}\right) = 0 \quad F = \frac{2\sqrt{3}}{3}W$$

12.13. $F_1 + F_2 = 800\,\text{N}$, and the torque about the point of application of F_1 is zero.

$$(F_2)(1.8\cos 30°) - (800\,\text{N})(0.9\cos 30° - 0.36\sin 30°) = 0$$

$$F_2 = 308\,\text{N} \quad F_1 = 492\,\text{N}$$

Upper sphere

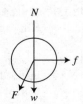

Lower sphere

12.14. Draw the force diagram for each sphere. The upward normal force of the table must support the weight of the two spheres, so $N = 2w$. Since the cylinder is not moving sideways, the two horizontal contact forces f must be equal. The contact force between the spheres is F, directed at an angle θ to vertical, where $\sin\theta = 0.5$. Applying $\sum F_y = 0$ to the upper sphere yields $F\cos\theta - w = 0$, so $F = 1.15w$. Applying $\sum F_x = 0$ yields $F\sin\theta - f = 0$, so $f = 0.58\,w$. If w is increased, the cylinder will tend to tip about point P. The weight of the cylinder acts downward from its CM on its axis with a lever arm of $3\,R/4$ about point P. The lower contact force has lever arm R about P, and the upper contact force has lever arm $R + 2R\cos\theta$ about P. Thus for no tipping, $W(3R/4) + fR - f(R + 2R\cos\theta) = 0$, where $f = 0.58w$, and thus $w = 0.75W$.

12.15. The CM of the plank (at its center) is 1.2 m from the edge. The torque due to it must balance the torque due to the child, who has walked a distance x beyond the edge. Here $g = 9.8\,\text{m/s}^2$. $20\,g(1.2\,\text{m}) - 15\,gx = 0$, so $x = 1.6\,\text{m}$.

12.16. If the string is not to unwind, the torque about point P must vanish. The friction force creates a counterclockwise torque with lever arm $R + R\cos\theta$, and the normal force creates a clockwise torque about P with lever arm $R\sin\theta$. Thus $F_f(R + R\cos\theta) - NR\sin\theta = 0$. Also $F_f = \mu N$. Thus

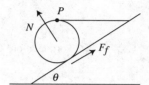

$$\mu = \frac{\sin\theta}{1 + \cos\theta}$$

12.17. Look at a small segment of rope that subtends a small angle $d\theta$. Because of friction the tension at one end is T and at the other end slightly larger, $T + dT$. Apply $\sum F_y = 0$.

$$N - T\sin\frac{d\theta}{2} - (T + dT)\sin\frac{d\theta}{2} = 0$$

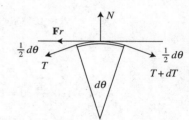

For $d\theta \ll 1$,

$$\sin\frac{d\theta}{2} \simeq \frac{d\theta}{2}$$

Neglect the very small term $2\,dT\sin d\theta/2$. Thus $N = 2T\,d\theta/2 \simeq T\,d\theta$. The friction force is $F_f = \mu N = \mu T\,d\theta$. Apply $\sum F_x = 0$:

$$(T + dT)\cos\frac{d\theta}{2} - F_f - T\cos\frac{d\theta}{2} = 0$$

$$\cos\frac{d\theta}{2} \simeq 1 \quad \text{if} \quad \frac{d\theta}{2} \ll 1, \quad \text{so} \quad dT = F_f = \mu T\,d\theta \quad \text{or} \quad \frac{dT}{T} = \mu\,d\theta$$

If the tensions at the two ends are T_1 and T_2, then

$$\int_{T_1}^{T_2} \frac{dT}{T} = \mu \int_{\theta_1}^{\theta_2} d\theta \quad \text{or} \quad \ln\frac{T_2}{T_1} = \mu(\theta_2 - \theta_1)$$

$$\theta_2 - \theta_1 = \frac{1}{\mu}\ln\frac{T_2}{T_1} = \frac{1}{0.4}\ln\frac{24{,}000}{400} = 10.2\,\text{rad} = \frac{10.2}{2\pi}\,\text{rev} \quad \theta_2 - \theta_1 = 1.63\,\text{revs of rope}$$

12.18. If normal forces on front and back wheels are N_1 and N_2, then $N_1 + N_2 = mg$.

$$L = 3.2\,\text{m} \qquad h = 0.60\,\text{m} \qquad \mu = 0.80$$

(a) Torque about CM is zero.

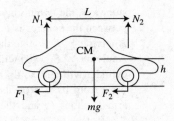

$$N_2\left(\frac{L}{2}\right) - N_1\left(\frac{L}{2}\right) - F_1 h - F_2 h = 0$$

$$F_1 = \mu N_1 \qquad F_2 = \mu N_2$$

Solve by finding $N_1 = 0.35\, mg$ and $N_2 = 0.65\, mg$.

$$F_{net} = ma = F_1 + F_2 = \mu N_1 + \mu N_2$$

so

$$a = \frac{F_{net}}{m} = \frac{\mu N_1 + \mu N_2}{m} = 0.8g = 7.84\,\text{m/s}^2$$

(b) Torque about CM is zero: $N_2(L/2) - N_1(L/2) - \mu N_2 h = 0$. Solve by finding

$$N_1 = 0.41\, mg \qquad N_2 = 0.59\, mg \qquad a = \frac{\mu N_2}{m} = 4.63\,\text{m/s}^2$$

(c) Torque about CM is zero: $N_2(L/2) - N_1(L/2) - \mu N_1 h = 0$. Solve:

$$N_1 = 0.43\, mg \qquad N_2 = 0.57\, mg \qquad a = \frac{\mu N_1}{m} = 3.37\,\text{m/s}^2$$

12.19.
$$\sum F_y = 0 \qquad N_1 - W_L - W = 0$$

$$N_1 - 120 - 800 = 0 \qquad N_1 = 920 N$$

$$\sum F_x = 0 \qquad F - N_2 = 0$$

$$F = \mu N_1 = (0.5)(920\,N) = 460\,N, \quad \text{so } N_2 = 460\,N$$

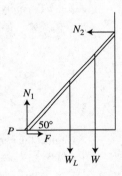

Torque about base of ladder, point P, is zero. The worker climbs distance x up ladder.

$$N_2(L\sin 50°) - W_L\left(\frac{L}{2}\cos 50°\right) - Wx\cos 50° = 0. \quad \text{Solve } x = 3.67\,\text{m}$$

12.20. Torque about pivot P is zero.

$$FR\cos\theta = mgR\sin\theta$$

$$F = mg\tan\theta$$

$$\cos\theta = \frac{R - h}{R} = \frac{28 - 15}{28}$$

$$\theta = 62.3°$$

$$F = (20\,\text{kg})(9.8\,\text{m/s}^2)\tan(62.3°)$$

$$F = 374\,\text{N}$$

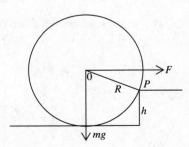

12.21. (a) For tipping require $mg(L/2) = FL$

$$F = \frac{mg}{2} = \frac{(40\,\text{kg})(9.8\,\text{m/s}^2)^2}{2}$$

$$F = 196\,\text{N}$$

(b) If force is applied a distance y above the floor,

$$F_y = mg\frac{L}{2} \qquad F = \frac{mgL}{2y}$$

Friction force is $F_f = \mu mg$
For tipping with no sliding require

$$F_f \geq F \text{ or } \mu mg \geq \frac{mgL}{2y}, \quad y \geq \frac{L}{2y}$$

$$y \geq \frac{1.2\,\text{m}}{(2)(0.8)} = 0.75\,\text{m}$$

12.22. $Y = \dfrac{F/A}{\Delta L/L} \quad \Delta L = \dfrac{1}{Y}\dfrac{F}{A}L \quad F = mg$

$$\Delta L = \frac{1}{0.38 \times 10^{10}\,\text{N/m}^2}\left[\frac{(60\,\text{kg})(9.8\,\text{m/s}^2)}{\pi(0.003\,\text{m})^2}\right]$$

$$\Delta L = 0.082\,\text{m} = 8.2\,\text{cm}$$

CHAPTER 13

Oscillations

When a particle or a system repeats the same motion again and again at regular time intervals, the motion is **periodic**. We also use the word **oscillatory** to describe periodic motion, usually in the context of a vibrating object or small system, such as a mass on a spring or a violin string or an electric circuit, whereas we describe the motion of a planet around the Sun as "periodic." Many kinds of oscillatory behavior are analogous to the motion of a mass attached to a spring, and it is important to understand thoroughly this system.

13.1 Simple Harmonic Motion

Consider a mass attached to a spring with spring constant k. The spring exerts a force $-kx$, where x is the displacement of the mass from equilibrium. The law $F = ma$ is thus

$$F = m\frac{d^2x}{dt^2} = -kx \tag{13.1}$$

We can integrate this equation twice to obtain the solution, which is

$$x = A\cos(\omega t + \theta) \tag{13.2}$$

where

$$\omega = 2\pi f = \sqrt{\frac{k}{m}} \tag{13.3}$$

Equation 13.2 describes **simple harmonic motion** (SHM). A is the **amplitude**. f is the frequency in vibrations per second. One vibration per second is called **1 hertz** (Hz). ω is the angular frequency, in radians per second. θ is the **phase constant**. The quantity in parenthesis $(\omega t + \theta)$ is the **phase** of the oscillation. A and θ are determined by the initial conditions of the problem. All of these terms are very important. You can confirm that Eq. 13.2 is indeed the solution of Eq. 13.1 by differentiating. Thus

$$v = \frac{dx}{dt} = -A\omega\sin(\omega t + \theta) \tag{13.4}$$

$$a = \frac{d^2x}{dt^2} = -A\omega^2\cos(\omega t + \theta) \tag{13.5}$$

or

$$a = -\omega^2 x = -\frac{k}{m}x, \quad \text{so } F = ma = -kx$$

Graphs of displacement, velocity, and acceleration are shown in Fig. 13-1 for the case $\theta = 0$. The time repeat interval T is the **period**.

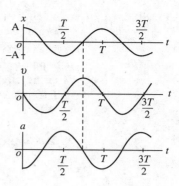

$$T = \frac{1}{f}$$

(13.6)

Note that the displacement curve and the acceleration curve are similar, but a is shifted by $T/2$ from x. a is "out of phase" with x, or "180° out of phase" with x. Similarly, v is 90° out of phase with x, and v leads x by 90° in phase, because the peak of v occurs at an earlier time than the nearest peak of x.

The maximum value of x is A, the maximum value of v is $A\omega$, and the maximum value of a is $A\omega^2$. When the displacement is large, the mass stops and $v = 0$. At this point the spring is fully stretched and F and a are both large (but negative).

If the mass on a spring is displaced a distance x_0 and released ($v = 0$ at $t = 0$), then $\theta = 0$ and $x_0 = A$. This is the case we will most often encounter.

PROBLEM 13.1. A beam oscillates according to $x = (0.002 \text{ m}) \cos(\pi t)$. What are the amplitude, maximum velocity, maximum acceleration, frequency, and period?

Solution By comparison with Eqs. 13.2, 13.4, and 13.5, $A = 0.002 \text{ m}$, $\omega = \pi$, $v_{\max} = A\omega = 0.002\pi \text{ m/s}$, $a_{\max} = A\omega^2 = 0.002\pi^2 \text{ m/s}^2$, $f = \omega/2\pi = 0.5 \text{ Hz}$, and $T = 1/f = 2.0 \text{ s}$.

We sometimes encounter a mass hanging from a spring. The equilibrium position thus corresponds to the spring being initially stretched somewhat. However, we can show that the same equations as above still apply.

PROBLEM 13.2. Determine the motion of a mass m hung from a spring with spring constant k.

Solution Let $y = 0$ at the end of the unstretched spring (that is, before the mass is attached), taking y as positive downward. With the mass attached, $F = ma$ becomes

$$m\frac{d^2y}{dt^2} = mg = ky$$

The mass stretched the spring a distance of $\Delta y = mg/k$ to the new equilibrium point, so transform to a new variable y', where $y' = y - mg/k$. Now $F = ma$ becomes

$$m\frac{d^2y'}{dt^2} = -ky'$$

This is the same equation as Eq. 13.1, and once again the solution is SHM. The only change is that the equilibrium point has been shifted downward by the action of the gravity force.

PROBLEM 13.3. When a mass of 0.050 kg is suspended from a spring, it stretches the spring by 0.012 m. If now the mass is displaced slightly and allowed to oscillate, what will be its frequency?

Solution The spring constant is $k = mg/x$ (k is always positive, and here x stands for the amount of stretch, a positive distance).

$$f = \frac{1}{2\pi}\sqrt{\frac{k}{m}} = \frac{1}{2\pi}\sqrt{\frac{mg}{mx}} = \frac{1}{2\pi}\sqrt{\frac{g}{x}} = \frac{1}{2\pi}\sqrt{\frac{9.8}{0.012}} \text{ s}^{-1} = 4.5 \text{ Hz}$$

PROBLEM 13.4.   When a mass is attached to a spring, it is observed to oscillate at 10 Hz. Suppose the spring is cut in half and again the same mass is attached. At what frequency will the mass now oscillate?

Solution   Suppose the force mg stretched the whole spring a distance x. This means half of the spring stretched a distance $x/2$ due to the force mg. Thus when the spring is cut in half, its new spring constant is $k' = 2mg/x = 2k$. Thus the new frequency is

$$f_2 = \frac{1}{2\pi}\sqrt{\frac{k'}{m}} = \frac{1}{2\pi}\sqrt{\frac{2k}{m}} = \sqrt{2}f_1 = \sqrt{2}(10\,\text{Hz}) = 14\,\text{Hz}.$$

13.2   Energy and SHM

We saw previously (Eq. 8.10) that the potential energy of a spring is

$$U = \tfrac{1}{2}kx^2 = \tfrac{1}{2}kA^2\cos^2(\omega t + \theta) \tag{13.7}$$

The kinetic energy is

$$K = \tfrac{1}{2}mv^2 = \tfrac{1}{2}m\omega^2 A^2 \sin^2(\omega t + \theta) \tag{13.8}$$

The total energy of the simple harmonic oscillator is

$$E = K + U = \tfrac{1}{2}kA^2[\sin^2(\omega t + \theta) + \cos^2(\omega t + \theta)] = \tfrac{1}{2}kA^2 \tag{13.9}$$

since $\sin^2(\omega t + \theta) + \cos^2(\omega t + \theta) = 1$.

This result makes sense, since when the spring is fully stretched, $v = 0$, $x = A$, and the energy is all potential energy. When the mass passes through the equilibrium position ($x = 0$), the potential energy is zero and the energy is all kinetic. Then $E = \tfrac{1}{2}mv_{max}^2$. Since $v_{max} = A\omega$, $E = \tfrac{1}{2}mA^2\omega^2 = \tfrac{1}{2}mA^2k/m = \tfrac{1}{2}kA^2$, the same result as above. As the mass oscillates, its energy switches back and forth between kinetic energy and potential energy, with the sum of the two remaining constant.

PROBLEM 13.5.   A mass is attached to a spring and displaced and then released from rest. Determine the time when the KE and PE are first equal.

Solution

$$\frac{1}{2}mv^2 = \frac{1}{2}kx^2, \quad \text{so } m(-A\omega\sin\omega t)^2 = k(A\cos\omega t)^2$$

$$\tan^2\omega t = \frac{k}{m\omega^2} \quad \text{and} \quad \omega^2 = \frac{k}{m}, \quad \text{so } \tan^2\omega t = 1$$

$$\omega t = \frac{\pi}{4}, \qquad t = \frac{\pi}{4\omega} = \frac{T}{8}$$

13.3   SHM and Circular Motion

SHM can be related to circular motion in the following way. Imagine a peg P attached to a wheel oriented with its axis perpendicular to the plane of Fig. 13-2. The peg is a distance A from the axis, and the wheel rotates with constant angular velocity ω. Light is directed down from above so that the peg casts a shadow on the horizontal plane (the x-axis in Fig. 13-2a).

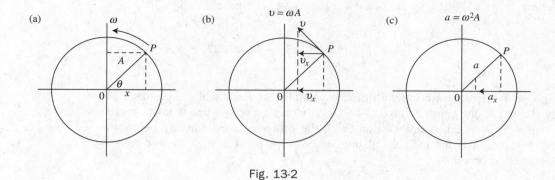

Fig. 13-2

At $t = 0$, the peg is all the way to the right and the shadow is at $x = A$. At a later time the position of the shadow is $x = A \cos \theta = A \cos \omega t$. The tangential velocity of the peg is of magnitude $A\omega$, and its projection on the x-axis is $v = -A\omega \sin \omega t$ as shown in Fig. 13-2b. The acceleration of the peg (centripetal) is $A\omega^2$ directed as shown in Fig. 13-2c. The projection of the acceleration on the x-axis is $a = -A\omega^2 \cos t$. Thus we see that the x position of the shadow exhibits SHM since the equations for x, v, and a are the same as obtained above. If instead of setting $t = 0$ when the shadow was all the way to the right, we had chosen a different starting point with $\omega t = \theta$, our equations would have included the phase angle θ.

From the above discussion you can see why ω is sometimes called the *angular velocity*, as well as the *angular frequency*.

13.4   Pendulum

A simple pendulum consists of a mass m suspended from a light string of length L as in Fig. 13-3. If the linear displacement s is measured along the arc, $F = ma$ becomes

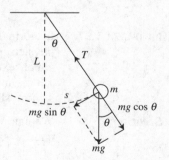

$$m\frac{d^2s}{dt^2} = -mg \, \sin \theta \qquad (13.10)$$

Since $s = L\theta$, this may be written

$$\frac{d^2\theta}{dt^2} = -\frac{g}{L}\sin \theta \simeq -\frac{g}{L}\theta \qquad (13.11)$$

where I made the approximation $\sin \theta \simeq \theta$ for small angles. This is of the same form as Eq. 13.1, so the solution is

Fig. 13-3

$$\theta = \theta_0 \cos (\omega t + \theta) \qquad (13.12)$$

θ_0 is the maximum angular displacement. The angular frequency and period are

$$\boxed{\omega = \sqrt{\frac{g}{L}} \quad \text{and} \quad T = \frac{2\pi}{\omega} = 2\pi\sqrt{\frac{L}{g}}} \qquad (13.13)$$

Somewhat surprisingly, the frequency depends only on g and on the length of the pendulum, not on its mass or on θ_0 (as long as the oscillations are small).

PROBLEM 13.6.   The period of a simple pendulum is 2 s. What will be the period if the mass and the length of the pendulum string are both doubled?

Solution   Changing the mass will have no effect. From Eq. 13.13,

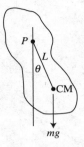

$$\frac{T_1}{T_2} = \sqrt{\frac{L_1}{L_2}}, \quad \text{so } T_2 = \sqrt{\frac{L_2}{L_1}}T_1 = \sqrt{2}T_1 = 2.8 \text{ s}$$

The **physical** (or **compound**) **pendulum** is a pendulum consisting of an extended rigid body pivoted at one point, P (Fig. 13-4). Gravity exerts a torque that tends to restore the object to its equilibrium position (with the center of mass directly below the point of support). If the moment of inertia about the pivot P is I, the equation of motion $\tau = I\alpha$ becomes

Fig. 13-4

$$-mgL \sin\theta = I\frac{d^2\theta}{dt^2} \qquad (13.14)$$

The minus sign indicates that the torque tends to decrease θ. If we again limit the oscillations to small angles, $\sin\theta \simeq \theta$. Equation 13.14 becomes

$$\frac{d^2\theta}{dt^2} = -\frac{mgL}{I}\theta = -\omega^2\theta \qquad (13.15)$$

This is the same as Eq. 13.1, and the solution is SHM, $\theta = \theta_0 \cos(\omega t + \theta)$. The period and angular frequency are

$$\omega = \sqrt{\frac{mgL}{I}} \quad \text{and} \quad T = \frac{2\pi}{\omega} = 2\pi\sqrt{\frac{I}{mgL}} \qquad (13.16)$$

PROBLEM 13.7.   In normal walking, the legs of a human or animal swing more or less freely like a physical pendulum. This observation has enabled scientists to estimate the speed at which extinct creatures such as dinosaurs traveled. If a giraffe has a leg length of 1.8 m, and a step length of 1 m, what would you estimate to be the period of its leg swing? How fast would it then travel when walking?

Solution   We can model the giraffe's leg as a physical pendulum of length L pivoted about one end.

$$I = \tfrac{1}{3}mL^2$$

for a rod pivoted at one end, from Table 10.2.

$$T = 2\pi\sqrt{\frac{mL^2}{3\,mgL}} = 2\pi\sqrt{\frac{L}{3g}} = 1.6 \text{ s} \quad v = \frac{\text{step length}}{\text{period}} = \frac{1 \text{ m}}{1.6 \text{ s}} = 0.6 \text{ m/s}$$

PROBLEM 13.8.   A physical pendulum consists of a sphere of mass m and radius R attached to the end of a string of length L and negligible mass. What is the period of this pendulum?

Solution   By the parallel-axis theorem, Eq. 10.9, the moment of inertia of the sphere about the point where the string is attached to the ceiling is

$$I = I_{CM} + m(R + L)^2 = \tfrac{2}{5}mR^2 + m(R + L)^2$$

From Eq. 13.16 the period is

$$T = 2\pi\sqrt{\frac{I}{mg(L + R)}} = 2\pi\sqrt{\frac{2R^2 + 5(R + L)^2}{5g(R + L)}}$$

In the limit $R \ll L$, this reduces to the period of a simple pendulum of length L. Note also that the period of this physical pendulum is longer than that of a simple pendulum of length L or of length $L + R$.

A **torsional pendulum** can be constructed by attaching a mass to a watch spring or by suspending a mass from a thin fiber that can exert a restoring torque when twisted. If the torque is proportional to the angular displacement, as is often true for small displacements, then $\tau = -\kappa\theta$, where κ is the torsional constant of the spring or fiber. Proceeding as above, we conclude that SHM results, with

$$\omega = \sqrt{\frac{\kappa}{I}} \tag{13.17}$$

Here I is the moment of inertia of the mass attached to the spring or fiber.

13.5   Damped Oscillations and Forced Oscillations

Real oscillators experience dissipative forces such as friction that damp the motion. Frequently such damping forces may be approximated by a term $-bv$ in the force equation. In this case the equation $F = ma$ becomes

$$m\frac{d^2x}{dt^2} = -kx - bv \tag{13.18}$$

The solution is

$$x = Ae^{-(b/2m)t}\cos(\omega t + \theta) \tag{13.19}$$

where

$$\omega = \sqrt{\frac{k}{m} - \left(\frac{b}{2m}\right)^2} = \sqrt{\omega_0^2 - \left(\frac{b}{2m}\right)^2} \tag{13.20}$$

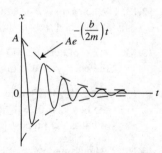

Fig. 13-5

Here $\omega_0 = \sqrt{k/m}$, the "natural frequency." Fig. 13-5 is a representative graph of x versus t. The exact shape of the curve depends on the size of the damping parameter b. In all cases there is exponential decay. Frequently an oscillator is driven with a force $F_0 \cos \omega t$. When this is done, it is found that the system oscillates at the driving frequency ω, but the displacement has a difference in phase from the driving force (Fig. 13-6). The amplitude of oscillation A is strongly dependent on how close ω is to the natural frequency ω_0. When $\omega = \omega_0$, the amplitude can become very large. This condition is called **resonance**. A large driving force at the resonant frequency can cause a large structure to collapse, as happened to the Tacoma Narrows bridge in 1940 due to vibrations driven by a light wind.

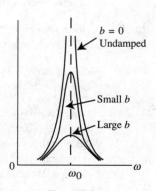

Fig. 13-6

13.6   Summary of Key Equations

SHM:
$$F = -kx$$
$$x = A\cos(\omega t + \theta) \qquad v = -A\omega\sin(\omega t + \theta) \quad a = -A\omega^2\cos(\omega t + \theta)$$
$$E = PE + KE = \frac{kA^2}{2} \quad PE = U = \frac{kx^2}{2} \quad \omega = \sqrt{\frac{k}{m}} = 2\pi f = \frac{2\pi}{T}$$

Pendulum:
$$\omega = \sqrt{\frac{g}{L}}$$

Damped oscillator: $x = Ae^{-(b/2m)t}\cos(\omega t + \theta)$

SUPPLEMENTARY PROBLEMS

13.9.   A 0.20-kg mass sliding on a horizontal plane is attached to a spring, stretched 0.12 m, and released from rest. After 0.40 s the speed of the mass is zero. What is the maximum speed of the mass?

13.10.   The piston in a car engine undergoes SHM, moving back and forth a distance of 0.084 m at 2400 rev/min. The piston mass is 1.25 kg. What are the maximum speed and acceleration of the piston? What is the maximum force on it?

13.11.   A spring rests on a plane inclined at angle θ above horizontal. The upper end of the spring is fixed, and the lower end is attached to a mass m that can slide on the plane without friction. What is the angular frequency for oscillations of the mass?

13.12.   A piston undergoes SHM with a frequency of 2 Hz. A coin rests on top of the piston. What is the maximum amplitude for which the coin will always remain in contact with the piston?

13.13.   When a mass m is attached to a particular spring, it oscillates with a period T. With what period will it oscillate if attached to two such springs connected side by side (in "parallel")?

13.14.   A car of mass 1200 kg oscillates on its springs at a frequency of 0.50 Hz with an amplitude of 0.04 m. What is the energy of this motion?

13.15.   A 0.24-kg block rests on a table, and a 0.12-kg block is placed on top of it. The coefficient of friction between the blocks is 0.25. The lower block is now moved back and forth horizontally in SHM with amplitude 0.05 m. What is the highest frequency for which the upper block will not slip with respect to the lower block?

13.16.   A mass undergoes SHM with amplitude A. What fraction of the energy is kinetic energy when $x = A/2$?

13.17.   A grandfather's clock is to be designed so that on every half swing a small gear is moved one notch to indicate the passage of 1 s (that is, the period of the simple pendulum is 2 s). What length pendulum is required?

13.18.   A student attempts to use a simple pendulum to measure g, the acceleration due to gravity. She observes that a pendulum of length 1.50 m makes 24 oscillations in 60 s. What is the value of g at her location?

13.19.   A disk of mass m and radius R is pivoted at a point on its perimeter and allowed to swing freely parallel to its plane. What is the period of such motion for small oscillations?

13.20.   An engineer wishes to determine the moment of inertia of a machine part of mass 1.20 kg about a particular axis (the X–X' axis) through the CM. He locates the CM by suspending the object motionless from several different points around its periphery. He then suspends the object from a pivot a distance 0.25 m from the CM and observes that it undergoes small oscillations with a period of 1.50 s about an axis parallel to the X–X' axis. What is the moment of inertia of the object about the X–X' axis?

13.21.   A damped simple harmonic oscillator is characterized by $m = 0.2\,\text{kg}$, $k = 80\,\text{N} \cdot \text{m}$, and $b = 0.072\,\text{kg/s}$. What is the period? How long does it take for the amplitude to decrease to half its original value?

13.22. A mass suspended from a spring is observed to oscillate at a frequency of 2 Hz. Now the spring is cut in half and the two pieces are placed side by side. The mass is now suspended from this composite spring consisting of the two smaller springs arranged in parallel. At what frequency will the mass now oscillate?

13.23. An object undergoing SHM has a maximum velocity of 25 m/s and a maximum acceleration of 1500 m/s^2. What is the amplitude and frequency of this motion?

13.24. A pendulum consists of a small brass weight attached to a string 1.20 m long. The pendulum is displaced 6° from vertical and released from rest. How long does it take for the pendulum to swing to a displacement of 3° on the opposite side?

13.25. Two springs of spring constants k_1 and k_2 are attached end to end (in "series"). What is the effective spring constant of this composite spring?

SOLUTIONS TO SUPPLEMENTARY PROBLEMS

13.9. $\quad T = 0.8\,\text{s} \qquad A = 0.12\,\text{m} \qquad v_{max} = A\omega = A\left(\dfrac{2\pi}{T}\right) = 0.94\,\text{m/s}$

13.10. $\quad f = \dfrac{2400}{60}\,\text{s}^{-1} = 40\,\text{s}^{-1} \qquad a_{max} = A\omega^2 = (0.042\,\text{m})(2\pi)^2(40\,\text{s}^{-1})^2 = 2650\,\text{m/s}^2$

$\qquad v_{max} = A\omega = 10.5\,\text{m/s} \qquad F_{max} = ma_{max} = (1.25\,\text{kg})(2650\,\text{m/s}^2) \qquad F_{max} = 3310\,\text{N}$

13.11. The force along the plane is $-ks + mg\sin\theta$, measuring displacements as positive down the plane, where $s = 0$ at equilibrium. The equation of motion is $m(d^2s/dt^2) = -ks + mg\sin\theta$. Let

$$s' = s - mg\sin\theta \qquad \text{and} \qquad m\frac{d^2s'}{dt^2} = -ks', \quad \text{so} \quad \omega = \sqrt{\frac{k}{m}}$$

13.12. If the coin leaves the piston, it will do so just as the piston is starting its downward motion, where its acceleration is a maximum. This acceleration should not exceed g if the coin is not to lose contact.

$$a_{max} = A\omega^2 = g, \quad A = \frac{g}{\omega^2} = \frac{9.8\,\text{m/s}^2}{(2\pi)^2(2\,\text{s}^{-1})^2} = 0.062\,\text{m}$$

13.13. If force F stretches one spring a distance x, it will require force $2F$ to stretch two parallel springs the same amount, so the effective spring constant is

$$k' = 2k \quad \text{and} \quad T' = 2\pi\sqrt{\frac{m}{k'}} = 2\pi\sqrt{\frac{m}{2k}} = \frac{1}{\sqrt{2}}T$$

13.14. From Eq. 13.9, $E = \frac{1}{2}kA^2$ and $\omega^2 = k/m$. Thus

$$E = \frac{1}{2}m\omega^2 A^2 = \frac{1}{2}(1200\,\text{kg})(2\pi)^2(0.50\,\text{s}^{-1})^2(0.04\,\text{m})^2, \quad E = 9.5\,\text{J}$$

13.15. Friction acts to move the upper block, so for that block, $m_1 a_1 = F_f = \mu m_1 g$; thus $a_1 = \mu g$. The acceleration of the lower block should not exceed this if the upper block is not to slip; therefore,

$$a_{max} = A\omega^2 = \mu g \quad \omega^2 = \frac{\mu g}{A} = \frac{(0.25)(9.8\,\text{m/s}^2)}{(0.05\,\text{m})} \quad \omega = 7\,\text{rad/s} \quad f = \frac{\omega}{2\pi} = 1.1\,\text{Hz}$$

13.16. From Eq. 13.9, the total energy is $E = \frac{1}{2}kA^2$. When $x = \frac{1}{2}A$, $\text{PE} = \frac{1}{2}kx^2 = \frac{1}{8}kA^2$. $\text{PE} + \text{KE} = E$, so $\frac{1}{8}kA^2 + \text{KE} = \frac{1}{2}kA^2$, and $\text{KE} = \frac{3}{8}kA^2 = \frac{3}{4}E$.

13.17. $\quad T = 2\pi\sqrt{\dfrac{L}{g}} \qquad L = \left(\dfrac{T}{2\pi}\right)^2 g = \left(\dfrac{2\mathrm{s}}{2\pi}\right)^2 (9.8\,\mathrm{m/s}^2) = 0.99\,\mathrm{m}$

13.18. $\quad T = \dfrac{60\,\mathrm{s}}{24} = 2.5\,\mathrm{s} \qquad T = 2\pi\sqrt{\dfrac{L}{g}} \qquad g = L\left(\dfrac{2\pi}{T}\right)^2 = 9.47\,\mathrm{m/s}^2$

13.19. From the parallel-axis theorem, Eq. 10.9, $I = I_{CM} + mR^2$. From Table 10.2, $I_{CM} = \frac{1}{2}mR^2$, so $I = \frac{3}{2}mR^2$. From Eq. 13.16,

$$T = 2\pi\sqrt{\frac{I}{mgL}} = 2\pi\sqrt{\frac{3/2mR^2}{mgR}} = 2\pi\sqrt{\frac{3R}{2g}}$$

13.20. $\quad T = 2\pi\sqrt{\dfrac{I}{mgL}} \qquad I = mgL\left(\dfrac{T}{2\pi}\right)^2 = (1.2\,\mathrm{kg})(9.8\,\mathrm{m/s}^2)(0.25\,\mathrm{m})\left(\dfrac{1.5\,\mathrm{s}}{2\pi}\right)^2 = 0.17\,\mathrm{kg\cdot m}^2$

$\quad I = I_{CM} + mL^2, \quad \text{so } I_{CM} = I - mL^2 = 0.17\,\mathrm{kg\cdot m}^2 - (1.2\,\mathrm{kg})(0.25\,\mathrm{m})^2 = 0.10\,\mathrm{kg\cdot m}^2$

13.21. From Eq. 13.18,

$$A = A_0 e^{-(b/2m)t} = \frac{1}{2}A_0, \quad \text{so } e^{-(b/2m)t} = 2 \quad \text{and} \quad \ln^{-(b/2m)t} = \ln 2$$

$$\frac{bt}{2m} = \ln 2, \quad \text{since } \ln e = 1 \quad \text{and} \quad t = \frac{2m\ln 2}{b} = \frac{(2)(0.2\,\mathrm{kg})(\ln 2)}{0.072\,\mathrm{kg/s}} = 3.9\,\mathrm{s}$$

$$\frac{k}{m} = \frac{80\,\mathrm{N\cdot m}}{0.20\,\mathrm{kg}} = 400\,\mathrm{s}^{-2} \qquad \left(\frac{b}{2m}\right)^2 = \left(\frac{0.072\,\mathrm{kg/s}}{(2)(0.2\,\mathrm{kg})}\right)^2 = 0.03\,\mathrm{s}^{-2}$$

$$\text{so } \frac{k}{m} \gg \left(\frac{b}{2m}\right)^2 \quad \text{and} \quad \omega = \sqrt{\frac{k}{m} - \left(\frac{b}{2m}\right)^2} \simeq \sqrt{\frac{k}{m}} \qquad f = \frac{\omega}{2\pi} = \frac{1}{2\pi}\sqrt{\frac{k}{m}} = 3.18\,\mathrm{Hz}$$

13.22. When a spring is cut in half the spring constant of each half is double that of the original spring. You can see this by recognizing that when a given force is applied to the original spring, and also to the smaller piece, the separation between coils will be the same for both springs, but the total extension will be only half as great for the shorter spring. This means that its spring constant is twice as large as that for the long spring. When the two small springs are connected side by side, their effective spring constant is double that of one of these springs. Thus if k is the spring constant of the original spring, $4k$ is the effective spring constant of the two halves placed side by side.

$$\omega_1^2 = 2\pi f_1 = \sqrt{\frac{k}{m}} \qquad \omega_2^2 = 2\pi f_2 = \sqrt{\frac{4k}{m}} = 2\sqrt{\frac{k}{m}}$$

Thus,

$$f_2 = 2f_1 = (2)(2\,\mathrm{Hz}) = 4\,\mathrm{Hz}$$

13.23. $\quad v_{max} = A\omega$

$\quad a_{max} = A\omega^2$

$\quad \omega = \dfrac{v_{max}}{a_{max}} = \dfrac{1500\,\mathrm{m/s}^2}{25\,\mathrm{m/s}} = 60\,\mathrm{rad/s}$

$\quad \omega = 2\pi f, \quad f = \dfrac{\omega}{2\pi} = \dfrac{60\,\mathrm{rad/s}}{2\pi} = 9.55\,\mathrm{Hz}$

$\quad A = \dfrac{v_{max}}{\omega} = \dfrac{25\,\mathrm{m/s}}{60\,\mathrm{rad/s}} = 0.42\,\mathrm{m}$

13.24. $\theta = \theta_0 \cos{(\omega t + \varphi)}$

Take $\varphi = 0$ and $\theta_0 = 6°$

We wish to find t when $\theta = -3°$

$\theta = -3° = 6° \cos{\omega t}$

$\cos{\omega t} = -\dfrac{3°}{6°} = -0.5$

So, $\omega t = 120°$

$\omega = \sqrt{\dfrac{g}{L}} = \left(\dfrac{9.8 \text{ m/s}^2}{1.2 \text{ m}}\right)^{\frac{1}{2}} = 2.86 \text{ s}^{-1}$

$\omega = 2\pi f = \dfrac{2\pi}{T} \quad T = \dfrac{2\pi}{\omega} = \dfrac{2\pi}{2.86} \quad \text{s} = 2.20 \text{ s}$

$\omega t = 120°$ is one-third of a period,

So, $t = \dfrac{T}{3} = \dfrac{2.20 \text{ s}}{3} \qquad t = 0.73 \text{ s}$

13.25. Note that if force F is applied to the springs, each spring experiences this force. If the two springs stretch an amount x, $x = x_1 + x_2$ where

$x_1 = $ distance spring one stretches
$x_2 = $ distance spring two stretches
$F = k_{\text{eff}}^x$
$F = k_1 x_1$
$F = k_2 x_2$
$x = x_1 + x_2 = \dfrac{F}{k_1} + \dfrac{F}{k_2}$

$F = k_{\text{eff}} x_1 = k_{\text{eff}}(x_1 + x_2)$

$F = k_{\text{eff}}\left(\dfrac{F}{k_1} + \dfrac{F}{k_2}\right)$

$1 = k_{\text{eff}}\left(\dfrac{1}{k_1} + \dfrac{1}{k_2}\right)$

$k_{\text{eff}} = \dfrac{k_1 k_2}{k_1 + k_2}$

Gravity

Gravity is one of the fundamental forces of nature. The other forces discussed previously, such as friction, tension, and the normal force, are derived from the electric force, another of the fundamental forces. Gravity is a rather weak force unless the objects involved are large, like planets or stars. The electric force between two protons is much, much stronger than the gravitational force between them. Most of what we know about gravity came from studies of the motion of the Moon and the planets.

14.1   The Law of Gravity

Isaac Newton deduced that two particles of masses m_1 and m_2, separated by a distance r, will attract each other with a force

$$F = G\frac{m_1 m_2}{r^2}$$

(14.1)

This is the **universal law of gravity**. The same law applies to two spherically symmetric masses, when the mass is treated as if it were all concentrated at the center of the sphere. The universal gravity constant G is $6.67 \times 10^{-11}\,\text{N} \cdot \text{m}^2/\text{kg}^2$.

PROBLEM 14.1.   A man of mass 80 kg stands 2 m from a woman of mass 50 kg. What gravitational force acts on each?

Solution   By Newton's third law, the same magnitude force acts on each.

$$F = (6.67 \times 10^{-11}\,\text{N} \cdot \text{m}^2/\text{kg}^2)\frac{(80\,\text{kg})(50\,\text{kg})}{(2\,\text{m})^2} = 6.7 \times 10^{-8}\,\text{N}$$

Note that an 80-kg person has weight $W = mg = 784\,\text{N}$, so the gravitational force of attraction here is much smaller than a person's weight.

The force of gravity on a person of mass m due to the Earth's mass M_E is called the person's weight. Thus

$$W = mg = G\frac{mM_E}{r^2}, \quad \text{so} \quad \boxed{g = G\frac{M_E}{r^2}}$$

(14.2)

At the Earth's surface, $r \simeq R_E$ and $g = 9.80\,\text{m/s}^2$. Equation 14.2 shows that g decreases as we go away from the Earth. We can measure g and figure out R in various ways and thereby determine the mass of the Earth. Amazing!

PROBLEM 14.2. A 100-kg astronaut weighs 980 N on Earth. How much would he weigh on the Moon (mass 7.36×10^{22} kg, radius 1.74×10^6 m)?

Solution On the Moon,

$$g_m = G\frac{M_M}{R_M^2} = (6.67 \times 10^{-11})\frac{(7.36 \times 10^{22})}{(1.74 \times 10^6)} = 1.62\,\text{m/s}^2 \simeq \frac{1}{6}g$$

So

$$W = mg_m = (100\,\text{kg})(1.62\,\text{m/s}^2) = 162\,\text{N}$$

PROBLEM 14.3. How far from the Earth, in terms of the Earth–Moon separation d, should a satellite be positioned if it is to experience no net gravitational force from the Earth and the Moon? The Moon's mass is $M = 0.012\,M_E$.

Solution $$G\frac{M_E m}{x^2} = G\frac{M_M m}{(d-x)^2}, \quad \text{so } (d-x)^2 = \frac{M_M}{M_E}x^2$$

Take the square root,

$$d - x = \pm\sqrt{\frac{M_M}{M_E}}x + \text{applies since } x < d$$

$$x = \frac{1}{1 + \sqrt{M_M/M_E}}d$$

$$= \frac{1}{1 + \sqrt{0.012}}d = 0.90d$$

An interesting and useful result (that I won't prove mathematically) is that the gravity force vanishes on an object placed inside a spherically symmetric shell of material. Outside a spherically symmetric mass distribution, the gravity force acts as if all of the mass were concentrated at the center of the sphere.

PROBLEM 14.4. What is the gravitational force on a particle of mass m at a point within the earth, that is, where $r < R_E$?

Solution Break the Earth into two parts: a spherical shell extending from r to R_E and an inner sphere of radius r and mass M'. Since the particle is inside the spherical shell, it feels no gravity force due to the shell. Due to the inner sphere the particle experiences a force of

$$F = G\frac{mM'}{r^2}, \quad \text{where } \frac{M'}{M_E} = \frac{4/3\,\pi r^3}{4/3\,\pi R_E^3}$$

Thus

$$F = G\frac{mM_E}{R_E^3}r \tag{14.3}$$

14.2 Gravitational Potential Energy

Gravity is a conservative force, and we may define a potential energy associated with it. Recall that the work you must do to lift a mass m from one point to another is equal to the gain in potential energy. Work is done against gravity only when the displacement is radial. Going sideways to r requires no work. Suppose a mass m starts a distance r_1 from a mass M and we lift it out to $r_2 > r_1$. The work done is the gain in potential energy $U(r)$.

$$W = U(r_2) - U(r_1) = \int_{r_1}^{r_2} F \, dr = \int_{r_1}^{r_2} G \frac{mM}{r^2} \, dr = -GmM\left(\frac{1}{r_2} - \frac{1}{r_1}\right) \qquad (14.4)$$

The zero point of potential energy is arbitrary. Only differences in potential energy matter. If we choose $U = 0$ at $r = \infty$, then the **gravitational potential energy** is

$$\boxed{U = -G\frac{mM}{r}} \qquad (14.5)$$

Frequently, we are interested in objects a small distance h above the surface of the Earth, where $h \ll R_E$. If we take $r_1 = R_E$ and $r_2 = R_E + h$, Eq. 14.4 becomes

$$\Delta U = U(R_E + h) - U(R_E) = -GM_E m\left(\frac{1}{R_E + h} - \frac{1}{R_E}\right)$$

$$= -GM_E m\left(\frac{R_E - R_E - h}{R_E^2 - hR_E}\right) \simeq m\frac{GM_E}{R_E^2} h = mgh$$

Here I neglected $hR \ll R^2$ and used Eq. 14.2 for g. If we choose $U = 0$ at $r = R_E$, then

$$\boxed{U = mgh} \qquad (14.6)$$

This approximation is useful near the surface of the Earth.

PROBLEM 14.5.   Determine the minimum speed that a rocket at the surface of the Earth must have to escape completely the Earth's gravity field.

Solution   The rocket's energy is conserved. We want the rocket to go far away ($r = \infty$), and barely reach there ($v = 0$). Thus $U_1 + \text{KE}_1 = 0 + 0$.

$$-G\frac{M_E m}{R_E} + \frac{1}{2} mv^2 = 0, \quad \text{so } v_{\text{esc}} = \sqrt{\frac{2GM_E}{R_E}} \qquad (14.7)$$

v_{esc} is the **escape velocity**. Using the appropriate values for R and M, the escape velocity from the sun (that is, from the solar system) or from another planet could be found. Calculation shows that the escape velocity from the Earth is 11.2 km/s and from the solar system, 618 km/s.

14.3   The Motion of Planets

Much of what we know about gravity resulted from observing the motion of the planets. Using the observational data of Tycho Brahe (1546–1601), Johannes Kepler (1571–1630) deduced the following empirical laws (**Kepler's laws**):

1.   The planets move in elliptical orbits with the Sun at a focal point of the ellipse.
2.   The radius vector drawn from the sun to a planet sweeps out equal areas in equal time intervals.
3.   The square of the orbital period of a planet is proportional to the cube of the semimajor axis of the elliptical orbit.

Newton postulated the universal law of gravity (Eq. 14.1) and proceeded to solve $F = ma$, thereby obtaining the mathematical description of the orbits as ellipses with the Sun at a focal point (Kepler's first law). This is slightly complicated, and I won't reproduce his derivation here. Law 2 follows from the observation that gravity is a **central force** and thus exerts no torque. This is so because the gravity force is directed along the line joining the two masses and hence has zero lever arm. We have seen that if no torque acts, the angular momentum of the system remains constant. Since the Sun is so much more massive than a planet, it remains essentially stationary and the planet goes around it with angular momentum $\mathbf{L} = \mathbf{r} \times m\mathbf{v} =$ constant. In time dt the planet has displacement $d\mathbf{r} = \mathbf{v}\,dt$ and sweeps out an area dA (Fig. 14-1). But the magnitude of $\mathbf{r} \times \mathbf{v}\,dt$ is $rv\,dt\,\sin\theta$, the area of the parallelogram formed by $\mathbf{r}$ and $\mathbf{v}\,dt$. dA is just half of this, so

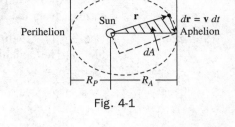

Fig. 4-1

$$dA = \frac{1}{2}|\mathbf{r} \times \mathbf{v}\,dt| = \frac{L}{2m}\,dt$$

or

$$\frac{dA}{dt} = \frac{L}{2m} = \text{constant} \tag{14.8}$$

Here

$$m = \text{mass of the planet}$$

$$L = \text{orbital angular momentum of the planet}$$

L and m are both constants, so dA/dt is constant, and equal areas are swept out in equal times (Kepler's second law).

All of the planet orbits, except those of Mars, Mercury, and Pluto, are almost circular. For the case of a circular orbit one can prove Kepler's third law as follows: The gravity force provides the needed centripetal force to move a planet in a circle. Thus

$$G\frac{M_3 m}{r^2} = \frac{mv^2}{r}$$

The orbital speed is $v = 2\pi r/T$, where T is the period of the motion.

$$G\frac{M_3}{r^2} = \frac{(2\pi r/T)^2}{r} \quad \text{or} \quad T^2 = \left(\frac{4\pi^2}{GM_3}\right)r^3 \quad \text{and} \quad v_{\text{orbit}} = \sqrt{\frac{GM_3}{r}} \tag{14.9}$$

Thus Kepler's third law is proved for circular orbits. It can be shown that the same equation is true for elliptical orbits, provided r is replaced by the semimajor axis distance.

The mathematical solutions for possible orbits under the gravity force are ellipses, hyperbolas, and parabolas. It is possible to show that for a bound orbit (an ellipse) the total energy is negative. That is, the potential energy (which is negative in a bound orbit) has a larger magnitude than does the kinetic energy (which is positive). To illustrate this, consider a circular orbit. Gravity provides the centripetal force, so

$$G\frac{Mm}{r^2} = \frac{mv^2}{r} \quad \text{and} \quad E = \frac{1}{2}mv^2 - G\frac{Mm}{r}$$

Substitute the first equation in the second and obtain

$$E = -G\frac{Mm}{2r} \tag{14.10}$$

Thus the total energy is negative, and the magnitude of the potential energy is twice that of the kinetic energy. The same expression applies to elliptical orbits, but r is replaced by the value of the semimajor axis.

PROBLEM 14.6. A satellite is in a circular orbit around the Earth at a radius of $2R_E$. How much energy is required to move it out to an orbit of radius $4R_E$?

Solution

$$\Delta E = -\frac{GM_E m}{2(4R_E)} - \left(-\frac{GM_E m}{2(2R_E)}\right) = \frac{GM_E m}{8R_E}$$

PROBLEM 14.7.   Comet Halley orbits the sun with a period of 76 years. Its perihelion distance (when it is closest to the Sun) is 8.9×10^{10} m. What is the comet's aphelion distance (when it is farthest from the Sun)?

Solution   Use Eq. 14.9, with r replaced by a, the semimajor axis length of the elliptical orbit, to find a. The aphelion distance R_A and the perihelion distance R_P are related to a by $R_A + R_P = 2a$.

$$a = \left(\frac{GMT^2}{4\pi^2}\right)^{1/3}$$

Substitute $M_{\text{sun}} = 1.99 \times 10^{30}$ kg, $T = 76\,\text{y} = 2.4 \times 10^9$ s. Find $a = 2.7 \times 10^{12}$ m:

$$R_A = 2a - R_P = 2(2.7 \times 10^{12}\,\text{m}) - 8.9 \times 10^{10}\,\text{m} = 5.3 \times 10^{12}\,\text{m}$$

14.4   Summary of Key Equations

$$F = G\frac{m_1 m_2}{r^2} \qquad W = mg \quad g = G\frac{M_E}{r^2}$$

$$U = -G\frac{m_1 m_2}{r} \qquad \text{with } U(\infty) = 0$$

$$U \simeq mgh \qquad \text{with } U(R_E) = 0$$

$$E = -G\frac{Mm}{2r} \qquad \text{Circular orbit}$$

$$T^2 = \left(\frac{4\pi^2}{GM_3}\right)r^3 \qquad \text{Planetary orbit}$$

SUPPLEMENTARY PROBLEMS

14.8.   People who believe in astrology imagine that the Moon, planets, and stars have an effect on each of us. The only known way for such objects to influence us is through the gravity force. Calculate the force exerted by the Moon on a person of mass 50 kg. The Moon is 3.8×18^8 m from the Earth, and its mass is 7.4×10^{22} kg. For comparison, estimate the weight of a paper clip of mass 0.10 g.

14.9.   What is the weight of a 70 kg person at the surface of the Earth? What is the weight and mass of such a person at a point two earth radii from the center of the Earth?

14.10.   It is difficult to measure the universal gravitational constant in the laboratory because the gravity force between terrestrial objects is so small. To confirm this, calculate the gravitational force between two 2-kg masses separated by 1 m.

14.11.   Engineers have proposed building a rail transit system using the following idea: A straight tunnel would be drilled between two cities, for example, Boston and Washington, D.C. The tunnel would be evacuated, and the cars could be magnetically levitated so that frictional drag is negligible. If there were no frictional losses, a car released from rest at one end of the tunnel would fall freely for half of the journey and then coast up to the exit at the opposite end. Show that such a car is executing simple harmonic motion. Determine the time to go from one end of the tunnel to the other. You will find that this time is independent of the length of the tunnel.

14.12.   Mass m is divided into two parts, xm and $(1 - x)m$. For a given separation, what value of x will maximize the gravitational attraction between the two pieces?

14.13. A rocket engine burns out at an altitude h above the Earth's surface, at which point its speed is v_0, where v_0 exceeds the escape velocity v_{esc} at the burnout altitude. Determine the speed of the rocket far from Earth in terms of v_{esc} and v_0.

14.14. When nuclear chain reactions were first created in connection with the development of the atom bomb, there was some concern that a chain reaction could be set off that would totally blow apart the Earth. Calculate the energy needed to totally explode the Earth, that is, to separate all of the mass out to infinity. (*Hint:* Consider the work required to peel layers off the Earth like layers of skin on an onion.)

14.15. An experimental rocket burns up all of its fuel when it is at an altitude of 180 km above the earth's surface, at which point it is moving vertically at 8.2 km/s. How high above the Earth's surface will it rise? Ignore friction.

14.16. Geosynchronous satellites are placed in an orbit in the equatorial plane such that they are always above a given point on the Earth's surface; that is, their period is 24 h. How far from the center of the Earth is such a satellite?

14.17. Three identical stars each of mass m positioned at the vertices of an equilateral triangle of side a move in a circular orbit. For each of the stars the needed centripetal force is provided by the gravitational force due to the other two stars. What is the period of such a star system?

14.18. A certain binary star consists of two identical stars of mass M, each moving in a circular orbit of radius R about the center of mass. Determine (a) the gravitational force exerted on each star, (b) the orbital speed of each star, (c) the period of the motion, and (d) the energy needed to separate the stars to infinity.

14.19. An Earth satellite in low orbit experiences a drag force due to the Earth's atmosphere, and this reduces the mechanical energy of the satellite. We are used to thinking of the friction force as slowing down moving objects, but a satellite actually speeds up as a result of the presence of friction. To see how this happens, consider a satellite in circular orbit of radius r. Because of friction, the orbit radius decreases very slightly to $r - \Delta r$. Show that this results in an increase in speed of $\Delta v = (\Delta r/2)\sqrt{GM/r^3}$, where M is the mass of the Earth. Calculate the work done by friction, the change in kinetic energy, and the change in potential energy in this process.

14.20. The Moon goes around the Earth every 27.3 days. The mass of the Earth is 5.98×10^{24} kg. Determine the Earth–Moon distance.

14.21. Two planets of masses M and m are initially at rest and very far apart. They fall toward each other. What is the speed of one relative to the other when they are a distance d apart?

14.22. Geosynchronous satellites play an important role in communications systems. Such satellites circle the Earth in a circular orbit in the equatorial plane with a period equal to that of the Earth's rotation. Thus they always stay directly above a certain point on the Earth's surface. Such satellites have many uses, including reflecting television signals back to Earth. How high above the surface of the Earth should a synchronous satellite be placed? Take one day to be 8.64×10^4 s.

14.23. A mass m is placed at the origin and a second mass $2m$ is placed on the x-axis at $x = L$. A third mass M is placed between them at position x. Determine x such that its presence doubles the gravitational force on each of the other two masses.

SOLUTIONS TO SUPPLEMENTARY PROBLEMS

14.8. $F = G\dfrac{Mm}{r^2} = \dfrac{(6.65 \times 10^{-11}\,\text{N}\cdot\text{m}^2/\text{kg}^2)(7.4 \times 10^{-4}\,\text{kg})(50\,\text{kg})}{(3.8 \times 10^8\,\text{m})^2} = 0.0017\,\text{N}$

For a paper clip

$$W = mg = 9.8 \times 10^{-4}\,\text{N} \simeq 0.001\,\text{N}$$

14.9. If $g = 9.8 \, \text{m/s}^2$ at the surface of the Earth,

$$g' = G\frac{M_E}{r^2} = G\frac{M_E}{(2R_E)^2} = \frac{1}{4}\frac{GM_E}{R_E^2} = \frac{1}{4}g$$

At the Earth's surface, a person weighs $W = mg = (70 \, \text{kg})(9.8 \, \text{m/s}^2) = 696 \, \text{N}$. At $r = 2R_E$, $W' = mg' = 176 \, \text{N}$.

14.10. $F = G\dfrac{m^2}{r^2} = \dfrac{(6.67 \times 10^{-11} \, \text{N} \cdot \text{m}^2/\text{kg}^2)(2 \, \text{kg})^2}{(1 \, \text{m})^2} = 2.7 \times 10^{-10} \, \text{N}$

14.11. $F_x = m\dfrac{d^2x}{dt^2} = -F_G \cos\theta$
where from Eq. 14.3,

$$F = G\frac{M_E m}{R_E^3}r$$

$$\cos\theta = \frac{x}{r}$$

so

$$m\frac{d^2x}{dt^2} = -k'x \qquad\qquad\qquad (i)$$

where

$$k' = G\frac{M_E m}{R_E^3}$$

Equation (i) is the same as Eq. 13.1, which results in simple harmonic motion with

$$\omega = \frac{2\pi}{T} = \sqrt{\frac{k'}{m}}$$

Travel from one end to the other requires time $t = T/2$.

$$T = 2\pi\sqrt{\frac{m}{k'}} = 2\pi\sqrt{\frac{R_E^3}{GM_E}} \cong 5060 \, \text{s}, \qquad t = 2530 \, s \cong 42 \, \text{min}$$

14.12. $F = G\dfrac{(xm)(1-x)m}{r^2}$
For maximum

$$\frac{d}{dx}(1-x)x = 0 \quad \text{and} \quad 1 - 2x = 0$$

so

$$x = \frac{1}{2} \quad \text{and} \quad \frac{d^2}{dx^2}(1-x)x = -2$$

14.13. Energy conserved: $\dfrac{1}{2}mv_0^2 - G\dfrac{M_E m}{R_E + h} = \dfrac{1}{2}mv^2 - 0 \quad$ as $r \to \infty$
From Eq. 13.7,

$$v_{\text{esc}} = \sqrt{\frac{2GM_E}{R_E}}$$

to escape from Earth's surface, so to escape from position $R = R_E + h$,

$$v_{\text{esc}} = \sqrt{\frac{2GM_E}{R_E + h}}, \quad \text{so } v^2 = v_0^2 - v_{\text{esc}}^2$$

14.14. KE is zero initially and finally, so work done = gain in PE. PE of a shell is $dU = -G(M/\tau)\,dm$, where the mass of outer shell of thickness dr is $4\pi r^2\,dr\,\rho$. Here ρ = density. M = mass of sphere inside the shell = $\frac{4}{3}\pi r^3\rho$. Thus total work required is

$$W = \int_0^{R_E} \frac{GM}{r}\,dm = \int_0^{R_E} \frac{G}{r}\left(\frac{4}{3}\pi r^3\rho\right)(4\pi r^2\rho\,dr) = \frac{(4\pi\rho)^2 G}{15} R_E^5$$

$$M_E = \frac{4\pi}{3} R_E\rho, \quad \text{so } W = \frac{3}{5}\frac{GM_E}{R_E}$$

14.15. $\frac{1}{2}mv_1^2 - G\dfrac{M_E m}{r_1} = 0 - G\dfrac{M_E m}{r^2}$

Solve for r_2, and find $r_2 = 183\,\text{km}$.

14.16. From Eq. 14.9,

$$T^2 = \left(\frac{4\pi^2}{GM_E}\right)r^3$$

Substitute $T = (24\,\text{h})(3600\,\text{s/h})$. Find $r = 42{,}300\,\text{km}$.

14.17. The gravitational force acting toward the center is

$$2F_G \cos 30° = \frac{mv^2}{r} \quad \text{and} \quad F_G = G\frac{m^2}{a^2}$$

$$r = \frac{2}{\sqrt{3}}a, \quad T = \frac{2\pi r}{v} = \sqrt{\frac{8a^3\pi^2}{3Gm}}$$

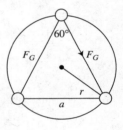

14.18. (a) $F = G\dfrac{M^2}{(2R)^2} = \dfrac{GM^2}{4R^2}$ (b) $\dfrac{Mv^2}{R} = \dfrac{GM^2}{4R^2}$ $v = \sqrt{\dfrac{GM}{4R}}$

(c) $T = \dfrac{2\pi R}{v} = 4\pi\sqrt{\dfrac{R^3}{GM}}$ (d) $E = -U = G\dfrac{M^2}{R}$

14.19. From Eq. 14.9, the orbital velocity is

$$v = \sqrt{\frac{GM}{r}}$$

Thus

$$dv = -\frac{1}{2}\sqrt{\frac{GM}{r^3}}\,dr \quad \text{if } dr < 0, \quad dv > 0$$

Initially $E = K + U$ and $-dE = dW$ = work against friction.

$$K = \frac{1}{2}mv^2 = \frac{1}{2}\frac{GMm}{r}, \quad \text{so } dK = \frac{GMm}{2r^2}\,dr \quad \text{and} \quad U = -\frac{GMm}{r}$$

$$dU = \frac{GMm}{r^2}\,dr = 2|dk|, \quad dE = dK + dU = -\frac{GMm}{2r}\,dr + \frac{GMm}{r}\,dr = \frac{GMm}{r}\,dr$$

So

$$dW = -\frac{Gmm}{r}\,dr \quad (\text{remember, } dr < 0)$$

For example, if a decrease in r causes KE to increase by 1 J, U decreases by 2 J, and the total energy E decreases by 1 J. Work against friction is thus 1 J. This 2:1 ratio for $|dU|$ and $|dK|$ is true in general for all orbits.

14.20. From Eq. 14.9,

$$T^2 = \left(\frac{4\pi^2}{GM}\right)r^3$$

Find $r = 3.83 \times 10^8$ m.

14.21. No external force acts, so CM doesn't move. In the CM reference frame, momentum is conserved, so $Mv = mv$. Energy is conserved and initially $E = 0$ (when $r = \infty$).

$$0 = \frac{1}{2}MV^2 + \frac{1}{2}mv^2 - G\frac{Mm}{d}$$

Solve and find

$$v = \sqrt{\frac{2GM^2}{(m+M)d}} \qquad V = \sqrt{\frac{2Gm^2}{(m+M)d}}$$

The relative velocity is

$$r_{12} = v + V = \sqrt{\frac{2G}{(m+M)d}}\,(m+M)$$

14.22. $T^2 = \left(\frac{4\pi^2}{GM}\right)r^3$, $\quad r = \left(\frac{GMT^2}{4\pi^2}\right)^{\frac{1}{3}}$

$$r = \left[\frac{(6.67 \times 10^{-11}\,\text{N}\cdot\text{m}^2/\text{kg}^2)(5.97 \times 10^{24}\,\text{kg})(8.64 \times 10^4\,\text{s})}{4\pi^2}\right]^{\frac{1}{3}}$$

$r = 4.23 \times 10^7$ m

Distance above Earth's surface is

$h = r - R_E = 4.23 \times 10^7\ - 0.67 \times 10^7$ m

$h = 3.59 \times 10^7$ m 22,300 mi

14.23. $G\dfrac{2m^2}{L^2} + G\dfrac{mM}{x^2} = 2G\dfrac{2m^2}{L^2}$ (1) For mass m

$G\dfrac{2m^2}{L^2} + G\dfrac{2mM}{(L-x)^2} = 2G\dfrac{2m^2}{L^2}$ (2) For mass $2m$

These are two simultaneous equations for the two unknowns x and M. Solve the first equation for M and substitute the result into the second equation.

$$\frac{M}{x^2} = \frac{m}{L^2} \tag{1}$$

$$\frac{2M}{(L-x)^2} = \frac{m}{L^2} \tag{2}$$

So,

$$\frac{M}{x^2} = \frac{2M}{(L-x)^2}$$

$$(L-x)^2 = 2x^2$$

$$L - x = \pm\sqrt{2}x$$

$$L = (1 \pm \sqrt{2})x$$

We want the solution $x < L$, so $x = L/(1 + \sqrt{2})$

$$x = 0.41\,L$$

CHAPTER 15

Fluids

15.1 Pressure in a Fluid

Matter occurs in three phases: solid, liquid, and gas. A solid has a definite shape and volume. Liquids and gases are **fluids**. When a fluid is placed in a container, it assumes the shape of the container. A gas is distinguished from a fluid in that it fills the entire container, whereas the liquid forms a definite interface (as when water half fills a bottle, leaving an air–water interface).

Pressure plays an important role in the behavior of fluids. When a force of magnitude F acts normal to a surface of area A, the pressure on the surface is defined as

$$P = \frac{F}{A} \tag{15.1}$$

More precisely, pressure is the force per unit area when the area A is vanishingly small. Pressure is measured in newtons per square meter, where $1\,\text{N/m}^2 = 1$ pascal (Pa). Pressure is also measured in pounds per square inch (lb/in or psi), where $1\,\text{psi} = 6.9 \times 10^3$ Pa; in atmospheres (the pressure due to the atmosphere at sea level), where $1\,\text{atm} = 1.01 \times 10^5\,\text{Pa} = 14.7\,\text{psi}$; the bar, where $1\,\text{bar} = 10^5$ Pa; and the torr, where $1\,\text{torr} = 133$ Pa. Note that pressure is a scalar quantity, not a vector. In a fluid at rest the pressure is the same in all directions at a given point.

An important property of matter is its **mass density** ρ, defined as the mass per unit volume. If mass m occupies a small volume V, then

$$\rho = \frac{m}{V} \tag{15.2}$$

The density of water is $1000\,\text{kg/m}^3$. Solids and liquids have comparable densities, but the density of gas is normally much less than that of a solid or liquid.

The pressure in a liquid can be calculated by considering the open container in Fig. 15-1. Consider an imaginary cylinder of fluid of height h and area A. Pushing down on the top surface is atmospheric pressure P_0, and pushing up on the bottom of the cylinder is P. The piece of fluid is in equilibrium, so $F_{\text{up}} = F_{\text{down}}$. The weight of the fluid is mg, so $PA = P_0 A + mg$, where $m = \rho V = \rho A h$ (since $V = Ah$). Thus

$$PA = P_0 A + \rho g h A \quad \text{or} \quad \boxed{P = P_0 + \rho g h} \tag{15.3}$$

The pressure due to the fluid alone is $\rho g h$, and it depends *only on the depth below the surface, not on the shape or size of the cylinder*. In Fig. 15-2 several containers of water are illustrated. The pressure at point P is the same in each case, since point P is at the same depth below the surface in each container.

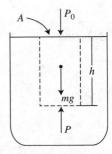

Fig. 15-1

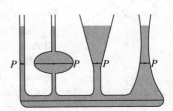

Fig. 15-2

Pressure in excess of atmospheric pressure is called **gauge pressure**: $P_g = P - P_0$. If you use a pressure gauge to measure the pressure in a flat tire, you get a reading of zero for the gauge pressure. The absolute pressure is still P_0. If P_0 is increased, P is increased at every point in the fluid. Thus **any change in pressure is transmitted undiminished throughout the fluid (Pascal's principle)**.

Liquids are virtually incompressible, so their density does not change with depth. Equation 15.3 applies to them. The pressure in a gas can also be deduced, using the line of reasoning above. However, gases are easily compressed, so density is a function of depth, and we must recognize this in calculating the mass contained in a vertical cylinder. This can be done by considering thin layers of the gas and integrating to find the total mass in the cylinder. As for liquids, the pressure increases with depth, but not in a linear fashion.

PROBLEM 15.1. What depth of (a) water (density $1000 \, \text{kg/m}^3$) and of (b) mercury (Hg, density $13,600 \, \text{kg/m}^3$) is required to produce a pressure of 1 atm ($1.01 \times 10^5 \, \text{N/m}^2$)?

Solution

(a) $P = \rho g h = (10^3 \, \text{kg/m}^3)(9.8 \, \text{m/s}^2)h = 1.01 \times 10^5 \, \text{N/m}^2$; $h = 10.3 \, \text{m}$.

(b) $P = (13,600 \, \text{kg/m}^3)(9.8 \, \text{m/s}^2)h = 1.01 \times 10^5 \, \text{N/m}^2$; $h = 0.76 \, \text{m}$.

We sometimes encounter pressures measured in "millimeters of mercury" or in "feet of water." Using the result of Problem 15.1, we can convert such units to pascals. However, for many applications it is more convenient simply to use these funny units of depth of a column of liquid. Thus we say that 1 atm of pressure is equal to 10.3 m of water or 76 cm of mercury. For example, to test a pipe joint, a building inspector will require that it does not leak when subjected to a certain head of water so many feet tall. A standpipe can be installed and filled with water to check the joint. Cerebral spinal fluid has a density about equal to that of water, so if a tube is inserted into the spinal column and the fluid is observed to rise 15 cm, the pressure may be recorded as "15 cm of H₂O." The pressure of a gas can be measured using a column of liquid supported by the gas pressure. Such manometers use the very dense liquid mercury in order to reduce the length of the column needed.

PROBLEM 15.2. A hydraulic jack consists of a large cylinder of area A connected to a small cylinder of area a. Both cylinders are filled with oil. When force f is applied to the small cylinder, the resulting pressure is transmitted to the large cylinder, which then exerts an upward force F. Suppose a car of weight 12,000 N rests on the large cylinder of area $0.10 \, \text{m}^2$. What force must be applied to the small cylinder of area $0.002 \, \text{m}^2$ in order to support the car?

Solution $F/A = f/a$ so

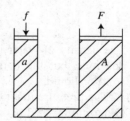

$$f = \frac{Fa}{A} = \frac{(12,000 \, \text{N})(0.002 \, \text{m}^2)}{0.10 \, \text{m}^2}$$

$$= 240 \, \text{N}$$

Thus the jack has a mechanical advantage of 50.

PROBLEM 15.3. What is the total net force acting on the surface of a dam of height h and width w?

Solution Observe that the pressure increases with depth (which is why dams are built thicker at the bottom). Since the pressure varies linearly with depth, the average pressure is $\rho g h/2$ and the total net force on the dam is $F = PA = \rho g h^2 w/2$. Note that atmospheric pressure acts on both sides of the dam and so does not contribute to the net force. We could also obtain this result by breaking the surface area into thin horizontal strips of height dy. The force on a strip is then $dF = (\rho g y)(w \, dy)$ where y is to be

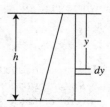

integrated from 0 to *h*. Thus

$$F = \int_0^h \rho g w y \, dy = \frac{\rho g w h^2}{2}$$

PROBLEM 15.4. An experimenter wishes to determine the density of an oil sample she has extracted from a plant. To a glass U tube open at both ends she adds some water colored with food coloring (for visibility). She then pours a small amount of the oil sample on top of the water on one side of the tube and measures the heights h_1 and h_2, as shown in the drawing. What is the density of the oil in terms of the density of water and h_1 and h_2?

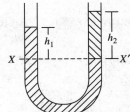

Solution The pressure at the level $X–X'$ is the same on both sides of the tube. Thus

$$\rho_w g h_1 = \rho g h_2 \quad \text{and} \quad \rho = \frac{h_1}{h_2} \rho_w$$

15.2 Buoyancy

When an object is immersed in a fluid (either a liquid or a gas), it experiences an upward buoyancy force because the pressure on the bottom of the object is greater than on the top. The great Greek scientist Archimedes (287–212 B.C.) made the following careful observation, now called **Archimedes' principle**.

> **Any object completely or partially immersed in a fluid is buoyed up by a force equal to the weight of the displaced fluid.**

To see that this is true, consider a small piece of water in a beaker of water (Fig. 15-3). The water above this piece acts downward on the piece, as does its weight. The water under the piece pushes up. Since the piece of water is in equilibrium, the upward force balances the downward forces.

$$F_1 + W = F_2$$

The net upward force due to the fluid is called the *buoyancy force*, $F_B = F_2 - F_1$. Thus

Fig. 15-3

$$F_B = F_2 - F_1 = W \quad \boxed{F_B = W} \tag{15.4}$$

Here *W* is the weight of the fluid displaced by the object. If the piece of water of weight *W* is replaced by an object of the same shape and size, this object would also feel the upward buoyant force $F = W$. If the weight of the object is greater than *W* (the weight of the displaced fluid), the object will sink (but it still experiences a buoyant force, which is why a rock does not feel as heavy when it is submerged as it does when it is lifted out of the water). If the weight of the object is less than the weight of water displaced when it is totally immersed, it will experience a net upward force and will float to the surface. Some of the object will protrude above the surface, so that the portion still submerged will displace a weight of fluid equal to the weight of the object.

PROBLEM 15.5. A man fishing in the Aegean Sea hooks an ancient gold artifact. The density of gold is $19.3 \times 10^3 \, \text{kg/m}^3$, and the density of seawater is $1.03 \times 10^3 \, \text{kg/m}$. While he is pulling up the treasure, the tension in his line is 120 N. What will be the tension when he lifts the object out of the water? *Note*: If you hook a treasure or a big fish, don't lift it out of the water. Your line may break.

Solution   If the object of weight mg is pulled up slowly, it is in equilibrium and $mg = T_1 + F_B$, where T_1 is the tension in the line while in the water and F_B is the buoyant force $m_w g = \rho_w V g$, where V is the volume of the object and $m = \rho V$. Thus

$$\rho V g = T_1 + \rho_w V g \quad \text{or} \quad V = \frac{T_1}{(\rho - \rho_w)g}$$

When the object is in air, the tension is equal to the weight mg.

$$W = mg = \rho V g = \frac{\rho g T_1}{(\rho - \rho_w)g} = \frac{\rho/\rho_w}{(\rho/\rho\omega) - 1} T_1 = \frac{19.3}{19.3 - 1}(120\,\text{N}) = 127\,\text{N}$$

PROBLEM 15.6.   A block of wood of specific gravity 0.8 floats in water. What fraction of the volume of the block is submerged?

Solution   If V is the volume of the block and xV is the submerged volume, then

$$mg = F_B \quad \text{or} \quad \rho V g = \rho_w x V g, \quad \text{so } x = \frac{\rho}{\rho_w} = 0.8$$

15.3   Fluid Flow

We can visualize the motion of a flowing fluid by means of **stream lines**. A stream line traces out the path followed by a particle of the fluid. For example, we can put a drop of ink in a stream of flowing water and see a stream line traced out. The velocity of the fluid at any point is tangent to the stream line at that point. Where stream lines are close together, the fluid is flowing fast. In the following I will limit the discussion by the following assumptions:

1.   The flow is steady. The velocity is not time dependent.
2.   The flow is **laminar**, as opposed to **turbulent**. In turbulent flow the fluid has whirlpools, whereas in laminar flow the fluid flows smoothly in layers ("laminae"). Turbulent flow is very complicated and is not yet well understood.
3.   The fluid is incompressible (that is, like a liquid). Despite this limitation, the following results apply approximately to gases as well.
4.   The temperature of the fluid is constant.
5.   Friction is neglected; that is, the fluid has zero viscosity.

Suppose a fluid flows through a pipe whose cross-sectional area decreases from A_1 to A_2 (Fig. 15-4). When this happens, the fluid will speed up in going from one section to the next. If a fluid volume dV enters the big pipe in time dt, it must flow out of the small pipe in the same time, since no fluid can accumulate inside.

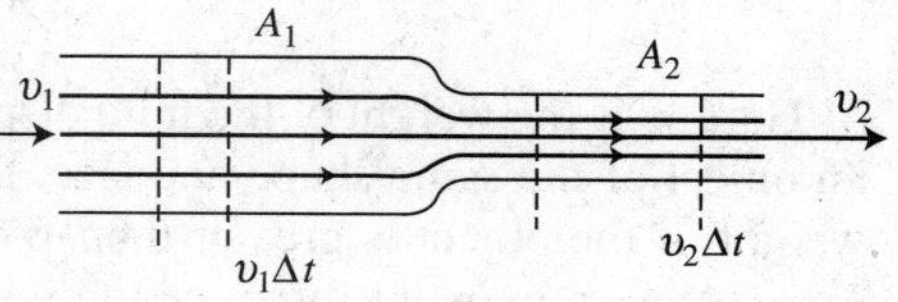

In time dt the fluid moves a distance of $dx = vdt$, so $dV = Adx = Avdt$. Thus $A_1 v_1 dt = A_2 v_2 dt$, or

Fig. 15-4

$$\boxed{A_1 v_1 = A_2 v_2} \qquad (15.5)$$

This is the **equation of continuity**.

PROBLEM 15.7.   Water flows through a garden hose of inside diameter 2 cm at a speed of 1.2 m/s. At what speed will it emerge from a nozzle of diameter 0.5 cm?

Solution $$v_2 = \frac{A_1}{A_2} v_1 = \frac{\pi(0.01\,\text{m})^2}{\pi(0.025\,\text{m})^2}(1.2\,\text{m/s}) = 4.8\,\text{m/s}$$

15.4 Bernoulli's Equation

When fluid in horizontal flow speeds up, its kinetic energy increases, and this means work was done on the fluid. This work is done by the pressure in the fluid, which then decreases when work is done. Similarly, when a fluid flows uphill at constant speed, its potential energy increases, and this also requires that work be done on the fluid.

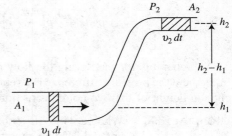

Consider a liquid flowing through the pipe shown here (Fig. 15-5). When the liquid at the lower end moves a distance dx, work $dW_1 = F_1dx_1 = P_1A_1dx_1 = P_1dV$ is done on the liquid. Here dV is the volume of liquid pushed out of the lower pipe in time dt. In this time the pressure in the upper pipe does work $dW_2 = P_2dV$ on the liquid to the right in the upper pipe. Thus the net work done on the liquid moving from the lower pipe up into the upper pipe is

$$dW = dW_1 - dW_2 = P_1dV - P_2dV = (P_1 - P_2)dV$$

Fig. 15-5

This work increases the kinetic energy and potential energy of the mass of liquid pushed from the lower pipe to the upper pipe in time dt. Calling this mass dm,

$$dW = (P_1 - P_2)dV = \Delta KE + \Delta PE = \tfrac{1}{2}dm(v_2^2 - v_1^2) + mg(h_2 - h_1)$$

Divide by dV and note that the density $\rho = dm/dV$. Thus

$$P_1 - P_2 = \tfrac{1}{2}\rho v_2^2 - \tfrac{1}{2}\rho v_1^2 + \rho gh_2 - \rho gh_1$$

Rearranging terms,

$$P_1 + \tfrac{1}{2}\rho v_1^2 + \rho gh_1 = P_2 + \tfrac{1}{2}\rho v_2^2 + \rho gh_2$$

This is *Bernoulli's equation*, and it may be written

$$\boxed{P + \tfrac{1}{2}\rho v^2 + \rho gh = \text{constant}} \tag{15.6}$$

The variation of fluid pressure with velocity has many important applications. For horizontal flow, where the velocity is high, the pressure is low, and vice versa. This effect occurs in gases as well as in liquids, and it explains the lift of an airplane wing, why the cloth roof of a convertible car bulges outward, and why houses can "explode" in a typhoon when air moves rapidly over a peaked roof. Sometimes the "wind" involved results from the "ball of air" that sticks like cotton candy to a moving object. This is why a thrown baseball can curve or an oncoming large truck can "suck" your car toward it. Don't stand near the edge of a subway platform. You may be sucked onto the tracks when a train rushes by.

The Bernoulli effect provides a means of measuring the velocity of a fluid. For a flowing liquid, pressure differences can be measured by means of a standpipe, as in Fig. 15-6a. Where the velocity is high, the pressure is low, and the liquid does not rise very high in the standpipe.

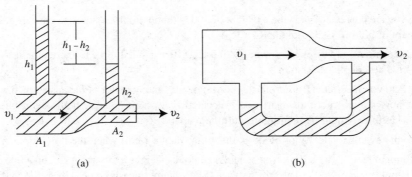

(a)   (b)

Fig. 15-6

The pressure difference between the standpipes is $P_1 - P_2 = \rho g h_1 - \rho g h_2$. Using Eq. 15.6,

$$P_1 + \tfrac{1}{2}\rho v_1^2 = P_2 + \tfrac{1}{2}\rho v_2^2$$

From the equation of continuity (Eq. 15.5), $A_1 v_1 = A_2 v_2$. Thus

$$\rho g h_1 + \frac{1}{2}\rho\left(\frac{A_2}{A_1}v_2\right)^2 = \rho g h_2 + \frac{1}{2}\rho v_2^2, \quad v_2 = A_1\sqrt{\frac{2\rho g(h_1 - h_2)}{\rho(A_1^2 - A_2^2)}}$$

An apparatus for measuring the flow rate of a gas is shown in Fig. 15-5b. The U tube is filled with a liquid, and the pressure difference between the two sections of tubing is determined by measuring the difference in height of the two columns of liquid.

PROBLEM 15.8.   When a wind blows between two large buildings, a significant drop in pressure can be created. The air pressure is normally 1 atm inside the building, so the drop in pressure just outside can cause a plate glass window to pop out of the building and crash to the street below. (I saw this happen once in Boston.) What pressure difference would result from a 27-m/s wind (about 60 mi/h)? What force would be exerted on a 2×3 m plate glass window? The density of air is $1.29\,\text{kg/m}^3$ at $27°\text{C}$ and 1 atm.

Solution   Far from the buildings the pressure is 1 atm, and the wind velocity is approximately zero. Thus $P + 1/2\rho v^2 = P_0 + 0$.

$$P - P_0 = \tfrac{1}{2}\rho v^2 = (0.5)(1.29\,\text{kg/m}^3)(27)^2 = 470\,\text{Pa}$$

and

$$F = PA = (470\,\text{Pa})(2\,\text{m})(3\,\text{m}) = 2820\,\text{N} = 634\,\text{lb}$$

GOOD-BYE WINDOW!!!

15.5   Summary of Key Equations

Density: $\rho = \dfrac{m}{v}$

Pressure: $P = \dfrac{F}{A} = \rho g h$

Buoyancy force: $F_B = \text{weight of displaced fluid}$

Bernoulli's equation: $P + \tfrac{1}{2}\rho g h + \tfrac{1}{2}\rho v^2 = \text{constant}$

Equation of continuity for flow: $A_1 v_1 = A_2 v_2$

SUPPLEMENTARY PROBLEMS

15.9.   Suppose you are capable of carrying a weight of 400 N (about 90 lb). What size cube of gold could you carry? The density of gold is $19{,}300\,\text{kg/m}^3$.

15.10.   A deep sea exploration vessel has a window of area $0.10\,\text{m}^2$. What force is exerted on it by sea water (density $1030\,\text{kg/m}^3$) at a depth of 5000 m?

15.11.   In 1654, Otto von Guericke, Burgermeister of Magdeburg and inventor of the air pump, demonstrated that two teams of horses could not pull apart two evacuated brass hemispheres. If the hemisphere diameters were 0.30 m, what force would be required to pull them apart?

15.12.   Calculate the average velocity of blood in the aorta (radius 1 cm) when the flow rate is 5 l/min.

15.13.   In an attempt to identify a rock specimen, a field geologist weighs a sample in air and while immersed in water, using a simple improvised equal arm balance. She obtains measurements of 120 g and 78 g. What is the specific gravity of the sample?

15.14. A beaker is partially filled with water. Oil of density $750\,\text{kg/m}^3$ is poured on top of the water, and it floats on the water without mixing. A block of wood of density $820\,\text{kg/m}^3$ is placed in the beaker, and it floats at the interface of the two liquids. What fraction of the block's volume is immersed in water?

15.15. An ice floe (density $917\,\text{kg/m}^3$) floats in sea water (density $1030\,\text{kg/m}^3$). If the surface area of the ice is $20\,\text{m}^2$ and it is $0.20\,\text{m}$ thick, what is the mass of the heaviest polar bear that can stand on the ice without causing it to go below the surface of the water?

15.16. A siphon is a device for removing liquid from a container that is inaccessible or that cannot readily be tipped. The outlet C must be lower than the inlet A, and the tube must initially be filled with liquid (this is usually accomplished by sucking on the tube at point C). The density of the liquid is ρ. (a) With what speed does the fluid flow out at point C? (b) What is the pressure at point B? (c) What is the maximum height H that the siphon can lift water?

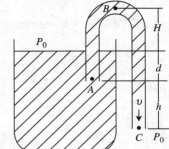

15.17. A large open storage tank is filled with water. A small hole is punctured in the side of the tank at a depth h below the surface of the water. At what speed will water flow out of the hole?

15.18. Firemen use a hose of inside diameter $6.0\,\text{cm}$ to deliver $1000\,\text{l}$ of water per minute. A nozzle is attached to the hose, and they want to squirt the water up to a window $30\,\text{m}$ above the nozzle. (a) With what speed must the water leave the nozzle? (b) What is the inside diameter of the nozzle? (c) What pressure in the hose is required?

15.19. A geode is a hollow rock much prized by rock hounds because such stones often contain beautiful crystals in the interior void. Suppose you find such a treasure. From previous measurements you are fairly confident that the density of the stone material itself is about 2.4 times the density of water. You weigh your geode, and find that it weighs three times as much when weighed in air as when weighed while immersed in water. From this information, deduce what fraction of the total geode volume is occupied by the interior air pocket.

15.20. A balloon and payload of mass M are to be lifted by a gas-filled balloon. The gas in the balloon has density ρ and the surrounding air has density ρ_a. Show that the minimum volume of the balloon required is $V = M/(\rho_a - \rho)$.

15.21. Water flows through a garden hose at a rate of $0.60\,\text{l/s}$. The inner diameter of the hose is $1.6\,\text{cm}$. The water squirts out through a nozzle of inner diameter $0.8\,\text{cm}$. At what velocity does water flow in the hose? At what velocity does water leave the nozzle?

SOLUTIONS TO SUPPLEMENTARY PROBLEMS

15.9. $\quad W = mg = \rho V g = \rho a^3 g, \quad a = \left(\dfrac{W}{\rho g}\right)^{1/3} = \left[\dfrac{400}{(19{,}300)(9.8)}\right]$

$$\text{Side of cube} = a = 0.13\,\text{m} = 13\,\text{cm}$$

15.10. $\qquad\qquad F = PA = \rho g h A = (1030)(9.8)(5000)(0.1) = 5.05 \times 10^6\,\text{N}$

15.11. Consider the hemisphere oriented with its axis along the x-axis. Consider a narrow strip of width ds that circles the hemisphere. The x component of the force on this strip is

$$dF_x = P_0\,dA\,\cos\theta = P_0(2\pi r \sin\theta)\,ds\,\cos\theta \quad \text{and} \quad ds = r\,d\theta$$

so

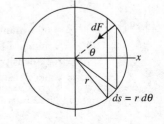

$$F_x = \int_0^{\pi/2} 2\pi r P_0 \sin\theta \cos\theta\, r\,d\theta$$

$$= 2\pi r^2 P_0 \left(\tfrac{1}{2}\sin^2\theta\right)\Big|_0^{\pi/2} = \pi r^2 P_0$$

$$= \pi (0.115)^2 (1.01 \times 10^5)\,\text{N}$$

$$= 7100\,\text{N} \simeq 600\,\text{lb}$$

Note that this result could have been obtained by observing that $F_x = (P_0)$ (area seen looking along x-axis) $= P_0 \pi r^2$.

15.12. $Flow = Av$, $\quad v = \dfrac{Flow}{A} = \left(\dfrac{5000\,\text{cm}^3}{60\,\text{s}}\right)\left[\dfrac{1}{\pi(1\,\text{cm})^2}\right] = 27\,\text{cm/s}$

15.13. $m = \rho V = 120\,\text{g}$ and $78 = 120 - \rho_w V$, so

$$\rho_w V = 42 \quad \text{or} \quad \rho_w\dfrac{120}{\rho} = 42, \qquad \dfrac{\rho}{\rho_w} = \dfrac{120}{42} = 2.86$$

15.14. Volume xV is in water, and volume $(1-x)V$ is in oil.

$$\rho Vg = \rho_w xVg + \rho_0(1-x)Vg$$

Solve:

$$x = \dfrac{\rho - \rho_0}{\rho_w - \rho_0} = \dfrac{820 - 750}{1000 - 750} = 0.28$$

15.15. $m_B g + m_i g = m_w g$, $\qquad V = (20\,\text{m}^2)(0.2\,\text{m}) = 4\,\text{m}^3$

$m_B = m_w - m_i = \rho_w V - \rho_i V = (1000 - 917)(4)$, $\qquad m = 332\,\text{kg}$

15.16. (a) Compare the surface (where pressure is atmospheric pressure P_0 and velocity is approximately zero) with point C. From Eq. 15.6,

$$P_0 + 0 + \rho g(h+d) = P_0 + \tfrac{1}{2}\rho v^2 + 0, \qquad v = \sqrt{2\rho g(h+d)}$$

(b) Compare the surface with point B: $P_0 + \rho g(h+d) = P + 1/2\rho v^2 + \rho g(h+d+H)$. From (a), $\tfrac{1}{2}\rho v^2 = \rho g(h+d)$, so $P = P_0 - \rho g(h+d+H)$.

(c) When H is a maximum, the velocity and pressure there approaches zero, so comparing the surface and point B yields $P_0 + 0 + \rho g(h+d) = 0 + 0 + \rho g(h+d+H)$, or

$$\rho g H = P_0, \quad H = \dfrac{P_0}{\rho g} = \dfrac{1.01 \times 10^5}{(10^3)(9.8)} = 10.3\,\text{m}$$

15.17. At the surface $P = P_0$ and $v \simeq 0$. At the hole $P = P_0$ and $v = v$, so

$$P_0 + 0 + \rho g h = P_0 + \tfrac{1}{2}\rho v^2 + 0, \quad v = \sqrt{2gh}$$

15.18. (a) As the water leaves the nozzle, $P = P_0$ and $v = v$. At the highest point $v = 0$, so from Eq. 15.6, $P_0 + \tfrac{1}{2}\rho v^2 + 0 = P_0 + 0 + \rho g h$ and $v = \sqrt{2gh} = \sqrt{2(0.8)(30)} = 24.2\,\text{m/s}$.

(b) Flow $Q = Av = \dfrac{\pi d^2 v}{4}$, $\quad d^2 = \dfrac{4Q}{\pi v} = \dfrac{4Q}{\pi\sqrt{2gh}} = \dfrac{(4)(1000 \times 10^{-3}\,\text{m}^3)}{(60\,\text{s})(\pi)\sqrt{2(9.8\,\text{m/s}^2)(30\,\text{m})}}$, $\quad d = 0.03\,\text{m}$

(c) The velocity in the hose is V, $A_H V = $ flow, and

$$V = \dfrac{\text{flow}}{A_H} = \dfrac{(1000 \times 10^{-3}\,\text{m}^3)}{(\pi)(0.03\,\text{m})^2(60\,\text{s})} = 5.9\,\text{m/s}$$

$$P + \tfrac{1}{2}\rho V^2 + 0 = P_0 + \tfrac{1}{2}\rho v^2 + 0, \qquad P = P_0 + \tfrac{1}{2}\rho(v^2 - V^2)$$

$$P - P_0 = \tfrac{1}{2}(1000\,\text{kg/m}^3)\big[(24.2\,\text{m/s})^2 - (5.9\,\text{m/s})^2\big] = 2.7 \times 10^5\,\text{Pa} \simeq 2.7\,\text{atm}$$

15.19. Let $V = $ geode volume $= V_R + V_A$

$V_R = $ volume of rock, $V_A = $ volume of cavity

$\rho_R = $ rock density $= 2400\,\text{kg/m}^3$

$\rho_W = $ water density $= 1000\,\text{kg/m}^3$

Weight in air $= 3$ (weight in water)

$$Mg = 3(Mg - mg) \tag{15.7}$$

Here mg is the buoyant force and $m = $ mass of displaced water

$$M = \rho_R V_R, m = \rho_W V$$

From Eq. 1, $2Mg = 3mg$

$$2\rho_R V_R = 3\rho_W V$$
$$2\rho_R V_R = 3\rho_W(V_R + V_A)$$
$$(2\rho_R - 3\rho_W)V_R = 3\rho_W V_A$$

$$\frac{V_A}{V_R} = \frac{2\rho_R - 3\rho_W}{3\rho_W} = \frac{2(2400) - 3(1000)}{(3)(1000)} = 0.6$$

Fraction of geode volume that is air is

$$R = \frac{V_A}{V_A + V_R} = \frac{V_A}{V_A + (1/0.6)V_A}, \quad R = 0.38$$

15.20. Buoyant force = weight of gas in balloon + weight of load

$$\rho_A Vg = (\rho V + M)g$$
$$(\rho_A - \rho)V = M$$
$$V = M/(\rho_A - \rho)$$

15.21. Flow $Q = Av$, $v = Q/A$, $Q = \text{constant} = 0.60\,l/s$

For the hose,

$$v_h = \frac{Q}{\pi d^2/4} = \frac{4Q}{\pi d^2}$$

$$v_h = \frac{(4)(0.60 \times 10^{-3}\,\text{m}^3/\text{s})}{\pi(1.6 \times 10^{-2}\,\text{m})^2} = 3.0\,\text{m/s}$$

For the nozzle the diameter is half that of the hose, hence the area is one-fourth that of the hose, so the water squirts out with velocity four times as great as in the hose.

$$v_N = 12.0\,\text{m/s}$$

CHAPTER 16

Waves and Sounds

A wave is a periodic disturbance that travels through space. Examples are water waves, sound waves, electromagnetic waves (for example, radio waves, microwaves, light, and x-rays), and vibrational waves on a stretched string. In quantum mechanics we encounter probability waves that tell us the likelihood of finding an electron at one place or another. We can learn about the basic properties of waves by studying the waves that propagate on a stretched string, and from there we can go on to understand other kinds of waves.

16.1 Transverse Mechanical Waves

Suppose one end of a string is tied to something, and you hold the other end, pulling the string taut. If you now give a sudden jerk on your end of the string, a pulse will travel along the string, as shown in Fig. 16-1. Shown there are "snapshots" of the string at successively later time intervals. The pulse travels at speed v, the **wave velocity**, in the positive x direction.

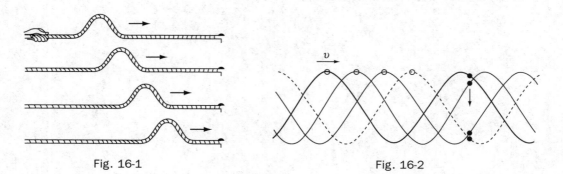

Fig. 16-1   Fig. 16-2

If you move your hand up and down in simple harmonic motion, you will generate a **transverse sinusoidal traveling wave** on the string, as shown in Fig. 16-2. In this drawing the open circles indicate the progress of a certain feature of the wave (a peak), and the solid circles indicate the up and down motion of a piece of the string. In your mind's eye imagine a sine wave (the darkest curve) moving to the right at speed v. Its position at three subsequent times is described by the curves in the drawing. Observe that although the wave is moving to the right, no matter is moving in this direction. The particles of the string just move up and down, transverse (that is, perpendicular) to the direction of propagation of the wave. Be careful not to confuse the particle velocity dy/dt (in the y direction) with the wave velocity v (in the x direction).

Although complicated waveforms are encountered in nature, I will focus on the properties of sinusoidal waves since more complex waves can be described in terms of combinations (superpositions) of sinusoidal

178

waves. We can obtain an equation of motion (a "wave equation") for a particle on a stretched string by applying $F = ma$ to a little piece of the string. When we do this, we find that a solution is any function whose argument is $x + vt$ or $x - vt$, that is, $f(x + vt)$ or $f(x - vt)$. The exact nature of the function $f(x - vt)$ depends on how you wiggle the end of the string. When you wiggle the end of the string in simple harmonic motion at frequency f, the transverse displacement of a piece of the string is given by

$$y(x, t) = A \sin \frac{2\pi}{\lambda}(x - vt) \tag{16.1}$$

This describes a wave moving to the right along the x-axis with speed v. The quantity $\phi = (2\pi/\lambda)(x - vt)$ is the **phase of the wave**. When the phase has a given value, y has a given value. Thus a constant value of y requires that the phase $(2\pi/\lambda)(x - vt)$ be constant. Suppose $(2\pi/\lambda)(x - vt) = \phi = $ constant. If we take the time derivative of the phase, we see that $dx/dt - v = 0$, or $v = dx/dt$, which shows that the quantity v is indeed the velocity of the wave. v is called the **phase velocity** (or wave velocity) because it describes the speed of a point of constant phase on the wave. Observe that a function of the form $f(x + vt)$, that is, with argument $(x + vt)$, instead of $(x - vt)$, describes a wave traveling toward the negative x direction because for it we find that $dx/dt = -v$, and v is a positive number.

The form of Eq. 16.1 is reminiscent of the way we described the y coordinate of a point on the rim of a rotating disk. As the disk rotated, the angle of rotation (the argument of the sine function) varied from 0 to 2π rad, or from 0 to 360°. Thus we sometimes refer to phase in terms of degrees or radians. A picture of a wave at $t = 0$ is shown in Fig. 16-3. Observe that moving a distance of $\lambda/4$ in space along the x-axis causes the phase to change by 90°. Moving a distance $\lambda/2$ results in a phase change of 180°.

The maximum value y reaches is the **amplitude** A in meters. The separation in space between two adjacent points with the same phase (for example, between two crests) is the *wavelength* λ, in meters, of the wave. Amplitude and wavelength are indicated in Fig. 16-4. From Eq. 16.1 we see that when x increases by an amount λ, the phase increases by 2π rad, and since $\sin(\phi + 2\pi) = \sin \phi$, y has the same value at points x and at $x + \lambda$.

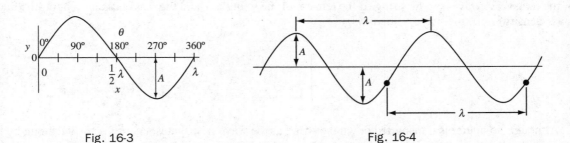

Fig. 16-3   Fig. 16-4

The number of crests passing a given point in space each second is the **frequency** f. One crest (or vibration) per second is **1 hertz (Hz)**, so $1\,\text{Hz} = 1\,\text{s}$. Consider a length of the wave of length $L = n\lambda$, where n is the number of crests in the length L. In t seconds it moves a distance $L = vt = n\lambda$ or $v = \lambda n/t$. But n/t is the number of crests passing a given point per second, which is the frequency f. Thus

$$v = f\lambda = \frac{\lambda}{T} \tag{16.2}$$

Here the time in seconds between adjacent passing crests is the **period** $T = 1/f$. Thus if 10 crests pass each second ($f = 10\,\text{Hz}$), the period is $T = 1/f = 1/10 = 0.1\,\text{s} = $ time between crests or for oscillations to repeat.

We frequently encounter the **wave number** k and the **angular frequency** ω (in radians per second) defined as follows:

$$\omega = 2\pi f \quad \text{and} \quad k = \frac{2\pi}{\lambda}, \quad \text{so } v = f\lambda = \frac{\omega}{k} \tag{16.3}$$

Using this notation, Eq. 16.1 can be conveniently written as

$$y(x, t) = A \sin \frac{2\pi}{\lambda}(x - vt) = A \sin 2\pi\left(\frac{x}{\lambda} - \frac{t}{T}\right) = A \sin(kx - \omega t) \tag{16.4}$$

In the above I assumed the vertical displacement $y(x, t)$ is zero at $t = 0$ and $x = 0$. This need not be the case (we can start $t = 0$ whenever we want), so a more general form for $y(x, t)$ is $y = A \sin(kx - \omega t + \phi)$, where ϕ, the **phase constant**, is determined from the given initial conditions. If we choose the zero of time and the x-axis origin so that $\phi = 90°$, then $y = A \sin(kx - \omega t + 90°) = A \cos(kx - \omega t)$. In some books we see the sine function and in others the cosine function. Either form can be used.

PROBLEM 16.1.   A string wave is described by $y = 0.002 \sin(0.5x - 628t)$. Determine the amplitude, frequency, period, wavelength, and velocity of the wave.

Solution   From Eq. 16.4,

$$A = 0.002 \, \text{m} \qquad \frac{2\pi}{\lambda} = 0.5 \quad \lambda = 12.6 \, \text{m}$$

$$\frac{2\pi}{T} = 628 \qquad T = 0.01 \, \text{s} \qquad f = \frac{1}{T} = 100 \, \text{Hz} \qquad v = f\lambda = 1260 \, \text{m/s}$$

16.2   Speed and Energy Transfer for String Waves

By applying $F = ma$ to a small piece of a vibrating string, we can deduce the wave equation and the speed of the transverse waves on the string. If the tension of the string is T and the mass per unit length (the linear mass density) is μ, the wave velocity is

$$v = \sqrt{\frac{T}{\mu}} \tag{16.5}$$

Although no matter is transported down the string as the wave propagates, energy is carried along by the wave with velocity v. As a piece of the string moves up and down executing simple harmonic motion, it has kinetic energy as well as potential energy (because the string is stretched like a spring). In Eq. 13.9 we saw that the total energy of a mass m that oscillates with amplitude A and angular frequency ω is $E = \frac{1}{2} kA^2 = \frac{1}{2} m\omega^2 A^2$, where k is the spring constant and $k = m\omega^2$. Consider a small length dx of the string. The mass of this piece is $dm = \mu dx$, where μ is the mass per unit length of the string. This infinitesimal mass of string thus has a small energy $dE = \frac{1}{2} dm^2 A^2 = \frac{1}{2} \mu dx \omega^2 A^2$. As this small mass moves up and down, it pulls on the piece of string to its right and does work on it, thereby transferring energy down the string to the right as the wave moves in that direction. If energy dE is transferred in time dt, the rate of energy transfer (the power) is $P = dE/dt = \frac{1}{2} \mu\omega^2 A^2 dx/dt$ where $dx/dt = v = $ wave velocity. Thus the power transmitted by the wave is

$$P = \tfrac{1}{2}\mu v\omega^2 A^2 \tag{16.6}$$

PROBLEM 16.2.   A string of linear mass density 480 g/m is under a tension of 48 N. A wave of frequency 200 Hz and amplitude 4.0 mm travels down the string. At what rate does the wave transport energy?

Solution
$$\omega = 2\pi f = 2\pi(200) = 400\pi\,\text{s}^{-1}, \qquad v = \sqrt{\frac{T}{\mu}} = \sqrt{\frac{48\,\text{N}}{0.48\,\text{kg/m}}} = 10\,\text{m/s}$$

$$P = \frac{1}{2}\mu v \omega^2 A^2 = (0.5)(0.480\,\text{kg/m})(10\,\text{m/s})(400\pi\,\text{s})^2(4\times10^{-3}\,\text{m})^2 = 61\,\text{W}$$

Equation 16.6 proves to be qualitatively true for all kinds of waves, including electromagnetic waves such as light. The power transmitted is proportional to the wave velocity and to the square of the frequency and to the square of the amplitude.

16.3 Superposition of Waves

Consider a stretched string tied at one end (Fig. 16-5a). If you jerk the other end, a pulse will travel along the string. When the pulse reaches the tied end, it will be turned upside down (180° phase shift) and be reflected. If the far end of the string is free to move up and down (perhaps it slides on a pole, as in Fig. 16-5b), the reflected pulse does not turn upside down (no phase shift). Suppose two pulses are sent down the string. The first one is reflected, and on its way back, it encounters the second oncoming pulse. The two will interact (they are said to **interfere**). As they pass each other, their displacements will add. The pulses in Fig. 16-5a are upside down from each other, and as they pass, they cancel each other out (**destructive interference**). The two pulses in Fig. 16-5b reinforce each other (**constructive interference**). This adding together of wave displacements is called **superposition**. By adding together very many sine waves, very complicated wave forms can be constructed.

Conversely, a complicated wave pattern can be decomposed into many sine waves.

If two waves of the same velocity and wavelength are traveling in the same direction on a string, they will interfere. If they are in phase (Fig. 16-6a), they interfere constructively and result in a stronger wave. If they are out of phase (Fig. 16-6b) and have the same amplitude, they cancel each other out (destructive interference).

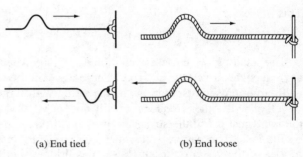

(a) End tied   (b) End loose

Fig. 16-5

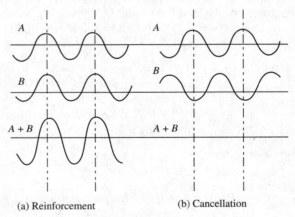

(a) Reinforcement   (b) Cancellation

Fig. 16-6

16.4 Standing Waves

Consider two sinusoidal traveling waves with the same amplitude and wavelength moving in opposite directions on a string. The resultant combination for the two waves is obtained by superposition; thus

$$y(x, t) = y_1(x, t) + y_2(x, t) = A\sin(kx - \omega t) + A\sin(kx + \omega t) \tag{16.7}$$

Use a trig identity to simplify this: $\sin \alpha + \sin \beta = 2 \sin \frac{1}{2}(\alpha + \beta) \cos \frac{1}{2}(\alpha - \beta)$. Thus

$$y(x, t) = (2A \sin kx) \cos \omega t \qquad (16.8)$$

Here $y(x, t)$ is a **standing wave**. We think of the magnitude of the quantity $2A \sin kx$ as the amplitude for simple harmonic motion of a small piece of the string at the position x. A point where the amplitude of the standing wave is zero is called a **node**. A point where the amplitude is a maximum is an **antinode**.

Now consider a string of length L with both ends fixed, so $y = 0$ at $x = 0$ and at $x = L$. Imagine that one end jiggles slightly, so that a wave travels down the wave and is reflected back. These two oppositely traveling waves can interfere and set up standing waves, as illustrated in Fig. 16-7. There we see that possible standing waves are those for which the length of the string is an integer multiple of one-half wavelength.

$$L = n\frac{\lambda}{2} \quad \text{or} \quad \lambda = \frac{2L}{n} \quad \text{or} \quad \boxed{f = \frac{v}{\lambda} = n\frac{v}{2L} \text{ where } n = 1, 2, 3, \dots} \qquad (16.9)$$

This makes

$$\sin kx = \sin \frac{2\pi}{\lambda} = 0 \quad \text{at } x = L$$

The standing waves described here result only when the string oscillates at frequencies given by Eq. 16.9. These are called **resonant frequencies**, and they represent oscillations of the string with large amplitude. Waves traveling with other frequencies will not set up standing waves. Instead, they will just cause the string to vibrate with very small or imperceptible oscillations. The patterns shown in Fig. 16-7 are examples of **resonant modes** of the system. Structures such as bridges, buildings, and freeways have many possible resonant modes. If a structure is driven at one of its resonant frequencies, large amplitude oscillations can result, and the structure may fall down.

The lowest resonant frequency ($n = 1$ in Eq. 16.9) is the **fundamental frequency** or the **first harmonic** f_1. The second harmonic is the mode with $n = 2$ and frequency $f_2 = 2f_1$, and so on for the higher harmonics f_3, f_4, and so on.

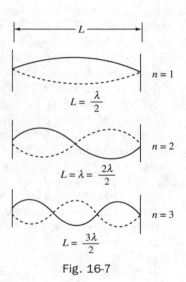

Fig. 16-7

PROBLEM 16.3.   The G string of a mandolin is 0.34 m long and has a linear mass density of 0.004 kg/m. The thumbscrew attached to the string is adjusted to provide a tension of 71.1 N. What then is the fundamental frequency of the string?

Solution

$$f_1 = \frac{v}{2L} = \frac{1}{2L}\sqrt{\frac{T}{\mu}} = \frac{1}{(2)(0.34 \text{ m})}\sqrt{\frac{71.1 \text{ N}}{0.004 \text{ kg/m}}} = 196 \text{ Hz}$$

A stringed instrument such as a guitar is tuned by adjusting the tension in a string by means of a thumbscrew. The length of the string is fixed, so adjusting the tension adjusts the fundamental frequency. Other fundamental frequencies can be achieved by shortening the string length by pressing on a fret. Finally, several strings of different mass densities are used to give a range of wave velocities, thereby providing access to a greater range of fundamental frequencies.

16.5   Sound Waves

A sound wave is a longitudinal pressure wave. *Longitudinal* means that the pressure variations are parallel to the direction of travel, whereas in vibrating string waves, the variations in displacement are transverse to the wave velocity. We can envision what happens by placing a long coiled spring on a horizontal table. When one end is moved back and forth harmonically, regions of compression and rarefaction travel along the spring, as sketched in Fig. 16-8. We can derive the speed of a sound wave using $F = ma$, and the result is

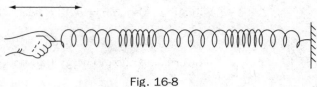

Fig. 16-8

$$v = \sqrt{\frac{B}{\rho}}$$

(16.10)

Here B is the bulk modulus and ρ is the mass density of the medium in which the sound is traveling. This is similar in form to the expression for the speed of the transverse waves on a stretched spring, $v = \sqrt{T/\mu}$. In fact, the velocity of any mechanical wave is of the form $v = \sqrt{\text{elastic property/inertial property}}$ or $\sqrt{\text{"stiffness"/density}}$.

As for all waves, $f\lambda = v$. Representative values for sound velocities are 343 m/s in air at 20°C, 1493 m/s in water at 25°C, and 5130 m/s in iron.

PROBLEM 16.4. For copper the bulk modulus is 14×10^{10} N/m² and the density is 8920 kg/m³. What is the speed of sound in copper?

Solution
$$v = \sqrt{\frac{B}{\rho}} = \sqrt{\frac{14 \times 10^{10}\,\text{N/m}^2}{8920\,\text{kg/m}^3}} = 3960\,\text{m/s}$$

A sound wave can transport energy since as it moves along it causes molecules to vibrate with kinetic energy. When we hear a sound wave, we detect the pitch and the loudness. The **pitch** of a sound is its **frequency**, and its **loudness** is proportional to the **power intensity** of the wave. Humans can typically hear a frequency range of 20–20,000 Hz (when you're 16 years old, not when you are over the hill at 25). As you get older, the high-frequency response gets worse and worse. The average power per unit area perpendicular to the direction of travel of a sound wave is the **intensity**. Humans can detect power intensities ranging from $I_0 = 10^{-12}$ W/m² up to about 1 W/m². Any higher intensities are very painful to the ear. Because sound intensities vary over such a wide range, it is convenient to use a different quantity as a measurement of intensity. A dimensionless quantity β is defined, measured in units of **decibels** (dB).

$$\beta = 10 \log_{10} \frac{I}{I_0} \quad \text{where } I_0 = 10^{-12}\,\text{W/m}^2$$

(16.11)

PROBLEM 16.5. Normal conversation is carried on at about 60 dB. To what intensity level does this correspond?

Solution
$$60 = 10 \log_{10} \frac{I}{I_0}, \quad \text{so } 10^6 = \frac{I}{I_0} \quad \text{and} \quad I = 10^6 I_0 = 10^{-6}\,\text{W/m}^2$$

I can deduce the power carried by a sound wave as follows: Suppose the wave is traveling along the x-axis of a cylinder of material of cross-sectional area A and density ρ. A piece of mass dm occupies volume dV and is undergoing simple harmonic motion along the x-axis. The average energy dE of the mass is equal to its maximum kinetic energy $1/2(dm)v_{\max}^2$ where $v_{\max} = \omega x_{\max}$ is the maximum particle velocity (not the sound wave velocity) and $x_{\max}$ is the maximum amplitude of vibration. Also, $dm = \rho dV = \rho A\, dx$. Thus

$$dE = \tfrac{1}{2}(dm)v^2 = \tfrac{1}{2}\rho A\, dx(\omega x_{max})^2$$

Power is the rate of energy transport, so

$$P = \frac{dE}{dt} = \frac{1}{2}\rho A\omega^2 x_{max}^2 \frac{dx}{dt} \quad \text{and} \quad \frac{dx}{dt} = v \text{ is the wave velocity,}$$

so

$$\boxed{P = \tfrac{1}{2}\rho A\omega^2 x_{max}^2 v} \tag{16.12}$$

Here v is the wave velocity, x_{max} is the maximum displacement, A is the cross-sectional area through which the sound is propagating, ρ is the density of the material, and $\omega = 2\pi f$, where f is the sound wave frequency. The sound intensity is defined as

$$\boxed{I = \frac{\text{power}}{\text{area}} = \frac{1}{2}\rho(\omega x_{max})^2 v} \tag{16.13}$$

It can be shown that the variation ΔP_{max} in pressure amplitude can be expressed as

$$\Delta P_{max} = \rho v\omega x_{max}, \quad \text{so} \quad \boxed{I = \frac{(\Delta P_{max})^2}{2\rho v}} \tag{16.14}$$

PROBLEM 16.6. A source emits sound uniformly in all directions at a power level of 60 W. What is the intensity at a distance of 4 m from the source?

Solution The power is distributed over the surface area of a sphere: $A = 4\pi r^2$.

$$I = \frac{P}{4\pi r^2} = \frac{60\,\text{W}}{4\pi(4\,\text{m})^2} = 0.30\,\text{W/m}^2$$

PROBLEM 16.7. At a distance of 5 m from a source the sound level is 90 dB. How far away has the level dropped to 50 dB?

Solution
$$I_1 = \frac{P}{4\pi r_1^2} \quad \text{and} \quad I_2 = \frac{P}{4\pi r_2^2}, \quad \text{so} \quad \frac{I_2}{I_1} = \frac{r_1^2}{r_2^2}$$

$$\beta_1 = 10\log\frac{I_1}{I_0} = 90\,\text{dB}, \quad \text{so} \quad \frac{I_1}{I_0} = 10^9$$

Similarly,

$$\frac{I_2}{I_0} = 10^5$$

Thus

$$\frac{I_2}{I_1} = \frac{10^5}{10^9} = 10^{-4} = \frac{r_1^2}{r_2^2}, \quad \text{so} \quad r_2 = 10^2 r_1 = 500\,\text{m}$$

16.6 Standing Sound Waves

Standing sound waves can be set up whenever sound is reflected back and forth in an enclosure. In particular, standing sound waves are set up in a column of air, such as in an organ pipe or in a horn. Longitudinal pressure waves are reflected back when they hit an obstruction (for example, the closed end of a pipe) or when they encounter any change in the nature of the structure in which they are propagating. Thus when a sound wave in a pipe encounters the open end of the pipe, it is reflected back. At the closed end of a pipe, the molecules cannot move longitudinally, so this point is a node for displacement (zero displacement). Conversely, a closed pipe end is a point where the pressure variations are large (an antinode). The open end of a pipe is an antinode for displacement and a node for pressure variations. The latter is plausible since the open end is in contact with atmospheric pressure, and this is constant. The above ideas are only approximately true, and some corrections have to be made for very accurate calculations. Also, it is assumed that the pipe diameter is small compared to its length. In Fig. 16-9 the pressure variation for several modes is shown for a pipe open at both ends and for a pipe closed at one end. The frequencies (resonances) are related to the pipe length. For a pipe open at both ends the resonant frequencies are integer multiples of the first harmonic (the fundamental frequency), just as for a string fixed at both ends.

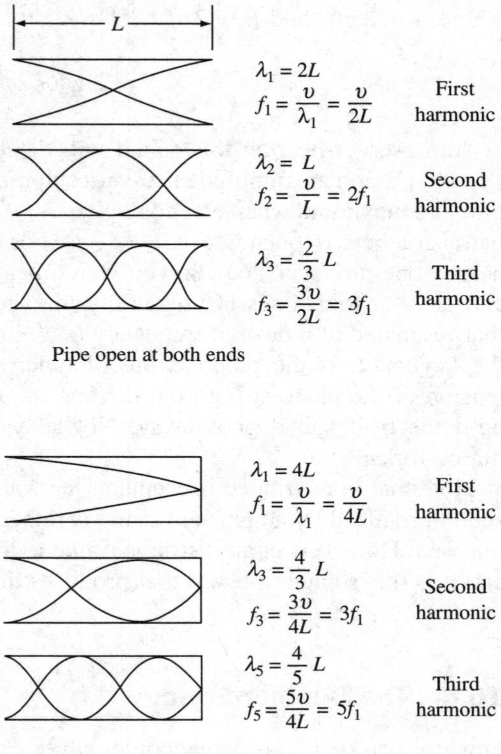

Pipe open at both ends:

$$\boxed{f_n = n\frac{v}{2L} \quad (n = 1, 2, 3, \ldots)} \tag{16.15}$$

In a pipe closed at one end, only odd harmonics are present.

Pipe closed at one end:

$$\boxed{f_n = n\frac{v}{4L} \quad (n = 1, 3, 5, \cdots)} \tag{16.16}$$

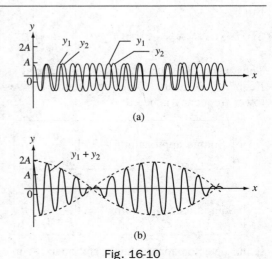

$$\lambda_1 = 2L$$
$$f_1 = \frac{v}{\lambda_1} = \frac{v}{2L} \qquad \text{First harmonic}$$

$$\lambda_2 = L$$
$$f_2 = \frac{v}{L} = 2f_1 \qquad \text{Second harmonic}$$

$$\lambda_3 = \frac{2}{3}L$$
$$f_3 = \frac{3v}{2L} = 3f_1 \qquad \text{Third harmonic}$$

Pipe open at both ends

$$\lambda_1 = 4L$$
$$f_1 = \frac{v}{\lambda_1} = \frac{v}{4L} \qquad \text{First harmonic}$$

$$\lambda_3 = \frac{4}{3}L$$
$$f_3 = \frac{3v}{4L} = 3f_1 \qquad \text{Second harmonic}$$

$$\lambda_5 = \frac{4}{5}L$$
$$f_5 = \frac{5v}{4L} = 5f_1 \qquad \text{Third harmonic}$$

Pipe open at one end

Fig. 16-9

16.7 Beats

We have seen that two traveling waves of the same frequency and velocity can interfere. They can reinforce each other, or, if they have equal amplitudes and are 180° out of phase, they can cancel each other. If they have slightly different frequencies, they interfere and produce a phenomenon called **beats**. In Fig. 16-10a two waves of slightly different frequency are shown, and their superposition $y = y_1 + y_2$ is graphed in Fig. 16-10b. Suppose you are at a fixed point in space and the sound wave train shown passes by you. For simplicity, suppose you are at the origin, and the time dependence of the two superimposed waves is $y_1 = A\cos 2\pi f_1 t$ and $y_2 = A\cos 2\pi f_2 t$. The resultant wave is $y = y_1 + y_2$. Use the trigonometric identity

$$\cos\alpha + \cos\beta = 2\cos\left(\frac{a-b}{2}\right)\cos\left(\frac{a+b}{2}\right)$$

(a)

(b)

Fig. 16-10

where $\alpha = 2\pi f_1 t$ and $\beta = 2\pi f_2 t$. Thus

$$y = 2A \cos 2\pi \left(\frac{f_1 - f_2}{2}\right) t \cos 2\pi \left(\frac{f_1 + f_2}{2}\right) t \qquad (16.17)$$

An observer hearing this sound will detect a predominant frequency equal to the average frequency $(f_1 + f_2)/2$ and an amplitude that varies in time at frequency $(f_1 - f_2)/2$, as seen in Fig. 16-10b. The amplitude is a maximum whenever $\cos 2\pi[(f_1 - f_2)/2] = \pm 1$. Thus there are two maxima per cycle, so one hears **beats** at a beat frequency of $f = f_2 - f_1$. For example, if one tuning fork produces sound at 260 Hz and a nearby one produces 264 Hz, you would hear something that sounded like 262 Hz, but it would get loud and quiet at four times per second. Long ago piano tuners would tune a piano by striking a tuning fork that resonated at a desired frequency (say, low C, 65 Hz) and listen to it while striking the same note on the keyboard. If the piano is out of tune, a beat sound will result. The piano tuner then adjusts the tension of the piano string (and thereby the wave velocity on the string and the fundamental frequency) until the beat sound goes away. Nowadays electronic signal generators are usually used in place of tuning forks.

Note that in a stringed instrument, the sound you hear is not a direct result of the vibrating string. The sound produced by an isolated string is rather weak, as you may have noticed when you hear a bow string released. However, in an instrument like a guitar, the vibrating string causes a sounding board to vibrate, and it is this sounding board that produces the pressure waves we hear as sound.

16.8 The Doppler Effect

Suppose you stand on a street corner while a race car is speeding toward you. The sound you hear is something like this: "Eeeeeeeeeeeuuuuuuuunnnnhhhhh." As the car approaches, you hear a high-pitched sound, and as it passes you and moves away, the pitch drops noticeably. Here is what happens: If the source is stationary, it emits crests of the pressure wave at a rate of f times per second (f is the frequency). The crests are separated by a distance λ, the wavelength. They travel toward you at v, the speed of sound. If now the source moves toward you at speed v_s, the spacing between crests is reduced because the source chases after each crest it emits, moving a distance $v_s T$ before emitting the next crest, where $T = 1/f =$ the period of the sound wave (see Fig. 16-11). Thus the spacing of the crests coming toward you is $\lambda' = vT - v_s T = \lambda - v_s T = \lambda - v_s/f$. The waves will pass you with frequency f'.

$$f' = \frac{v}{\lambda'} = \frac{v}{\lambda(-v_s/f)}$$

But

$$\lambda = \frac{v}{f}, \quad \text{so } f' = \left(\frac{v}{v - v_s}\right) f$$

Thus when the source approaches at speed v_s, a higher frequency is heard. When the source moves away, a lower frequency is heard.

$$\boxed{\text{Source approaching } f' = \left(\frac{v}{v - v_s}\right) f} \qquad (16.18)$$

$$\boxed{\text{Source receding from } f' = \left(\frac{v}{v + v_s}\right) f} \qquad (16.19)$$

If the detector approaches a stationary source, he will hear a frequency higher than normal (Fig. 16-11). Suppose T is the time between crests an observer

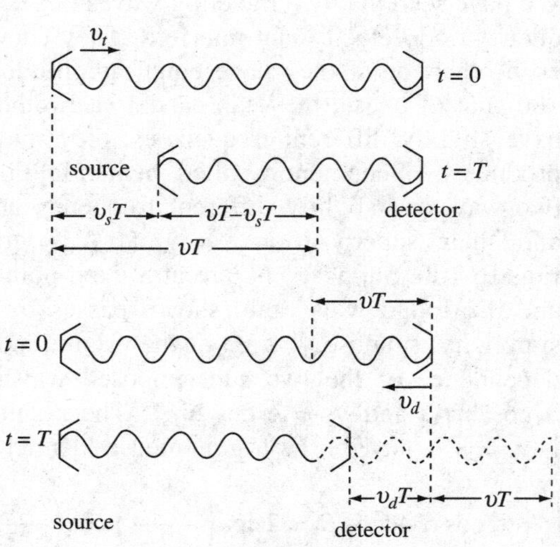

Fig. 16-11

detects when he and the source are both stationary, where $T = 1/f$ and $v = f\lambda$. If the detector moves with speed v_d toward an oncoming crest, the detector approaches the crest with speed $v + v_d$, so the time to cover the distance from one crest to the next will be shorter; that is, $T' = \lambda/(v + v_d) = 1/f'$. Substitute $\lambda = v/f$ and we obtain $1/f' = v/f(v + v_d)$ or $f' = [(v + v_d)/v]\,f$. Similar reasoning shows that when the detector moves away from the source, he detects a lower frequency. Thus

$$\text{Detector approaching } f' = \left(\frac{v + v_d}{v}\right)f \qquad (16.20)$$

$$\text{Detector receding from } f' = \left(\frac{v - v_d}{v}\right)f \qquad (16.21)$$

If both the source and the detector are moving, the frequency detected is

$$f' = \left(\frac{v \pm v_d}{v \mp v_s}\right)f \qquad (16.22)$$

Here use the upper sign ($+v_d$ and $-v_s$) if the source and detector are approaching. Use the lower sign ($-v_d$ and $+v_s$) if the source and detector are moving away from each other. f is the frequency when both detector and source are stationary.

PROBLEM 16.8.   A stationary police car siren emits a sound at 1200 Hz. Under conditions when the velocity of sound in air is 340 m/s, what frequency will you hear when stationary if the siren is approaching at 30 m/s? What frequency will you hear when the siren is moving away at 30 m/s?

Solution   Approaching: $f' = \dfrac{v}{v - v_s}f = \left(\dfrac{340}{340 - 30}\right)(1200\,\text{Hz}) = 1316\,\text{Hz}$

                 Going away: $f' = \dfrac{v}{v + v_s}f = \left(\dfrac{340}{340 + 30}\right)(1200\,\text{Hz}) = 1103\,\text{Hz}.$

16.9   Summary of Key Equations

Wave traveling toward $+x$-axis:	$y = A\sin 2\pi\left(\frac{x}{\lambda} - \frac{t}{T}\right)$ or $y = A\sin(kx - \omega t)$
Angular frequency:	$\omega = 2\pi f = 2\pi\frac{1}{T}$
Wave number:	$k = \frac{2\pi}{\lambda}$
For all waves:	$v = f\lambda$
Transverse string wave:	$v = \sqrt{\frac{T}{\mu}}, \quad T = \text{tension}, \ \mu = \text{mass/length}$
Power transmitted by a string wave:	$P = \frac{1}{2}\mu v\omega^2 A^2$
Standing waves in a string fixed at both ends:	$f = n\frac{v}{2L} \quad (n = 1, 2, 3, \dots)$
Standing waves in a pipe open at both ends:	$f = n\frac{v}{2L} \quad (n = 1, 2, 3, \dots)$
Standing waves in a pipe open at one end:	$f = n\frac{v}{4L} \quad (n = 1, 3, 5, \dots)$
Intensity:	$I = \text{power/area}$
Decibel level:	$\beta = 10\log_{10}\frac{I}{I_0}, \quad \text{where } I_0 = 10^{-12}\,\text{W/m}^2$

SUPPLEMENTARY PROBLEMS

16.9. A transverse string wave travels in the negative x direction with amplitude 0.002 m, frequency 200 Hz, and wavelength 0.20 m. The displacement of the wave is $y = 0$ at $t = 0$ and $x = 0$. Write an expression for the displacement y.

16.10. Humans can hear a range of frequencies from 20 to 20,000 Hz. To what range of wavelengths in air does this correspond (assume the sound velocity is 340 m/s)?

16.11. A copper wire (density $8920 \, \text{kg/m}^3$) has diameter 2.4 mm. With what velocity will transverse waves travel along it when it is subjected to a tension of 20 N?

16.12. The G string on a guitar has a length of 0.64 m and a fundamental frequency of 196 Hz. We can effectively shorten the length of the string by pressing it down on a fret (a small ridge on the neck of the guitar). How far should the fret be from the end of the string if you are to produce a fundamental frequency of 262 Hz (the C note)?

16.13. A tuning fork vibrates at 462 Hz. An untuned violin string vibrates at 457 Hz. How much time elapses between successive beats?

16.14. Energy is transmitted at a power level P_1 with frequency f_1 on a string under tension T_1. (a) What will be the power transmitted if the tension is increased to $2T_1$? (b) What will be the power transmitted if the tension remains T_1 but the frequency is increased to $2f_1$?

16.15. A long pipe is closed at one end and open at the other. If the fundamental frequency of this pipe for sound waves is 240 Hz, what is the length of the pipe? Assume sound velocity is 340 m/s in air.

16.16. Suppose a source of sound radiates uniformly in all directions. By how many decibels does the sound level decrease when the distance from the source is doubled?

16.17. What is the intensity level in decibels of a sound whose intensity is $4.0 \times 10^{-7} \, \text{W/m}^2$? What is the pressure amplitude of such a wave? Assume sound velocity is 340 m/s in air.

16.18. In my town a siren is sounded at the fire station to call volunteer firemen to duty. If the frequency of the siren is 300 Hz, what frequency would you hear when driving toward the siren at 20 m/s?

16.19. You drop a stone off a bridge from a height of 20.0 m above the water below. How long after dropping the stone will you hear the splash as the stone hits the water? Assume the speed of sound in air is 343 m/s.

16.20. A string fixed at both ends vibrates with a fundamental frequency of 320 Hz. How long does it take for a transverse wave to travel the length of this string?

SOLUTIONS TO SUPPLEMENTARY PROBLEMS

16.9. $y = 0.002 \sin 2\pi \left(\dfrac{x}{0.20} + 200t \right)$

16.10. $\lambda_1 = \dfrac{v}{f_1} = \dfrac{340 \, \text{m/s}}{20/\text{s}} = 17 \, \text{m}, \qquad \lambda_2 = \dfrac{340 \, \text{m/s}}{20{,}000/\text{s}} = 0.017 \, \text{m}$

16.11. Consider 1 m of wire:

$$\mu = \frac{m}{L} = \frac{\rho v}{L} = \rho \frac{AL}{L} = \rho A$$

$$v = \sqrt{\frac{T}{\mu}} = \sqrt{\frac{T}{\rho A}} = \sqrt{\frac{20 \, \text{N}}{(8920 \, \text{kg/m}^3)(\pi)(0.0012 \, \text{m})^2}} = 22.3 \, \text{m/s}$$

16.12. $f = n \dfrac{v}{2L}, \quad n = 1, \quad \text{so } f_G = \dfrac{v}{2L_G}, \qquad f_c = \dfrac{v}{2L_c}$

$\dfrac{f_c}{f_G} = \dfrac{v/2L_c}{v/2L_G} = \dfrac{L_G}{L_c}, \quad L_c = \dfrac{f_G}{f_c} L_G = \dfrac{196}{262}(0.64 \, \text{m})$

$L_c = 0.48 \, \text{m}, \quad \text{so } \Delta L = L_G - L_c = 0.64 \, \text{m} - 0.48 \, \text{m} = 0.16 \, \text{m}$

16.13. Beat frequency:

$$f_B = f_2 - f_1 = 462 - 457 \, \text{Hz} = 5 \, \text{Hz}$$

$$T_B = \frac{1}{f_B} = \frac{1}{5} \, \text{s} = 0.2 \, \text{s}$$

16.14. $P = \frac{1}{2}\mu\omega^2 v$, $v = \sqrt{\frac{1}{\mu}}$, so $P = \frac{1}{2}\mu\omega^2\sqrt{\frac{T}{\mu}} = \frac{1}{2}\omega^2\sqrt{\mu T}$

(a) $\frac{P_2}{P_1} = \frac{1/2\,\omega^2\sqrt{\mu(2T_1)}}{1/2\,\omega^2\sqrt{\mu T_1}} = \sqrt{2}$, $P_2 = \sqrt{2}P_1 = 1.4P_1$

(b) $\frac{P_3}{P_1} = \frac{1/2\,\omega_3^2\sqrt{\mu T_1}}{1/2\,\omega_1^2\sqrt{\mu T_1}} = \frac{f_3^2}{f_1^2} = \frac{(2f_1)^2}{f_1^2} = 4$, $P_3 = 4P_1$

16.15. $f_n = n\frac{v}{4L}$, $n = 1$, so $L = \frac{v}{4f_1} = \frac{340 \, \text{m/s}}{(4)(240/\text{s})} = 0.35 \, \text{m}$

16.16. Intensity $= \dfrac{\text{power}}{\text{area}} = \dfrac{P}{4\pi r^2}$, so $\dfrac{I_1}{I_2} = \dfrac{P/4\pi r_1^2}{P/4\pi r_2^2} = \left(\dfrac{r_2}{r_1}\right)^2$

$$\beta_1 = 10\log\frac{I_1}{I_0}, \qquad \beta_2 = 10\log\frac{I_2}{I_0}$$

$$\beta_1 - \beta_2 = 10\log\frac{I_1}{I_0} - 10\log\frac{I_2}{I_0} = 10\log I_1 - 10\log I_0 - 10\log I_2 + 10\log I_0$$

$$= 10\log I_1 - 10\log I_2 = 10\log\frac{I_1}{I_2} = 10\log\left(\frac{r_2}{r_1}\right)^2 = 20\log\frac{r_2}{r_1}$$

If

$$r_2 = 2r_1, \text{ then } \beta_1 - \beta_2 = 20\log 2 = 6.02 \, \text{dB}$$

16.17. $\beta = 10\log\dfrac{I_1}{I_0} = 10\log\dfrac{4\times 10^{-7} \, \text{W/m}^2}{10^{-12} \, \text{W/m}^2} = 44 \, \text{dB}$

16.18. $f' = f\left(1 + \dfrac{v_d}{v}\right) = (300 \, \text{Hz})\left(1 + \dfrac{20 \, \text{m/s}}{340 \, \text{m/s}}\right) = 318 \, \text{Hz}$

16.19. The time for the stone to fall is given by $h = \frac{1}{2}gt^2$

Thus, $t_1 = \left(\dfrac{2h}{g}\right)^{\frac{1}{2}}$

The time for sound to come back is $t_2 = \dfrac{h}{v}$

Total time is $t = t_1 + t_2 = \left(\dfrac{2h}{g}\right)^{\frac{1}{2}} + \dfrac{h}{v}$

$$t = \left[\frac{(2)(20 \, \text{m})}{8.5 \, \text{m/s}^2}\right]^{\frac{1}{2}} + \frac{20 \, \text{m}}{343 \, \text{m/s}} = 2.020 \, \text{s} + 0.058 \, \text{s}$$

$$t = 2.08 \, \text{s}$$

16.20. $v = f\lambda = f(2L)$

$$t = \frac{L}{v} = \frac{L}{2fL} = \frac{L}{2f} = \frac{1}{2(320 \, \text{s}^{-1})} = 1.56 \times 10^{-3} \, \text{s}$$

CHAPTER 17

Temperature, Heat, and Heat Transfer

Thermodynamics deals with the internal energy of systems (thermal energy) and how this energy is transferred from one system to another. A central concept in thermodynamics is that of temperature. The temperature of an object determines or influences many of its properties. No branch of science has greater consequences for human life than does thermodynamics.

17.1 Temperature

Temperature is a measure of how hot something is. We will see later that temperature is proportional to the average kinetic energy of the atoms in a substance. At high temperature the atoms jump around more vigorously. Be sure to recognize the distinction between temperature and heat: **Heat** is energy that flows between two objects because of their temperature difference.

When you touch something, you can tell if it is hot or cold, but your senses are not reliable, and psychological factors can mislead you. To measure temperature accurately, we need to measure some physical property that varies with temperature in a repeatable way. If two objects are placed in contact, they will come to the same temperature. **Two objects at the same temperature are in thermal equilibrium.** The basis for making a thermometer is the **zeroth law of thermodynamics**.

> **If body *A* is in thermal equilibrium with body *C*, and body *B* is in thermal equilibrium with body *C*, then *A* is in thermal equilibrium with *B*.**

Suppose we want to measure the temperature of two beakers of liquid, *A* and *B*. We place a thermometer (body *C*) in contact with *A* and record a reading, say, 20°C. Now we place the thermometer in contact with liquid *B* and again obtain the reading 20°C. Then we know that *A* and *B* are at the same temperature and are in thermal equilibrium.

In order to attach numbers to temperature, we arbitrarily call the temperature at the triple point of water 273.19 K. There are historical reasons for this funny choice of a number, but I will not explain them here. The **triple point of water** occurs when the liquid, solid (ice), and vapor phases of water all coexist. This temperature scale is the **Kelvin** or **absolute** temperature scale. We say that the water triple point is "273.19 Kelvin," not "273.19 degrees Kelvin." On this scale 0 K is absolute zero, that is, the point at which classically all motion of atoms stops. This is not quite an accurate statement, since there are quantum mechanical effects that result in a "zero-point motion" at a temperature of 0 K, but this need not concern us for most ordinary purposes.

A standard thermometer is made using a small amount of gas contained in a flask. The pressure of such a gas is proportional to its temperature on the Kelvin scale, and it is calibrated so that the triple point of water is 273.19 K. On the Kelvin scale water boils at 1 atm at 100 K above the triple point, or 373.19 K.

In everyday work we often use the Celsius temperature scale. This scale is defined so that

$$T_C = T_K - 273.15 \tag{17.1}$$

Thus ice water is at about 0°C or 273 K, and boiling water is at 100°C or 373 K.

In some backward countries (such as the United States) the Fahrenheit scale is still used. A Fahrenheit degree is only 5/9 as large as a Celsius degree, and ice water is said to be at 32°F. Thus

$$T_F = \frac{9}{5}T_C + 32° \quad \text{or} \quad T_C = \frac{5}{9}(T_F - 32°) \tag{17.2}$$

PROBLEM 17.1. Express the following temperatures using the other temperature scales: 98°C, −40°F, and 77 K.

Solution For 98°C,

$$T_K = T_C + 273 = 98°C + 273°C = 371°F$$
$$T_F = \tfrac{9}{5}T_C + 32 = \tfrac{9}{5}(98) + 32 = 208°F$$

For −40°F,

$$T_C = \tfrac{5}{9}(T_F - 32) = \tfrac{5}{9}(-40 - 32) = -40°C$$

$$T_K = T_C + 273 = -40 + 273 = 233 \text{ K}$$

For 77 K,

$$T_C = T_K - 273 = 77 - 273 = -196°C$$
$$T_F = \tfrac{9}{5}T_C + 32 = -321°F$$

The variation of a physical property with temperature can be used to make a thermometer, and many types are used. The thermal expansion of solids and liquids, the variation in electrical resistance, and changes in color are all used as the bases for thermometers.

17.2 Thermal Expansion

When the temperature of a solid or liquid is increased, the atoms vibrate more vigorously, and they tend to push each other farther apart. There are some exceptions to this behavior; for example, because of its unique structure, water contracts between 0°C and 4°C. This has important consequences for the temperature distribution in lakes. If at an initial temperature T_0 an object has length L_0 along some dimension, its length will change by an amount $\Delta L = L - L_0$ when the temperature is changed to T, where $\Delta T = T - T_0$. If $T < T_0$, the object becomes smaller. Experimentally it is found that if ΔT is small, then

$$\Delta L = \alpha L_0 \Delta T \tag{17.3}$$

α is the **coefficient of linear expansion**, with units °C^{-1}.

Area and volume V also change with temperature. Thus for a small square of material originally of area $A_0 = L_0 \times L_0$ at temperature T_0, the area at temperature T is

$$A = L \times L = L_0(1 + \alpha \Delta T)L_0(1 + \alpha \Delta T) = L_0^2\big[1 + 2\alpha \Delta T + \alpha^2(\Delta T)^2\big]$$

Since $\alpha \ll 1$, I neglect the last term, and $A \simeq L_0^2(1 + 2\alpha \Delta T) = A_0(1 + 2\alpha \Delta T)$. Thus

$$\Delta A = 2\alpha A_0 \Delta T \tag{17.4}$$

In a similar fashion a small cube initially of side L_0 and volume V_0 changes to volume V when the temperature is changed. $V = L_0^3(1 + \alpha \Delta T)^3 \simeq V_0(1 + 3\alpha \Delta T)$. Thus

$$\boxed{\Delta V = 3\alpha V \Delta T = \beta V \Delta T} \tag{17.5}$$

Here β is the **coefficient of volume expansion**.

PROBLEM 17.2.   In order to keep concrete slabs from buckling or cracking, felt or wood spacers are often placed between sections. Temperature variations might range from $-10°C$ in winter to $35°C$ in summer. If a slab is 10 m long at $-10°C$, how much will its length increase at $35°C$? For concrete, $\alpha = 1 \times 10^{-5}°C^{-1}$.

Solution $\Delta L = L_0 \alpha \Delta T = (10\,\text{m})(1 \times 10^{-5}°C^{-1})(45°C) = 4.5 \times 10^{-3}\,\text{m}$

PROBLEM 17.3.   Aluminum rivets used in airplane construction are made slightly larger than the holes into which they fit and then cooled in dry ice (CO_2) to $-78°C$ before being inserted. When they then warm up to room temperature ($23°C$), they fit very tightly. If a rivet at $-78°C$ is to be inserted into a hole of diameter 3.20 mm, what should the diameter of the rivet be at $23°C$? For aluminum, $\alpha = 2.4 \times 10^{-5}°C^{-1}$.

Solution $L = L_0(1 + \alpha \Delta T) = (3.20\,\text{mm})\left[1 + (2.4 \times 10^{-5}°C^{-1})(23°C + 78°C)\right] = 3.21\,\text{mm}$

PROBLEM 17.4.   Here is a problem I've encountered a few times, and it could lead to a dangerous accident and fire. I filled the gas tank of my truck (25 gal) all the way to the top when the temperature was $23°C$. After sitting in the Sun all day, the temperature of the gasoline and the steel tank rose to about $35°C$. The coefficient of volume expansion for gasoline ($96 \times 10^{-5}°C^{-1}$) is greater than that for steel ($1.1 \times 10^{-5}°C^{-1}$), and so some gas overflowed out of the tank and on to the ground. Fortunately no one threw a lighted cigarette on the ground by my rig. About how much gas would overflow under these conditions?

Solution   Let $\Delta V_G =$ the volume increase of gas and $\Delta V_T =$ the volume increase of tank. Then the overflow is

$$\Delta V = \Delta V_G - \Delta V_T = V_0 \beta_G \Delta T - V_0 \beta_T \Delta T = V_0 \Delta T(\beta_G - \beta_T)$$

$$= (25\,\text{gal})(12°C)(96 \times 10^{-5} - 1.1 \times 10^{-5})(°C^{-1}) = 0.28\,\text{gal}$$

Note: Observe that a hollow steel tank expands in volume just as if it were a solid piece of steel (not hollow). By the same reasoning, a hole in a sheet of metal expands in diameter when the sheet is heated. To convince yourself of this, imagine drawing a circle on a sheet with no hole in it. Now heat the sheet. The circle expands, as does the rest of the sheet. Now cut out the circle. This is where the hole would have been had you cut it out initially before heating the sheet. The hole becomes bigger when the sheet is heated.

Thermal expansion can result in very large forces when an object is constrained so that its expansion is limited, as illustrated by the following problem.

PROBLEM 17.5.   In an engine a cylinder of steel has a cross-sectional area of $1.2\,\text{cm}^2$ and a length of 2.00 cm at $23°C$. When it is heated to $90°C$, what compressive force would have to be applied to it to keep it from expanding? For steel, $\alpha = 1.2 \times 10^{-5}°C^{-1}$ and Young's modulus is $Y = 20 \times 10^{10}\,\text{Pa}$ (see Eq. 12.6).

Solution

$$Y = \frac{F/A}{\Delta L/L_0} \quad \text{and} \quad \Delta L = L_0 \alpha \Delta T, \quad \text{so} \quad F = AY\frac{\Delta L}{L_0} = AY\alpha\,\Delta T$$

$$F = (1.2 \times 10^{-4}\,\text{m}^2)(20 \times 10^{10}\,\text{Pa})(1.2 \times 10^{-5}\,{}^{\circ}\text{C}^{-1})(67{}^{\circ}\text{C})$$

$$= 1.9300 \times 10^4\,\text{N} \quad (1\,\text{Pa} = 1\,\text{N/m}^2)$$

17.3   Heat and Thermal Energy

The **internal energy** (also called **thermal energy**) of a system is the collective kinetic and potential energy associated with the random motion of the atoms and molecules comprising the system. When a system at temperature T is placed in contact with surroundings at a different temperature, energy will be transferred into or out of the system.

> **Heat is the energy transferred between a system and its surroundings because of their temperature difference.**

Heat flow Q is positive when energy flows into a system and negative when heat flows out. Heat is measured in joules. Other units used are calories and British thermal units (Btu's); $1\,\text{cal} = 4.186\,\text{J}$ and $1\,\text{Btu} = 252\,\text{cal} = 1054\,\text{J}$.

Note that the internal energy of a system can change if heat is added to the system or if work is done on the system. Whereas pressure, volume, and temperature are intrinsic properties of a system, heat and work are not. Do not say that a system "contains" heat or work.

17.4   Heat Capacity and Latent Heat

When heat is added to a substance, it will become hotter unless it changes from one phase (for example, solid, liquid, or gas) to another phase. The resulting rise in temperature ΔT depends on the mass of the substance, on the heat added, and on the kind of material. The amount of heat required to raise a substance by $1{}^{\circ}\text{C}$ is called the **heat capacity**. The amount of heat required to raise 1 kg of a substance by $1{}^{\circ}\text{C}$ is the **specific heat** c. Thus if heat Q causes mass m to rise in temperature by ΔT, then $c = Q/m\Delta T$, or

$$\boxed{Q = mc\,\Delta T} \tag{17.6}$$

Note that for water, $c = 1\,\text{cal/g} \cdot {}^{\circ}\text{C} = 4190\,\text{J/kg} \cdot {}^{\circ}\text{C}$. The specific heat of water is much larger than that of most other substances.

PROBLEM 17.6.   Eighty grams of brass (specific heat 0.092 cal/g-K) at $292{}^{\circ}\text{C}$ is added to 200 g of water (specific heat 1 cal-K) at $14{}^{\circ}\text{C}$ in an insulated container of negligible heat capacity. What is the final temperature of the system?

Solution   The heat lost by brass equals the heat gained by water, so

$$m_c c_c(292 - T) = m_w c_w(T - 14)(80\,\text{g})(0.092\,\text{cal/g} \cdot {}^{\circ}\text{C})(292 - T)$$

$$= (200\,\text{g})(1\,\text{cal/g} \cdot {}^{\circ}\text{C})(T - 14) \quad T = 23.9{}^{\circ}\text{C}$$

PROBLEM 17.7.   To 160 g of water at $10{}^{\circ}\text{C}$ is added 200 g of iron ($c = 0.11\,\text{cal/g} \cdot {}^{\circ}\text{C}$) at $80{}^{\circ}\text{C}$ and 80 g of marble ($c = 0.21\,\text{cal/g} \cdot {}^{\circ}\text{C}$) at $20{}^{\circ}\text{C}$. What is the final temperature of the mixture?

Solution   I can see that the water will get hotter and the iron will get colder, but I can't tell if the marble will get hotter or colder. No matter. Just call the final temperature T. Suppose the marble heats up. Then heat loss of iron equals the heat gain of water plus the heat gain of marble. Thus

$$m_I c_I(80 - T) = m_w c_w(T - 10) + m_M c_M(T - 20)$$

If I am mistaken in assuming that the marble heats up (that is, if T is actually less than 20°C), the heat "gained" by the marble will turn out to be negative; that is, the marble will cool off. Substitute numerical values and solve.

$$(200)(0.11)(80 - T) = (160)(1)(T - 10) + (80)(0.21)(T - 20), \quad T = 18.6°C$$

PROBLEM 17.8. A solar collector placed on the roof of a house consists of a black plastic sheet of area $5.0\,m^2$ behind which are copper coils through which water runs. The collector faces the Sun, and the intensity of sunlight incident on it is approximately $1000\,W/m^2$. The water circulating through the coils is to be heated by 38°C. Assuming all of the sunlight energy goes into heating the water, at what rate, in liters per minute, should water circulate through the coils?

Solution The energy absorbed per minute is

$$Q = (1000\,W/m^2)(5\,m^2)(60\,s/min)$$

$$= 3 \times 10^5\,J/min$$

Thus

$$m = \frac{Q}{c\,\Delta T} = \frac{3 \times 10^5\,J/min}{4200\,J/kg \cdot °C}(38°C) = 1.9\,kg/min = 1.9\,l/min$$

Solid, liquid, and gas (or vapor) are called **phases** of matter. A gas in contact with the liquid form of the same substance is called a **vapor**. Energy must be added to a substance in order to change it from solid to liquid (to melt it) or from liquid to gas (to boil it). Conversely, energy must be removed to change from gas to liquid (to condense it) or from liquid to solid (to freeze it). The energy that must be added (or removed) to cause the solid–liquid transition in 1 kg of a given material is called the **latent heat of fusion** L_f. The energy that must be added (or removed) to cause the liquid–gas transition in 1 kg of material is the **latent heat of vaporization** L_v.

PROBLEM 17.9. An espresso stand prepares steamed milk by bubbling steam at 140°C through a cup of milk at 30°C, raising the temperature of the milk and its container to 50°C. What mass of steam is required to heat 220 g of milk (essentially water) in a cup of mass 100 g given the following data: The specific heat of milk $= 1\,kcal/kg \cdot °C$, the specific heat of the cup $= 0.20\,kcal/kg \cdot °C$, the specific heat of the steam $= 0.48\,kcal/kg \cdot °C$, the heat of vaporization of water $= 540\,kcal/kg$, and the boiling temperature of water $= 100°C$.

Solution Key idea: Thermal energy lost by steam = energy gained by milk plus cup. First the steam cools from 140°C to 100°C, releasing energy $Q_1 = m_s c_s T1$. Now the steam turns into water, releasing energy $Q_2 = m_s L_v$. Finally, the steam (which is now liquid water) cools from 100°C to 50°C, releasing energy $Q_3 = m_s c_w \Delta T_2$. The energy gained by the milk is $Q_4 = m_m c_m \Delta T_3$. The energy gained by the cup is $Q_5 = m_c c_c \Delta T_4$. Thus

$$Q_1 + Q_2 + Q_3 = Q_4 + Q_5$$

$$m_s(0.48\,kcal/kg \cdot °C)(140°C - 100°C)$$
$$+ m_s(540\,kcal/kg) + m_s(1\,kcal/kg \cdot °C)(100°C - 50°C)$$
$$= (0.220\,kg)(1\,kcal/kg \cdot °C)(50°C - 30°C)$$
$$+ (0.100\,kg)(0.20\,kcal/kg)(50°C - 30°C)$$

Solve:

$$m_s = 0.0079\,kg = 7.9\,g$$

Observe that the energy given up during the phase transformation is much greater than that associated with the temperature change.

PROBLEM 17.10.   Heat is added at the rate of 200 W to an isolated sample of mass 0.40 kg. The temperature is measured as a function of time, and the curve shown here is obtained. From this information determine

(a) The melting and boiling temperatures of the substance
(b) The nature of the substance at $t = 7.5$ min
(c) The latent heat of fusion
(d) The latent heat of vaporization
(e) The specific heat of the liquid phase
(f) Whether the specific heat of the solid is greater than or less than the specific heat of the liquid

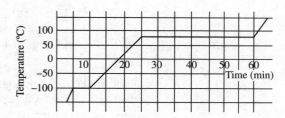

Solution   Up until $t = 5$ min, the substance is a solid. It then starts to melt (melting temperature $-100°C$). At $t = 10$ min, the material is all liquid and then rises in temperature until it starts to boil at $80°C$. At $t = 60$ min, the material is all vapor (gas).

(a) $T_M = -100°C$ and $T_B = 80°C$.
(b) At $t = 7.5$ min, half of the material has melted, so the mass is 50 percent liquid and 50 percent solid.
(c) The material melts in 5 min, so

$$L_f = \frac{(5 \text{ min })(60 \text{ s/min })(200 \text{ J/s})}{0.4 \text{ kg}} = 1.5 \times 10^5 \text{ J/kg}$$

(d) Thirty-five minutes is required to boil the material completely into vapor, so

$$L_v = \frac{(35 \text{ min })(60 \text{ s/min })(200 \text{ J/s})}{0.4 \text{ kg}} = 1.05 \times 10^6 \text{ J/kg}$$

(e) The liquid rises from $-100°C$ to $80°C$ in 15 min, so

$$c = \frac{Q}{m\Delta T} = \frac{(15 \text{ min })(60 \text{ s/min })(200 \text{ J/s})}{(0.4 \text{ kg})(180°C)} = 2500 \text{ J/kg} \cdot °C$$

(f) The specific heat of the liquid is greater than that of the solid because the temperature of the liquid rises more slowly than does that of the solid.

17.5   Heat Transfer

When two interacting systems or objects are at different temperatures, heat energy will flow from the hot one to the cold one. There are three mechanisms of heat transfer, as described below.

Conduction: If you heat one end of a metal rod with a flame, the atoms there will begin to vibrate more vigorously as the temperature rises. Since they interact with their neighbors, the neighbors will also vibrate more vigorously. This process continues, much in the same way a sound wave is transmitted along a rod (except that here the vibrations are random), with the atoms remaining more or less fixed, but passing energy to their neighbors. In metals there are many free electrons that can move out of hot regions, thereby contributing another aspect to conduction.

Consider a slab of material of cross-sectional area A and thickness Δx. One face is maintained at temperature T_1 and the other at T_2 (Fig. 17-1). Experimentally we find that the thermal energy ΔQ that flows through the slab in time Δt is $\Delta Q = kA(\Delta T/\Delta x)\Delta t$, where $\Delta T = T_2 - T_1$ and k is the **thermal conductivity** of the material.

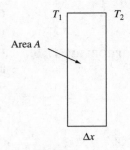

For a slab of infinitesimal thickness dx and temperature dT, this becomes

$$H = \frac{dQ}{dt} = -kA\frac{dT}{dx} \qquad (17.7)$$

Fig. 17-1

The minus sign in Eq. 17.7 indicates that heat flows from high to low temperature.

PROBLEM 17.11.   A windowpane is 1.2 m high and 0.6 m wide. The glass is 4 mm thick and has thermal conductivity $k = 0.78$ W/m $\cdot$ °C. The temperature of the inner face of the glass is 12°C and that of the outer surface is -12°C. What is the rate of heat flow through the window?

Solution $H = kA\dfrac{\Delta T}{\Delta x} = -\dfrac{(0.78\,\text{W/m} \cdot \text{°C})(1.2\,\text{m})(0.6\,\text{m})(12\text{°C} - 12\text{°C})}{0.004\,\text{m}}$, $H = 3.4\,\text{kW}$

PROBLEM 17.12.   Two slabs, of thicknesses L_1 and L_2 and area A, are in thermal contract with their surfaces at temperatures T_1 and T_2, respectively. What is the temperature at the interface between the two slabs? What is the rate of heat flow?

Solution   The heat flow through each slab is the same, so

$$H = -k_1 A\frac{(T - T_1)}{L_1} = -k_2 A\frac{(T_2 - T)}{L_2}$$

Solve:

$$T = \frac{k_1 L_2 T_1 + k_2 L_1 T_2}{k_1 L_2 + k_2 L_1}$$

and

$$H = \frac{A(T_2 - T_1)}{L_1/k_1 + L_2/k_2}$$

Convection: The transfer of thermal energy by motion of material is called **convection**. Natural convection results from the fact that when a gas or liquid is heated it expands and floats upward, carrying thermal energy with it. This is a major factor in determining the weather. This is also the mechanism for circulation of water in lakes, a process essential for animal and plant life. Forced convection of hot air is used to heat most buildings.

Radiation: All objects emit electromagnetic radiation, and this radiation carries energy. The power radiated from a surface area A at temperature T is given by the **Stefan-Boltzmann law**,

$$P = e\sigma AT^4 \qquad (17.8)$$

The emissivity is e, a dimensionless constant between 0 and 1 that depends on the nature of the surface. The constant $\sigma = 5.57 \times 10^{-8}$ W/m^2 $\cdot$ K. Note that T must be expressed in Kelvin. A hot object radiates electromagnetic waves at all frequencies (or wavelengths), but the frequency at which the most intense radiation is emitted shifts to higher values as the temperature is increased.

If an object is at temperature T and its surroundings are at temperature T_0, the net rate of energy loss is $P = e\sigma A(T^4 - T_0^4)$.

PROBLEM 17.13. The Sun's surface has a temperature of about 5800 K, and the radius of the Sun is about 7×10^8 m. Calculate the total energy radiated by the Sun each day, assuming the emissivity is $e = 1$. By way of comparison, the total energy consumed worldwide each year by humans is about 10^{21} J.

Solution
$$E = Pt = e\sigma AT^4 t = e\sigma(4\pi R^2)T^4 t$$

$$= (1)(5.57 \times 10^{-8}\,\text{W/m}^2 \cdot \text{K})(4\pi)(7 \times 10^8\,\text{m})^2(5800\,\text{K})^4(365\,\text{day})(24\,\text{h/day})(3600\,\text{s/h})$$

$$= 1.75 \times 10^{25}\,\text{J}$$

17.6 Summary of Key Equations

$$T_C = T_K - 273.2$$
$$T_F = \frac{9}{5}T_C + 32 \quad \text{or} \quad T_C = \frac{5}{9}T_F - 32$$
$$\Delta L = \alpha L_0\,\Delta T$$
$$Q = mc\Delta T, \quad Q_f = mL_f, \quad Q_v = mL_v$$
$$H = -kA\frac{dT}{dx}$$
$$P = e\sigma AT^4$$

SUPPLEMENTARY PROBLEMS

17.14. At what temperature is the Fahrenheit scale reading equal to (a) the reading on the Celsius scale, (b) half that of the Celsius scale, and (c) twice that of the Celsius scale?

17.15. Gold melts at $1064°C$ and boils at $2660°C$. Express these temperatures in Kelvin.

17.16. A steel beam is 12 m long when installed at $23°C$. By how much does its length change when it changes temperature from $-32°C$ to $55°C$? For steel, $\alpha = 1.1 \times 10^{-5}°C^{-1}$.

17.17. A 100-cm^3 beaker made of Pyrex glass ($\alpha = 3.2 \times 10^{-6}°C^{-1}$) is filled to the brim with water at $12°C$. What volume of water will overflow when the temperature is raised to $60°C$?

17.18. A brass sleeve of inside diameter 1.995 cm at $20°C$ is to be heated so that it will just barely slide over a shaft of diameter 2.005 cm. To what temperature must the sleeve be heated? For brass, $\alpha = 1.9 \times 10^{-6}°C^{-1}$.

17.19. A new engine design incorporates a piston that contains 0.60 kg of steel (specific heat $= 470\,\text{J/kg} \cdot \text{K}$) and 1.20 kg of aluminum (specific heat $= 910\,\text{J/kg} \cdot \text{K}$). How much heat is required to raise the piston from an initial temperature of $20°C$ to its operating temperature of $160°C$?

17.20. While resting, a person has a metabolic rate of about 4.2×10^5 J/h. This energy flows out of the body as heat. Suppose the person is immersed in a tub containing 1200 kg of water, initially at $27.0°C$ when the person gets in the tub. If the heat generated in the person goes only into the water, by how much will the water temperature rise in 1 h?

17.21. An ice cube of mass 60 g is taken from a freezer at $-15°C$ and dropped into 120 cm^3 of coffee at $80°C$. What will be the final temperature of the mixture? The specific heat of ice is $0.5\,\text{cal/g} \cdot °C$.

17.22. A 4.0-g lead bullet traveling at 350 m/s strikes a block of ice at 0°C. If all of the heat generated goes into melting the ice, how much ice is melted? The latent heat of fusion of ice is 80 kcal/kg, and its specific heat is 0.50 kcal/kg · °C.

17.23. A copper rod 24 cm long has cross-sectional area of 4 cm². One end is maintained at 24°C and the other is at 184°C. What is the rate of heat flow in the rod? The heat conductivity of copper is 397 W/m · °C.

17.24. We frequently encounter layers of thermally conducting material, for example, in a double-pane glass window. Derive an expression for the heat flow through three sheets of material, two of which are of the same material (conductivity k_1, thickness L_1, area A). The third sheet (with conductivity k_2, thickness L_2, area A) is sandwiched between the other two, in terms of the temperatures T_1 and T_4 on the outer surfaces.

17.25. We can estimate the temperature at the inner surface of the Earth's crust as follows: The temperature at the surface is about 10°C. The average thermal conductivity of rocks near the surface is about 2.5 W/m · K. In North America heat flows through the surface layer of the Earth at about 54 W/m². Neglecting any heat generated by radioactive decay within the crust, estimate the temperature at the inner surface of the Earth's crust, a depth of 35 km below the surface.

17.26. The surface of the Sun is at 6000 K, and the radius of the Sun is 6.96×10^8 m. The distance of the Sun from the Earth is 1.5×10^{11} m. Assuming the Sun is a perfect emitter (emissivity $e = 1$), find (a) the total power radiated by the Sun and (b) the intensity of sunlight in watts per square meter at the top of the Earth's atmosphere.

17.27. The tungsten filament of an incandescent light bulb is at 3200 K. The filament wire has a diameter of 0.08 mm and is 6 cm long. Assuming the emissivity $e = 1$, calculate the power radiated by the filament.

17.28. Thermal expansion can have serious consequences for steel beams used in constructing large buildings. To gain a feeling for the magnitude of this effect, calculate the fractional change in length for a beam of length L when its temperature changes from very hot, 50°C, to very cold, −50°C. For steel, the coefficient of thermal expansion is 1.2×10^{-5}°C^{-1}.

SOLUTIONS TO SUPPLEMENTARY PROBLEMS

17.14. (a) $T_F = 9/5T_C + 32$. If $T_F = T_C$, then $T_C = 9/5T_F + 32$ and $T_F = -40°$ F.

(b) If $T_F = 1/2T_C$, then $T_F = 9/5(2T_F) + 32$ and $T_F = 12.3°$ F.

(c) If $T_C = 1/2T_F$, then $T_F = 9/5(1/2T_F) + 32$ and $T_F = 320°$ F.

17.15. $1064°C = (1064 + 273)$ K $= 1337$ K, $2660°C = (2660 + 273)$ K $= 2933$ K

17.16. $\Delta L = \alpha L_0 \Delta T = (1.1 \times 10^{-5}°C^{-1})(12\,m)[55°C - (-32°C)] = 0.011$ m

17.17. $\Delta V = 3\alpha V_0 \Delta T = (3)(3.2 \times 10^{-6}°C^{-1})(100)(0.01\,m)^3(60°C - 12°C)$

$= 4.6 \times 10^{-8}\,m^3 = 0.046\,cm^3$

17.18. $\Delta V = \alpha L_0 \Delta T$, $\Delta T = \dfrac{\Delta L}{\alpha L_0} = \dfrac{(2.005 - 1.9995)\,cm}{(1.9 \times 10^{-6}°C^{-1})(1.9995\,cm)}$

$\Delta T = 263°C$,   so $T = T_0 + \Delta T = 20°C + 263°C = 283°C$

17.19. $Q = m_1 c \Delta T + m_2 c_2 \Delta T = (10.60\,kg)(470\,J/kg \cdot K)(160°C - 20°C)$

$+ (1.20\,kg)(910\,J/kg \cdot K)(160°C - 20°C) = 1.9 \times 10^5$ J

17.20. Heat lost by person = heat gained by person

$(4.2 \times 10^5\,J/h)(1\,h) = mc\Delta T = (1200\,kg)(4.2\,kJ/kg)(T - 27),$ $T = 100°C$

17.21. Heat gained by coffee = heat lost by ice

$$m_c c_c (80 - T) = m_w L_f + m_w c_w (T - 0)(0.120\,\text{kg})(1\,\text{kcal/kg} \cdot {}^\circ\text{C})(80 - T)$$
$$= (0.060\,\text{kg})(80\,\text{kcal/kg}) + (0.060\,\text{kg})(1\,\text{kcal/kg} \cdot {}^\circ\text{C})(T - 0) \quad T = 26.7^\circ$$

17.22. Energy lost by bullet = energy gained by ice

$$1/2 m_B v^2 = m_i L_f, \quad m_i = \frac{(0.004\,\text{kg})(350\,\text{m/s})^2 (1\,\text{kcal}/4200\,\text{J})}{(2)(80\,\text{kcal/kg})(4.2\,\text{kJ/kcal})} \quad m_i = 0.17 \times 10^{-3}\,\text{kg} = 0.17\,\text{g}$$

17.23. $H = kA \dfrac{\Delta T}{\Delta x} = (397\,\text{W/m} \cdot {}^\circ\text{C})(4 \times 10^{-4}\,\text{m}^2)\left(\dfrac{184^\circ\text{C} - 24^\circ\text{C}}{0.24\,\text{m}}\right) \quad H = 106\,\text{W}$

17.24. The heat flow through each sheet is the same:

$$
\begin{array}{lll}
T_1 - - - - - - - - & & \\
& k_1 & L_1 \\
T_2 - - - - - - - - & & \\
& k_2 & L_2 \\
T_3 - - - - - - - - & & \\
& k_1 & L_1 \\
T_4 - - - - - - - - & &
\end{array}
$$

$$H = k_1 A \frac{T_1 - T_2}{L_1} = k_2 A \frac{T_2 - T_3}{L_2} \quad \text{and} \quad k_2 A \frac{T_2 - T_3}{L_2} = k_1 A \frac{T_3 - T_4}{L_1}$$

Observe that

$$T_1 - T_4 = (T_1 - T_2) + (T_2 - T_3) + (T_3 - T_4) = \frac{HL_1}{k_1 A} + \frac{HL_2}{k_2 A} + \frac{HL_1}{k_1 A}$$

So

$$H = \frac{T_1 - T_4}{L_1/k_1 A + L_2/k_2 A + L_1/k_1 A} = \frac{A(T_1 - T_4)}{2L_1/k_1 + L_2/A_2}$$

17.25.
$$H = kA \frac{\Delta T}{\Delta x}, \text{ so } \Delta T = T - 10 = \frac{H \Delta x}{kA}$$

For $A = 1\,\text{m}^2$,

$$T = 10^\circ\text{C} + \frac{(54 \times 10^{-3}\,\text{W/m}^2)(35 \times 10^3\,\text{m})}{(2.5\,\text{W/m} - \text{K})(1\,\text{m}^2)}$$
$$= 766^\circ\text{C}$$

17.26. (a) $P = eA\sigma T^4 = 4\pi R^2 + T^4 = (1)(4\pi)(6.96 \times 10^8\,\text{m})^2 (5.67 \times 10^{-8}\,\text{W/m}^2 \cdot \text{K}^4)(600\,\text{K})^4 = 4.5 \times 10^{26}\,\text{W}$
(b) The power is spread over the surface of a sphere of radius R_{SE}, so the intensity is

$$I = \frac{P}{\text{area}} = \frac{4.5 \times 10^{26}\,\text{W}}{4\pi(1.5 \times 10^{11}\,\text{m})^2} = 1600\,\text{W/m}^2$$

17.27. $P = eA\sigma T^4 = e(\pi dL)\sigma T^4$

$$= (1)(\pi)(0.08 \times 10^{-3}\,\text{m})(0.06\,\text{m})(5.67 \times 10^{-8}\,\text{W/m}^2 \cdot \text{K}^4)(3200\,\text{K})$$
$$= 90\,\text{W}$$

17.28. $\Delta L = \alpha L_0 \Delta T$

$$\frac{\Delta L}{L} = \alpha \Delta T = (1.2 \times 10^{-5}\,{}^\circ\text{C}^{-1})(50^\circ\text{C} - (-50^\circ\text{C}))$$

$$\frac{\Delta L}{L} = 1.2 \times 10^{-3}$$

CHAPTER 18

The Kinetic Theory of Gases

Gases play an important role in many thermodynamic processes, and before investigating thermodynamics further, it is helpful to consider an ingenious way to understand the properties of gases. This idea, called the *kinetic theory of gases*, attempts to explain the macroscopic properties of a gas by examining the behavior of individual atoms or molecules making up the gas. At first this seems to be an overwhelming challenge, because the numbers of atoms involved are so large (there are more than 10^{27} atoms in a room). However, by using a statistical approach, we can predict with very high accuracy the characteristics of a gas. In the following I will assume we are dealing with an **ideal gas** with the following properties: The number of molecules is large and the gas is very dilute (that is, it is mostly empty space, so the average distance between molecules is small compared to the molecular size). I will treat the molecules as point objects. I assume that the molecules obey Newton's laws and that they move randomly. I assume that the molecules do not interact except when they undergo collisions, and I assume that the collisions are elastic. Real gases do not behave exactly like this, but this is a good place to start.

18.1   The Ideal Gas Law

Because there are so many particles in a gas, it is useful to have a large unit with which to measure their number. One **mole** is defined as 6.02×10^{23} particles (the particles can be electrons, helium atoms, or golf balls). This number is arrived at in the following way: The nucleus of ordinary carbon contains 12 particles (six protons and six neutrons). We say that 12 g of carbon is 1 mole of carbon, and 12 is the **atomic mass** of carbon. It turns out that 12 g of carbon contains 6.02×10^{23} atoms (this number is called **Avogadro's number** N_A). The atomic mass of oxygen is 16, so 16 g of oxygen contains N_A atoms. The same idea applies to molecules. **Molecules** are stable clusters of atoms, for example, H_2O or CO_2. The molecular mass is the sum of the atomic mass of the atoms that comprise the molecule. Thus the atomic mass of hydrogen is 1, the atomic mass of oxygen is 16, and the molecular mass of H_2O is $1 + 1 + 16 = 18$.

Experimentally it is found that the pressure, volume, and absolute temperature of a gas obey approximately the following equation of state, called the **ideal gas law**:

$$pV = nRT \tag{18.1}$$

Here n is the number of moles of the gas and $R = 8.31 \, \text{J/K}$, the **universal gas constant**. *Caution*: In using this equation, be sure to express temperature in K, not °C or °F.

PROBLEM 18.1.   A motorist starts a trip on a cold morning when the temperature is 4°C, and she checks her tire pressure at a gas station and finds the pressure gauge reads 32 psi. After driving all day, her tires heat up, and by afternoon the tire temperature has risen to 50°C. Assuming the tire volume is constant, to what pressure will the air in the tires have risen?

Solution   Note that a pressure gauge measures **gauge pressure**, that is, the pressure in excess of atmospheric pressure (1 atm = 15 psi = 15 lb/in). Thus the initial pressure is 47 psi at a temperature of 273 K + 4 K = 276 K. The final temperature is 323 K.

$$p_1V_1 = nRT_1 \quad \text{and} \quad p_2V_2 = nRT_2, \quad V_1 = V_2$$

Divide these equations:

$$\frac{p_1V_1}{p_2V_2} = \frac{nRT_1}{nRT_2}, \qquad p_2 = \frac{T_2}{T_1}p_1 = \left(\frac{323\,\text{K}}{277\,\text{K}}\right)(47\,\text{psi}) = 54.8\,\text{psi absolute}$$

or

$$P = 39.8\,\text{psi gauge pressure}$$

PROBLEM 18.2.   **Standard temperature and pressure (stp)** for a gas is defined as 0°C (273 K) and 1 atm (1.013×10^5 Pa). What volume does 1 mol of an ideal gas occupy?

Solution $$V = \frac{nRT}{P} = (1)\frac{(8.31\,\text{J/K})(273\,\text{K})}{1.013 \times 10^5\,\text{Pa}} = 22.4 \times 10^{-3}\,\text{m}^3 = 22.4\,\text{L}$$

PROBLEM 18.3.   How many molecules are in $1\,\text{cm}^3$ of helium gas at 300 K?

Solution $$N = nN_A = \frac{pV}{RT}N_A = \frac{(1.013 \times 10^5\,\text{Pa})(10^{-6}\,\text{m}^3)(6.02 \times 10^{23})}{(8.31\,\text{J/K})(300\,\text{K})} = 2.4 \times 10^{19}$$

18.2   Molecular Basis of Pressure and Temperature

An approximate expression for the pressure due to an ideal gas can be obtained as follows: Consider N molecules contained in a cubical container of side L, with edges aligned parallel to the xyz axes. A molecule moving along the positive x-axis with velocity v_x will collide elastically with a wall and bounce back with velocity $-v_x$. Its momentum in the x direction will change from $+mv_x$ to $-mv_x$, a change of $-2mv_x$. After striking the wall, the molecule will bounce back and travel in the negative x direction until it strikes the opposite wall and rebounds. Again it moves in the positive x direction until it strikes the first wall a second time. In traversing the box twice, the molecule travels a distance $2L$ in the x direction, and this requires time $2L/v_x$. Hence the time between collisions with the first wall is $2L/v_x$. The force exerted on the molecule by the wall in order to change its momentum by $-2mv_x$ is, from Newton's second law,

$$F' = \frac{\text{change in momentum}}{\text{time of change}} = \frac{-2mv_x}{2L/v_x} = -\frac{mv_x^2}{L}$$

By Newton's third law, the force exerted on the wall by the molecule is $F = -F'$. The total force exerted on the wall is the sum of the forces exerted by each molecule,

$$F = \frac{m}{L}\left(v_{x1}^2 + v_{x2}^x + \cdots + v_{Nx}^2\right)$$

But the average value of v_x^2 for N molecules is

$$\overline{v_x^2} = \frac{1}{N}\left(v_{1x}^2 + v_{2x}^2 + \cdots + v_{Nx}^2\right)$$

Thus

$$F = \frac{Nm}{L}\overline{v_x^2}$$

If one molecule has velocity components v_x, v_y, and v_z, then by the Pythagorean theorem, $v^2 = v_x^2 + v_y^2 + v_z^2$ and the average values are related by

$$\overline{v^2} = \overline{v_x^2} + \overline{v_y^2} + \overline{v_z^2}$$

Since the motion is random,

$$\overline{v_x^2} = \overline{v_y^2} = \overline{v_z^2} = \tfrac{1}{3}\overline{v^2}$$

The total force on the wall is thus

$$F = \frac{N}{3}\left(\frac{m\overline{v^2}}{L}\right)$$

The pressure on the wall is

$$p = \frac{F}{A} = \frac{F}{L^2} \quad \text{or} \quad \boxed{p = \frac{2}{3}\left(\frac{N}{V}\right)\left(\tfrac{1}{2}m\overline{v^2}\right)} \tag{18.2}$$

It is useful to define **Boltzmann's constant**, $k_B = 1.38 \times 10^{-23}$ J/K, where $R = N_A k_B$. The ideal gas law, Eq. 18.1, then becomes $pV = nRT = nN_A k_B T = Nk_B T$. Comparing this with Eq. 18.2, I see that they are the same, provided that

$$\boxed{T = \frac{2}{3k_B}\left(\tfrac{1}{2}m\overline{v^2}\right)} \tag{18.3}$$

This interesting result gives us added insight into the nature of temperature. **The absolute temperature of a gas is proportional to the average molecular kinetic energy.** Further, since

$$\overline{v_x^2} = \overline{v_y^2} = \overline{v_z^2} = \tfrac{1}{3}\overline{v^2}$$

Then

$$\tfrac{1}{2}m\overline{v_x^2} = \tfrac{1}{2}m\overline{v_y^2} = \tfrac{1}{2}m\overline{v_z^2} = \tfrac{1}{3}\left(\tfrac{1}{2}\overline{v^2}\right) = \tfrac{1}{2}k_B T \tag{18.4}$$

Equation 18.4 illustrates a general result called the **theorem of equipartition of energy** that says that each "degree of freedom" of a gas contributes an amount of energy $\tfrac{1}{2}k_B T$ to the total internal energy. A *degree of freedom* is an independent motion that can contribute to the total energy. For example, a molecule such as O_2 has, in principle, seven degrees of freedom. Three are associated with translation along the x, y, and z axes, three are associated with rotations about the x, y, and z axes, and one is associated with vibrations of the molecule along the $O-O$ axis (like masses vibrating on the ends of a spring). However, since the moment of inertia I for rotations about the $O-O$ axis is approximately zero, rotations about this axis add almost nothing to the energy (KE $= \tfrac{1}{2}I\omega^2$). Further, quantum mechanics shows that vibrational modes are not appreciably excited until the gas temperature is high, so for most purposes we assume a diatomic molecule has five degrees of freedom. A monatomic gas like helium has three degrees of freedom.

The total internal energy of n moles of a monatomic gas (with three degrees of freedom) is

$$\boxed{E = N\left(\tfrac{1}{2}m\overline{v^2}\right) = \tfrac{3}{2}Nk_B T = \tfrac{3}{2}nRT} \tag{18.5}$$

Equation 18.3 can be solved to find the root mean square (rms) molecular speed.

$$\boxed{v_{\text{rms}} = \sqrt{\overline{v^2}} = \sqrt{\frac{3k_B T}{m}} \quad \text{or} \quad v_{\text{rms}} = \sqrt{\frac{3RT}{M}}} \tag{18.6}$$

The last expression I obtained using $R = N_A k_B$ and $M = N_A m =$ molar mass in grams, where $m =$ mass of one molecule.

This *root mean square* speed v_{rms}, is an average speed, and some molecules move slower and some faster than this.

PROBLEM 18.4. What is the rms speed of a nitrogen molecule (N_2) in air at 300 K? The atomic mass of nitrogen is 14.

Solution

$$v_{rms} = \sqrt{\frac{3RT}{M}} = \sqrt{\frac{(3)(8.31 \, \text{J/K})(300 \, \text{K})}{28 \times 10^{-3} \, \text{kg}}} = 517 \, \text{m/s}$$

In the above discussion I assumed the molecules were point masses. If we make the more realistic assumption that the molecules are spheres of diameter d, it is possible to calculate the mean free path λ between collisions of the molecules. Using a statistical approach yields the result

$$\lambda = \frac{1}{\sqrt{2}\pi d^2 N/V} = \frac{RT}{\sqrt{2}\pi d^2 N_A P} \tag{18.7}$$

PROBLEM 18.5. Estimate the mean free path of an air molecule at 273 K and 1 atm, assuming it to be a sphere of diameter 4.0×10^{-10} m. Estimate the mean time between collisions for an oxygen molecule under these conditions, using $v = v_{rms}$.

Solution

$$\lambda = \frac{(8.31 \, \text{J/K})(273 \, \text{K})}{\sqrt{2}\pi(4 \times 10^{-10} \, \text{m})^2(6.02 \times 10^{23})(1.01 \times 10^5 \, \text{Pa})} = 5.2 \times 10^{-8} \, \text{m}$$

$$t = \frac{\lambda}{v_{rms}}$$

where $v_{rms} = 517 \, \text{m/s}$ from Problem 18.4. Thus

$$t \cong \frac{5.2 \times 10^{-8} \, \text{m}}{517 \, \text{m/s}} = 1.0 \times 10^{-10} \, \text{s}.$$

18.3   The Maxwell-Boltzmann Distribution

The molecules in a gas travel at a wide range of speeds. Using the methods of statistical mechanics (which I won't go into here), we can deduce the number of particles dN in a gas with speed between v and $v + dv$ to be

$$dN = N f(v) \, dv, \quad \text{where} \quad f(v) = 4\pi \left(\frac{m}{2\pi k_B T} \right)^{3/2} v^2 e^{-mv^2/2k_B T} \tag{18.8}$$

$f(v)$ is the **Maxwell-Boltzmann distribution function**, and N is the number of gas particles, each with mass m. Fig. 18-1 shows the distribution function for several temperatures. Note that $f(v)$ has units of reciprocal speed, s/m. The most probable speed is that corresponding to the peak of the distribution, that is, where $df/dv = 0$. The result is

$$v_{mp} = \sqrt{\frac{2k_B T}{m}} = \sqrt{\frac{2RT}{M}} \tag{18.9}$$

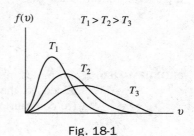

Fig. 18-1

$M = N_A m$ is the molar mass. Using the distribution function, we can calculate other quantities of interest. For example, the mean speed is

$$\bar{v} = \int_0^\infty v f(v)\, dv = \sqrt{\frac{8k_B T}{\pi m}} = \sqrt{\frac{8RT}{\pi M}} \tag{18.10}$$

The mean square speed is

$$\overline{v^2} = \int_0^\infty v^2 f(v)\, dv = \frac{3k_B T}{m} = \frac{3RT}{M} \tag{18.11}$$

The above integrals are tricky to work out. Look them up in tables.

18.4   Molar Specific Heat and Adiabatic Processes

The molar specific heat of a gas is the amount of energy that must be added to 1 mole of the gas to increase its temperature by 1 degree. Consider first 1 mole of an ideal monatomic gas with fixed volume V (that is, in a rigid container). The internal energy of the gas is given by Eq. 18.5 with $n = 1$, and $E = \frac{3}{2} RT$. Since the specific heat is defined as $C_v dT = dE$, $C_v = dE/dT = 3R/2$ (taking the derivative of E in Eq. 18.5). Thus for a monatomic ideal gas at constant volume,

$$C_v = \tfrac{3}{2} R \quad \text{monatomic gas} \tag{18.12}$$

If the gas is not monatomic and has f effective degrees of freedom, each degree of freedom contributes $\frac{1}{2} RT$ to the internal energy, and then $C = fR/2$. For example, a diatomic molecule like N_2 or O_2 has effectively five degrees of freedom, and so for it, $C = 5R/2$ per mole.

If the gas is not held at constant volume when heat is added to it, it will expand. Imagine a gas contained in a cylinder, above which is a piston on which rests a fixed weight (Fig. 18-2). The piston maintains a constant pressure on the gas. As the gas expands, it pushes the piston up and does work on it. If the area of the piston is A and it is pushed up a distance dx, the work done by the gas is $dW = F\, dx = PA\, dx = F\, dV$, where the increase in volume of the cylinder is $dV = A\, dx$ and $P = F/A$. Using the idea of the conservation of energy, heat added equals gain in internal energy plus work done, or

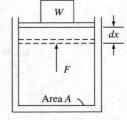

Fig. 18-2

$$dQ = dE + dW = dE + p\, dV$$

Dividing by dT yields $dQ/dT = dE/dT + p\, dV/dV$. The molar specific heat is $C_p = dQ/dT$. From the ideal gas law, $pV = nRT$, so $p\, dV/dT = R$ (with $n = 1$). From Eq. 18.5, $dE/dT = 3R/2$. Thus

$$C_p = 3\frac{R}{2} + R = 5\frac{R}{2} \quad \text{monatomic gas} \tag{18.13}$$

In general, for *any* ideal gas,

$$C_p = C_v + R \tag{18.14}$$

PROBLEM 18.6.   4.0 moles of argon gas is contained in a cylinder at 300 K. How much heat must be added to the gas to raise its temperature to 600 K at (a) constant volume and (b) constant pressure?

Solution

(a) $Q = nC_v\Delta T = n\frac{3}{2}R(T_2 - T_1) = (4)\left(\frac{3}{2}\right)(8.31\,\text{J/K})(600\,\text{K} = 30\,\text{K})$

$= 1.50 \times 10^4\,\text{J}$

(b) $Q = nC_p\Delta T = n\left(\frac{3}{2}R + R\right)(\Delta T) = (4)\left(\frac{5}{2}\right)(8.31\,\text{J/K})(600\,\text{K} - 300\,\text{K})$

$= 2.5 \times 10^4\,\text{J}$

An *adiabatic process* in a gas is one in which no heat flows into or out of the gas. Experimentally we can approximate this process by changing the volume of a gas very quickly, or by enclosing the gas in a well-insulated container, so that little heat can flow in or out. For such a process it is possible to show (trust me) that if $\gamma = C_p/C_v$, then

$$pV^\gamma = \text{constant} \quad \text{and} \quad TV^{\gamma-1} = \text{constant adiabatic} \qquad (18.15)$$

It is instructive to examine a graph of p versus V for the adiabatic case (no heat flow, $Q = 0$, $pV^\gamma = \text{constant}$) and for the isothermal case ($pV = nRT = \text{constant}$). Such curves are shown in Fig. 18-3. Note that the adiabat is steeper than the isotherms.

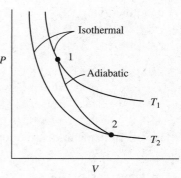

Fig. 18-3

PROBLEM 18.7 Consider an adiabatic process and an isothermal process that pass through a common point p_0, V_0 in a graph of p versus V (see Fig. 18-3). Show that the slope of the adiabatic curve is greater than that of the isothermal curve at this point.

Solution For the adiabat,

$$pV^\gamma = p_0V_0^\gamma = \text{constant}$$

$$P = \frac{p_0V_0^\gamma}{V^\gamma}, \quad \left(\frac{dp}{dV}\right)_{\text{ad}} = -\gamma\frac{p_0V_0^\gamma}{V^{\gamma+1}} = -\gamma\frac{p_0V_0^\gamma}{V_0^{\gamma+1}} = -\frac{\gamma p_0}{V_0} \text{ at } V = V_0$$

For the isotherm,

$$PV = nRT = p_0V_0, \quad p = \frac{p_0V_0}{V}$$

$$\left(\frac{dp}{dV}\right)_{\text{iso}} = -\frac{p_0V_0}{V^2} = -\frac{p_0V_0}{V_0^2} \text{ at } V = V_0, \quad \text{so} \left(\frac{dp}{dV}\right)_{\text{iso}} = -\frac{p_0}{V_0}$$

Since $\gamma > 1$, the slope of the adiabat is greater than the slope of the isotherm.

PROBLEM 18.8. Helium gas (a monatomic gas) at 400 K and 1 atm is compressed adiabatically from 20 to 4 L. What are the final temperature and pressure?

Solution

$$C_v = \frac{3}{2}R, \quad C_p = \frac{5}{2}R, \quad \gamma = \frac{C_p}{C_v} = \frac{5}{3} = 1.67$$

$$p_1V_1^\gamma = p_2V_2^\gamma, \quad p_2 = \left(\frac{V_1}{V_2}\right)^\gamma, \quad p_1 = \left(\frac{20\,\text{L}}{1\,\text{L}}\right)^{1.67}(1\,\text{atm}) = 14.6\,\text{atm}$$

$$pV^2 = pV, \quad V^{\gamma-1} = (nRT)V^{\gamma-1}$$

So

$$nRT_1 V_1^{\gamma-1} = nRT_2 V_2^{\gamma-1}, \qquad T_2 = \left(\frac{V_1}{V_2}\right)^{\gamma-1} T_1$$

$$T_2 = \left(\frac{20\,\text{L}}{1\,\text{L}}\right)^{0.67} (400\,\text{K}) = 1170\,\text{K}$$

18.5   Summary of Key Equations

Ideal gas law: $pV = nRT$

Kinetic theory: $p = \dfrac{2}{3}\left(\dfrac{N}{V}\right)\left(\dfrac{1}{2}m\overline{v^2}\right) \quad T = \dfrac{2}{3k_B}\left(\dfrac{1}{2}m\overline{v^2}\right) \quad E = N\left(\dfrac{1}{2}m\overline{v^2}\right) = \dfrac{3}{2}nRT$

$$v_{\text{mp}} = \sqrt{\frac{2RT}{M}} \quad \overline{v} = \sqrt{\frac{8RT}{\pi M}} \quad v_{\text{rms}} = \sqrt{\overline{v^2}} = \sqrt{\frac{3RT}{M}}$$

Mean free path: $\lambda = \dfrac{RT}{\sqrt{2}\,\pi d^2 N_A p}$

Maxwell-Boltzmann distribution: $f(v) = 4\pi\left(\dfrac{m}{2\pi k_B T}\right)^{3/2} v^2 e^{-(mv^2/2k_B T)}$

Monatomic gas: $C_v = \dfrac{3}{2}RT$

Diatomic gas: $C_v = \dfrac{5}{2}RT \quad C_p = C_v + R$

Adiabatic process: $(Q = 0) \quad pV^\gamma = \text{constant} \quad \text{or} \quad TV^{\gamma-1} = \text{constant}$

SUPPLEMENTARY PROBLEMS

18.9. An air compressor used for spraying paint has a tank of capacity 0.40 m that contains air at 27°C and 6.0 atm. How many moles of air does the tank hold?

18.10. In a certain lab experiment a small vessel of volume V holds an ideal gas at 300 K and 5 atm. The small vessel is connected to a large reservoir of volume 6 V that contains the same kind of gas at a pressure of 1 atm and a temperature of 600 K. A stopcock in the tube connecting the two containers keeps the gases separate. The temperature of each container is held constant. When the stopcock is opened, what will be the final pressure in each container?

18.11. 6×10^{22} molecules of an ideal gas are stored in a tank at 0.5 atm at 37°C. Determine (a) the pressure in Pascals and the temperature in Kelvin, (b) the volume of the tank, and (c) the pressure when the temperature is increased to 152°C.

18.12. The best vacuum attainable in the laboratory is about 5.0×10^{-18} Pa at 293 K. How many molecules are there per cubic centimeter in such a vacuum?

18.13. A hot air balloon stays aloft because it experiences a buoyant lift force equal to the weight of the air it displaces. Suppose a balloon of volume 420 m^3 in air at 13°C, the density of which is 1.24 kg/m. To what temperature must the hot air be raised if the balloon is to support a payload of 240 kg (not including the mass of the hot air in the balloon)?

18.14. In helium gas (molar mass 4 g) at 300 K, the rms speed of an atom is 1350 m/s. What is the rms speed of an oxygen molecule (molar mass 32 g) at this temperature?

18.15. Natural uranium occurs in two common forms, ^{238}U and ^{235}U. The latter isotope constitutes only about 0.7 percent of the uranium mined, and it is the uranium commonly used in power reactors and bombs. The technical problem of separating the two isotopes (and thereby obtaining "enriched" fuel) played a key role in World War II. The United States was able successfully to separate significant amounts of ^{235}U, whereas Germany could not do so, and this accomplishment was probably pivotal in determining the outcome of the war. Even today the difficulty in obtaining enriched uranium is an important deterrent in the use of nuclear weapons by terrorists. ^{238}U and ^{235}U are very similar chemically, so an ingenious diffusion technique is used to separate them. Both isotopes are made into a gas of uranium hexafluoride (UF$_6$), of molecular masses 349 and 352, respectively. They are allowed to diffuse through a porous membrane, and the lighter ^{235}UF$_6$ molecules diffuse more rapidly and are then harvested. What is the percentage difference in their rms speeds at a given temperature?

18.16. Argon gas (with an atomic diameter of about 3.1×10^{-10} m) is used in a laboratory vessel maintained at 300 K. To what pressure must the container be evacuated in order that the mean free path be 1 cm?

18.17. A room in a well-insulated building has a volume of 120 m^3. The air in the room is at 21°C. How much heat must be added to the air in order to increase its temperature by 1°C?

18.18. Two moles of air ($C_v = 5R/2$) at 300 K are contained under a heavy piston in a cylinder of volume 6.0 L. If 5.2 kJ of heat is added to the gas, what will be the resulting volume of the gas?

18.19. During the compression stroke of an internal combustion engine, the pressure changes adiabatically from 1 to 18 atm. Assuming the gas is ideal and has $\gamma = 1.4$, by what factor does the temperature change? By what factor does the volume change?

18.20. It is not uncommon where I live in Idaho for a warm west wind to blow down from the Cascade Mountains and melt all of the snow right in the midst of the ski season. This wind, called a "Chinook," a Native American word meaning "snow eater," is encountered all over the world. In southern California where I grew up, we called it a "Santa Ana" (I think because it came from the direction of the city by that name). When I lived in Zurich, Switzerland, they called the wind a "foehn" when it blew down out of the Alps. People said it gave them headaches and a reason for leaving work early. One guy said it impelled him to kill his wife. If one considers the wind dropping down from a high elevation to undergo an adiabatic compression, it is possible to calculate the resulting temperature rise. Suppose a Chinook wind blows down from the Rockies (elevation 4300 m, temperature −20°C, pressure 5.5×10^4 Pa) eastward into Denver (elevation 1630 m, temperature 2°C, and pressure 9.10×10^4 Pa before the wind arrives). By how many degrees will the temperature rise when the Chinook arrives?

18.21. Air containing gasoline vapor is injected into the cylinder of an internal combustion engine. The initial pressure and temperature are 1 atm and 300 K. The piston rapidly compresses the gas from 400 cm^3 to 50 cm^3. Since the compression is very rapid, there is no time for heat to flow into or out of the gas, so we may approximate the compression as adiabatic. Also, air is a mixture of nitrogen and oxygen, two diatomic gases, so we use the value $\gamma = 1.4$ for the gas constant. What are the final temperature and pressure of the gas?

SOLUTIONS TO SUPPLEMENTARY PROBLEMS

18.9. $n = \dfrac{pV}{RT} = \dfrac{(6 \text{ atm})(1.013 \times 10^5 \text{ Pa/atm})(0.4 \text{ m}^3)}{(8.31 \text{ J/K})(300 \text{ K})} = 97.5 \text{ mol}$

18.10. Initially,

$$n_1 = \frac{p_1 V_1}{RT_1} \quad \text{and} \quad n_2 = \frac{p_2 V_2}{RT_2}$$

Finally,

$$p_1 = p_2 = p \quad \text{and} \quad n_1' = n_1 = \frac{pV_1}{RT_1} \quad \text{and} \quad n_2' = \frac{pV_2}{RT_2}$$

$$n_1 + n_2 = n_1' + n_2' \qquad \frac{p_1V_1}{RT_1} + \frac{p_2V_2}{RT_2} = \frac{pV_1}{RT_1} + \frac{pV_2}{RT_2}$$

$$V_2 = 6V_1 \qquad p = \frac{p_1T_2 + 6p_2T_1}{T_2 + 6T} = \frac{(5\,\text{atm})(600\,\text{K}) + 6(1\,\text{atm})(300\,\text{K})}{600\,\text{K} + 300\,\text{K}} \qquad P = 5.3\,\text{atm}$$

18.11. (a) $p = 0.50\,\text{atm} = (0.50\,\text{atm})(1.013 \times 10^5\,\text{Pa/atm}) = 5.07 \times 10^4\,\text{Pa}$

$T = 37°\text{C} = (37 + 273)\,\text{K} = 310\,\text{K}$

(b) $V = \dfrac{nRT}{p} = \left(\dfrac{N}{N_A}\right)\dfrac{RT}{p} = \left(\dfrac{6 \times 10^{22}}{6.02 \times 10^{23}}\right)\dfrac{(8.31\,\text{J/K})(310\,\text{K})}{5.07 \times 10^4\,\text{Pa}} = 5.1 \times 10^{-3}\,\text{m}^3$

(c) $\dfrac{p_1}{T_1} = \dfrac{p_2}{T_2} = \dfrac{nR}{V} \qquad p_2 = \dfrac{T_2}{T_1} \qquad p_1 = \left(\dfrac{425\,\text{K}}{310\,\text{K}}\right)(0.5\,\text{atm}) = 0.69\,\text{atm}$

18.12. $pV = nRT = \dfrac{N}{N_A}RT \qquad \dfrac{N}{V} = \dfrac{N_A p}{RT} = \dfrac{(6.02 \times 10^{23})(5 \times 10^{-18}\,\text{Pa})}{(8.31\,\text{J/K})(293\,\text{K})} = 1240/\text{m}^3 = 1.2/\text{cm}^3$

18.13. $m_L g + \rho_H g V = \rho_C g V \qquad \rho_H = \text{density of hot air}$

$\rho_H = \dfrac{\rho_C V - m_L}{V} = \rho_C - \dfrac{m_L}{V} \qquad \rho_C = \text{density of cold air}$

Also,

$$\rho = \frac{nM}{V}$$

where n = number of moles and M = molar mass.

$$\rho = \frac{nM_p}{mRT} = \frac{M_p}{RT}, \quad \text{so} \quad \frac{\rho_H}{\rho_C} = \frac{T_C}{T_H}$$

Thus

$$\frac{\rho_H}{\rho_C} = \frac{T_C}{T_H} = 1 - \frac{m_L}{\rho_C V} = 1 - \frac{240\,\text{kg}}{(1.24\,\text{kg/m}^3)(420\,\text{m}^3)} = 0.54 \qquad T_H = \frac{T_C}{0.54} = \frac{286\,\text{K}}{0.54} = 530\,\text{K} = 257°\text{C}$$

18.14. $v_{\text{rms}} = \sqrt{\dfrac{3RT}{M}}, \quad \text{so} \quad \dfrac{v_{\text{rms}} - 0}{v_{\text{rms-He}}} = \dfrac{\sqrt{3RT/M_0}}{\sqrt{3RT/M_{\text{He}}}} = \sqrt{\dfrac{M_{\text{He}}}{M_0}} = \sqrt{\dfrac{4}{32}} = 0.35$

$v_{\text{rms-0}} = 0.35 v_{\text{rms-He}} = (0.35)(1350\,\text{m/s}) = 477\,\text{m/s}$

18.15. Let

$$\alpha = \frac{^{235}v_{\text{rms}}}{^{238}v_{\text{rms}}} = \frac{\sqrt{3RT/^{235}M}}{\sqrt{3RT/^{238}M}} = \sqrt{\frac{^{238}M}{^{235}M}} = \sqrt{\frac{352}{249}} = 1.0043$$

$$\text{Percent difference} = \frac{^{235}v_{\text{rms}} - ^{238}v_{\text{rms}}}{^{238}v_{\text{rms}}} = \alpha - 1 = 0.0043 = 0.43 \text{ percent}$$

18.16. $\lambda = \dfrac{RT}{\sqrt{2}\pi d^2 N_A p}, \quad p = \dfrac{RT}{\sqrt{2}\pi d^2 N_A \lambda}$

$p = \dfrac{(8.31\,\text{J/K})(300\,\text{K})}{(\sqrt{2}\pi)(3.1 \times 10^{-10}\,\text{m})^2(6.02 \times 10^{23})(0.01\,\text{m})} = 3.0 \times 10^{-10}\,\text{Pa} = 3 \times 10^{-15}\,\text{atm}$

18.17. $Q = nC_v\Delta T, \quad n = \dfrac{pV}{RT}, \quad C_v = \tfrac{5}{2}R, \text{ so } Q = \left(\dfrac{pV}{RT}\right)\left(\tfrac{5}{2}R\right)\Delta T$

$Q = \dfrac{(5pV)(\Delta T)}{T} = \dfrac{(5)(1.01 \times 10^5\,\text{Pa})(120\,\text{m}^3)(1\,\text{K})}{294\,\text{K}} = 2.06 \times 10^5\,\text{J}$

18.18. $Q = nC_p(T_2 - T_1), \quad C_p = C_v + R = \tfrac{5}{2}R + R = \tfrac{7}{2}R$

$Q = \dfrac{7nR}{2}(T_2 - T_1), \quad T_2 = T_1 + \dfrac{2Q}{7nR}, \quad T_2 = 300\,\text{K} + \dfrac{(2)(5.2 \times 10^3\,\text{J})}{7(2)(8.31\,\text{J/K})} = 389\,\text{K}$

$pV_2 = nRT_2, \quad V_2 = \dfrac{nRT_2}{p} = \dfrac{nRT_2}{nRT_1}V_1 \quad \text{since } p = \dfrac{nRT_1}{V_1}, \quad V_2 = \dfrac{T_2}{T_1}V_1 = \dfrac{389\,\text{K}}{300\,\text{K}}(6.0\,\text{L}) = 7.8\,\text{L}$

18.19. $p_1 V_1^\gamma = p_2 V_2, \quad V_2 = \left(\dfrac{p_1}{p_2}\right)^{1/\gamma} = \left(\dfrac{1}{18}\right)^{1/1.4} = 0.13 V_1 \qquad pV^\gamma = p^{1-\gamma}(pV)^\gamma = p^{1-\gamma}(nRT)^\gamma$

so

$$p_1^{1-\gamma}T_1^\gamma = p_2^{1-\gamma}T_2^\gamma, \quad T_2 = \left(\dfrac{P_1}{P_2}\right)^{(1-\gamma)/\gamma} T_1 \quad T_2 = \left(\dfrac{1}{18}\right)^{(1-1.4)/1.4} T_1 = 2.3 T_1$$

18.20. From 18.11,

$$T_2 = \left(\dfrac{P_1}{P_2}\right)^{(1-\gamma)/\gamma} T_1 \quad T_2 = \left(\dfrac{0.7\,\text{atm}}{0.9\,\text{atm}}\right)^{(1-1.4)/1.4} (253\,\text{K}) = 272\,\text{K}$$

$$\Delta T = T_2 - T_1 = 272\,\text{K} - 253\,\text{K} = 19\,\text{K} = 19°\text{C}$$

18.21. $p_1 V_1^\gamma = p_2 V_2^\gamma$

$p_2 = \left(\dfrac{V_1}{V_2}\right)^\gamma p_1 = \left(\dfrac{400}{50}\right)^{1.4} 1\,\text{atm} = 18.4\,\text{atm}$

$pV = nRT$

So,

$$\dfrac{p_1 V_1}{T_1} = \dfrac{p_2 V_2}{T_2}$$

$$T_2 = \dfrac{p_2 V_2}{p_1 V_1} T_1 = \left(\dfrac{18.4}{1}\right)\left(\dfrac{50}{400}\right)300\,\text{K}$$

$$T_2 = 690\,\text{K} = 417°\text{C}$$

CHAPTER 19

The First and Second Laws of Thermodynamics

Thermodynamics is concerned with how energy flows between different systems. As a result, it has important consequences for fields as diverse as engineering, biology, meteorology, and, of course, chemistry and physics. I will focus on applying thermodynamics to gases since this is a good way to make the ideas clear. For the ideal gases I will discuss a **state** of the system as characterized by the values of temperature, pressure, and volume for the system. I will discuss different **thermodynamic processes** by means of which a system changes from one state to another.

19.1   The First Law of Thermodynamics

Consider a gas at pressure p in a cylinder sealed with a movable piston. If the gas pushes against the piston and moves it a small distance dx, the gas does work $dW = F\,dx$. Since $F = PA$, $dW = PA\,dx$. In this process the volume of the gas increases by $dV = A\,dx$, so

$$\boxed{dW = p\,dV}$$

(19.1)

Note carefully that dW is positive when the gas does work on its surroundings, that is, when it expands. If dW is negative, the surroundings do work on the gas and its volume decreases. The total work done in a process is equal to the area under the curve representing the process in a p–V diagram, as indicated in Fig. 19-1.

PROBLEM 19.1.   An ideal gas expands from p_1, V_1 to p_2, V_2 isothermally (at constant temperature). How much work does it do?

Solution

$$pV = nRT \quad p = \frac{nRT}{V}$$

$$W = \int dW = \int_{V_1}^{V_2} \frac{nRT}{p}\,dV$$

$$\boxed{W = nRT \ln\frac{V_2}{V_1}}$$

(19.2)

Fig. 19-1

Observe that the work done in going from one state to another in a thermodynamic process depends on the path followed in the p–V diagram.

PROBLEM 19.2. How much work is done by a gas in going from p_1, V_1 to p_2, V_2 for the paths a, b, and c shown here?

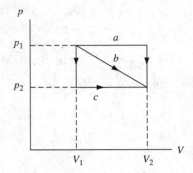

Solution The work is the area under the curve.

$$W_a = p_1(V_2 - V_1)$$
$$W_b = p_{ave}(V_2 - V_1) = \tfrac{1}{2}(p_1 + p_2)(V_2 - V_1)$$
$$W_c = p_2(V_2 - V_1)$$

There are an infinite number of different paths by means of which a system can go from one state to another. Some commonly encountered paths (processes) are the following: **isothermal** (constant temperature), **isobaric** (constant pressure), **isochoric** (constant volume), and **adiabatic** (no heat flows in or out).

Since the work done by a system (or on a system) depends on the path followed from one state to another, it makes no sense to speak of the "work in a system." Similarly, the heat added to a system when it goes from one state to another depends on the path followed, so it makes no sense to speak of the "heat in a system" (even though the unfortunate term "heat capacity" is used for historical reasons). A system does, however, have a definite internal energy. For an ideal gas we saw that the internal energy depends *only* on the temperature and on the amount of gas (Eq. 18.5). For a monatomic gas, $E = 3nRT/2$. Changes in internal energy depend only on the initial and final states and hence are *path independent*.

When heat is added to a system, two things can happen. The internal energy of the system may increase, or the system may do work on its surroundings. Either or both may occur. Applying the conservation of energy law to this situation,

$$\boxed{dQ = dE + dW} \tag{19.3}$$

This statement of energy conservation is the **first law of thermodynamics**. If heat is added to the system, dQ is positive. If heat is removed, dQ is negative. If the internal energy increases, dE is positive, and dE is negative when the internal energy decreases. When the system does work on its surroundings, dW is positive. When work is done on the system, dW is negative.

PROBLEM 19.3. 22.0 J of heat is added to 2 moles of a monatomic ideal gas. In the process the gas does 10 J of work. By how much does the temperature of the gas increase?

Solution From Eq. 18.5,

$$E = \frac{3}{2}nRT \qquad \Delta Q = \frac{3}{2}nR\Delta T + \Delta W$$

$$\Delta T = \frac{2}{2nR}(\Delta Q - \Delta W) = \frac{2(22\,\text{J} - 10\,\text{J})}{(2)(2)(8.31\,\text{J/K})} = 0.72\,\text{K}$$

We frequently encounter processes that are closed paths on the p–V diagram. Such cyclic processes are basic to all engines. Suppose, for example, a system goes from p_1, V_1 to p_2, V_2 along the upper path in Fig. 19-2. It completes the cycle by returning to p_1, V_1 along the lower path. The work done along the upper path is positive and equal to the area under the upper curve. The work done along the lower path is negative and equal to the area under that curve. Thus the net work done in each cycle is equal to the area enclosed by the complete cyclic path.

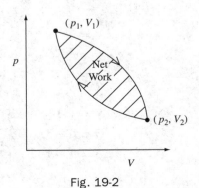

Fig. 19-2

PROBLEM 19.4.   When 1 g of water is boiled at a constant pressure of 1 atm, it becomes 1670 cm of steam. The latent heat of vaporization of water at this pressure is $L_v = 2.26 \times 10^6$ J/kg. Calculate (a) the work done by the water in this process and (b) the increase in the internal energy of the water.

Solution $Q = mL_v = (0.001 \text{ kg})(2.26 \times 10^6 \text{ J/kg}) = 2260 \text{ J}$

(a) $W = p(V_2 - V_1) = (1.013 \times 10^5 \text{ Pa})(1670 - 1)(10^{-6} \text{ m}^3) = 169 \text{ J}$

(b) $\Delta E = Q - W = 2260 \text{ J} - 169 \text{ J} = 2091 \text{ J}$

In an adiabatic process no heat enters or leaves the system, so $Q = 0 = \Delta E + W$, and thus $W = -\Delta E$. For an ideal gas, $E = nC_v(T_2 - T_1)$. Thus

$$\boxed{W = nC_v(T_1 - T_2) \quad \text{adiabatic, ideal gas}} \tag{19.4}$$

Using $pV = nRT$ and $\gamma = C_p/C_v = (C_v + R)/C_v$, Eq. 19.4 may be written as

$$\boxed{W = \frac{C_v}{P}(p_1V_1 - p_2V_2) = \frac{1}{\gamma - 1}(p_1V_1 - p_2V_2) \quad \text{adiabatic, ideal gas}} \tag{19.5}$$

PROBLEM 19.5.   One mole of an ideal monatomic gas moves through the cycle shown here. (a) For each segment of the cycle calculate the work done, the heat flow, and the change in internal energy. Do this for the complete cycle also. (b) Find the pressure and volume at points 2 and 3 if the pressure at point 1 is 2 atm.

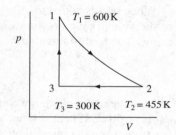

Solution

(a) $1 \to 2$ $W = nC_v(T_1 - T_2) = \frac{3}{2}R(T_1 - T_2)$

$= (1.5)(8.31 \text{ J/K})(600 \text{ K} - 455 \text{ K}) = 1810 \text{ J}$

$Q = 0$ $\Delta E = Q - W = 0 - 1810 \text{ J} = -1810 \text{ J}$

$2 \to 3$ $W = p(V_3 - V_2) = nR(T_3 - T_2)$ using $pV = nRT$

$= (1)(8.31 \text{ J/K})(300 \text{ K} - 455 \text{ K}) = -1290 \text{ J}$

$Q = nC_p\Delta T = n(\frac{5}{2}R)(T_3 - T_2)$

$= (1)(2.5)(8.31 \text{ J/K})(300 \text{ K} - 455 \text{ K}) = -3220 \text{ J}$

$\Delta E = Q - W = -3220 \text{ J} - (-1290 \text{ J}) = -1930 \text{ J}$

$3 \to 1$ $\Delta V = 0$,   so $W = 0$

$Q = nC_v(T_1 - T_3) = (1)(\frac{3}{2}R)(T_1 - T_3)$

$= (1.5)(8.31 \text{ J/K})(600 \text{ K} - 300 \text{ K}) = 3740 \text{ J}$

$\Delta E = Q - W = 3740 \text{ J} - 0 = 3740 \text{ J}$

Cycle $W = 1810 \text{ J} - 1290 \text{ J} + 0 = 520 \text{ J}$

$Q = 0 - 3220 \text{ J} + 3740 \text{ J} = 520 \text{ J}$

$\Delta E = -1810 \text{ J} - 1930 \text{ J} + 3740 \text{ J} = 0$

(b) $p_1V_1 = nRT$ $p_3V_3 = nRT_3$ $V_1 = V_3$

So

$$\frac{p_1V_1}{p_3V_3} = \frac{nRT_1}{nRT_3}   p_3 = \frac{T_3}{T_1}p_1   p_3 = \frac{300\,\text{K}}{600\,\text{K}}(2\,\text{atm}) = 1\,\text{atm} = p_2$$

$$V_3 = V_1 = \frac{nRT_1}{p_1} = \frac{(1)(8.31\,\text{J/K})(600\,\text{K})}{(2)(1.013 \times 10^5\,\text{Pa})} = 0.025\,\text{m}^3$$

$$p_2V_2 = nRT_2   V_2 = \frac{nRT_2}{p_2} = \frac{(1)(8.31\,\text{J/K})(455\,\text{K})}{1.013 \times 10^5\,\text{Pa}} = 0.037\,\text{m}^3$$

19.2   The Second Law of Thermodynamics

The first law of thermodynamics is a statement of the conservation of energy. It is a very useful statement of how the universe is observed to behave, but in itself it does not give a complete description of how things behave. For example, according to the first law, there is no reason heat can flow from cold objects to hot ones spontaneously, or why all of the molecules in the room don't just gather in a tiny space in the corner leaving an absolute vacuum everywhere else. To understand such matters, and to gain insight into such profound concepts as time itself, we rely on another observation concerning how our universe behaves. This second law of thermodynamics can be stated in several ways. First I will tell you how it came about historically with respect to heat engines and heat flow. This never seemed clear to me, intuitively, but you can accept it as true based simply on experimental evidence. This "heat engine" approach is very useful for engineering applications, and you need to understand this for practical reasons (sort of like being able to use your computer, even though you don't know exactly what is inside the mysterious device). In the next section I'll give you a better insight into what it all means by introducing the concept of entropy and statistical mechanics. In this chapter I will just sketch the results, rather than derive them, in order to save space. Two equivalent statements of the **second law of thermodynamics** are the following:

> **The Kelvin form: It is impossible to build a cyclic engine that converts thermal energy completely into mechanical work.**

> **The Clausius form: It is impossible to build an engine whose only effect is to transfer thermal energy from a colder body to a hotter body.**

Before seeing how this law applies to engines, note that two kinds of thermodynamic processes can be identified. Real processes are **irreversible**. They go in one direction only. If a box is slid across the floor against friction, mechanical energy is converted into heat. One never observes the reverse process where a box spontaneously decides to start sliding while the floor cools down. If you put a drop of ink in a glass of water, the ink spreads throughout. You never see an inky glass of water become clear. However, for the sake of understanding what is happening in nature, we consider **reversible processes** that can go either way. Such processes involve the interaction between two systems very nearly in equilibrium, such that an infinitesimal change in the properties of one system can change the direction of the process. For example, if heat is flowing from one object to another, and the temperatures of the two objects differ by only an infinitesimal amount, a slight change in the temperature of one object can reverse the direction of the heat flow.

A device that transforms heat partly into work is a **heat engine**. An engine utilizes a **working substance** (for example, gasoline and air in a car engine or water in a steam engine). Most engines use a **cyclic process** in which the working substance returns to the same state at periodic intervals.

The operation of an engine can be represented with an energy-flow diagram (Fig. 19-3). Heat Q is removed from a hot reservoir. Some of this heat is rejected as exhaust into a cold reservoir, and some is used to do work. Thus $Q_H = Q_C + W$. The efficiency of an engine is defined as

$$\boxed{e = \frac{W}{Q_H} = \frac{Q_H - Q_C}{Q_H} = 1 - \frac{Q_C}{Q_H}}$$ (19.6)

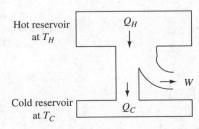

Fig. 19-3

PROBLEM 19.6.   Each cycle an engine gains 120 J of heat from fuel and does 30 J of work. What is the efficiency of the engine?

Solution

$$e = \frac{W}{Q_H} = \frac{30\,\text{J}}{120\,\text{J}} = 0.25 = 25 \text{ percent}$$

19.3   The Carnot Engine

According to the second law, no engine can be 100 percent efficient, no matter how much you reduce friction and other heat losses. The most efficient engine possible is an idealized one called the *Carnot engine*. The working material is carried through *reversible* processes (not possible in real engines). A Carnot cycle using an ideal gas is shown in Fig. 19-4. Two legs of the Carnot cycle are isothermals (*AB* and *CD*), and two are adiabats (*BC* and *DA*). If T_H and T_C are the temperatures of the hot and cold reservoirs, the efficiency of the Carnot engine is

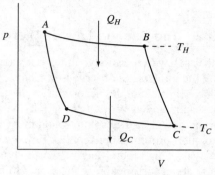

Fig. 19-4

$$\boxed{e = 1 - \frac{T_C}{T_H}} \qquad (19.7)$$

PROBLEM 19.7.   Derive Eq. 19.7.

Solution   The temperature is constant for *AB* and *CD*, so $\Delta E = 0$ for these portions. Thus $Q_{\Delta B} = \Delta E_{AB} + W_{AB} = 0 + W_{AB}$. Similarly, $Q_{CD} = W_{CD}$. From Eq. 19.2, the work done in an isothermal expansion is

$$W_{AB} = nRT_H \ln \frac{V_B}{V_A} = Q_{AB} \quad \text{and} \quad W_{CD} = Q_{CD} = nRT_C \ln \frac{V_D}{V_C}.$$

Dividing,

$$\frac{Q_{CD}}{Q_{AB}} = \frac{nRT_C \ln(V_D/V_C)}{nRT_H \ln(V_B/V_A)} = \frac{T_C \ln(V_D/V_C)}{T_H \ln(V_B/V_A)} \qquad (i)$$

From Eq. 18.15, $PV^\gamma = \text{constant}$ and $TV^{\gamma-1} = \text{constant}$, so $T_H V_B^{\gamma-1}$ and $T_H V_A^{\gamma-1} = T_C V_D^{\gamma-1}$. Divide:

$$\frac{T_H V_B^{\gamma-1}}{T_H V_A^{\gamma-1}} \quad \text{or} \quad \frac{V_B}{V_A} = \frac{V_C}{V_D}. \qquad (ii)$$

Substitute (ii) into (i),

$$\frac{Q_{CD}}{Q_{AB}} = \frac{T_C \ln(V_A/V_B)}{T_H \ln(V_B/V_A)} = -\frac{T_C}{T_H}$$

Here I used $\ln(V_A/V_B) = -\ln(V_B/V_A)$. Also, if we choose $Q_C > 0$, then $Q_C = |Q_{CD}|$, so $Q_C/Q_H = T_C/T_H$. The efficiency of the Carnot engine is thus

$$e = 1 - \frac{Q_C}{Q_H} = 1 - \frac{T_C}{T_H}$$

19.4   The Gasoline Engine

The gasoline engine cycle can be approximated by the Otto cycle, shown in Fig. 19-5. No work is done for processes BC and DA, since here there is no volume change. The heat input is $Q_H = C_V(T_C - T_B)$, and the heat out is $Q_C = C_V(T_D - T_A)$. The efficiency is thus

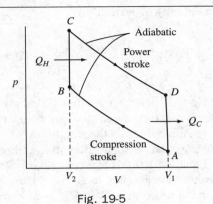

Fig. 19-5

$$e = 1 - \frac{Q_C}{Q_H} = 1 - \frac{T_D - T_A}{T_C - T_B}$$

This can be simplified using the relation for adiabatics, Eq. 18.5, $TV^{\gamma-1} = $ constant, yielding

$$\boxed{e = 1 - \frac{1}{(V_1/V_2)^{\gamma-1}} = 1 - \frac{T_A}{T_B} = 1 - \frac{T_D}{T_C}} \qquad (19.8)$$

In the last step, $T_A V_1^{\gamma-1} = T_B V_2^{\gamma-1}$, so

$$\frac{V_2^{\gamma-1}}{V_1} = \frac{T_A}{T_B} = \frac{T_D}{T_C}$$

$V_1/V_2 = r$, the compression ratio, so $e = 1 - (1/r^{\gamma-1})$.

19.5   Refrigerators and Heat Pumps

A refrigerator is essentially an engine operating in reverse. In Fig. 19-3, imagine the arrows on $Q_H, Q_C,$ and W reversed. Work is done by a motor, heat Q_C is sucked out of the stuff inside the refrigerator, and heat Q_H is kicked out into the surrounding room. A sort of upside down efficiency is defined to describe how well a refrigerator works. This *coefficient of performance* (COP) is defined as

$$\boxed{\text{COP} = \frac{Q_H}{W} = \frac{Q_H}{Q_H - Q_C} \quad \text{refrigerator}} \qquad (19.9)$$

For a Carnot refrigerator (the best possible), the COP $= T_H/(T_H - T_C)$. A good commercial refrigerator for home use has COP $= 6$. A heat pump is just a refrigerator in which the outdoors is the cold reservoir to be cooled, and the inside of the building is where the heat is dumped. A heat pump run in reverse becomes an air conditioner to cool a building.

19.6   Entropy

Recall that for the reversible Carnot cycle,

$$\frac{Q_H}{Q_C} = \frac{T_H}{T_C} \quad \text{or} \quad \frac{Q_H}{T_H} - \frac{Q_C}{T_C} = 0$$

It is possible to approximate any reversible cycle by a series of Carnot cycles, and this leads us to the conclusion that

$$\boxed{\oint \frac{dQ}{T} = 0 \quad \text{for a reversible cycle}} \qquad (19.10)$$

This is reminiscent of what we found for conservative forces, where $\oint \mathbf{F} \cdot \mathbf{ds} = 0$ for a closed path. This led us to define a potential energy U where $U_B - U_A = \int_A^B \mathbf{F} \cdot \mathbf{ds}$. In this case a state of the system was characterized by a definite value of U, the potential energy. In the same way, we define a new state variable, the **entropy** S, such that

$$dS = \frac{dQ}{T} \quad \text{and} \quad S(B) - S(A) = \int_A^B \frac{dQ}{T} \tag{19.11}$$

Notice that although a definite value of Q does not characterize a state (that is, a point on a p–V diagram), every point on the p–V diagram *does* have a definite value of S. It is curious that although the heat flow into a system depends on the path followed between the two states, the change in S is path independent. We say that dQ is an *inexact* differential, and dS is an *exact* differential.

Equation 19.10 is true for a reversible cycle. One can reason that $\oint (dQ/T) > 0$ for an irreversible cycle. Further, it is possible to extend this reasoning to any process that carries a system from state A to state B, with the result that $\Delta S = S(B) - S(A) = \int_A^B dQ/T$. For an isolated system ($dQ = 0$), this becomes

$$\boxed{\Delta S = 0 \ \text{ for a reversible cycle and } \ \Delta S > 0 \ \text{ for an irreversible cycle}} \tag{19.12}$$

This means that **the entropy of an isolated system either remains constant or increases**. Since real processes are all irreversible, this means that the entropy of the universe always increases in *every* process.

One can obtain a feeling for the nature of entropy as a state function by calculating the change in entropy for an ideal gas undergoing a process that carries it from p_1, T_1, V_1 to p_2, T_2, V_2, as shown in Fig. 19-6. No matter what path I follow, the entropy change will be the same since S is a state function. To simplify the calculation, I will choose the reversible path shown, first traveling along an isothermal path, and then along a path with constant volume. Along the isotherm the temperature doesn't change, and so there is no change in internal energy ($E = nC_vT$). Thus $dQ = dW$ for this process, and

$$S(B) - S(A) = \int_A^B \frac{dQ}{T_1} = \int_A^B \frac{dW}{T_1} = \int_{V_1}^{V_2} \frac{p\,dV}{T_1}$$

$$pV = nRT, \text{ so } S(B) - S(A) = \int_{V_1}^{V_2} \frac{nRT_1\,dV}{VT_1}$$

$$S(B) - S(A) = nR \ln \frac{V_2}{V_1}$$

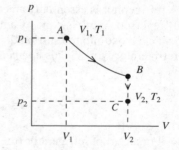

Fig. 19-6

For $B \to C$, no work is done, so $dQ = dE = nC_v\,dT$:

$$S(C) - S(B) = \int_B^C \frac{dQ}{T} = \int_{T_1}^{T_2} C_v \frac{dT}{T} = nC_v \ln \frac{T_2}{T_1}$$

The total entropy change is thus $S(B) - S(A) + S(C) - S(B)$:

$$\boxed{S(p_2, V_2, T_2) - S(p_1, V_1, T_1) = nC_v \ln \frac{T_2}{T_1} + nR \ln \frac{V_2}{V_1} \quad \text{ideal gas}} \tag{19.13}$$

PROBLEM 19.8.   An insulated container is partitioned into two compartments, one of volume V and the other of volume $2V$. Three moles of an ideal gas are contained in the smaller compartment, and the larger compartment is evacuated. A thin membrane separates the two regions. When the membrane is broken, the gas undergoes a **free expansion**. Calculate the change in entropy.

Solution   This is an irreversible adiabatic expansion (no heat flows in or out). Further, since the gas expands against no outside force, it does no work. Since no heat flows and no work is done, the internal energy of the gas doesn't change. This makes sense since the molecules don't bump into anything new. The internal energy of an ideal gas depends only on temperature, so I reason that the temperature doesn't change. Since the entropy is a state function, and I am going from the state (p_1, V, T) to $(p_2, 3V, T)$, I can calculate the entropy by considering any path I want (not just the real irreversible adiabatic one). A simple choice is an isothermal. For such a path the internal energy E doesn't change and $dQ = dW$, so

$$Q = W = \int_{V_1}^{V_2} p\, dV = nRT \int \frac{dV}{V} = nRT \ln \frac{V_2}{V_1}$$

$$\Delta S = \int \frac{dQ}{T} = \frac{1}{T} \int dQ = \frac{Q}{T},$$

so

$$\boxed{\Delta S = nR \ln \frac{V_2}{V_1} \quad \text{free expansion}} \tag{19.14}$$

Here $V_2 = 3V_1$ and $n = 3$, so

$$\Delta S = (3)(8.31 \text{ J/K})(\ln 3) = 24.9 \text{ J/K}$$

We have seen that *energy* has a fairly simple meaning. It is the ability of a system to do work. Entropy is a little trickier. It increases in all irreversible processes (that is, in all real processes), and in a way it is *a measure of energy that is not available to do work*. That is to say, it is a measure of the *disorder* of a system. To see what I mean by this, think of all the molecules rattling around in a gas. They are going every which way, and the kinetic energy associated with this random motion constitutes the thermal energy of the system. By contrast, suppose the system is highly organized and all of the molecules are moving together in the same direction, like the atoms in a thrown baseball. Then we say that the system has a net kinetic energy, and this organized system can hit into an external object and do more work on it than can the disordered randomly moving molecules. In this sense we can view entropy as a measure of the disorder of a system. For example, consider 10 coins in a cup. If you shake the cup and then slam the coins down on the table, some will be heads up and some will be tails up. There is only one way all 10 can be heads (each and every coin must be a head), but there are lots of ways to get, say, 4 heads and 6 tails. For example, some possible sequences of coins could be *HHHHTTTTTT* or *HHTTHHTTTT* or *THTHTHTHTT*, and so on. You can see the idea. These arrangements are called **microstates** of the system, and the combination $(10H)$ or $(4H6T)$ are **macrostates**. It is easy to see that the $(4H6T)$ macrostate is much more likely than the $(10H)$ macrostate because it is composed of many more microstates. In statistical mechanics we define the entropy of a system of N particles in a given macrostate (which consists of Ω microstates) in terms of Boltzmann's constant k_B and the number of accessible microstates as

$$\boxed{S = Nk_B \ln \Omega} \tag{19.15}$$

Now we can see why entropy always increases. If you take a collection of coins initially in the macrostate $(7H3T)$ and shake them or do something to them, they are more likely to end up in a more disordered state like $(5H5T)$ than in a highly ordered state like $(10H)$ or $(9H1T)$. This may not seem like a big deal when you have only 10 coins, but when you are dealing with billions of billions of billions of atoms, the probabilities of ending up in the most probable state is so high that we call it a law, the second law of thermodynamics. The air molecules in the room *could* all gather in one cubic centimeter, but the smart money is always going to bet they don't.

This entropy business can be confusing when you first meet it at an elementary level, but if you go on to investigate it carefully with statistical mechanics, it becomes simple and crystal clear. Statistical mechanics is the greatest thing since sliced bread. I urge you to look into it, since it will lead you to an exciting and fascinating exploration of everything from chaos, linguistics, and genetics to information theory, computers, and quantum field theory.

19.7   Summary of Key Equations

Work: $$dW = p\,dV$$

Ideal gas (isothermal): $$W = nRT\,\ln\frac{V_2}{V_1}$$

Ideal gas (adiabatic): $$W = nC_v(T_1 - T_2) = \frac{C_v}{R}(p_1V_1 - p_2V_2) = \frac{1}{\gamma - 1}(p_1V_1 - p_2V_2)$$

First law of thermodynamics: $$dQ = dE + dW$$

Engine efficiency: $$e = \frac{W}{Q_H} = 1 - \frac{Q_c}{Q_H}$$

Carnot engine efficiency: $$e = 1 - \frac{T_C}{T_H}$$

Gasoline engine (Otto cycle): $$e = 1 - \frac{1}{(V_1/V_2)^{\gamma-1}}$$

Refrigerator coefficient of performance: $$\text{COP} = \frac{Q_H}{Q_H - Q_C}$$

Entropy: $$\Delta S = \int_A^B \frac{dQ}{T} \quad \text{or} \quad S = Nk_B\,\ln\Omega$$

Second law of thermodynamics: $$\Delta S \geq 0$$

SUPPLEMENTARY PROBLEMS

19.9. A gas expands at constant pressure of 3 atm from a volume of 2 to 5 L. How much work is done?

19.10. One mole of an ideal gas initially at p_1, V_1, T_1 is taken through the cycle shown here. Calculate the work done per cycle. Obtain an expression for the efficiency of an engine using this process.

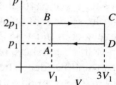

19.11. Two moles of an ideal gas at 600 K are compressed until the pressure triples. How much work does the gas do?

19.12. Calculate the efficiency of the Stirling engine that uses the cycle shown here with an ideal gas as the working substance.

19.13. A power plant generates 640 MW of electric power. It has an efficiency of 38 percent. At what rate does it dissipate heat? If the waste heat is carried away by a river, and environmental concerns limit the temperature rise of the water to 3°C, what flow rate is required in the river?

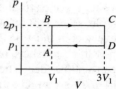

19.14. A coal-fired power plant operates between 490°C and 38°C. What is the maximum efficiency possible under these conditions?

19.15. Calculate the change in entropy if 2 kg of water at 80°C is mixed with 3 kg of water at 20°C.

19.16. Three moles of ideal gas expand from 0.02 to 0.06 m³ in a free expansion. What is the change in entropy of the gas?

19.17. One mole of a monatomic ideal gas initially at p_1, V_1 is heated and expands at constant pressure to a volume $4V$. It is then cooled at constant volume until the pressure has dropped to $0.5p$. Calculate (a) the work done, (b) the change in internal energy, (c) the heat flow, and (d) the net change in entropy.

19.18. An ideal (Carnot) engine operates between a hot reservoir at 360 K and a cold reservoir at 270 K. It absorbs 600 J of heat per cycle at the hot reservoir. (a) How much work does it do each cycle? (b) If the same engine is operated in reverse as a refrigerator, how much work must be done each cycle in order to remove 1200 J of heat from the cold reservoir each cycle?

19.19. What is the change in entropy of 1 kg of water when it is heated from 0°C to 100°C?

19.20. The latent heat of fusion of a substance is L_T, and its melting temperature is T_m. By how much does the entropy of mass m change when it melts?

19.21. An engine operates at 1800 RPM. It extracts 1800 J of energy from a hot reservoir and exhausts 1400 J to a cold reservoir. (a) What is the efficiency of the engine? (b) What is the power output of the engine? The engine undergoes one thermodynamic cycle per revolution.

19.22. Consider a two-engine arrangement in which the exhaust energy from one engine is used as the input energy for a second engine. Such engines are said to be operating in *series*. (a) What is the efficiency of such a combination if the first engine operates between temperatures T_H and T with efficiency e_1 and the second operates between T and T_C with efficiency e_2? (b) Would such a series engine be more efficient than a single engine operating between T_H and T_C? To gain a feeling for the answer, calculate the efficiency that would result for two Carnot engines operating between 800 K and 600 K, and 600 K and 400 K, as compared to a single Carnot engine operating between 800 K and 400 K.

SOLUTIONS TO SUPPLEMENTARY PROBLEMS

19.9. $W = \int_{v_1}^{V_2} p\, dV = p(V_2 - V_1) = (3)(1.013 \times 10^5\,\text{Pa})(5 - 2)(10^{-3}\,\text{m}^3) = 912\,\text{J}$

19.10.

$$W_{AB} = 0 \quad W_{CD} = 0 \quad W_{BC} = 2p_1(3V_1 - V_1) = 4p_1V_1$$
$$W_{DA} = p_1(V_1 - 3V_1) = -2p_1V_1$$
$$W = W_{AB} + W_{BC} + W_{CD} + W_{DA} = 0 + 4p_1V_1 + 0 - 2p_1V_1$$
$$p_1V_1 = RT_A \quad 2p_1V_1 = RT_B, \text{ so } T_g = 2T_A$$

Similarly,

$$(2p_1)(3V_1) = RT_C \quad T_C = 6T_A$$
$$p_1(3V_1) = RT_D \quad T_D = 3T_A \qquad Q_{AB} = C_v(T_B - T_A) = \tfrac{3}{2}R(2T_A - T_A) = \tfrac{3}{2}RT_A$$
$$Q_{BC} = C_p(T_C - T_B) = \tfrac{5}{2}R(6T_A - 2T_A) = 10RT_A$$
$$Q_{CD} = C_v(T_D - T_C) = \tfrac{3}{2}R(3T_A - 6T_A) = -\tfrac{9}{2}RT_A$$
$$Q_{DA} = C_v(T_A - T_D) = \tfrac{5}{2}R(T_A - 3T_A) = -5RT_A$$

$$Q = Q_{AB} + Q_{BC} + Q_{CD} + Q_{DA} = 2RT_A \qquad e = \frac{W}{Q_{1N}} = \frac{W}{Q_{AB} + Q_{BC}} = \frac{2RT_A}{(3/2)RT_A + 10RT_A} = 0.17$$

19.11. $W = \int_{V_1}^{V_2} p\, dV = \int_{V_1}^{V_2} \frac{nRT\, dV}{V} = nRT \ln\frac{V_2}{V_1} = nRT \ln\frac{p_1}{p_2}$ since $p_1V_1 = p_2V_2 = nRT$

$$W = (2)(8.31\,\text{J/K})(600\,\text{K}) \ln\frac{1}{3} = -1.1 \times 10^4\,\text{J}$$

19.12. For $1 \to 2$,

$$\Delta E = 0 \qquad Q_{12} = W_{12} = \int p \, dV = RT_H \int_{V_1}^{V_2} \frac{dV}{V} = RT_H \ln \frac{V_2}{V_1}$$

For $4 \to 1$,

$$W_{41} = 0 \qquad Q_{14} = \Delta E_{41} = C_v(T_H - T_C)$$

Thus the total heat is

$$Q_{in} = RT \frac{V_2}{V_1} + C_v(T_H - T_C)$$

The total work is

$$W = RT_H \ln \frac{V_2}{V_1} + RT_C \ln \frac{V_1}{V_2} = R(T_H - T_C) \ln \frac{V_2}{V_1}$$

$$e = \frac{W}{Q_{in}} = \frac{R(T_H - T_C) \ln(V_2/V_1)}{RT_H \ln(V_2/V_1) + [C_V(T_H - T_C)]}$$

19.13. Useful power $P = eP_H$, so waste power is $P_C = (1 - e)P_H = [(1 - e)/e]P$.

$$P_C = \frac{1 - 0.38}{0.38}(640 \, \text{MW}) = 1044 \, \text{MW}$$

If the flow rate is mass m per second, then $P_C = mc\Delta T$.

$$m = \frac{P_C}{c\Delta T} = \frac{1044 \times 10^6 \, \text{J/s}}{(4200 \, \text{J/kg} \cdot \text{K})(3 \, \text{K})} = 8.3 \times 10^5 \, \text{kg/s}$$

19.14. $e_{max} = 1 - \dfrac{T_C}{T_H} = 1 - \dfrac{313 \, \text{K}}{763 \, \text{K}} = 0.59$

Be careful always to express temperature in degrees Kelvin!

19.15. The final temperature is T, $m_1 c(T - T_1) = m_2 c(T_2 - T)(3 \, \text{kg})(T - 20°\text{C}) = (2 \, \text{kg})(80°\text{C} - T)$, $T = 44°\text{C} = 317 \, \text{K}$. Replace the irreversible process (the actual one) with a reversible one in which the hot water, say, is successively placed in contact with an infinitesimally cooler reservoir and thereby cooled in an infinite number of small steps. A similar procedure is followed with the cold water. Finally, the desired state of $44°\text{C}$ is reached. For masses m_1 and m_2 with specific heats c_1 and c_2,

$$\Delta S = \int_1 \frac{dQ}{T} + \int_2 \frac{dQ}{T} = m_1 c_1 \int_{T_1}^{T} \frac{dT}{T} + m_2 c_2 \int_{T_2}^{T} \frac{dT}{T}$$

$$= m_1 c_1 \ln \frac{T}{T_1} + m_2 c_2 \ln \frac{T}{T_2}$$

This is the entropy change for mixing, where mass m_1 at T_1 is mixed with mass m_2 at T_2, with a final temperature T. Here $c_1 = c_2 = 4.2 \, \text{kJ/kg} \cdot \text{K}$, so

$$\Delta S = (3 \, \text{kg})(4.2 \, \text{kJ/kg} \cdot \text{K}) \ln\left(\frac{317}{293}\right) + (2 \, \text{kg})(4.2 \, \text{kJ/kg} \cdot \text{K}) \ln\left(\frac{317}{353}\right) \qquad \Delta S = 0.88 \, \text{J/K}$$

19.16. From Eq. 19.14,

$$\Delta S = nR \ln \frac{V_2}{V_1} = (3)(8.31 \, \text{J/K}) \ln\left(\frac{0.06}{0.02}\right) = 27.4 \, \text{J/K}$$

19.17. From $pV = RT$, conclude $T_B = 4T_A$ and $T_C = 2T_A = 2T_0$.

$$Q_{AB} = C_p(T_B - T_C) = \frac{5}{2}R(4T_0 - T_0) = \frac{15}{2}RT_0 = \frac{15}{2}p_0V_0$$

$$Q_{BC} = C_v(T_C - T_B) = \frac{3}{2}R(2T_0 - 4T_0) = -3RT_0 = -3p_0V_0$$

$$W_{AB} = p_0\Delta V = 3p_0V_0 \quad W_{BC} = 0$$

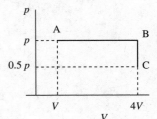

(a) $W = 3p_0V_0$

(b) $\Delta E = Q - W = \frac{15}{12}p_0V_0 - 3p_0V_0 - 3p_0V_0 = \frac{3}{2}p_0V_0$

(c) $Q = \frac{15}{2}p_0V_0 - 3p_0V_0 = \frac{9}{2}p_0V_0$

(d) $\Delta S = \int_A^B \frac{C_p\,dT}{T} + \int_B^C \frac{C_v\,dT}{T} = C_p\ln\frac{2T_0}{T_0} + C_v\ln\frac{2T_0}{4T_0} \quad \Delta S = \frac{5}{2}R\ln 2 + \frac{3}{2}R\ln 2 = 0.69R$

19.18. (a) $e = 1 - \frac{T_C}{T_H} = 1 - \frac{270\,\text{K}}{360\,\text{K}} = 0.25 \quad 0.25 = \frac{W}{Q} = \frac{W}{600\,\text{J}} \quad W = 150\,\text{J}$

(b) $\text{COP} = \frac{T_C}{T_H - T_C} = \frac{270\,\text{K}}{360\,\text{K} - 270\,\text{K}} = 3 \quad \text{CIO} = \frac{Q_C}{W} \quad 3 = \frac{1200\,\text{J}}{W}$

19.19. $\Delta S = \int \frac{dQ}{T} = \int_{273\,\text{K}}^{373\,\text{K}} \frac{mc\,dT}{T} = mc\ln\left(\frac{373}{273}\right) = (1\,\text{kg})(4.2\,\text{kJ/kg}\cdot\text{K})\ln\left(\frac{373}{273}\right) = 1.3\,\text{kJ/K}$

19.20. $\Delta S = \frac{Q}{T} = \frac{mL_f}{T}$

19.21. (a) $e = \frac{W}{Q_H} = \frac{Q_h - Q_c}{Q_H} = 1 - \frac{Q_c}{Q_H}$

$e = 1 - \frac{1400\,\text{J}}{1800\,\text{J}} = 0.22$

(b) Period is $t = \frac{1}{f} = \frac{1}{1800/60}\,\text{s} = 0.033\,\text{s}$

Power $P = \frac{W}{t} = \frac{Q_H - Q_c}{t} = \frac{1800\,\text{J} - 1400\,\text{J}}{0.033\,\text{s}}$

$P = 12,000\,\text{W} = 12\,\text{kW}$

19.22. (a) $W = W_1 + W_2$

$eQ_H = e_1Q_H + e_2Q$

$e_1 = 1 - \frac{Q}{Q_H}$ so, $Q = (1 - e_1)Q_H$

$e_2 = 1 - \frac{Q_C}{Q}$ so, $Q = (1 - e_2)Q_C$

$eQ_H = e_1Q_H + e_2(1 - e_1)Q_H$

$e = e_1 + e_2 - e_1e_2$ Overall efficiency

(b) $e = 1 - \dfrac{T_C}{T_H} = 1 - \dfrac{400\,\text{K}}{800\,\text{K}} = 0.5$ Single engine

$e_1 = 1 - \dfrac{T}{T_H} = 1 - \dfrac{600\,\text{K}}{800\,\text{K}} = \dfrac{1}{4}$

$e_2 = 1 - \dfrac{T_C}{T} = 1 - \dfrac{400\,\text{K}}{600\,\text{K}} = \dfrac{1}{3}$

$e' = e_1 + e_2 - e_1 e_2 = \dfrac{1}{4} + \dfrac{1}{3} - \left(\dfrac{1}{4}\right)\left(\dfrac{1}{3}\right) = \dfrac{3}{12} + \dfrac{4}{12} - \dfrac{1}{12}$

$e' = \frac{1}{2}$, so $e' = e$ both cases have the same efficiency

Electric Fields

20.1 Properties of Electric Charge

Electric charge and electric forces play a major role in determining the behavior of the universe. The basic building blocks of matter, electrons and protons, have a property called *electric charge*. Electric charge is observed to have the following characteristics:

An electric charge has a polarity; that is, it is either positive or negative. Like charges repel each other, and opposite charges attract.

The force between charges is proportional to their magnitudes and varies as the inverse square of their separation.

An electric charge is conserved. It cannot be created or destroyed. We obtain charge by separating neutral objects into a negative piece and a positive piece.

An electric charge is quantized. It is always observed to occur as an integer multiple of e, the fundamental quantity of charge. We choose the unit of electric unit of electric charge as the coulomb, where $e = 1.60 \times 10^{-19}$ **coulomb (C)**. The charge on an electron is $-e$, and on a proton, $+e$. Subnuclear particles called *quarks* have charges that are multiples of $e/3$, but individual *quarts* have not been detected.

Conductors are materials in which charge can move relatively freely and in which there are some free charges. Examples of good conductors are metals, plasmas (ionized gases), liquids containing ions (for example, sulfuric acid, blood, salt water), and some semiconductors. **Insulators** are materials that do not readily transport charge. Examples are a vacuum, glass, distilled water, paper, and rubber. There is not a sharp demarcation between conductors and insulators. Some materials (for example, **semi-conductors** like silicon or indium antinimide) have properties intermediate between a good conductor and a good insulator.

The force between two charges of magnitudes q_1 and q_2 separated by a distance r is given by Coulomb's law:

$$\boxed{F = k\frac{q_1 q_2}{r^2} = \frac{1}{4\pi\epsilon_0}\frac{q_1 q_2}{r^2}} \tag{20.1}$$

Here

$$k = \frac{1}{4\pi\epsilon_0} = 8.99 \times 10^9 \,\text{N}\cdot\text{m}^2/\text{C}^2 \simeq 9 \times 10^9 \,\text{N}\cdot\text{m}^2/\text{C}^2$$

ϵ_0 is the **permittivity of free space**, $\epsilon_0 = 8.85 \times 10^{-11} \,\text{C}^2/\text{N}\cdot\text{m}^2$. The force between two point charges is directed along the line joining them. When more than two charges are present, the force on any one of them is the vector sum of the forces due to each of the others. This **principle of superposition** will play an important role in all of our analysis.

PROBLEM 20.1. Table salt (sodium chloride) is a crystal with a simple cubic structure with Na^+ ions and Cl^- ions alternating at adjacent lattice sites. The distance between ions is $a = 2.82 \times 10^{-10}$ m $= 0.282$ nm(1 nm $= 10^{-9}$ m). (a) What force does an Na^+ ion experience due to one of its nearest Cl^- neighbors? (b) What force does a Cl^- ion experience due to a neighboring Na^+? (c) What force does an Na^+ ion at the origin experience due to Cl^- ions at $(a, 0, 0)$ and $(0, a, 0)$? (d) What is the weight of an Na^+ ion of mass 3.82×10^{-26} kg?

Solution

(a) $F_1 = k\dfrac{e^2}{r^2} = 9 \times 10^9$ N$\cdot$m^2/C$^2 \dfrac{(1.6 \times 10^{-19}\,\text{C})^2}{(0.282 \times 10^{-9}\,\text{m})^2}$

$= 2.90 \times 10^{-9}$ N

(b) By Newton's third law, the force on the Cl^- due to the Na^+ is the same as the force on the Na^+ due to the Cl^-.

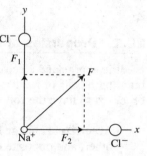

(c) $F = F_1 + F_2 = 2.90 \times 10^{-9}(\mathbf{i} + \mathbf{j})$ N

$= \sqrt{F_1^2 + F_2^2} = 4.10 \times 10^{-9}$ N

(d) $W = mg = (3.82 \times 10^{-26}\,\text{kg})(9.8\,\text{m/s}^2) = 3.7 \times 10^{-25}$ N

Since the electric forces acting on small charged objects are so much larger than their weights, I will often neglect the latter.

PROBLEM 20.2. A charge $q_1 = +4\,\mu$C($1\,\mu$C $= 10^{-6}$ C) is positioned at the origin. A charge $q_2 = +9\,\mu$C is positioned on the x-axis at $x = 4$ m. Where on the x-axis can a negative charge q_3 be placed so that the force on it is zero? Is there any position off the x-axis where the force on q_3 will be zero?

Solution Let q_3 be placed at position x in between the two positive charges, which will then pull in opposite directions on q_3.

$$k\frac{q_1 q_3}{x^2} = k\frac{q_2 q_3}{(4-x)^2} \quad q_2 x^2 = q_1(4-x)^2 \quad 4x^2 = 9(4-1)^2$$

Take the square root:

$$\pm 2x = 3(4-x) \quad x = 2.4\,\text{m}$$

Choose the positive root, because x must lie between 0 and 4 m.

There is no position off the x-axis where the force on q_3 would be zero since in order to cancel, the forces due to q_1 and q_2 must be directed antiparallel (that is, exactly opposite).

When charge is uniformly distributed over the surface of a spherical object or symmetrically throughout the volume of a sphere, then for charges outside the sphere, the spherical charge acts as if it is all concentrated at the center of the sphere.

PROBLEM 20.3. Two identical Styrofoam spheres, each of mass 0.030 kg, are each attached to a thread 30 cm long and suspended from a point. Each sphere is given a charge q (perhaps by rubbing it on a piece of cloth), and the two spheres repel each other and hang with each thread making an angle of 7 degrees with vertical. What is the charge on each sphere?

Solution   Draw the force diagram for one of the spheres. The sphere is in equilibrium, so $F_{up} = F_{down}$ and $F_L = F_R$.

$$T \cos \theta = mg \quad T \sin \theta = F$$

Divide: $\dfrac{T \sin \theta}{T \cos \theta} = \tan \theta = \dfrac{F}{mg}$, where $F = k\dfrac{q^2}{r^2} = k\dfrac{q^2}{(2L \sin \theta)^2}$

Solve: $q^2 = \dfrac{(mg \tan \theta)(2L \sin \theta)^2}{k} = \dfrac{(0.03 \, \text{kg})(9.8 \, \text{m/s}^2)(\tan 7°)[(2)(0.3 \, \text{m}) \sin 7°]^2}{(9 \times 10^9 \, \text{N} \cdot \text{m}^2/\text{C}^2)}$

$$q = 0.146 \times 10^{-6} \quad \text{C} = 0.146 \, \mu\text{C}$$

20.2   The Electric Field

A good way to describe electric effects is by means of the **electric field**. Imagine that around every charge there is a sort of "aura" that fills all space. This aura is the electric field due to the charge. The electric field is a **vector field**, and at every point in space it has a magnitude and direction. The total electric field at any point is the vector sum of the electric fields due to all charges that are present. Analytically, I define the electric field at a point (x, y, z) to be

$$\boxed{\mathbf{E}(x, y, z) = \dfrac{\mathbf{F}}{q_0}} \tag{20.2}$$

Here $\mathbf{F}$ is the force that acts on a test charge q_0 placed at the point (x, y, z). The electric field points in the direction of the force on a positive test charge. If a negative charge is placed at (x, y, z), the force on it will be opposite in direction to $\mathbf{E}$. We can visualize electric fields by means of **electric field lines**. Imagine these lines sprout out of positive charges (I think of positive charges as **sources** of the electric field), and they end on negative electric charges (the **sinks** of the electric field). At a given point, the electric field direction is tangent to the electric field line passing through that point, and the magnitude of the electric field at a point is proportional to the density of lines (the number per unit area measured in a plane perpendicular to the E line). Some examples of E line distributions are illustrated in Fig. 20-1.

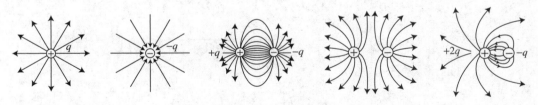

Fig. 20-1

By comparing Eqs. 20.1 and 20.2, you can see that the electric field of a point charge q at a "field point" at (x, y, z), a distance r from the charge, is

$$\boxed{\mathbf{E} = k\dfrac{q}{r^2}\hat{\mathbf{r}} = \dfrac{1}{4\pi\epsilon_0}\dfrac{q}{r^2}\hat{\mathbf{r}} \quad \text{point charge}} \tag{20.3}$$

Here $\hat{\mathbf{r}}$ is a unit vector directed radially away from the charge and directed along the line joining the position of the charge and the "field point" (x, y, z) where the field is to be found.

PROBLEM 20.4.   Find the force on a Ca^{2+} ion placed in an electric field of 800 N/C directed along the positive z-axis. For a doubly charged ion, $q = 2e$.

Solution $F = qE = (2)(1.6 \times 10^{-19} \, \text{C})(800 \, \text{N/C}) = 2.56 \times 10^{-16} \, \text{N}$

PROBLEM 20.5.   Four identical positive charges $+q$ are placed at the corners of a square of side L. Determine the magnitude and direction of the electric field due to them at the midpoint of one side of the square.

Solution   By symmetry I see that the fields due to the charges at B and D cancel. Further, the y components of the fields due to the charges at A and C cancel. Hence the resultant field points in the x direction and has magnitude

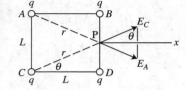

$$E = 2E_A \cos \theta$$

where

$$E_A = k\frac{q}{L^2 + (L/2)^2} \quad \text{and} \quad \cos\theta = \frac{L}{\tau} = \frac{L}{\sqrt{L^2 + L^2/4}} = \frac{2}{\sqrt{5}}$$

Hence

$$E = \frac{16\,kg}{5\sqrt{5}L^2}$$

PROBLEM 20.6.   An **electric dipole** consists of charges $+q$ and $-q$ separated by a distance $2a$. If the charges are positioned at $(0, 0, a)$ and $(0, 0, -a)$ on the z-axis (the axis of the dipole), determine the electric field of the dipole at a point a distance z from the origin on the z-axis, where $z \gg 2a$. Express the result in terms of the **electric dipole moment**, defined as $p = 2aq$.

Solution

$$E = k\frac{q}{(z-a)^2} + k\frac{(-q)}{(z-a)^2} = kq\left(\frac{1}{z^2 - 2az + a^2} - \frac{1}{z^2 + 2az + q^2}\right)$$

$$a^2 \ll 2az \ll z^2$$

So drop a^2:

$$\frac{1}{z^2 - 2az} = \frac{1}{z^2(1 - 2a/z)} \simeq \frac{1}{z^2}\left(1 + \frac{2a}{z^2}\right) \quad \text{using } \frac{1}{1-\delta} \simeq 1+\delta \text{ if } \delta \ll 1$$

$$\frac{1}{z^2 + zaz} = \frac{1}{z^2(1 + 2a/z)} \simeq \frac{1}{z^2}\left(1 - \frac{2a}{z}\right)$$

Thus

$$E = \frac{kq}{z^2}\left[1 + \frac{2a}{z} - \left(1 - \frac{2a}{z}\right)\right] = \frac{4kaq}{z^3}$$

or

$$\boxed{E = k\frac{2p}{z^3} \quad \text{dipole on axis}} \tag{20.4}$$

Matter in nature is generally electrically neutral, so often we do not encounter situations where the forces are due to net charge on an object. However, essentially everything can acquire a dipole moment when placed in an electric field because the negative charges in the atoms are pulled one way and the positive charges in the nucleus are pulled in the opposite direction. Thus electric dipoles play a very important role in our understanding of matter. When an electric dipole is placed in an electric field, it tends to line up with its axis parallel to the field. If it is not parallel to the field, it experiences a torque $\vec{\tau}$. Associated with this torque is a potential energy U, where

$$\boxed{\vec{\tau} = \mathbf{p} \times \mathbf{E} \quad \text{and} \quad U = -\mathbf{p} \cdot \mathbf{E} = -pE\cos\theta} \tag{20.5}$$

Here θ is the angle between the electric field and the dipole axis and $p = 2aq$ is the dipole moment.

20.3   Motion of a Charged Particle in a Uniform Electric Field

Consider a charge q subject to a uniform electric field in the z direction. Neglect the gravity force since it is normally much smaller than the electric force. This force qE gives rise to acceleration $a_z = F/m = qE/m$ and $a_x = 0$. We can determine the motion using the kinematic equations developed previously.

PROBLEM 20.7.   Two very large parallel metal plates separated by a small distance d are given opposite uniform charges, creating a uniform electric field E in the space between them. An electron of charge $-e$ is projected with initial velocity v_0 through a small hole in the positive plate. It travels halfway across the gap between the plates before stopping and reversing direction. What is E in terms of the initial velocity v_0?

Solution
$$a_z = \frac{F}{m} = \frac{-eE}{m} \quad v^2 = v_0^2 + 2az = 0 \quad \text{at } z = \frac{d}{2}$$

So
$$0 = v_0^2 - 2\left(\frac{eE}{m}\right)\left(\frac{d}{2}\right) \quad E = \frac{mv_0^2}{ed}$$

One could also solve this using energy principles. Thus, loss in KE = work done against the E field:

$$\frac{1}{2}mv_0^2 = (eE)\left(\frac{d}{2}\right) \quad E = \frac{mv_0^2}{ed}$$

PROBLEM 20.8.   Two large charged parallel plates are often used to create a uniform electric field E. A charged particle fired between the plates will be deflected by the electric field. This technique is used to deflect electrons in a cathode ray tube (as in an oscilloscope) or to deflect ink droplets in an ink jet printer. Suppose a particle of mass m, charge q, and initial velocity v_0 is projected parallel to two plates where the electric field is E. The length of the plates is L. Through what angle will the particle be deflected?

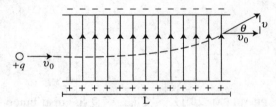

Solution   There is no acceleration in the x direction, so $L = v_0 t$ where $t =$ time between plates. $v_x = v_0 =$ constant:

$$v_y = v_{0y} + at = 0 + \frac{qE}{m}\left(\frac{L}{v_0}\right) \quad \tan\theta = \frac{v_y}{v_x} = \frac{1}{v_0}\left(\frac{qEL}{mv_0}\right) = \frac{qEL}{mv_0^2}.$$

20.4   Electric Field of a Continuous Charge Distribution

One way to find the electric field due to a continuous charge distribution is to break the charge into infinitesimal pieces of charge dq and then sum the contribution of each dq using integral calculus. *CAUTION:* The contributions must be added vectorially, so be careful to separate the components, as illustrated in the following problems.

PROBLEM 20.9.   Charge Q is distributed uniformly along the x-axis over a distance L. Determine the electric field on the axis a distance d from one end of the charged segment.

Solution
$$E = \int k\frac{dq}{r^2} = \int_0^L k\frac{\lambda\, dx}{(d+x)^2}$$

where

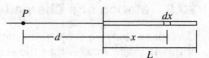

$$\lambda = \frac{L}{Q} = \text{linear charge density}$$

Thus

$$E = k\lambda\left[-\frac{1}{d+x}\right]_0^L = k\frac{Q}{L}\left(\frac{1}{d} - \frac{1}{d+L}\right)$$

$$= \frac{kQ}{d(d+L)}$$

PROBLEM 20.10.   Charge Q is distributed uniformly along a rod. The rod is then bent to form a semicircle of radius R. What is the electric field at the center of the semicircle?

Solution   By symmetry, the y components of the field due to each dq cancel. So $E = E_x$.

$$dq = \lambda\, ds = \lambda R\, d\theta$$

where

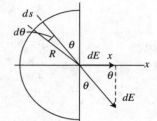

$$\lambda = \frac{Q}{\pi R}$$

Thus

$$E_x = \int k\frac{dq}{R^2}\sin\theta = \frac{k\lambda R}{R^2}\int_0^\pi \sin\theta\, d\theta = \frac{k\lambda}{R}(-\cos\theta)_0^\pi$$

$$E = E_x = \frac{kQ}{\pi R^2}(1+1) = \frac{2kQ}{\pi R^2}$$

PROBLEM 20.11.   Charge Q is distributed uniformly over the surface of a disk of radius R. Determine the electric field at a point on the axis of the disk at a distance z from the center of the disk.

Solution   Let σ = surface charge density = $Q/\pi R^2$. Also, $dq = \sigma\, dA$ and dA = area of angular ring = $2\pi r_1 dr_1$. I use a ring for dA since every point on it is at the same distance r from point P. By symmetry, components of E parallel to the plane of the disk cancel, so $E = E_z$. Thus

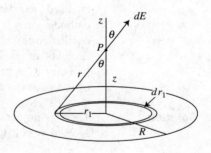

$$E = E_z = \int \frac{k\, dq}{r^2}\cos\theta = \int \frac{k\cos\theta\,\sigma\, dA}{r^2} = \int_0^R \frac{k\cos\theta\,\sigma 2\pi r_1\, dr_1}{r^2}$$

$$r = \sqrt{r_1^2 + z^2}\quad \cos\theta = \frac{z}{r}\quad E = \frac{2\pi kQ}{\pi R^2}\int_0^R \frac{z r_1\, dr_1}{\left(r_1^2 + z^2\right)^{3/2}} = \frac{2kQ}{R^2}\left(1 - \frac{z}{\sqrt{R^2 + z^2}}\right)$$

Far from the disk,

$$z \gg R,\quad \frac{z}{\sqrt{R^2 + z^2}} \simeq \frac{z}{z\sqrt{1 + R^2/z^2}} \simeq \frac{1}{1 + \frac{1}{2}R^2/z^2}$$

$$\simeq \left(1 - \frac{R^2}{2z^2}\right)\quad \text{and}\quad E \simeq \frac{Q}{4\pi\epsilon_0 z^2}$$

20.5 Summary of Key Equations

Coulomb's law: $F = k\dfrac{q_1 q_2}{r^2} = \dfrac{1}{4\pi\epsilon_0}\dfrac{q_1 q_2}{r^2}$

Electronic field: $\mathbf{E} = \dfrac{\mathbf{F}}{q_0}$

Point charge: $\mathbf{E} = k\dfrac{q}{r^2}\hat{\mathbf{r}} = \dfrac{1}{4\pi\epsilon_0}\dfrac{q}{r^2}\hat{\mathbf{r}}$

Dipole moment: $\mathbf{p} = 2aq\mathbf{k}$

Torque on a dipole: $\vec{\tau} = \mathbf{p} \times \mathbf{E}$

Energy of a dipole: $U = -\mathbf{p} \cdot \mathbf{E}$

SUPPLEMENTARY PROBLEMS

20.12. In the hydrogen atom the electron and the proton are separated by 0.53×10^{-10} m. What electric field does the electron experience?

20.13. Charges are placed on the x-axis as follows: $q_1 = +2\,\mu\text{C}$ at $x = 0$, $q_2 = -3\,\mu\text{C}$ at $x = 2$ m, $q_3 = -4\,\mu\text{C}$ at $x = 3$ m, and $q_4 = +1\,\mu\text{C}$ at $x = 3.5$ m. What is the magnitude and direction of the force on q_3?

20.14. Three identical positive charges q are placed at the corners of an equilateral triangle of side L. What force does one of the charges experience?

20.15. Identical charges are placed at the vertices of an n-sided regular polygon. What is the magnetic field at the center of the polygon? This question requires thinking, not calculating.

20.16. Three small identical spheres, each of mass m, are each attached to light strings of length L. They are each given charge q and suspended from a common point. What angle does each string make with the vertical? $m = 0.02$ kg, $L = 0.10$ m, and $q = 8 \times 10^{-8}$ C.

20.17. Alkali halide crystals such as NaCl have a cubic structure, with alternate lattice sites occupied by different kinds of ions. Consider a single cube of this structure, of side a, with alternating charges $+q$ and $-q$ at the vertices. What force does one of the charges experience due to the other seven charges?

20.18. A particle of charge q and mass m is fired with initial velocity v_0 into an opposing uniform electric field E set up by two parallel plates of separation d. The particle enters the field through a small hole in one plate, at which point its velocity makes angle θ with the plane of the plate. Where will the particle strike the opposite plate (assuming it is moving fast enough to do so)?

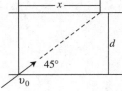

20.19. A uniform electric field is set up between two oppositely charged parallel plates. An electron (mass 9.11×10^{-31} kg) is released from rest at the negatively charged plate and strikes the positive plate a distance 1.5 cm away after 1.2×10^{-8} s. Determine the speed of the electron when it strikes the field and the magnitude of the electric field.

20.20. Charge is distributed uniformly, starting at the origin, along the positive semi-infinite x-axis with linear charge density. Determine the electric field at the point $(0, 0, z)$.

20.21. Charge Q is distributed uniformly along the x-axis from $-L$ to L. Determine the electric field at the point $(0, 0, z)$.

20.22. Three thin plastic rods, each of length 2L, carry charges Q, Q, and, −Q, respectively, distributed uniformly on each rod. The rods are arranged to form an equilateral triangle. Determine the electric field at the center of the triangle. You may use the result of Problem 20.20.

20.23. An infinite plane positioned in the xy plane has uniform surface charge density. Determine the electric field a distance z from the plane.

20.24. A hemispherical shell of radius R carries a uniform surface charge density. Determine the electric field at the center of the hemisphere.

20.25. In a certain thundercloud charges of +45 C and −45 C build up, separated by 6 km. What is the dipole moment of these charges?

20.26. The HCl molecule has a dipole moment of 3.4×10^{-30} C · m. What torque does the molecule experience in an electric field of 4×10^6 N/C when the axis of the dipole makes an angle of 30° with the direction of the electric field?

20.27. A water molecule has a dipole moment of 6.1×10^{-30} C · m. This dipole moment plays a very important role in determining the properties of water, such as its amazing solvency properties and its use in heating food in microwave ovens. What energy is required to change the orientation of the water dipole moment from parallel to the electric field to antiparallel to the electric field when $E = 2.4 \times 10^5$ N/C?

20.28. When a point charge is placed on the x-axis it gives rise to an electric field of E_o at the origin and a field of $2E_o$ at the point where x = 4 cm on the x-axis. Both fields point in the positive x direction. What is the position and sign of the charge that produces these fields?

20.29. Eight identical charges q are placed at the corners of a cube of side L. What is the magnitude and direction of the electric field at the center of one face of the cube?

SOLUTIONS TO SUPPLEMENTARY PROBLEMS

20.12. $E = k\dfrac{q}{r^2} = (9 \times 10^9 \text{ N} \cdot \text{m}^2/\text{C}^2)\dfrac{(1.6 \times 10^{-19} \text{ C})}{(0.53 \times 10^{-10} \text{ m})^2} = 5.1 \times 10^{11}$ N/C

20.13. $F = kq_3\left(\dfrac{q_1}{r_{13}^2} + \dfrac{q_2}{r_{23}^2} + \dfrac{q_2}{r_{43}^2}\right)$

$= (9 \times 10^9 \text{ N} \cdot \text{m/C}^2)(4\,\mu\text{C})\left[-\dfrac{2\,\mu\text{C}}{(3\text{ m})^2} + \dfrac{3\,\mu\text{C}}{(1\text{ m})^2} + \dfrac{1\,\mu\text{C}}{(0.5\text{ m})^2}\right]$

$= 2.44 \times 10^{-10}$ N/C

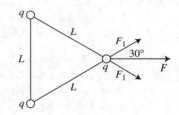

20.14. $F = 2F_1 \cos 30°$

$= 2k\dfrac{q^2}{L^2}\cos 30° = \sqrt{3}k\dfrac{q^2}{L^2}$

20.15. Suppose the field at the center is not zero. Rotate the n polygon by 360°/n. This leaves the charge distribution unchanged because of its symmetry. Thus the electric field vector at the center must also be unchanged, and the only vector that is unchanged by a rotation of 360°/n is one of magnitude zero, so E = 0.

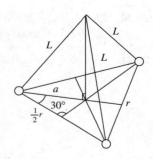

20.16. The charges will be positioned at the vertices of an equilateral triangle of side r. The electric force on each will be $F = \sqrt{3}k(q^2/r^2)$ (see Problem 20.14). As

in Problem 20.3,

$$\tan\theta = \frac{F}{mg} = \frac{\sqrt{3}kq^2}{r^2 mg} \qquad \frac{r}{2} = a\cos 30° \quad \text{and} \quad a = L\sin\theta$$

So

$$r = 2L\frac{\sqrt{3}}{2}\sin\theta = \sqrt{3}L\sin\theta \qquad \sin^2\theta\tan\theta = \frac{kq^2}{\sqrt{3}mgL}$$

Substitute numerical values and find, by trial and error, that $\theta = 6°$.

20.17. Find the force on charge 8 here. Charges 1, 3, and 6 are equivalent, as are 4, 5, and 7. By symmetry, the force is along the body diagonal of the cube. Thus

$$F = \frac{kq^2}{a^2}\left[-\frac{1}{(\sqrt{3})^2} - 3\frac{\cos\alpha}{(1)^2} + 3\frac{\cos\beta}{(\sqrt{2})^2}\right]$$

From the drawing I see that

$$\cos\alpha = \frac{1}{\sqrt{3}} \qquad \cos\beta = \frac{\sqrt{2}}{\sqrt{3}}$$

So

$$F = \frac{kq^2}{a^2}\left[-\frac{1}{3} - \frac{3}{\sqrt{3}} + \frac{3}{2}\left(\sqrt{\frac{2}{3}}\right)\right] = -0.84\frac{kq^2}{a^2}$$

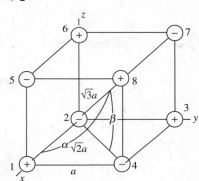

20.18. Here

$$ay = \frac{F}{m} = -\frac{qE}{m} \qquad y = v_{0y}t + \frac{1}{2}a_y t^2 \quad \text{and}$$

$$d = v_0\sin 45°t - \frac{qE}{2m}t^2$$

Solve the quadratic equation:

$$t = \frac{\sqrt{2}v_0 \pm \sqrt{2v_0^2 - 8a_y d}}{2a_y}$$

Choose the smallest solution for t (the minus sign) since this is when the particle first has $y = d$:

$$x = v_0\cos 45°t = \frac{1}{\sqrt{2}}v_0 t = \frac{1}{\sqrt{2}}v_0\left(\frac{\sqrt{2}v_0 - \sqrt{2v_0^2 - 8a_y d}}{2a_y}\right)$$

20.19. $v_0 = 0 \quad x = \frac{1}{2}at^2 \quad \text{and} \quad v = at$

So

$$a = \frac{2x}{t^2} \quad \text{and} \quad v = \frac{2st}{t^2} = \frac{2x}{t} = \frac{(2)(0.015\,\text{m})}{1.2\times 10^{-8}\,\text{s}} = 2.5\times 10^6\,\text{m/s}$$

20.20. Find E_z and E_x separately:

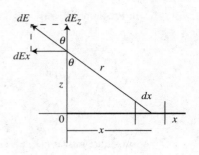

$$E_z = \int \frac{k\,dq}{r^2}\cos\theta = k\lambda \int \frac{dx}{r^2}\left(\frac{z}{r}\right)$$

$$= k\lambda z \int_0^\infty \frac{dx}{(z^2+x^2)^{3/2}} = k\lambda z \frac{x}{z^2(z^2+x^2)^{1/2}}\Bigg|_0^\infty.$$

$$= \frac{k\lambda}{z}$$

$$E_x = -\int \frac{k\,dq}{r^2}\sin\theta = -\int_0^\infty \frac{k\lambda\,dx}{r^2}\left(\frac{x}{r}\right)$$

$$= -k\lambda \int_0^\infty \frac{x\,dx}{(z^2+x^2)^{3/2}} = -k\lambda\left(-\frac{1}{\sqrt{z^2+x^2}}\right)_0^\infty = -\frac{k\lambda}{z}$$

20.21. By symmetry, $E_x = 0$. Thus

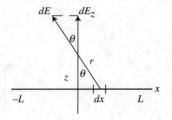

$$E_z = \int \frac{k\,dq}{r^2}\cos\theta = \int_{-L}^{L} \frac{k\lambda\,dx}{r^2}\left(\frac{z}{r}\right) = k\lambda z \int_{-L}^{L} \frac{dx}{(z^2+x^2)^{3/2}}$$

$$= k\lambda z\left(\frac{1}{z^2}\frac{x}{\sqrt{z^2+x^2}}\right)_{-L}^{L} = \frac{2k\lambda L}{L\sqrt{z^2+L^2}}$$

$$\lambda = \frac{Q}{2L}, \quad \text{so } E_z = \frac{kQ}{L\sqrt{z^2+L^2}}$$

20.22. The distance from a rod to the center is $z = L\tan 30° = L/\sqrt{3}$. If the field due to one rod is E_1, then $E = E_1 + 2E_1\cos 60° = E_1 + 2E_1\left(\frac{1}{2}\right) = 2E_1$. E_1 was found in Problem 20.21. Thus

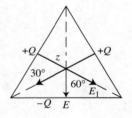

$$E = 2\frac{kQ}{L\sqrt{(L/\sqrt{3})^2 + L^2}}$$

$$= \frac{\sqrt{3}kQ}{L^2}$$

20.23. The solution is given by Problem 20.10 in which I let $R \to \infty$ and I substituted for the surface charge density $\sigma = Q/\pi R^2$. Thus

$$E = \frac{2kQ}{R^2}\left(1 - \frac{z}{\sqrt{R^2+z^2}}\right) = 2\pi k\sigma \quad \text{as } R \to \infty$$

Since

$$k = \frac{1}{4\pi\epsilon_0} \quad E = \frac{\sigma}{2\epsilon_0} \quad \text{field of uniformly charged infinite plane}$$

20.24. Break the surface into small rings of radius $2\pi R \sin\theta$ and width $R\,d\theta$.
$dq = \sigma\,dA = \sigma(2\pi R \sin\theta)(R\,d\theta)$:

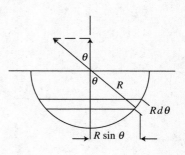

$E = E_z$ by symmetry

$$= \int \frac{k\,dq}{R^2}\cos\theta = \frac{k}{R^2}\int \sigma\,dA\cos\theta$$

$$= \frac{k\sigma}{R^2}\int_0^{\pi/2} 2\pi R^2 \sin\theta\cos\theta\,d\theta = 2\pi k\sigma\left(\frac{1}{2}\sin^2\theta\right)_0^{\pi/2} = \pi k\sigma$$

20.25. $p = 2aq = (6000\,\text{m})(45\,\text{C}) = 2.7 \times 10^4\,\text{C}\cdot\text{m}$

20.26. $\tau = |\mathbf{p}\times\mathbf{E}| = pE\sin\theta = (3.4\times10^{-30}\,\text{C}\cdot\text{m})(4\times10^6\,\text{N/C})(\sin30°) = 6.8\times10^{-24}\,\text{N}\cdot\text{m}$

20.27. $\Delta U = -pE\cos180° - (-pE\cos0°) = 2pE = (6.1\times10^{-30}\,\text{C}\cdot\text{m})(2.4\times10^5\,\text{N/C})$

$$= 1.5\times10^{-24}\,\text{N}\cdot\text{m}$$

20.28.

$$E_1 = k\frac{q}{x^2} = E_0$$

$$E_2 = k\frac{q}{(x-4)^2} = 2E_0$$

$$k\frac{q}{(x-4)^2} = 2k\frac{q}{x^2}$$

$$2(x-4)^2 = x^2$$

$$\sqrt{2}(x-4) = \pm x$$

$$(\sqrt{2}\pm1)x = 4\sqrt{2}$$

The charge must be negative and have $x > 4\,\text{cm}$, so choose negative sign.

$$x = \frac{4\sqrt{2}}{\sqrt{2}-1} = 13.7\,\text{cm}$$

20.29. By symmetry, the fields due to the four charges on the face cancel. The field due to the other four charges is

$$E = 4k\frac{q}{r^2}\cos\theta \quad r^2 = L^2 + \left(\frac{L}{2}\right)^2 + \left(\frac{L}{2}\right)^2 = \frac{3}{2}L^2$$

$$\cos\theta = \frac{L}{r}$$

The field is directed perpendicular to the face, hence the factor $\cos\theta$.

$$E = 4kq\frac{L}{r^3} = 4kqL\frac{1}{[(3/2)L^3]} = 2.18\frac{kq}{L^2}$$

CHAPTER 21

Gauss' Law

In the previous chapter I showed you how to find the electric field due to a charge distribution using straight-forward, brute-strength methods requiring evaluation of integrals that sometimes are complicated. Now I will show you a more elegant mathematical approach that is much simpler to use for symmetric distributions. Further, the mathematics involved has wide-ranging applications.

21.1 Electric Flux and Gauss' Law

A very small area dA can be represented by a vector directed perpendicular to the area and of a magnitude equal to the area. If dA is part of a closed surface, I choose the direction of $\mathbf{dA}$ to point outward. Notice that electric field lines look like flow lines in a fluid (even though nothing is flowing). The word *flux* means, roughly, "flow." Based on this analogy, I define the **flux** of the electric field through a surface $\mathbf{dA}$ as

$$d\Phi = \mathbf{E} \cdot \mathbf{dA} = E\, dA \cos \theta \qquad (21.1)$$

Here θ is the angle between $\mathbf{E}$ and $\mathbf{dA}$. The total flux through a closed surface S is

$$\Phi = \oint \mathbf{E} \cdot \mathbf{dA} \qquad (21.2)$$

The circle on the integral sign signifies a closed surface. *Flux* has a simple meaning. Recall that E can be measured pictorially as the number of lines per area perpendicular to E. Thus $(E)(dA \cos) = $ (number of lines per area perpendicular to E) (area perpendicular to E) = number of E lines passing through dA. Hence the **flux of E through dA** is just **the number of E lines passing through dA**. When E lines pass outward through a closed surface (like a sphere), the flux is positive. When E lines go into a closed surface, the flux is negative (because $\cos \theta < 0$ when $90° < \theta < 180°$). In Fig. 21-1a the flux is four lines. In Fig. 21-1b the flux through each of the two areas is 6, since both surfaces have the same projected area perpendicular to the electric field.

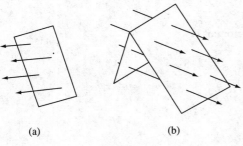

(a)   (b)

Fig. 21-1

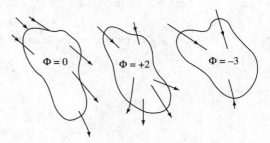

Fig. 21-2

In Fig. 21-2 are shown some examples of flux through closed surfaces. The net flux is the number of lines coming out minus the number of lines going in.

PROBLEM 21.1. Show that the flow of a fluid in cubic meters per second through a surface $d\mathbf{A}$ is $\mathbf{v} \cdot d\mathbf{A}$, where $\mathbf{v}$ is the fluid velocity.

Solution In time dt the fluid passing through dA moves forward a distance $v\,dt$, so the flow through dA is equal to the volume of the parallelopiped shown per time dt. Thus

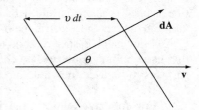

$$\text{Flow} = \frac{\text{volume}}{dt} = \frac{(v\,dt\,\cos\theta)(dA)}{dt} = v\,dA\cos\theta = \mathbf{v} \cdot d\mathbf{A}$$

Referring again to the analogy with fluids, suppose you surround a source of fluid (a mountain spring, perhaps) with a closed surface such as a sphere. If no water accumulates within the sphere, you can see that the total flow out through the spherical surface will equal the rate at which the source is producing water. We can apply this idea to static electric fields, the sources of which are electric charge. The sum of the sources within a closed surface is equal to the total flux through the surface. For electric fields we write this as **Gauss' law**:

$$\Phi = \oint_S \mathbf{E} \cdot d\mathbf{A} = \frac{q_{\text{in}}}{\epsilon_0} \qquad (21.3)$$

Let me emphasize that q_{in} is **the charge inside the surface S**. ϵ_0 is a constant needed to make this statement consistent with Coulomb's law. This simple approach works because E varies as $1/r^2$, just as the surface area of a sphere varies as $1/r^2$.

PROBLEM 21.2. Evaluate Gauss' law for a point charge q placed at the center of a sphere of radius r.

Solution For a sphere, $d\mathbf{A}$ points radially out, and by symmetry $\mathbf{E}$ must also be directed radially out and constant in magnitude over the surface of the sphere. Thus

$$\Phi = \oint_S \mathbf{E} \cdot d\mathbf{A} = \oint_S E\,dA \qquad \text{because } \mathbf{E} \text{ is parallel to } d\mathbf{A}$$

$$= E \oint_S dA \qquad \text{because } E \text{ is constant over the surface } S$$

$$= 4\pi r^2 E \qquad \text{because } da = \text{surface area of a sphere} = 4\pi r^2$$

Hence, Gauss' law yields

$$4\pi r^2 E = \frac{q}{\epsilon_0} \quad \text{or} \quad E = \frac{q}{4\pi\epsilon_0 r^2}$$

Thus we obtain Coulomb's law again.

PROBLEM 21.3. A point charge is placed at one corner of a cube of side L. What is the flux through each face of the cube?

Solution Imagine the charge is at the origin and that the cube faces lie in the xy, yz, and xz planes. The E lines radiate out symmetrically in all directions, and no lines pass through the faces in the xy, yz, or xz planes, so the flux through these faces is zero. By symmetry the flux through each of the remaining faces is the same. One-eighth of the total flux emanates out through the octant occupied by the cube, so the flux through each of the three faces that do not contain the origin is $\Phi = \left(\frac{1}{3}\right)\left(\frac{1}{8}\right)(q/\epsilon_0) = q/24\epsilon_0$.

21.2   Applications of Gauss' Law

There are two kinds of electric fields. The kind I am discussing here, the **electrostatic electric field**, is described by field lines that sprout out of positive charges (the "sources") and end on negative charges (the "sinks"). There is another kind of electric field, called an **induced electric field** (to be discussed later) in which the electric field lines loop back on themselves and hence have no beginning or end. Gauss' law is the easiest method to use when finding the electrostatic electric field for charge distributions that have high symmetry. The charge distributions that are easily solved, and the corresponding gaussian surfaces to be used in evaluating the integral for the flux, are the following:

> **For spherical charge distributions, the gaussian surface is a sphere.**
>
> **For charge distributions that have the symmetry of an infinitely long cylinder or wire, the gaussian surface is a coaxial cylinder of length L.**
>
> **For charge distributions that have the symmetry of an infinite uniform plane, the gaussian surface is a "coin-shaped" flat cylinder.**

Always use Gauss' law to find E for charge distributions with these symmetries, plus a few other problems that can be viewed as superpositions of the above charges. Gauss' law is true for any surface, but the above choices greatly simplify evaluation of the integral encountered. In choosing these surfaces, I was guided by a desire to find surfaces over which E is constant by symmetry, since then E can be taken outside the integral, and the remaining integral is easy to solve.

We encounter charges on or throughout insulators (like pieces of plastic) or on the surface of conductors (like metals). **Within the volume of a conductor, $E = 0$. Further, no net charge remains within the volume of a conductor in equilibrium, so all of the charge on a conductor resides on the surface.** This is so because if there were an electric field within a conductor, the relatively free charges there would move, and in so doing, they themselves would set up "reverse" fields that would cancel the original field. This process occurs very quickly, with the result that the field inside a conductor is zero and all of the net charges (which repel each other) push to the outer surface of the conductor. **At the surface of a conductor the electric field lines are perpendicular to the surface.** If this were not the case, the component of the electric field parallel to the surface would cause charges to move until finally static equilibrium was achieved with the parallel component of E being reduced to zero. Notice that in general the surface charge density on a conductor will not be uniform. The charge tends to accumulate most densely where there are sharp protrusions on the surface, and it is there that the electric field will be greatest.

Here is the problem-solving method to be used to find the electric field of a continuous charge distribution:

1. Identify that the symmetry is one of the above (spherical, long cylinder, or planar), in which case use Gauss' law. Otherwise break the charge into small pieces dq and add up the contributions to E using Coulomb's law.
2. To use Gauss' law, choose a gaussian surface of the type described above, and let one part of this surface pass through the point where you wish to find the electric field.
3. Evaluate the integral of Eq. 21.3, being careful to identify the charge q_{in} contained within the surface S.

The charge outside the gaussian surface enters the problem in a subtle way. It affects the symmetry of the charge distribution, and this in turn influences the value of the integral of the flux. These ideas are most easily understood by considering some examples.

PROBLEM 21.4.   Charge is distributed with uniform volume charge density ρ throughout the volume of a sphere of radius R. Determine E everywhere.

Solution   Inside:

$$q_{in} = \rho\left(\frac{4}{3}\pi r^3\right)$$

$$\oint_s \mathbf{E} \cdot \mathbf{dA} = \oint_s E \, dA = E \oint dA = 4\pi r^2 E$$

$$= \frac{q_{in}}{\epsilon_0}, \quad \text{so} \quad 4\pi r^2 E = \rho\left(\frac{4}{3}\pi r^3\right)$$

$$E = \frac{\rho r}{3\epsilon_0} \quad \text{and} \quad r < R$$

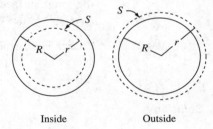

Inside   Outside

Outside:

$$q_{in} = \rho\left(\frac{4}{3}\pi R^3\right), \quad \text{so} \quad \oint_s \mathbf{E} \cdot \mathbf{dA} = 4\pi r^2 E = \rho\frac{(4/3\pi R^3)}{\epsilon_0}$$

and

$$E = \frac{\rho}{3\epsilon_0}\frac{R^3}{r^2} = \frac{Q}{4\pi\epsilon_0 r^2}, \quad r > R, \quad \text{where} \quad Q = \rho\left(\frac{4}{3}\pi R^3\right)$$

Observe that the field outside a spherical charge distribution is that resulting from a point charge equal to the total charge of the sphere and positioned at the center of the sphere.

PROBLEM 21.5. Charge Q is distributed uniformly over a hollow spherical surface of radius R. Determine E inside and outside the sphere.

Solution

$$\oint \mathbf{E} \cdot \mathbf{dS} = 4\pi r^2 E = \frac{q_{in}}{\epsilon_0}$$

Inside: $q_{in} = 0$, so $E = 0$ and $r < R$

Outside:
$q_{in} = q$, so $E = \frac{q}{4\pi\epsilon_0 \tau}$ and $r > R$

Inside   Outside

PROBLEM 21.6. Charge is distributed with constant density ρ throughout a sphere of radius R. There is a spherical void of radius $R/2$ within the large sphere, positioned as shown here. Show that the electric field within the void is constant in magnitude and direction.

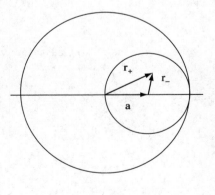

Solution This charge distribution can be viewed as a superposition of a small negative sphere on top of a big positive sphere. The field due to each is found using Gauss' law, with the result given by Problem 2.4. Thus

$$\mathbf{E}_+ = \frac{\rho}{3\epsilon_0}\mathbf{r}_+ \quad \text{and} \quad \mathbf{E}_- = -\frac{\rho}{3\epsilon_0}\mathbf{r}_-$$

The resultant field in the void is thus

$$\mathbf{E} = \mathbf{E}_+ + \mathbf{E}_- = \frac{\rho}{3\epsilon_0}(\mathbf{r}_+ - \mathbf{r}_-)$$

But if $\mathbf{a} = (R/4)\mathbf{i}$ is the vector joining the centers of the spheres, vector addition gives

$$\mathbf{a} + \mathbf{r}_- = \mathbf{r}_+, \quad \text{so} \quad \mathbf{r}_+ - \mathbf{r}_- = \mathbf{a} = \frac{R}{4}$$

Thus

$$\mathbf{E} = \frac{\rho}{3\epsilon_0}\left(\frac{R}{4}\mathbf{i}\right) = \frac{R\rho}{12\epsilon_0}\mathbf{i} \quad E \text{ is constant in the void}$$

This **principle of superposition** allows us to solve many problems whose lack of symmetry makes it appear that Gauss' law would not be a good way to find *E*.

PROBLEM 21.7. Three concentric hollow metallic spherical shells of radii *a*, *b*, and *c* carry charges $+2Q, -3Q$, and $+Q$, respectively. Determine the charge on the inner surface and on the outer surface of each sphere.

Solution First place a spherical gaussian surface within the metal of the inner sphere. Since $E = 0$ in the metal, $\oint E\, dA = 0$ for this surface, and the charge inside is zero. Thus no charge resides on the inner surface of the inner sphere. Since the total charge on this sphere is $2Q$, the charge on the outer surface of the inner sphere is $+2Q$. Repeat the process for the intermediate sphere, placing the gaussian surface within the metal. Again $E = 0$ here, so no net charge is enclosed. Since $2Q$ is on the inner sphere, the charge on the inner surface of the intermediate sphere must be $-2Q$. On the outer surface is $-Q$, yielding the total charge of $-3Q$ on the intermediate sphere. Finally, place a gaussian surface within the metal of the outer sphere. Again $E = 0$ and $q = 0$, so on the inner surface is charge $+2Q - 3Q = -Q$. The total charge on this sphere is $+Q$, so $+2Q$ resides on the outer surface of the outer sphere.

PROBLEM 21.8. Charge is distributed throughout a sphere of radius *R* with density $\rho(r) = \rho_0/r$. Find *E* everywhere. Express the field outside the sphere in terms of the total charge *Q* in the sphere.

Solution
$$\oint \mathbf{E} \cdot d\mathbf{A} = 4\pi R^2 E$$

Inside, use a gaussian sphere of radius $r < R$. There

$$q_{\text{in}} = \int \rho\, dV = \int_0^r \frac{\rho_0}{r}(4\pi r^2\, dr) = \frac{4\pi\rho_0 r^2}{2}$$

so Gauss' law yields

$$4\pi r^2 E = \frac{1}{\epsilon_0}\left(\frac{4\pi\rho_0 r^2}{2}\right) \quad E = \frac{\rho_0}{2\epsilon_0} \quad \text{and} \quad r \le R$$

Outside,

$$q_{\text{in}} = \int_0^R \frac{\rho_0}{r}(4\pi r^2\, dr) = 2\pi\rho_0 R^2 = Q \quad E = \frac{Q}{4\pi\epsilon_0 r^2} \quad \text{and} \quad r \ge R$$

PROBLEM 21.9. Charge is distributed with uniform density ρ throughout the volume of a very long cylinder of radius *R*. Determine *E* inside and outside the cylinder.

Solution Consider a gaussian cylinder *S* of length *L* and radius *r*. By symmetry **E** must be directed radially outward. **dA** is directed perpendicularly outward from the surface of S, so $\mathbf{E} \perp \mathbf{dA}$ on the ends of the cylinder and $\parallel \mathbf{dA}$ on the curved side of the cylinder.

$$\oint_s \mathbf{E} \cdot \mathbf{dA} = \int_{\text{curved side}} (E\,dA) = E \int_{\text{curved side}} dA = E(2\pi r L)$$

To see that $\int dA = 2\pi r L$, imagine you cut the cylinder along a line parallel to its axis and roll it out flat into a rectangle of dimensions $L \times 2\pi r$. When $r \leq R$,

$$q_{\text{in}} = \rho \pi r^2 L \quad \text{and} \quad 2\pi r L E = \frac{1}{\epsilon_0}(\rho \pi r^2 L)$$

$$E = \frac{\rho}{2\epsilon_0} r \quad \text{and} \quad r \leq R$$

When $r \geq R$,

$$q_{\text{in}} = \rho \pi R^2 \quad \text{and} \quad E = \frac{\rho}{2\epsilon_0}\frac{R^2}{r} \quad r \geq R$$

PROBLEM 21.10. An infinitely long straight line carries uniform linear charge density. Determine the electric field a distance r from the line.

Solution Use a gaussian cylinder of length L and radius r. As in Problem 21.9, $\oint_s \mathbf{E} \cdot \mathbf{dA} = 2\pi r L E$, and here $q_{\text{in}} = \lambda L$, so

$$E = \frac{\lambda}{2\pi\epsilon_0 r}$$

PROBLEM 21.11. Charge is distributed with uniform surface charge density σ over an infinite plane insulating sheet. Determine E outside the sheet.

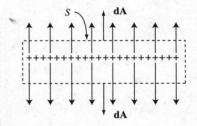

Solution The "coin-shaped" gaussian surface is oriented with its flat faces parallel to the plane. The plane passes through the center of the coin. By symmetry E must be directed perpendicular to the plane, so the magnitude of E does not decrease as you move away from the plane. This is not, of course, realistic, since there are no real "infinite planes." However, if you are close enough to any surface, curved or flat, it looks approximately like an infinite plane. On the flat faces of S of area A, $\mathbf{E}$ is parallel to $\mathbf{dA}$, so $\oint_s \mathbf{E} \cdot \mathbf{dA} = \int_{\text{top}} E\,dA + \int_{\text{bottom}} E\,dA = EA + EA = 2EA$ and $q_{\text{in}} = \sigma A$. Thus, Gauss' law yields

$$2EA = \frac{\sigma A}{\epsilon_0}, \quad E = \frac{\sigma}{2\epsilon_0} \quad \text{insulating sheet}$$

PROBLEM 21.12. Charge is distributed with uniform surface charge density over an infinite conducting plane sheet. Determine E outside the sheet.

Solution   Position the coin-shaped gaussian surface with one face inside the conductor, where $E = 0$. $\mathbf{E} \perp d\mathbf{A}$ on the curved side of the coin, so $\mathbf{E} \cdot d\mathbf{A} = 0$ there.

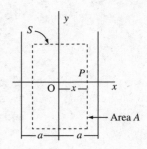

$$\oint_s \mathbf{E} \cdot d\mathbf{A} = \int_{\text{curved side}} \mathbf{E} \cdot d\mathbf{A} \int_{\text{bottom}} \mathbf{E} \cdot d\mathbf{A} + \int_{\text{top}} \mathbf{E} \cdot d\mathbf{A}$$

$$= 0 + 0 + EA = \frac{\sigma A}{\epsilon_0}$$

or

$$E = \frac{\sigma}{\epsilon_0} \quad \text{conducting sheet}$$

Thus for surface charge density on a conducting sheet, E is twice as strong as for the same charge density on an insulating sheet because all of the lines go out on one side (since $E = 0$ inside), whereas for the insulator the lines go out equally on both sides.

PROBLEM 21.13.   Charge is distributed with density ρ uniformly throughout the volume of an infinite plane slab of thickness $2a$ positioned in the yz plane with its center at the origin. Find E everywhere.

Solution   To find E at point P, position the gaussian coin as shown here. $\mathbf{E}$ is pointing parallel to $d\mathbf{A}$ on the flat faces of the coin, and $\mathbf{E}$ is perpendicular to $d\mathbf{A}$ on the curved side. The coin is positioned symmetrically with respect to the slab, so E has the same magnitude on each flat face.

$$\oint_s \mathbf{E} \cdot d\mathbf{A} = 0 + E \int_{\text{left side}} dA + E \int_{\text{right side}} dA = 2EA$$

$$q_{\text{in}} = \rho(2x)(A)$$

so Gauss' law yields

$$E = \frac{\rho x}{\epsilon_0} \quad |x| \leq a$$

If point P is outside, the flux calculation is unchanged, but $q_{\text{in}} = \rho(2a)(A)$. Thus

$$E = \frac{\rho a}{\epsilon_0}, \quad |x| \geq a$$

21.3   Summary of Key Equations

Flux: $$\Phi = \int \mathbf{E} \cdot d\mathbf{A}$$

Gauss' law: $$\oint_s \mathbf{E} \cdot d\mathbf{A} = \frac{q_{\text{in}}}{\epsilon_0}$$

Outside spherical charge: $$E = \frac{q}{4\pi\epsilon_0 r^2}$$

Surface of a conductor: $$E = \frac{\sigma}{\epsilon_0}$$

SUPPLEMENTARY PROBLEMS

21.14. Charges $+Q$, $-4Q$, and $+2Q$ are placed inside a cubic enclosure, but their positions are not specified. What is the total electric flux passing through the walls of the container?

21.15. Eight identical charges $+q$ are placed at the corners of a cube. What is the total flux through the faces of the cube?

21.16. A point charge q is placed on the axis of a disk of radius R a distance z from the center of the disk. What is the flux passing through the disk?

21.17. Lightning causes the Earth to acquire a negative charge and this charge creates an electric field in the air. Treating the Earth as a flat conductor, estimate the surface charge density on the ground when the field in the air is 100 N/C.

21.18. A point charge Q is placed at the center of a hollow conducting sphere of inner radius a and outer radius b. Determine the electric field everywhere and the charge on the inner and outer surfaces of the sphere. Sketch the field lines.

21.19. A coaxial cable consists of two long concentric conducting cylindrical shells of radii a and b. The linear charge density on the inner conductor is $+\lambda$ and on the outer conductor $-\lambda$. Determine E everywhere.

21.20. A very long cylinder of radius R has a uniform charge density ρ. It contains a long cylindrical void of radius $R/2$, positioned as shown here. Determine the electric field at point P.

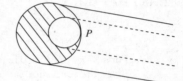

21.21. A sphere of radius r has charge density $\rho = \rho_0 r$. Determine E everywhere.

21.22. Two infinite, parallel nonconducting sheets carry surface charge densities $+\sigma$ and $-\sigma$. Find E in the space between the sheets and outside the sheets.

21.23. A slab of thickness $2a$ is parallel to the yz plane, with its midpoint at the origin. It has charge density $\rho = \rho_0(x/a)^2$. Such situations are encountered in constructing semiconductor devices', such as the chips used in computers. Determine the electric field everywhere.

21.24. A charge q is placed at the center of a regular tetrahedron of side a. What is the average value of the electric field over one face of the tetrahedron?

21.25. A solid sphere of radius R has an electric charge density $\rho = \rho_0 r$. Determine the electric field inside and outside the sphere.

SOLUTIONS TO SUPPLEMENTARY PROBLEMS

21.14. By Gauss' law,

$$\Phi = \oint_s \mathbf{E} \cdot d\mathbf{A} = \frac{q_{\text{in}}}{\epsilon_0}$$

so

$$\Phi = \frac{1}{\epsilon_0}(+Q - 4Q + 2Q) = -\frac{Q}{\epsilon_0}$$

21.15. In Problem 21.3 we found that charge Q provided flux of $1/8\,(1/\epsilon_0)$ through the sphere. Thus, charges on every corner (five in all) provide flux

$$(8)\left(\frac{1}{8}\frac{q}{\epsilon_0}\right) = \frac{q}{\epsilon_0}$$

Another way to see this is to imagine that each charge is like a small sphere, with 1/8 of its volume inside the sphere, so essentially one whole charge is inside, and Gauss' law yields $\Phi = q/\epsilon_0$.

21.16. Replace the disk with a spherical cap of radius $r = \sqrt{z^2 + R^2}$. The same flux passes through this cap as through the disk. Use the ratio

$$\frac{\Phi_{\text{cap}}}{\Phi_{\text{total}}} = \frac{\text{area of cap}}{\text{surface area of sphere}} = \frac{A}{4\pi r^2}$$

where

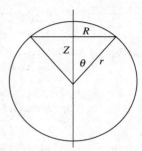

$$A = \int_0^{\sin^{-1} R/r} (r\,d\theta)(2\pi r \sin\theta) = 2\pi r^2 [-\cos\theta]^{\sin^{-1} R/r}$$

$$= (2\pi r^2)\left(-\frac{z}{r} + 1\right)$$

$$\Phi = \frac{q}{\epsilon_0}\frac{1}{4\pi r^2}(2\pi r^2)\left(1 - \frac{z}{r}\right) = \frac{q}{2\epsilon_0}\left(1 - \frac{z}{\sqrt{z^2 + R^2}}\right)$$

21.17. $E = \sigma/\epsilon_0$, so $\sigma = \epsilon_0 E = (8.85 \times 10^{-12}\ \text{C}^2\text{N}\cdot\text{m}^2)(100\ \text{N/C}) = 8.85 \times 10^{-10}\ \text{C/m}^2$.

21.18. For $r < a$,

$$q_{\text{in}} = q \quad \text{and} \quad E = \frac{q}{4\pi\epsilon_0 r^2}$$

For $a < r < b$, $E = 0$ in the conductor, so $q_{\text{in}} = 0$, and thus charge $-Q$ is induced on the inner surface at $r = a$. Since the conductor carries no net charge, charge $+q$ is induced on the outer surface.

21.19. Use a gaussian cylinder. For $r < a$ or $r > b$, $q_{\text{in}} = 0$, and thus $E = 0$ there. For $a < r < b$, $q_{\text{in}} = \lambda L$, so

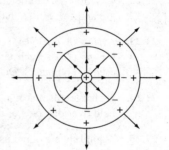

$$2\pi r L E = \frac{\lambda L}{\epsilon_0} \quad \text{and} \quad E = \frac{\lambda}{2\pi\epsilon_0 r}$$

21.20. Superimpose a cylinder of radius $R/2$ and charge density $-\rho$ on top of a solid cylinder of radius R and positive charge density $+p$. Add vectorially the field of each using the result of Problem 21.9. Thus, at point P,

$$E = \frac{\rho}{2\epsilon_0}R - \frac{\rho}{2\epsilon_0}\left(\frac{R}{2}\right) = \frac{\rho}{4\epsilon_0}R$$

21.21. For $r < R$,

$$q_{\text{in}} = \int \rho\,dv = \int_0^r \frac{\rho_0 r}{r}(4\pi r^2\,dr) = \frac{1}{2}\pi\rho_0 r^4$$

For $r > R$,

$$q_{\text{in}} = \int_0^R \frac{\rho_0 r}{r}(4\pi r^2\,dr) = \frac{1}{2}\pi\rho_0 R^4$$

$$\oint \mathbf{E}\cdot d\mathbf{A} = 4\pi r^2 E$$

Thus,

$$E = \frac{\rho_0 r^2}{8\epsilon_0} \quad r < R \quad \text{and} \quad E = \frac{\rho_0 R^4}{8\epsilon_0 r^2} \quad r > R$$

21.22. First let a gaussian coin surface enclose both sheets. Then $q_{in} = 0$, and thus, $E = 0$ outside the sheets. We can also see this result by superimposing the fields of the individual sheets. Now let a Gaussian coin enclose only one sheet. On outside face of the coin $E = 0$, so

$$\oint \mathbf{E} \cdot \mathbf{dA} = 0 + EA = \frac{\sigma A}{\epsilon_0}, \quad E = \frac{\sigma}{\epsilon_0}$$

between the sheets.

21.23. First position coin-shaped S as shown.

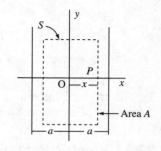

$$\oint \mathbf{E} \cdot \mathbf{dA} = 2EA \quad q_{in} = \int_{-a}^{a} \rho_0 \left(\frac{x}{a}\right)^2 A\, dx = \frac{1}{3} \rho_0 A \frac{x^3}{a^2} = \frac{2}{3} \rho_0 A a \left(\frac{x}{a}\right)^3$$

Thus,

$$E = \frac{\rho_0 a}{3\varepsilon_0} \left(\frac{x}{a}\right)^3 \quad |x| \le a$$

For $|x| > a$,

$$q_{in} = \frac{2}{3} \rho_0 A a \quad \text{and} \quad E = \frac{\rho_0 a}{3\varepsilon_0} \quad |x| > a$$

21.24.

$$\int \mathbf{E} \cdot \mathbf{dA} = \frac{q}{\epsilon_0}$$

$$4 E_{AV} A_1 = \frac{q}{\epsilon_0}, \quad \text{where } A_1 \text{ is the area of one face.}$$

Each face is an equilateral triangle of side a and area A_1,

$$A_1 = \left(\frac{a}{2}\right)\left(\frac{\sqrt{3}}{2} a\right) = \frac{\sqrt{3}}{4} a^2$$

Thus,

$$E_{AV} = \frac{q}{A_1 \epsilon_0} = \frac{q}{\sqrt{3}\epsilon_0}$$

21.25.

$$\int E\, dA = \frac{q}{\epsilon_0} = 4\pi r^2 E, \quad E = \frac{q}{4\pi\epsilon_0 r^2}$$

r is the radius of the Gaussian sphere passing through the point where E is to be determined. We must determine the enclosed charge q.

Case 1:

$$r \le R, \quad q = \int \rho\, dV = \int_0^r \rho_0 r(4\pi r^2\, dr)$$

$$q = \rho_0 \pi r^4$$

$$E = \frac{\rho_0}{\epsilon_0} r^2, \quad r \le R$$

Case 2:

$$r \ge R, \quad q = \int_0^R \rho_0 r(4\pi r^2)\, dr = \rho_0 \pi R^4$$

$$E = \frac{\rho_0 \pi R^4}{4\pi\epsilon_0 r^2} = \frac{\rho_0 R^4}{4\epsilon_0 r^2}, \quad r \ge R$$

CHAPTER 22

Electric Potential

The electrostatic force is conservative, and thus we can introduce a potential energy and use the powerful techniques of energy conservation in solving problems.

22.1 Electric Potential and Potential Energy

Consider a charge q in an electric field $\mathbf{E}$. The electric field exerts a force $q\mathbf{E}$ on the charge, and if you want to move the charge, you must exert a force $-q\mathbf{E}$ on it. If in so doing, you move the charge by a displacement $\mathbf{ds}$, you do work $dW = \mathbf{F} \cdot \mathbf{ds} = -q\mathbf{E} \cdot \mathbf{ds}$. The work done is equal to the gain in **electric potential energy**, so the gain in potential energy when a charge is moved from A to B is

$$\Delta U = U_A - U_B = -\int_A^B q\mathbf{E} \cdot \mathbf{ds} \tag{22.1}$$

It is useful to define the **electric potential** $V = U/q$. The change in electric potential between two points is

$$V = V_A - V_B = -\int_A^B \mathbf{E} \cdot \mathbf{ds}, \quad \text{where } \Delta U = q\Delta V \tag{22.2}$$

Potential energy U is measured in joules and potential V is measured in joules per coulomb. Potential is so important that a new unit, the volt, is used to describe it. We define 1 *volt* such that $1\text{ V} = 1\text{ J/C}$. The zero point of electric potential (or of electric potential energy) is arbitrary (just as the zero of elevation is arbitrary). Only differences in potential matter. In everyday usage the term *voltage* or *voltage difference* is often used for the correct term, *potential difference* (PD). Be sure to keep clear the distinction between the terms *electric potential V* (in volts), *electric potential energy U* (in joules), *electric field E* (in volts per meter or newtons per coulomb), and *electric charge Q* (in coulombs). From Eq. 22.2 we can see that when $\mathbf{ds} = dx\mathbf{i}$,

$$E_x = -\frac{dV}{dx} \tag{22.3}$$

In the case of a constant electric field, such as that between two large parallel charged plates, the magnitude of the potential difference between two points separated by a distance x is

$$\Delta V = Ex \quad \text{constant } E \text{ field} \tag{22.4}$$

If we choose the zero point of potential to be at $x = 0$, that is, $V(0) = 0$, then $V = Ex$ and the potential energy of a charge q at point x is $U = qEx = qV$. This is analogous to the potential energy of a mass m

in a gravitational field g. For the gravity case, $U = mgy$. Thus I find it helpful to make the following analogy: The amount of charge q (in coulombs) is like the amount of mass m. The electric field E is like the gravitational field g. The electric potential V is like gy, that is, proportional to the elevation. In this analogy, $-E$ is like the slope of a hill and potential is like the elevation. E points downhill, and positive charges tend to roll downhill to lower potential energy. Negative charges like electrons tend to move toward higher electric potential (but still lower potential energy).

In describing electrons and other small particles, it is useful to have a small unit of energy. When a particle of charge q falls through a potential difference ΔV, its potential energy changes by $\Delta U = q\Delta V$. When an amount of charge e (the charge on the proton) falls through 1 V, its potential energy changes by **1 electronvolt (eV)**, where

$$\boxed{1\,\text{eV} = e\Delta V = (1.60 \times 10^{-19}\,\text{C})(1\,\text{V}) = 1.6 \times 10^{-19}\,\text{J}} \tag{22.5}$$

PROBLEM 22.1   A proton initially at rest falls through a potential difference of 25,000 V. What speed does it gain?

Solution

$$\Delta\text{KE} = \Delta U \qquad \frac{1}{2}mv^2 = e\Delta V$$

$$v = \sqrt{\frac{2e\Delta V}{m}}$$

$$= \sqrt{\frac{(2)(1.6 \times 10^{-19}\,\text{C})(25,000\,\text{V})}{1.67 \times 10^{-27}\,\text{kg}}} = 2.19 \times 10^6\,\text{m/s}$$

Observe that if a charged particle moves perpendicular to electric field lines, no work is done on it, and its electric potential energy does not change. Along such a path the electric potential energy and the electric potential V remain constant. Such a path in two dimensions is called an **equipotential line**, and in three dimensions a particle can move over an **equipotential surface** or throughout an **equipotential volume**. An equipotential line is analogous to a contour line on a topographic map. Electric field lines are always perpendicular to equipotential lines. On a map, a contour line traces out points of equal elevation, and the direction of steepest descent (the direction water will run if you pour it on the ground) corresponds to the direction of the electric field lines. Where the equipotential lines are close together (like closely spaced contour lines), the E field is large (that is, the hill is steep). Equipotential lines can never intersect, nor can electric field lines. Equipotentials and E lines for a uniform field, a point charge, and a dipole are shown in Fig. 22-1.

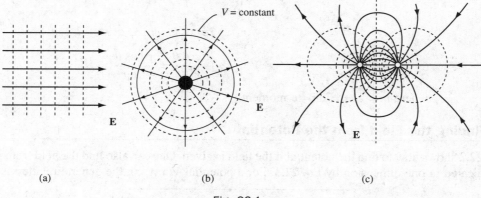

(a)   (b)   (c)

Fig. 22-1

22.2 Electric Potential of a Point Charge

We can integrate Eq. 22.2 for a point charge, where

$$\mathbf{E} = \frac{q}{4\pi\epsilon_0} \frac{\mathbf{r}}{r^2}$$

It is convenient to choose the zero of potential such that $V_A = 0$, and let $r_B = r$. Thus

$$\boxed{V = \frac{q}{4\pi\epsilon_0 r} \quad \text{point charge}} \tag{22.6}$$

For a collection of n point charges,

$$\boxed{V = \sum_{i=1}^{n} V_i = \frac{1}{4\pi\epsilon_0} \sum_{i=1}^{n} \frac{q_i}{r_i} \quad n \text{ point charges}} \tag{22.7}$$

Observe that positive charges always make the potential at every point more positive, and negative charges make the potential more negative.

PROBLEM 22.2.   A charge $-q$ is placed at one corner of a square of side a, and charges $+q$ are placed at each of the other corners. What is the potential at the center of the square?

Solution
$$V = \frac{1}{4\pi\epsilon_0} \left(\frac{q}{a/\sqrt{2}} + \frac{q}{a/\sqrt{2}} + \frac{q}{a/\sqrt{2}} - \frac{q}{a/\sqrt{2}} \right) = \frac{\sqrt{2}}{2\pi\epsilon_0} \frac{q}{a}$$

If a charge q_2 is placed in the field of a charge q_1 a distance r away, the potential energy of the pair is then

$$\boxed{U = q_2 V_1 = \frac{q_1 q_2}{4\pi\epsilon_0 r}} \tag{22.8}$$

If there are more than two charges present, the electrostatic energy of the system is obtained by finding the energy of each pair of charges and then summing to obtain the total energy. For three charges we obtain

$$\boxed{U = \frac{1}{4\pi\epsilon_0} \left(\frac{q_1 q_2}{r_{12}} + \frac{q_1 q_3}{r_{13}} + \frac{q_2 q_3}{r_{23}} \right)} \tag{22.9}$$

PROBLEM 22.3.   Find the potential due to an electric dipole far from the dipole.

Solution
$$V = kq \left(\frac{1}{r_+} - \frac{1}{r_-} \right) = kq \left(\frac{r_- - r_+}{r_- r_+} \right)$$

$$r_- - r_+ \simeq 2a\cos\theta, \quad r_- r_+ \simeq r^2$$

$$V \simeq kq \frac{2a\cos\theta}{r^2} = \frac{p\cos\theta}{4\pi\epsilon_0 r^2}$$

where $p = 2aq =$ dipole moment.

22.3 Finding the Field from the Potential

Equation 22.2 shows how to find the potential if the field is given. One can also find the field from the potential, as indicated in one dimension by Eq. 22.3. For a potential $V(x, y, z)$, the general relation is

$$\boxed{E_x = -\frac{\partial V}{\partial x} \qquad E_y = -\frac{\partial V}{\partial y} \qquad E_z = -\frac{\partial V}{\partial z}} \tag{22.10}$$

$\partial V/\partial x$ is a partial derivative that means we take the derivative of V with respect to x while holding y and z constant. In vector notation,

$$\mathbf{E} = -\left(\frac{\partial V}{\partial x}\mathbf{i} + \frac{\partial V}{\partial y}\mathbf{j} + \frac{\partial V}{\partial z}\mathbf{k}\right) \tag{22.11}$$

It is useful to introduce a differential vector operator ∇, called a *del*, such that

$$\nabla V = \frac{\partial V}{\partial x}\mathbf{i} + \frac{\partial V}{\partial y}\mathbf{j} + \frac{\partial V}{\partial z}\mathbf{k} \quad \text{Cartesian coordinates} \tag{22.12}$$

∇V is the **gradient of** V. In spherical coordinates this expression becomes

$$\nabla V = \frac{\partial V}{\partial r}\hat{\mathbf{r}} + \frac{1}{r}\frac{\partial V}{\partial \theta}\hat{\theta} + \frac{1}{r\sin\theta}\frac{\partial V}{\partial \theta}\hat{\phi} \tag{22.13}$$

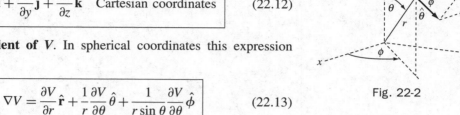

Fig. 22-2

Here $\mathbf{r}$, θ, and ϕ are the unit vectors shown in Fig. 22-2. Using this notation,

$$\mathbf{E} = -\nabla V \tag{22.14}$$

PROBLEM 22.4. Find the electric field of a dipole far from the dipole, using the potential found in Problem 22.3.

Solution Use spherical coordinates with

$$V = \frac{p\cos\theta}{4\pi\epsilon_0 r^2}$$

Then

$$E_r = -\frac{\partial V}{\partial r} = \frac{2p\cos\theta}{4\pi\epsilon_0 r^3} \qquad E_\theta = -\frac{1}{r}\frac{\partial V}{\partial \theta} = \frac{p\sin\theta}{4\pi\epsilon_0 r^3} \qquad E_\phi = -\frac{1}{r\sin\theta}\frac{\partial V}{\partial \phi} = 0$$

These E lines are shown in Fig. 22-1.

22.4 Potential of Continuous Charge Distributions

To find the potential due to a continuous charge distribution, break the charge into small pieces dq and find the potential due to each piece, treating it as a point charge. Then integrate to find the potential due to all dq's.

PROBLEM 22.5. A rod of length L has a uniform linear charge density λ. Determine the potential at a point P on the axis of the rod a distance d from one end.

Solution Consider small pieces of charge $dq = \lambda\,dx$ for which $dV = k\,dq/r = k\lambda\,dx/(d+x)$. Then

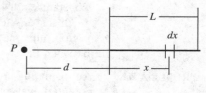

$$V = \int_0^L \frac{k\lambda\,dx}{d+x} = \frac{k\lambda}{4\pi\epsilon_0}\ln\left(\frac{d+\lambda}{d}\right)$$

PROBLEM 22.6.   Charge Q is uniformly distributed on a ring of
radius a. Determine the potential at a point on
the axis of the ring a distance x from the center.

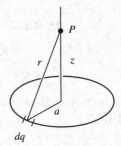

Solution   Break the ring into small pieces of charge dq,
and observe that every piece is the same
distance $r = \sqrt{z^2 + a^2}$ from point P.

$$V = k \int \frac{dq}{r} = k \oint_{\text{entire ring}} \frac{\lambda \, ds}{\sqrt{a^2 + z^2}} = \frac{k}{\sqrt{a^2 + z^2}} \oint_{\text{entire ring}} \lambda \, ds$$

$$\lambda \, ds = \frac{kQ}{\sqrt{a^2 + z^2}} \qquad V = \frac{Q}{4\pi\epsilon_0 \sqrt{a^2 + z^2}}$$

PROBLEM 22.7.   A disk of radius a carries a uniform surface charge
density σ. Find the potential on the axis at point P a
distance z from the center.

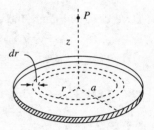

Solution   Break the disk into annular rings of radius r and width dr.
Each point on a given ring is the same distance r from
point P and hence makes a contribution dV to the
potential, where

$$dV = \frac{k \, dq}{\sqrt{r^2 + z^2}} = \frac{k\sigma 2\pi r \, dr}{\sqrt{r^2 + z^2}}$$

$$V = \int_0^a \frac{k\sigma 2\pi r \, dr}{\sqrt{r^2 + z^2}} = k\sigma 2\pi \int_0^a \frac{r \, dr}{\sqrt{r^2 + z^2}} = k\sigma 2\pi \left[\sqrt{r^2 + z^2} \right]_0^a$$

$$= k\sigma 2\pi \left[\sqrt{a^2 + z^2} - z \right] \qquad k = \frac{1}{4\pi\epsilon_0}$$

PROBLEM 22.8.   A spherical shell of radius R carries charge Q uniformly distributed over the surface.
Determine V inside and outside.

Solution   We saw in the previous chapter that outside a spherical shell the electric field is just
that due to a point charge, so the potential outside will be that of a point charge,
$V = Q/4\pi\epsilon_0 r$ for $r \geq R$. Inside Gauss' law showed that $E = 0$, so inside
$V = \text{constant}$, since

$$E_r = -\frac{\partial V}{\partial r} = 0$$

The constant must equal the value of V at the surface of the sphere where $r = R$, so the
constant $= Q/4\pi\epsilon_0 R = V$ inside, $r \leq R$.

Observe that since V has a definite value at every point, the solution for V in one
region must match the value of V in an adjacent region at every point on the interface
between the two regions. This is an example of what is called a *boundary condition*.

PROBLEM 22.9   An insulating sphere of radius R has uniform volume charge density ρ. Determine the
potential everywhere.

Solution   From Gauss' law we saw that outside the sphere the field is just that of a point charge, so for this case $V_{out} = Q/4\pi\epsilon_0 r$, where $Q = \rho(4\pi R^3/3)$. For $r \leq R$, consider first the charge within a sphere of radius r. The potential due to this charge is $V_1 = Q'/4\pi\epsilon_0 r = \rho r^2/3\epsilon_0$, where $Q = \rho(4\pi r^3/3)$.

For the charge between r and R, break the sphere into shells of radius r' and thickness dr'. The potential inside a shell is equal to the potential at the surface of the shell (see Problem 22.8). Thus the contribution of each shell is

$$dV_2 = \frac{dq}{4\pi\epsilon_0 r'} \quad \text{where } dq = \rho\left(4\pi r'^2 dr'\right)$$

Thus the potential due to the charge between r and R is

$$V_2 = \int_r^R \frac{\rho(4\pi r'^2 dr')}{4\pi\epsilon_0 r'} = \frac{\rho}{2\epsilon_0}(R^2 - r^2)$$

Hence the potential inside the sphere is

$$V = V_1 + V_2 = \frac{\rho r^2}{3\epsilon_0}(R^2 - r^2) = \frac{\rho}{3\epsilon_0}\left(R^2 - \frac{r^2}{3}\right)$$

In terms of the total charge

$$Q = \rho\left(\frac{4\pi R^3}{3}\right) \quad V = \frac{Q}{8\pi\epsilon_0 R}\left(3 - \frac{r^2}{R^2}\right)$$

An alternate way of finding V inside the sphere is to integrate the solution for E obtained using Gauss' law. Add a constant to the expression obtained for V so that the potential inside and outside match at the boundary $r = R$.

22.5   Potential of a Charged Conductor

We have seen that the charge on a conductor resides on the outer surface and that the electric field within the conductor is zero. This means that the potential everywhere in a conductor is constant (since $\mathbf{E} = -\nabla V$). The surface of a conductor is an equipotential surface, and the E lines are perpendicular to the surface at the surface. If there is an empty cavity in a conductor, $E = 0$ there. In general, charge is not uniformly distributed over the surface of a conductor. The charge concentrates where there are sharp protuberances (that is, small radii of curvature), and it is here that the electric field will be large. In some cases the electric field is large enough to cause electrical breakdown (corona) in the surrounding air.

PROBLEM 22.10.   Two conducting spheres of radii R_1 and R_2 are connected by a conducting wire of length $L \gg R_1, R_2$. A charge Q is placed on the spheres. Determine (a) the charge on each sphere, (b) the potential of each sphere, and (c) the electric field at the surface of each sphere.

Solution   Since the spheres are far apart, I assume the charge on each is uniformly distributed. Because they are joined by a wire, they constitute a single conductor, so both spheres are at the same potential, $V = V_1 = V_2$.

$$V_1 = k\frac{Q_1}{R_1} \quad \text{and} \quad V_2 = k\frac{Q_2}{R_2} \qquad V_1 = V_2 \quad \text{and} \quad Q = Q_1 + Q_2$$

so

$$Q_1 = \frac{R_1}{R_1 + R_2}Q \quad \text{and} \quad Q_2 = \frac{R_2}{R_1 + R_2}Q$$

and

$$V = k\frac{Q}{R_1 + R_2} \qquad E_1 = k\frac{Q_1}{R_1^2} = k\frac{Q}{R_1(R_1 + R_2)} \quad \text{and} \quad E_2 = k\frac{Q}{R_2(R_1 + R_2)}$$

Note that although $V_1 = V_2$, $E_1 > E_2$ if $R_1 < R_2$.

22.6 Summary of Key Equations

Electric potential energy: $\Delta U = U_A - U_B = -\int_A^B q\mathbf{E} \cdot \mathbf{ds}$

Electric potential: $V = V_A - V_B = -\int_A^B \mathbf{E} \cdot \mathbf{ds}, \quad \text{where } \Delta U = q\Delta V$

Electric field: $E_x = -\dfrac{dV}{dx} \quad \text{or} \quad \mathbf{E} = -\nabla V$

Potential of a point charge: $V = \dfrac{q}{4\pi\epsilon_0 r}$

SUPPLEMENTARY PROBLEMS

In the following, $k = 1/4\pi\epsilon_0 = 9 \times 10^9 \text{ N} \cdot \text{m}^2/\text{C}^2$.

22.11. What change in potential energy does a 4-$\bar{\mu}$C charge experience when it moves between two points that differ in potential by 80 V?

22.12. What is the energy in electronvolts of an electron with speed 4×10^6 m/s?

22.13. Two parallel metal plates separated by 2.4 cm have a potential difference of 100 V between them. An electron with speed 4×10^6 m/s passes perpendicularly through a hole in the positive plate. How far from the positive plate will it travel?

22.14. An infinite insulating sheet carries a uniform charge density of $\sigma = 0.40 \,\bar{\mu}$C/m^2. How far apart are equipotential surfaces that differ by 80 V?

22.15. How far from a point charge of 5 $\bar{\mu}$C is the potential 50 V? 100 V?

22.16. In a region of space the potential is given by $V(x,y) = 4xy + x^2$. Determine the strength of the electric field at the point (3, 2).

22.17. Considering the nucleus of platinum to be a uniformly charged sphere of charge 78e and radius 7.0×10^{-15} m, what is the potential energy of a proton when it reaches the surface of this nucleus?

22.18. Charge Q is uniformly distributed along a rod of length L. Find the potential a distance z from the midpoint of the rod along a line perpendicular to the rod.

22.19. A rod with uniform linear charge density λ is bent into the shape shown here. Find the potential at the center for this configuration.

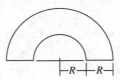

22.20. A hole of radius R is cut from the center of a uniformly charged disk of radius 2R. Determine the potential on the axis of the disk a distance z from the center.

22.21. 100 small identical spherical drops of mercury are each given enough charge to raise them to potential V_1. If the drops all coalesce into a single large spherical drop, what will be its potential?

22.22. In a certain region of space the electric potential is described by $V(x, y, z) = 3x - 2xy + 4yz^2$. Determine the components and the magnitude of the electric field at the point $(1, 1, 0)$.

22.23. In some cathode ray tubes, such as those used in old color TV sets, electrons fall from rest through a potential difference of 25 kV. What speed does such an electron attain in this case?

22.24. Charges $+q$ are placed at opposite corners of a square of side a. Charges $-q$ are placed at the other two opposite corners of the square. What is the potential energy of this arrangement of charges?

SOLUTIONS TO SUPPLEMENTARY PROBLEMS

22.11. $\Delta U = q\Delta V = (4 \times 10^{-6} \, \text{C})(80 \, \text{V}) = 3.2 \times 10^{-4} \, \text{J}$

22.12. $\text{KE} = \frac{1}{2}mv^2 = \frac{1}{2}(9.11 \times 10^{-31} \, \text{kg})(4 \times 10^6 \, \text{m/s})^2 = 7.29 \times 10^{-18} \, \text{J}$

$$= \frac{7.29 \times 10^{-18} \, \text{J}}{1.6 \times 10^{-19} \, \text{J/eV}} = 46 \, \text{eV}$$

22.13. $\frac{1}{2}mv^2 = eV$, and $V = Ex = \frac{V_0}{d}x$, so $x = \frac{mv^2 d}{2eV_0} = 0.011 \, \text{m}$

22.14. From Eq. 22.2,

$$E = \frac{\sigma}{2\epsilon_0} = \frac{\Delta V}{\Delta x} \quad \Delta x = \frac{2\epsilon_0 \Delta V}{\sigma}$$

$$= \frac{(2)(8.85 \times 10^{-12} \, \text{C}^2/\text{N} \cdot \text{m}^2)(80 \, \text{V})}{0.40 \times 10^{-6} \, \text{C/m}^2} = 0.0035 \, \text{m}$$

22.15. $V = k\dfrac{q}{r}$ and $r = k\dfrac{q}{V}$

For $V = 50 \, \text{V}$,

$$r = (9 \times 10^9 \, \text{N} \cdot \text{m}^2/\text{C}^2)\frac{(5 \times 10^{-6} \, \text{C})}{(50 \, \text{V})} = 900 \, \text{m}$$

For $V = 100 \, \text{V}$, $r = 450 \, \text{m}$.

22.16. $Ex = -\dfrac{\partial V}{\partial x} = -4y + 2x$ and $Ey = -\dfrac{\partial V}{\partial y} = -4x$

$$E = \sqrt{E_x^2 + E_y^2}$$

Thus at $(3, 2)$, $E = \sqrt{(-8 + 6)^2 + (-12)^2} = 18 \, \text{V/m}$.

22.17. $U = eV = ek\left(\dfrac{78e}{r}\right) = 2.57 \times 10^{-12} \, \text{J}$

22.18. $V = k\displaystyle\int_{-L/2}^{L/2} \frac{\lambda \, dx}{r} = k\int_{-L/2}^{L/2} \frac{\lambda \, dx}{\sqrt{x^2 + z^2}}$

$$= k\lambda \ln(x + \sqrt{x^2 + z^2}) \Big|_{-L/2}^{L/2}$$

$$= k\lambda \ln\left(\frac{\sqrt{L^2 + 4z^2} + L}{\sqrt{L^2 + 4z^2} - L}\right)$$

22.19. Using the result of Problem 22.5,

$$V = 2k\lambda \ln\left(\frac{R + R}{R}\right) + k\frac{\lambda \pi R}{R} + k\frac{\lambda \pi (2R)}{2R} = 2(\pi + 1)\lambda$$

22.20. The solution can be calculated as in Problem 22.7 with the integral limits changed to R to $2R$. An easier method is to superimpose a small disk with charge density $-\rho$ on top of a larger disk with charge density $+\rho$ and use the result of Problem 22.7.

$$V = V_+ + V_- = k\sigma 2\pi\left[\sqrt{(2a)^2 + x^2} - x\right] - k\sigma 2\pi\left[\sqrt{a^2 + x^2} - x\right]$$

$$= 2\pi k\sigma\left(\sqrt{4a^2 + x^2} - \sqrt{a^2 + x^2}\right)$$

22.21. Charge and total volume are unchanged.

$$100\left(\frac{4}{3}\pi r^3\right) = \frac{4}{3}\pi R^3 \quad \text{and} \quad V = k\frac{100q}{R}$$

where

$$V_1 = k\frac{q_1}{r}$$

Thus,

$$V = \frac{100V_1 r}{100^{1/3}r} = 21.5V_1$$

22.22. $E_x = -\dfrac{\partial V}{\partial x} = -3 + 2y^2 = -1$ at $(1, 1, 0)$

$E_y = -\dfrac{\partial V}{\partial y} = 4xy - 4z^2 = 4$

$E_z = -\dfrac{\partial V}{\partial z} = -8yz = 0$ at $(1, 1, 0)$

$E = \left(E_x^2 + E_y^2 + E_z^2\right)^{1/2} = (1 + 16 + 0)^{1/2} = \sqrt{17}$

22.23. $\dfrac{1}{2}mv^2 = qV$

$$v = \sqrt{\frac{2qV}{m}} = \left[\frac{(2)(1.6 \times 10^{-19}\text{ C})(2.5 \times 10^4\text{ V})}{9.11 \times 10^{-31}\text{ kg}}\right]^{1/2}$$

$$v = 9.4 \times 10^7\text{ m/s}$$

22.24. Imagine that a charge q is first placed at one corner. Now bring in an adjacent charge $-q$, the energy of this pair is $-qV$.

$$U_1 = -qV_1 = -k\frac{q^2}{a}$$

Now with these two charges in place bring in the charge $-q$. This requires energy U_2,

$$U_2 = -q\left(k\frac{q}{a} + k\frac{-q}{\sqrt{2}a}\right)$$

Finally, bring in the remaining charge q.

$$U_3 = q\left(k\frac{q}{\sqrt{2}a} - k\frac{q}{a} - k\frac{q}{a}\right)$$

The total energy is thus $U' = U_1 + U_2 + U_3$

$$U = k\frac{q^2}{a}\left(-1 - 1 + \frac{1}{\sqrt{2}} + \frac{1}{\sqrt{2}} - 1 - 1\right)$$

$$U = k\frac{q^2}{a}\left(-4 + \sqrt{2}\right) = -2.59k\frac{q^2}{a}$$

CHAPTER 23

Capacitance

Capacitance is a measure of the ability of two conductors to store charge when a given potential difference is established between them.

23.1   Calculation of Capacitance

Consider two conductors, on one of which is charge $+Q$ and on the other $-Q$. The **capacitance** of the conductors is defined as

$$C = \frac{Q}{V} \tag{23.1}$$

We measure capacitance in **farads** (F):

$$1 \text{ farad} = 1 \text{ coulomb/volt}$$

Note that the capacitance of two conductors does not depend on whether or not they are charged, just as the capacity of a 5-gal bucket does not depend on whether or not it contains water. To find the capacitance of two conductors, use the following procedure:

1. Place charge $+Q$ on one conductor and $-Q$ on the other.
2. Calculate the electric field in the space between the conductors. Usually this requires using Gauss' law.
3. Determine the potential difference between the conductors by integrating:

$$V = |V| = \int_{A}^{B} \mathbf{E} \cdot \mathbf{ds}$$

   Choose any convenient path from conductor A to conductor B. It is only the magnitude of V that counts, so I dropped the minus sign in front of the integral.
4. Since E depends on Q, you will obtain an expression of the form $V = (\text{constant}) \, Q$, where the constant is $1/C$.

PROBLEM 23.1.   Calculate the capacitance per unit area of two large parallel plates of area A separated by a distance d.

Solution   Using Gauss' law, I found that $E = \sigma/\epsilon_0$.

Thus

$$V = -\int_0^d E\,dx = E\int_0^d dx = -\frac{\sigma d}{\epsilon_0}$$

$$Q = \sigma A, \quad \text{so } V = \frac{Qd}{\epsilon_0 A}$$

and

$$\boxed{C = \frac{\epsilon_0 A}{d} \quad \text{parallel plates}} \tag{23.2}$$

PROBLEM 23.2. Calculate the capacitance of two plates of area $4\,\text{cm}^2$ separated by 1 mm.

Solution $C = \dfrac{\epsilon_0 A}{d} = \dfrac{(8.85 \times 10^{-12}\,\text{F/m})(0.0004\,\text{m}^2)}{0.001\,\text{m}} = 3.54 \times 10^{-12}\,\text{F} = 3.54\,\text{pF}$

Note: $1\,\text{pF} = 1$ picofarad $= 10^{-12}\,\text{F}$.

PROBLEM 23.3. Calculate the capacitance per unit length of a long, straight coaxial line made of two concentric cylinders of radii R_1 and R_2.

Solution Using Gauss' law I found that

$$E = \frac{\lambda}{2\pi\epsilon_0 r}$$

so

$$V = \int_{R_1}^{R_2} E\,dr = \frac{\lambda}{2\pi\epsilon_0} \ln\!\left(\frac{R_2}{R_1}\right) = \frac{\lambda}{C}$$

Thus

$$\boxed{C = \frac{2\pi\epsilon_0}{\ln R_2/R_1} \quad \text{coaxial line}} \tag{23.3}$$

PROBLEM 23.4. Calculate the capacitance of two concentric spheres of radii R_1 and R_2.
From Gauss' law I found that

$$E = \frac{Q}{4\pi\epsilon_0 r^2}, \quad \text{so } V = \int E\,dr = \frac{Q}{4\pi\epsilon_0}\left(\frac{1}{R_1} - \frac{1}{R_2}\right)$$

Thus

$$\boxed{C = 4\pi\epsilon_0 \frac{R_1 R_2}{R_2 - R_1} \quad \text{two spheres}} \tag{23.4}$$

When $R_2 \to \infty$, $C \to 4\pi\epsilon_0 R_1$. This describes the capacitance of an isolated sphere. Whenever we speak of the capacitance of an isolated conductor, we think of the second conductor as being at infinity.

23.2   Combinations of Capacitors

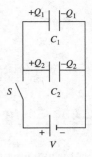

In Fig. 23-1 two capacitors connected in **parallel** to a battery by means of a switch are shown. The battery provides the same potential difference ("voltage") across each capacitor. The straight lines represent wires (conductors), and all of one wire is at the same potential. For the case shown, **the potential difference across each capacitor in parallel connection is the same**. Total charge Q flows out of the battery, with Q_1 going to the plates of C_1 and Q_2 going to C_2. Thus $Q = Q_1 + Q_2$. Since $V_1 = V_2 = V$, $Q = CV = C_1V_1 + C_2V_2 = (C_1 + C_2)V + C_pV$. Hence two capacitors in parallel are equivalent to a single capacitance C_p where $C_p = C_1 + C_2$. This reasoning can be extended to n capacitors in parallel, yielding the equivalent capacitance C_p.

Fig. 23-1

$$\boxed{C_p = C_1 + C_2 + C_3 + \cdots + C_n \quad \text{capacitors in parallel}} \qquad (23.5)$$

In Fig. 23-2 two capacitors connected in **series** are shown. Charge $+Q$ flows out of the positive terminal of the battery to the upper capacitor plate. This induces charge $-Q$ on the lower plate of the top capacitor and $+Q$ on the top plate of the lower capacitor. Finally, charge $-Q$ appears on the lower plate from the negative terminal of the battery. We see that **capacitors in series have the same amount of charge on their plates**. The potential drop from point A to point C is $V_{AC} = V_{AB} + V_{BC} = V_1 + V_2 = V$. Thus

$$V = \frac{Q_1}{C_1} + \frac{Q_2}{C_2} = \frac{Q}{C_s}$$

and

$$Q_1 + Q_2 = Q, \quad \text{so } \frac{1}{C_s} = \frac{1}{C_1} + \frac{1}{C_2}$$

Fig. 23-2

In general, for n capacitors in series, with equivalent capacitance C_s,

$$\boxed{\frac{1}{C_s} = \frac{1}{C_1} + \frac{1}{C_2} + \frac{1}{C_3} + \cdots + \frac{1}{C_n} \quad \text{capacitors in series}} \qquad (23.6)$$

In practice it is easiest to combine series capacitors two at a time, using the easily remembered equation $C_s = C_1C_2/(C_1 + C_2)$.

PROBLEM 23.5.   Find the equivalent capacitance of the capacitors shown here: $C_1 = 2\,\mu\text{F}$, $C_2 = 4\,\mu\text{F}$, $C_3 = 3\,\mu\text{F}$.

Solution   First combine C_1 and C_2 in parallel: $C_{12} = C_1 + C_2 = 2\,\mu\text{F} + 4\,\mu\text{F} = 6\,\mu\text{F}$. Next combine C_{12} and C_3 in series:

$$C = \frac{C_{12}C_3}{C_{12} + C_3} = \frac{(6\,\mu\text{F})(3\,\mu\text{F})}{6\,\mu\text{F} + 3\,\mu\text{F}} = 2\,\mu\text{F}$$

PROBLEM 23.6.   A 6-μF capacitor is charged by a 12-V battery and then disconnected. It is then connected to an uncharged 3-μF capacitor as shown here. What is the final potential difference across each capacitor?

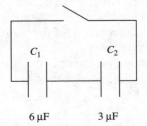

6 μF   3 μF

Solution   Initially, $Q = C_1 V_1 = (6 \mu F)(12 V) = 72 \mu C$. After reconnecting, $V_1 = Q_1/C_1$ and $V_2 = Q_2/C_2$ where $Q = Q_1 + Q_2$ and $V_1 = V_2$. Thus $VC_1 + VC_2 = V_1 C_1$.

$$V = \frac{C_1}{C_1 + C_2} V_1 = \frac{6 \mu F}{6 \mu F + 3 \mu F}(12 V) = 8 V$$

23.3 Energy Storage in Capacitors

An external agent, such as a battery, must do work to place charge on the plates of a capacitor (because the charge already on the plates repels the additional charge you are trying to add). If you add a small amount of charge dq to a capacitor across which a potential difference V exists, the work done is $dW = V\, dq$. The total work done to build up to a final charge Q is thus

$$W = \int_0^Q V\, dq = \int_0^Q \frac{q\, dq}{C} = \frac{1}{2}\frac{Q^2}{C}$$

Thus the stored energy in a capacitor can be written, using $Q = CV$, as

$$\boxed{U = \frac{1}{2}\frac{Q^2}{C} = \frac{1}{2}CV^2 = \frac{1}{2}QV} \tag{23.7}$$

Depending on what information is known, one of these expressions may be easier than the others to use. If you know Q and C, use the first one. If you know C and V, use the second one. If you know Q and V, use the third one.

PROBLEM 23.7.   Capacitors $C_1 = 5 \mu F$ and $C_2 = 8 \mu F$ are connected in series across a 9-V battery. How much energy do they store?

Solution

$$C_s = \frac{C_1 C_2}{C_1 + C_2} + \frac{(5)(8)}{5 + 8} \mu F = 3.08 \mu F$$
$$U = \tfrac{1}{2}C_s V^2 = \tfrac{1}{2}(3.08 \times 10^{-6} F)(9 V)^2 = 1.25 \times 10^{-4} J$$

23.4 Dielectrics

So far I considered capacitors filled with vacuum (air is approximately a vacuum for these purposes). However, in practical application the space between capacitor plates is filled with an insulator called a **dielectric**. Examples are paper, oil, or mica. Consider what happens when a dielectric is placed between the plates of a charged capacitor. Initially there is a free charge Q on the plates and a potential difference V between the plates. The electric field between the plates **polarizes** the dielectric. That is, it distorts slightly each molecule and thereby induces a dipole moment. If the molecules already had a dipole moment (**polar molecules**), the electric field tends to make them line up parallel to the electric field. In a uniform field the result is to cause a layer of induced charge to form on the surface of the dielectric, as illustrated in Fig. 23-3. This induced (polarization) charge tends to cancel the effect of some of the free charge on the plates, with the result that the field, and hence the potential difference, between the plates is reduced. Since $C = Q/V$, reducing V while maintaining the same free charge on the plates increases C. The amount of increase depends on the properties of the particular dielectric material, but in general the capacitance increases by a factor κ, **the dielectric constant**. If C_0 is the capacitance of the empty capacitor, the capacitance of the dielectric-filled capacitor is

$$\boxed{C = \kappa C_0} \tag{23.8}$$

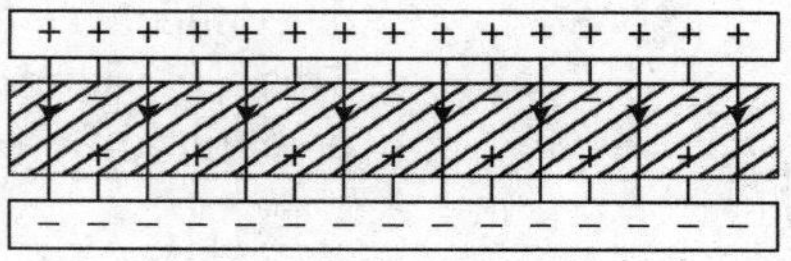

Fig. 23-3

When a capacitor is disconnected from the battery, $Q = $ constant, as in the above discussion. When the capacitor remains connected to the battery, V remains constant. Inserting a dielectric decreases the potential difference between the plates, and so the battery adds more charge to build the potential difference back up to the battery voltage. Again, increasing Q for given V increases C since $C = Q/V$.

PROBLEM 23.8.   A 200-μF parallel plate capacitor has a plate separation of 0.60 mm. A sheet of paper ($\kappa = 3.7$) 0.15 mm thick is slid between the plates. What then is the capacitance?

Solution   View this as two capacitors in series, one air-filled with plate separation $d_1 = 0.45$ mm and one of separation $d_2 = 0.15$ mm filled with paper. Let $C_0 = \epsilon_0 A/d$. Then

$$C_1 = \frac{\epsilon_0 A}{d_1} = \frac{d}{d_1} C_0 \qquad C_2 = \frac{\kappa \epsilon_0 A}{d_2} = \frac{\kappa d}{d_2} C_0$$

$$C = \frac{C_1 C_2}{C_1 + C_2} = \left(\frac{d C_0}{d_1}\right)\left(\frac{\kappa d C_0}{d_2}\right)\left(\frac{d C_0}{d_1} + \frac{\kappa d C_0}{d_2}\right)^{-1}$$

$$= \frac{\kappa C_0 d}{d_2 + \kappa d_1} = \frac{(3.7)(0.60 \text{ mm})(220\ \mu\text{F})}{0.15 \text{ mm} + (3.7)(0.45 \text{ mm})} = 269\ \mu\text{F}$$

Note that the exact location of the paper will not matter, as long as it is parallel to the plates. To see this, calculate the capacitance as if one had three capacitors in series, one filled with paper with $d_2 = 0.15$ mm, two air-filled with $d_1 = 0.45 - x$ and $d_3 = x$. The result will be independent of x.

PROBLEM 23.9.   A 6-μF capacitor filled with air is charged by a 12-V battery. It is disconnected from the battery, and the space between the plates is filled with oil ($\kappa = 2.5$). (a) What now is the potential difference between the plates? (b) What free charge is on one plate? (c) What induced charge is on the surface of the oil?

Solution   (a) $Q_f = C_0 V_0 = \kappa C_0 V$ and $V = \dfrac{1}{\kappa} V_0 = \dfrac{12 \text{ V}}{2.5} = 4.8 \text{ V}$

(b) $Q_f = C_0 V_0 = (6\ \mu\text{F})(12 \text{ V}) = 72\ \mu\text{F}$

(c) $\dfrac{Q_f - Q_i}{Q_f} = \dfrac{4.8 \text{ V}}{12 \text{ V}} = 0.6,$ so $Q_i = Q_f - 0.6 Q_f = 28.8\ \mu\text{C}$

PROBLEM 23.10.   A parallel plate capacitor with plate area A and separation d is filled with two equal slabs of dielectric of dielectric constants κ_1 and κ_2. What is the capacitance?

Solution   View this as two capacitors in parallel.

$$C = C_1 + C_2 = \frac{\kappa_1 \epsilon_0 A}{2d} + \frac{\kappa_2 \epsilon_0 A}{2d} = \frac{(\kappa_1 + \kappa_2)(\epsilon_0 A)}{2d}$$

23.5   Summary of Key Equations

Capacitance: $Q = CV$

Parallel plate capacitor: $C = \dfrac{\epsilon_0 A}{d}$

Parallel capacitors: $C_p = C_1 + C_2 + C_3 + \cdots + C_n$

Series capacitors: $\dfrac{1}{C_s} = \dfrac{1}{C_1} + \dfrac{1}{C_2} + \dfrac{1}{C_3} + \cdots + C_n$

Two capacitors in series: $C_s = \dfrac{C_1 C_2}{C_1 + C_2}$

Stored energy: $\qquad\qquad U = \dfrac{1}{2}QV = \dfrac{1}{2}CV^2 = \dfrac{1}{2}\dfrac{Q^2}{C}$

Capacitance with dielectric: $\qquad C = \kappa C_0$

SUPPLEMENTARY PROBLEMS

23.11. A parallel plate capacitor of plate area $0.04\ \text{m}^2$ and plate separation 0.25 mm is charged to 24 V. Determine the charge on a plate and the electric field between the plates.

23.12. Three capacitors $C_1 = 2\ \mu\text{F}$, $C_2 = 3\ \mu\text{F}$, and $C_3 = 5\ \mu\text{F}$ are connected as shown here. What is the equivalent capacitance of this arrangement?

23.13. Three capacitors $C_1 = 3\ \mu\text{F}$, $C_2 = 5\ \mu\text{F}$, and $C_3 = 8\ \mu\text{F}$ are each charged by a 24-V battery and then connected as shown here. Determine the final charge on each capacitor and the potential difference between points X and Y.

23.14. Capacitors $C_1 = C_0$ and $C_2 = 2C_0$ are given charges $4q$ and q, respectively. They are then connected in series, with the positive plate of one capacitor connected to the negative plate of the other. What is the final charge on C_2?

23.15. Tuning capacitors of the type used in radios have overlapping plates, and the capacitance is changed by varying the amount of overlap, as illustrated here. If each plate has area A and the spacing between the plates is d, what is the capacitance of the unit?

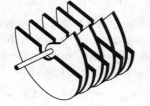

23.16. An aluminum plate of thickness L and area A is slid into a parallel plate capacitor of plate spacing d and area A, as shown here. The aluminum is a distance x from one of the plates. What is the capacitance of this arrangement?

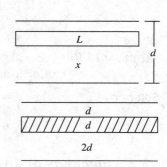

23.17. A parallel plate capacitor with plate area A and separation $4d$ is initially uncharged. A metal slab of thickness d and of the same area A is given charge Q and slid between the plates, as shown here. After this is done, what is the difference in potential between the plates?

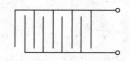

23.18. A parallel plate capacitor of area A and plate separation d is charged to a potential difference V and then disconnected from the battery. The plates are now pulled apart until their separation is $2d$. Determine the new potential difference, the initial and final stored energy, and the work required to pull the plates apart.

23.19. A parallel plate capacitor is half filled with a dielectric with $\kappa = 2.5$. What fraction of the energy stored in the charged capacitor is stored in the dielectric?

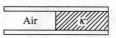

23.20. One frequently models real physical systems (for example, transmission lines or nerve axons) with an infinitely repeating series of discrete circuit elements such as capacitors. Such an array is shown here. What is the capacitance between terminals X and Y for such a line, assuming it extends indefinitely? All of the capacitors are identical and have capacitance C. (This requires thinking, not brute force.)

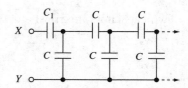

23.21. A 1.0-μF capacitor is charged to 100 V. (a) How much charge does it store? How much energy does it store? (b) A 2-μF capacitor is charged to 50 V. How much charge and energy does it store? Explain the reason for the difference in the result for (a) and (b).

23.22. In order to create a nuclear fusion reaction, of the type that generates energy in the Sun, it is necessary to raise the nucleons involved to very high temperatures. This is because the positively charged particles experience a large electrostatic repulsion. Many techniques have been tried to accomplish reaching the required high temperatures. One such approach is the National Ignition Facility experiment being carried out at the Lawrence Livermore Laboratory in California. Here, small pellets containing the nuclear reactants are irradiated simultaneously from all sides by many high powered lasers. The plan is to raise the pellets to such high temperatures that nuclear fusion occurs. A typical laser pulse generates about 10^{14} W of power for a duration of 10^{-9} s. This is a power level about 100 times as great as that of all the world's power plants. (a) How much energy is generated in one pulse of the lasers? (b) To what voltage must the capacitor bank supplying the energy be charged if the total capacitance is 0.25 F and 0.15% of the electrical energy appears as light in the lasers?

23.23. A 3-μF capacitor is connected in series with a 6-μF capacitor. A 4-μF capacitor is now connected in parallel with the series pair. Finally, a 3-μF capacitor is connected in series with the preceding combination. What is the capacitance of this array?

SOLUTIONS TO SUPPLEMENTARY PROBLEMS

23.11. $C = \dfrac{\epsilon_0 A}{d} = \dfrac{(8.55 \times 10^{-12}\,\text{F/m})(0.04\,\text{m}^2)}{(0.25 \times 10^{-3}\,\text{m})} = 1.42 \times 10^{-9}\,\text{F}$

$Q = CV = (1.42 \times 10^{-9}\,\text{F})(24\,\text{V}) = 3.4 \times 10^{-8}\,\text{C} \qquad E = \dfrac{\rho}{\epsilon_0} = \dfrac{Q}{A\epsilon_0} = 9.6 \times 10^4\,\text{V/m}$

23.12. $C_{12} = \dfrac{C_1 C_2}{C_1 + C_2} = \dfrac{(2)(3)}{2+3}\,\mu\text{F} = 1.2\,\mu\text{F}, \qquad C = C_{12} + C_3 = 9.2\,\mu\text{F}$

23.13. Let Q'_1, Q'_2, Q'_3 = final charges and V = final potential difference.
Then

$$Q_2 + Q_3 = Q'_2 + Q'_3 = \frac{C_1 C_2}{C_1 + C_2} V + C_3 V \qquad Q_2 = C_2 V_0 = (5)(24) = 120\,\mu\text{C}$$

$$Q_3 = C_3 V_0 = (8)(24) = 192\,\mu\text{C} \qquad Q_2 + Q_3 = Q'_2 + Q'_3 = 312\,\mu\text{C}$$

$$C_{eq} = \frac{C_1 C_2}{C_1 + C_2} + C_3 = 9.9\,\mu\text{F} \qquad Q'_2 + Q'_3 = C_{eq} V_{XY}, \quad \text{so } V_{XY} = 31.6\,\text{V}$$

$$Q'_3 = C_3 V_{XY} = 253\,\mu\text{C} \qquad Q_2 = Q_1 = C_{12} V_{XY} = \frac{(3)(5)}{3+5}(31.6) = 59\,\mu\text{C}$$

23.14. One has two capacitors in parallel with charge $3q$:

$$V = \frac{Q}{C} = \frac{3q}{C_0 + 2C_0} = \frac{q}{C_0}, \quad \text{so } Q_2 = C_2 V = 2C_0\left(\frac{q}{C_0}\right) = 2q$$

23.15. This is equivalent to eight capacitors in parallel, so

$$C = 8C_1 = 8\frac{\epsilon_0 A}{d}$$

23.16. The arrangement is effectively two capacitors in series, so

$$C = \frac{C_1 C_2}{C_1 + C_2}$$

where

$$C_1 = \frac{\epsilon_0 A}{x} \quad \text{and} \quad C_2 = \frac{\epsilon_0 A}{d - x - L}$$

Thus

$$C = \frac{(\epsilon_0 A / x)(\epsilon_0 A / d - x - L)}{(\epsilon_0 A / x) + (\epsilon_0 A / d - x - L)} = \frac{\epsilon_0 A}{d - L}$$

It is interesting to see that the result is independent of x; that is, it doesn't matter where the aluminum is placed, as long as it is parallel to the plates.

23.17. Charge will spread uniformly over both surfaces of the slab, so E at the face of the slab is $E = \rho / \epsilon_0 = Q / 2A\epsilon_0$. If we call the potential of the slab V_0, the potential of the top plate is $V_{\text{top}} = V_0 - Ed$ and of the bottom plate $V^{\text{bot}} = V_0 - 2Ed$. The potential difference between plates is

$$\Delta V = V_{\text{top}} - V_{\text{bot}} = Ed = \frac{Qd}{2A\epsilon_0}$$

23.18. E depends only on surface charge density, and this doesn't change, so the new potential difference is $V^1 = 2V$. Also, $U = \frac{1}{2}QV$, so $U^1 = \frac{1}{2}QV^1 = QV = 2U$, where $U = \frac{1}{2}CV^2$.

$$C = \frac{\epsilon_0 A}{d}, \quad \text{so } U = \frac{1}{2}\frac{\epsilon_0 A}{d}V^2$$

$$\text{Work} = U^1 - U = \frac{1}{2}\frac{\epsilon_0 A}{d}V^2$$

23.19. This arrangement is like two capacitors in parallel: $C = C_1 + C_2 = C_1 + \kappa C_1 = (1 + \kappa)C_1$.

The total stored energy is

$$U = \frac{1}{2}CV^2 = \frac{1}{2}(1 + \kappa)C_1 V^2$$

The energy in dielectric is

$$U_2 = \frac{1}{2}C_2 V^2 = \frac{1}{2}\kappa C_1 V^2$$

so the fraction is

$$\frac{U_2}{U} = \frac{\kappa}{1 + \kappa} = \frac{2.5}{1 + 2.5} = 0.71$$

23.20. Cut the array along the dashed line here. The capacitors to the right of this line are equivalent to the capacitance between X and Y. Thus we have the equivalent network below:

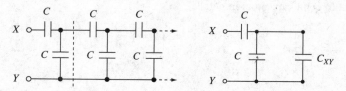

Thus

$$C_{XY} = \frac{(C)(C + C_{XY})}{C + (C + C_{XY})}$$

Solve: $2CC_{XY} + C_{XY}^2 = C^2 + CC_{XY}$ $C_{XY}^2 + CC_{XY} - C^2 = 0$

$$C_{XY} = \frac{-C \pm \sqrt{C^2 - 4(1)(-C^2)}}{2} = \frac{-1 + \sqrt{5}}{2}C = 0.62\,C \quad \text{choose the + sign since } C_{XY} > 0$$

23.21. (a) $Q_1 = C_1 V_1 = (1 \times 10^{-6}\,\text{F})(100\,\text{V}) = 10^{-4}\,\text{C}$

$U_1 = \frac{1}{2}C_1 V_1^2 = (0.5)(10^{-6}\,\text{F})(100\,\text{V})^2 = 5 \times 10^{-5}\,\text{J}$

(b) $Q_2 = C_2 V_2 = (2 \times 10^{-6}\,\text{F})(50\,\text{V}) = 10^{-4}\,\text{C}$

$U_2 = \frac{1}{2}C_2 V_2^2 = (0.5)(2 \times 10^{-6}\,\text{F})(50\,\text{V})^2 = 2.5 \times 10^{-3}\,\text{J}$

Both capacitors store the same charge, but energy storage depends on the voltage squared, so here the smaller capacitor stores more energy.

23.22. (a) $U = Pt = (10^{14}\,\text{W})(10^{-9}\,\text{S}) = 10^5\,\text{J}$

(b) $U = e\left(\frac{1}{2}CV^2\right)V^2 = \frac{2U}{eC} = \frac{(2)(10^5\,\text{J})}{(0.15 \times 10^{-2})(0.25\,\text{F})}\,V = 2.31 \times 10^{-6}\,\text{V} = 23.1\,\text{kV}$

23.23. $C_1 = \frac{(6\,\mu\text{F})(3\,\mu\text{F})}{6\,\mu\text{F} + 3\,\mu\text{F}} = 2\,\mu\text{F}$

$C_2 = 2\,\mu\text{F} + 4\,\mu\text{F} = 6\,\mu\text{F}$

$C_3 = \frac{(6\,\mu\text{F})(3\,\mu\text{F})}{6\,\mu\text{F} + 3\,\mu\text{F}} = 2\,\mu\text{F}$

CHAPTER 24

Current and Resistance

So far I have considered electric charges at rest (electrostatics). Now we turn to the interesting consequences of what happens when electric charge moves.

24.1 Electric Current

If charge dq passes a point in space in time dt, we define the **current** at that point as

$$I = \frac{dq}{dt} \quad \text{definition of current} \tag{24.1}$$

Current is measured in **amperes** (A): $1\,\text{A} = 1\,\text{C/s}$. The direction of current is positive in the direction of motion of positive charges. In fact, the charge carriers are often negative electrons. However, a flow of electrons to the left constitutes a current to the right, so this need not cause confusion. It is best simply to imagine the charge carriers as positive in all of your calculations (with a very few exceptions). Current can flow through a conductor like a copper wire or through a vacuum, as with the electron beam in a TV tube or in a linear accelerator.

Electrons move through a wire sort of like a swarm of bees. Each electron is moving at high speed every which way, but the swarm drifts slowly in the direction of the current. If there are n charge carriers per cubic meter, each with charge q, the amount of charge in a section of wire of cross-sectional area A and length Δx is $nqA\,\Delta x$. If this charge drifts with velocity v_d, then in time Δt the charge transported past a point is $\Delta Q = nqAv_d\,\Delta t$, so the current is

$$I = \frac{\Delta Q}{\Delta t} = nqv_dA \tag{24.2}$$

PROBLEM 24.1. Often no. 12 copper wire (0.205-cm diameter) is used in residential construction. What is the drift velocity of electrons in such a wire when it carries 12 A? Copper has a density of $8.95\,\text{g/cm}^3$ and an atomic mass of $63.5\,\text{g/mol}$. Each copper atom provides one free electron for conductivity.

Solution

$$n = (8.95\,\text{g/cm}^3)\left(\frac{1}{63.5}\,\text{mol/g}\right)(6.02 \times 10^{23}\,\text{mol}^{-1}) = 8.48 \times 10^{23}\,\text{cm}^{-3}$$

$$v_d = \frac{I}{nqA} = \frac{1\,\text{A}}{(8.48 \times 10^{22}\,\text{cm}^{-3})(1.6 \times 10^{-19}\,\text{C})(\pi/4)(0.205\,\text{cm})^2} = 0.027\,\text{cm/s}$$

The electrons drift very slowly!

24.2   Resistance, Resistivity, and Ohm's Law

If current I flows in a conductor of cross-sectional area A, the *current density J* is

$$\boxed{J = nqv_d} \tag{24.3}$$

For many materials the current density is proportional to the applied electric field, so

$$\boxed{\mathbf{J} = \sigma\mathbf{E}} \tag{24.4}$$

σ is the *conductivity* of the conductor. Assuming the electric field is constant in the conductor of length L, the potential difference over length L is $V = EL$. Since $I = JA$, $J = \sigma E$ becomes $I/A = \sigma V/L$, or $I = (\sigma A/L)V$. This is usually written

$$\boxed{V = IR} \tag{24.5}$$

where

$$\boxed{R = \frac{L}{\sigma A} = \rho\frac{L}{A}} \tag{24.6}$$

$V = IR$ is **Ohm's law**. R is the **resistance** (in **ohms**, Ω) of the conductor. $1\,\Omega = 1\,\text{V/A}$. $\rho = 1/\sigma$ is the **resistivity** of the material, measured in Ω-m. Resistance depends on the size, shape, kind of material, and temperature, whereas resistivity depends only on the kind of material and the temperature.

$$\boxed{\rho = \rho_0[1 + \alpha(T - T_0)]} \tag{24.7}$$

Here ρ_0 is the resistivity at temperature T_0 and ρ is the resistivity at temperature T. α is the temperature coefficient of resistivity. For metals, $\alpha > 0$, and for semiconductors, $\alpha < 0$.

Note that in previous chapters I stated that the electric field was zero in a conductor, and that all parts of a conductor were at the same potential. That is true for a *perfect* conductor, but for the resistive conductors I am discussing here, this is no longer true. Often, however, I will assume in circuits that the copper wire used is approximately a perfect conductor, and I will neglect changes in potential along a wire. Whether or not this assumption is justified depends on the circumstances.

PROBLEM 24.2.   An extension cord 20 m long uses no. 12 gauge copper wire (cross-sectional area $0.033\,\text{cm}^2$, resistivity $1.7 \times 10^{-8}\,\Omega$-m). What is its resistance?

Solution
$$R = \rho\frac{L}{A} = \frac{(1.7 \times 10^{-8}\,\Omega\text{-m})(20\,\text{m})}{0.033 \times 10^{-4}\,\text{m}^2} = 0.10\,\Omega$$

PROBLEM 24.3.   A tungsten wire of diameter 0.40 mm and length 40 cm is connected to a 36-V power supply. What current does it carry at 20°C and at 800°C? At 20°C the resistivity of tungsten is $5.6 \times 10^{-8}\,\Omega$-m and $\alpha = 4.5 \times 10^{-3}\,(°\text{C})^{-1}$.

Solution   At 20°C,
$$R_{20} = \rho\frac{L}{A} = \frac{(5.6 \times 10^{-8}\,\Omega\text{-m})(0.40\,\text{m})}{\pi(0.20 \times 10^{-3}\,\text{m})^2} = 0.18\,\Omega$$

At 800°C,
$$\rho = \rho_0(1 + \alpha\Delta T) = \rho_0[1 + (4.5 \times 10^{-3}°\text{C}^{-1})(780°\text{C})] = 4.5\rho_0$$

So
$$R_{800} = 4.5R_{20} = 0.81\,\Omega \qquad I_{20} = \frac{V}{R} = \frac{36\,\text{V}}{0.18\,\Omega} = 9.9\,\text{A} \qquad I_{800} = \frac{36\,\text{V}}{0.81\,\Omega} = 44\,\text{A}$$

PROBLEM 24.4.   A home experimenter wants to make a 1-Ω resistor from the 1-mm-diameter carbon rod from a mechanical pencil. What length rod does the experimenter need? For carbon, $\rho = 3.5 \times 10^{-5}$ Ω-m.

Solution
$$L = \frac{AR}{\rho} = \frac{\pi(0.5 \times 10^{-3} \text{ m})^2(1\,\Omega)}{3.5 \times 10^{-5}\,\Omega\text{-m}} = 0.022 \text{ m}$$

PROBLEM 24.5.   A conductor is machined to the shape of a truncated cone of circular cross section. One end has radius a and the other end radius b. The length of the cone is L, and its resistivity is ρ. What is its resistance for current flow along the axis?

Solution   Imagine the cone sliced into disks of thickness dx and radius r, where

$$r = a + \left(\frac{b-a}{L}\right)x$$

The total resistance is the sum of the resistance for each disk.

$$R = \int_0^L \rho \frac{dx}{\pi\{a + [(b-a)/L]x\}^2}$$

Let

$$u = a + \left(\frac{b-a}{L}\right)x$$

$$R = \frac{\rho}{\pi}\int_a^b \left(\frac{L}{b-a}\right)\frac{du}{u^2} = \frac{\rho}{\pi}\left(\frac{L}{b-a}\right)\left(-\frac{1}{u}\right)_a^b$$

$$= \frac{\rho}{\pi}\frac{L}{b-a}\left(-\frac{1}{b} + \frac{1}{a}\right) = \frac{\rho L}{\pi ab}$$

24.3 Electric Power and Joule Heating

When a charge q falls through a potential difference V, it loses potential energy qV. If it were falling freely through empty space, it would gain kinetic energy, but in a conductor there are strong frictional forces acting to keep the charge from gaining speed. It is as if the electron is falling through thick syrup. It falls very slowly at constant speed, and the loss in potential energy goes into heating the conductor. If charge dq falls through a potential difference V in time dt, the rate of generation of heat in the conductor (the power) is

$$\boxed{P = V\frac{dq}{dt} = VI = I^2R = \frac{V^2}{R} \quad \text{Joule heating}} \qquad (24.8)$$

PROBLEM 24.6.   A 12-V car battery powers a 70-W headlamp. What current does the lamp draw?

Solution
$$P = IV, \quad \text{so } I = \frac{P}{V} = \frac{70\,\text{W}}{12\,\text{V}} = 5.8 \text{ A}$$

PROBLEM 24.7. A car radio operating on 12 V draws 0.30 A. How much energy does it use in 4 h?

Solution
$$E = Pt = IVt = (0.30\,\text{A})(12\,\text{V})(240\,\text{s}) = 864\,\text{J}$$

PROBLEM 24.8. The South American eel *Electrophorus electricus* generates 0.80-A pulses of current at a voltage of about 640 V. At what rate does it develop power when giving a shock to its prey? If the shock pulse lasts 1 ms, how much energy is delivered?

Solution
$$P = IV = (0.80\,\text{A})(640\,\text{V}) = 512\,\text{W} \qquad E = Pt = (512\,\text{W})(10^{-3}\,\text{s}) = 0.5\,\text{J}$$

24.4 Summary of Key Equations

Definition of current: $\quad I = \dfrac{dq}{dt} \quad I = nqv_dA$

Ohm's law: $\qquad \mathbf{J} = \sigma\mathbf{E} \quad \text{and} \quad V = IR$

$$\rho = \rho_0(1 + \alpha\Delta T) \qquad R = \rho\frac{L}{A}$$

Joule heating: $\qquad P = IV = I^2R = \dfrac{V^2}{R}$

SUPPLEMENTARY PROBLEMS

24.11. A popular car battery is rated at 320 A-h. This indicates the amount of electric charge the battery can deliver. Express 320 A-h in coulombs.

24.12. An experimenter wishes to silver plate a microwave electronic component with a thickness of silver of 0.02 mm over an area of 6 cm². With a solution of Ag⁺ and a current of 1.8 A, how long will it take to deposit the desired amount of silver (density 10.5 g/cm³, atomic mass 108)?

24.13. A 6.0-MeV beam of protons has a density of $2.6 \times 10^{11}\,\text{m}^{-3}$ and an area of 1.9 mm². What is the beam current?

24.14. An immersion electric heater used in the lab will increase the temperature of 500 mL of water by 50°C in 6 min when operated with a 120-V supply. (a) What power does the heater deliver? (b) What current does it draw?

24.15. A spherical electrode of radius a is coated with a conducting layer of material with resistivity ρ. The thickness of the layer is d, and its outer surface is coated with a conducting electrode. What is the resistance of the spherical coating between the inner and outer electrodes?

24.16. A fuse is a conductor with a low melting point (usually a lead-tin alloy). It is inserted in a circuit (for example, in a stereo or in a car) so that if too much current is drawn, the fuse will get hot and melt, thereby stopping current flow, before a fire is started. Typically the temperature rise is proportional to the power P dissipated in the fuse; that is, $\Delta T = cP$, where c is a constant dependent on the geometry of how the fuse is constructed. c can be determined experimentally. The resistance of the fuse varies as $R = R_0(1 + \alpha\Delta T)$. Show that if α is large enough, the fuse will "blow" (ΔT will become very large) when the current reaches the value $I = 1/\sqrt{c\alpha R_0}$.

24.17. A lead wire of resistance R is drawn through a die so that its length is doubled, while its volume remains unchanged. What will be its new resistance in terms of its initial resistance?

24.18. An old streetcar draws 14 A from an overhead line at 480 V. What power does it draw?

24.19. The resistance of a certain component decreases as the current through it increases, as described by the relation $R = 80/(12 + bI^3)$, where $b = 4\,A^{-2}$. Determine the current that results in maximum power in the unit and the maximum power delivered to the unit.

24.20. A silicon solar cell of dimension 1×2 cm produces a current of 56 mA at 0.45 V when illuminated by sunlight of intensity $1000\,W/m^2$. What is the efficiency of the cell?

24.21. A certain flashlight operates on two 1.5-V batteries connected in series. The lamp draws a current of 0.50 A. (a) What electrical power is delivered to the lamp? (b) What is the resistance of the lamp filament?

24.22. A conducting block in the shape of a brick has side a, b, and c. It lies in the x–y plane with its axes along the coordinate axis. Side a is along the x-axis, side b is along the y-axis, and side c is along the z-axis. When a potential difference V is applies along the x-axis between opposite faces, a current I_x flows. When this same potential difference is applied along the y-axis between opposite faces, a current I_y flows. Determine I_x in terms of I_y and the dimensions of the block.

SOLUTIONS TO SUPPLEMENTARY PROBLEMS

24.11. $Q = 320\ \text{A-h} = (320\,\text{C/s})(3600\,\text{s}) = 1.15 \times 10^6\,\text{C}$

24.12. $Q = ne = \dfrac{(0.002\,\text{cm})(6\,\text{cm}^2)(10.5\,\text{g/cm}^3)(6.02 \times 10^{23})(1.6 \times 10^{-19}\,\text{C})}{108\,\text{g}} = 1.21 \times 10^4\,\text{C}$

$I = \dfrac{Q}{t}, \quad t = \dfrac{Q}{I} = \dfrac{1.21 \times 10^4\,\text{C}}{1.8\text{A}} = 6740\,\text{s} = 1.88\,\text{h}$

24.13. $\text{KE} = \dfrac{1}{2}mv^2, \quad v = \sqrt{\dfrac{2\text{KE}}{m}}, \quad I = nqv_dA$

$I = (2.6 \times 10^{11}\,\text{m}^{-3})(1.6 \times 10^{-19}\,\text{C})\left[\left(\dfrac{(2)(6 \times 10^6\,\text{eV})(1.6 \times 10^{-19})}{1.67 \times 10^{-27}\,\text{kg}}\,\text{J/eV}\right)\right]^{1/2}(1.9 \times 10^{-6}\,\text{m}^2)$

$= 2.7 \times 10^{-6}\,\text{A}$

24.14. The heat required is

$$H = mc\Delta T = (0.5\,\text{kg})(4{,}200\,\text{J/kg} \cdot {}^\circ\text{C})(50^\circ\text{C})$$

$$= 1.05 \times 10^5\,\text{J}$$

$$P = \dfrac{H}{T} = \dfrac{1.05 \times 10^5\,\text{J}}{300\,\text{s}} = 350\,\text{W}, \qquad I = \dfrac{P}{V} = \dfrac{350\,\text{W}}{120\,\text{V}} = 2.9\,\text{A}$$

24.15. Break the conductor into thin shells:

$$R = \int_a^{a+d} \rho\dfrac{dr}{4\pi r^2} = \dfrac{\rho}{4\pi}\left(-\dfrac{1}{r}\right)_a^{a+d} = \dfrac{\rho}{4\pi}\left(\dfrac{1}{a} - \dfrac{1}{a+d}\right) = \dfrac{\rho d}{4\pi a(a+d)}$$

24.16. $P = I^2R, \quad R = R_0(1 + \alpha\Delta T) \simeq R_0\alpha\Delta T$

$\Delta T = cP = cI^2R, \quad \text{so } \Delta T = cI^2R\alpha\Delta T \quad \text{and} \quad I \simeq \sqrt{\dfrac{1}{c\alpha\Delta T}}$

24.17. $R_1 = \rho\dfrac{L_1}{A_1}$ and $R_2 = \rho\dfrac{L_2}{A_2} = \rho\dfrac{2L_1}{A_1/2} = 4R_1$

since volume $= AL =$ constant.

24.18. $P = IV = (14\,\text{A})(480\,\text{V}) = 6720\,\text{W}$

24.19. $P = I^2R = I^2\left(\dfrac{80}{12 + 4I^3}\right)$

For maximum power,

$$\frac{dP}{dI} = 0 = 80\left(\frac{2I}{12 + 4I^3} - \frac{12I^4}{(12 + 4I^3)^2}\right)$$

Solve: $2I(12 + 4I^3) - 12I^4 = 0$, $I = 2\,\text{A}$ $P_{\text{max}} = I^2R = (4\,\text{A})^2\left(\dfrac{80}{12 + 32}\right) = 29\,\text{W}$

24.20. $P_{\text{out}} = 1\,\text{V} = (0.056\,\text{A})(0.45\,\text{V}) = 0.025\,\text{W}$

$P_{\text{in}} = 1\,\text{A} = (1000\,\text{W/m}^2)(2 \times 10^{-4}\,\text{m}^2) = 0.2\,\text{W}$ Efficiency $= \dfrac{P_{\text{out}}}{P_{\text{in}}} = \dfrac{0.025\,\text{W}}{0.20\,\text{W}} = 12.5$ percent

24.21. (a) $P = IV = (0.50\,\text{A})(3.0\,\text{V}) = 1.5\,\text{W}$

(b) $R = \dfrac{V}{I} = \dfrac{3.0\,\text{V}}{0.50\,\text{A}} = 6\,\Omega$

24.22. $I_y = \rho\dfrac{b}{ac}$

$I_y = \dfrac{1}{R_y}V = \dfrac{1}{\rho(b/ac)}V = \dfrac{ac}{b}\dfrac{V}{\rho}$

$I_x = \dfrac{1}{R_x}V = \dfrac{1}{\rho(a/bc)}V = \dfrac{bc}{a}\dfrac{V}{\rho}$

$\dfrac{I_x}{I_y} = \dfrac{bc/a}{ac/b}$

$I_x = \dfrac{b^2}{a^2}I_y$

CHAPTER 25

Direct Current Circuits

Charges are pushed through resistors by some kind of electric field. In the steady-state case we associated this electric field with a potential difference in volts, $\Delta V = -\int_A^B \mathbf{E} \cdot \mathbf{ds}$. We further identified the potential with the potential energy of a charge q as $U = qV$. However, as we shall see later, it is possible to create electric fields, such as by magnetic induction, that are not conservative, and for them there is no potential energy to be defined, since the work done in moving a charge in such a field depends on the path used. To avoid making the mistake of referring to $\int_A^B \mathbf{E} \cdot \mathbf{ds}$ as a "potential difference," I will define something called the **electromotive force**, or **EMF**, measured in volts.

$$\mathcal{E} = \int\limits_A^B \mathbf{E} \cdot \mathbf{ds} \quad \text{definition of EMF} \tag{25.1}$$

The integral is taken around a closed circuit. The EMF is not a force (we unfortunately use the term for historical reasons). For your purposes think of a source of EMF as something like a battery that provides the "push" to move charges around. The battery provides the energy that ends up as heat in resistors. Outside the battery in the circuit where the charges are moving the electric fields are conservative, and it is OK to speak of changes in potential there. At this point in your studies you don't have to make a big deal of these subtleties, but be aware that there is more involved here than first meets the eye.

In Fig. 25-1, I have illustrated a simple circuit in which a battery is causing a current to flow through a load resistor R. Current (imagined to be carried by positive charges) flows out of the positive terminal of the battery. Observe the following important point: **Current will flow only through a complete closed circuit.** If you create an **open circuit** by opening the switch S or breaking a wire, current flow will stop. The battery, indicated by the solid line, includes a seat of EMF $\mathcal{E}$ and some internal resistance r. If you were to open the switch so that no current flows, the voltage V measured between terminals X and Y would be

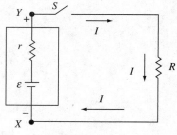

Fig. 25-1

equal to the EMF $\mathcal{E}$. However, when the switch is closed and current I flows, there is a drop in potential, $V = Ir$, across the resistance r, and so the potential difference between X and Y is only $V = \mathcal{E} - Ir$. The terminal voltage V_{XY} is equal to $\mathcal{E}$ when no current flows, so $\mathcal{E}$ is also called the **open circuit voltage**.

It is helpful to imagine a battery as analogous to a pump that raises the pressure (the voltage) needed to cause a flow of water (electric charge) through a system of pipes (resistors). The flow in gallons per minute is analogous to the electric current in coulombs per second.

PROBLEM 25.1. What current flows in the circuit of Fig. 25-1? For what value of the load resistor R will the maximum power be delivered to the load?

Solution The potential drops by Ir across the internal resistance and by IR across the load resistance, so the total potential drop around the circuit is $\mathcal{E} = Ir + IR$. Thus the current delivered by the battery is $I = \mathcal{E}/(r + R)$. The power delivered to the load is $P = I^2R = \mathcal{E}^2R(r + R)^2$. P is a maximum when $dP/dR = 0$.

$$\frac{dP}{dR} = \mathcal{E}^2\left[\frac{1}{(r+R)^2} - \frac{2R}{(r+R)^3}\right] = 0 \quad \text{or} \quad \frac{\mathcal{E}^2(r+R-2R)}{(r+R)^3} = 0$$

and **$R = r$ for maximum power to the load**. Under these conditions we say that the load is matched to the power source. Notice that equal amounts of power are delivered to the load and to the internal resistance under this condition.

In the following I will not display the internal resistance of the source explicitly. Instead, I will refer to a battery as providing a voltage V between terminals X and Y in Fig. 25-1.

25.1 Resistors in Series and Parallel

Two resistors are connected in series when one end, and only one end, of one resistor is connected to the second resistor, with nothing else connected to the junction. Two resistors connected in series are shown in Fig. 25-2. Since no charge accumulates in the circuit, the **current through two resistors in series is the same**. Imagine positive charges repelled by the positive terminal of the battery. They move in the direction of the current flow. It is convenient to imagine the potential to be zero at the negative terminal of the battery (this is an arbitrary choice). The battery raises the potential by an amount $\mathcal{E}$, and then the potential decreases in the direction of current flow. It is as if potential is like elevation, and the charges are like balls rolling downhill. To indicate a point of zero potential, we "ground" the circuit at any point we like. This is analogous to choosing the zero of elevation at sea level or at any arbitrary convenient point. We use the term "ground" because if we connect the circuit to a large conductor like the earth, it is not possible to transfer enough charge to the earth to raise its potential, so its potential remains at zero. In Fig. 25-2 the negative terminal of the battery has been grounded, as indicated by the circuit symbol there.

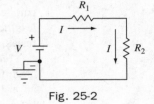

Fig. 25-2

As we move clockwise around the circuit of Fig. 25-2, starting at the grounded point, we see that the battery raises the potential by an amount $\mathcal{E}$. Then the potential drops by $V_1 = IR_1$ across R_1 and then by $V_2 = IR_2$ across R_2, bringing us back to our starting point. Thus $\mathcal{E} = IR_1 + IR_2 = I(R_1 + R_2)$. We would obtain the same current out of the battery if R_1 and R_2 were replaced by a single equivalent resistor $R_s = R_1 + R_2$. For several resistors in series the same reasoning leads us to conclude that the equivalent resistance is

$$\boxed{R_s = R_1 + R_2 + R_3 + \cdots + R_n \quad \text{resistors in series}} \tag{25.2}$$

PROBLEM 25.2. In Fig. 25-2, $\mathcal{E} = 12\,\text{V}$, $R_1 = 3\,\Omega$, and $R_2 = 1\,\Omega$. What current flows through each resistor, and what voltage drop occurs across each resistor?

Solution

$$R = R_1 + R_2 = 3\,\Omega + 1\,\Omega = 4\,\Omega$$

$$I = \frac{V}{R} = \frac{12\,\text{V}}{4\,\Omega} = 3\,\text{A in each resistor}$$

When two resistors have both ends connected together, with nothing intervening, they are connected in parallel, as in Fig. 25-3. The drop in potential when you go from X to Y is the same whether you go along R_1 or along R_2, so **two resistors in parallel**

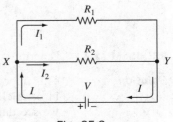

Fig. 25-3

have the same potential drop. Thus $V_1 = V_2 = V$. Further, the current I flowing from the battery splits when it meets point X (just as water flowing splits when it reaches a junction like this), with current I_1 flowing through R_1 and I_2 through R_2. Thus $I = I_1 + I_2$ or

$$\frac{V}{R_p} = \frac{V}{R_1} + \frac{V}{R_2}, \quad \text{so} \quad \frac{1}{R_p} = \frac{1}{R_1} + \frac{1}{R_2}$$

or

$$\boxed{R_p = \frac{R_1 R_2}{R_1 + R_2} \quad \text{two resistors in parallel}} \tag{25.3}$$

For many resistors in parallel,

$$\boxed{\frac{1}{R_p} = \frac{1}{R_1} + \frac{1}{R_2} + \frac{1}{R_3} + \cdots + \frac{1}{R_n}} \tag{25.4}$$

Here R_p is a single resistor equivalent to the parallel combination of resistors. R_p is always less than the resistance of any of the individual resistors connected in parallel.

You will find it easier to reduce parallel circuits by considering the resistors two at a time, using the easily remembered Eq. 25.3, rather than the cumbersome Eq. 25.4. Also, remember that the lines connecting resistors in a circuit represent

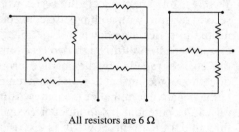

All resistors are 6 Ω

Fig. 25-4

flexible wires that may be pushed around into different shapes without changing the nature of the electrical circuit. Thus all of the circuits in Fig. 25-4 are equivalent and have $R_p = 2\,\Omega$.

PROBLEM 25.3.   Determine the equivalent resistance of the resistors shown here.

Solution   First combine 3 Ω and 6 Ω in parallel.

$$R_1 = \frac{(3)(6)}{3+6} = 2\,\Omega$$

Combine 2 Ω and R_1 in series,

$$R_2 = 2\,\Omega + 2\,\Omega = 4\,\Omega.$$

Combine R_2 and 4 Ω in parallel:

$$R = \frac{(4)(4)}{4+4}\,2\,\Omega$$

PROBLEM 25.4.   Determine the resistance between A and B here.

Solution   This appears complicated because the resistors are not connected either in series or in parallel. However, if we imagine a battery connected between terminals X and Y, we see that if the 5-Ω resistor is removed, we have a simple, readily soluble circuit. 1 Ω and 2 Ω are in series, so we have $1\,\Omega + 2\,\Omega = 3\,\Omega$ across which is 12 V, so current $I_1 = 12\,\text{V}/3\,\Omega = 4\,\text{A}$ flows through them. The potential at X is thus $12\,\text{V} - I_1(1\,\Omega) = 8\,\text{V}$. Similarly, the current I_2 through the $3\,\Omega$ and $6\,\Omega$ in series is $1.33\,\text{A}$, and so the potential at Y is also $8\,\text{V}$. Thus we see that X and Y are both at a potential of $8\,\text{V}$. This means that if we now insert a resistor between X and Y, no current will flow through it, since there will be no potential difference across it. Thus the 5-Ω resistor carries no current and may be removed with no effect. The equivalent resistance is just that of $3\,\Omega$ in parallel with $9\,\Omega$, so $R = (3)(9)/(3+9) = 2.25\,\Omega$. Often symmetry can be used to solve or simplify circuit problems.

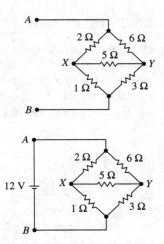

25.2 Multiloop Circuits

Some circuits do not involve only series and parallel connections. For such "multiloop" circuits one can find the current in each element by using **Kirchhoff's rules**.

> **The junction rule. The sum of the currents flowing into a junction is equal to the sum of the currents flowing out of the junction.**
>
> **The loop rule. The algebraic sum of the changes in potential around any closed loop is zero.**

In applying the loop rule, remember that when moving in the direction of current flow through a resistor (that is, downhill), the potential change is $-IR$; that is, the potential decreases. When moving opposite the direction of current through a resistor, the potential change is $+IR$. When moving across a battery from the negative to the positive terminal, the potential increases.

PROBLEM 25.5. Determine the current in each resistor in the circuit here.

Solution Draw arrows indicating the current in each circuit element. I choose the direction arbitrarily, since I do not yet know in what direction each current flows. If my initial guess is wrong for one of the currents, I will obtain a negative answer for that current.

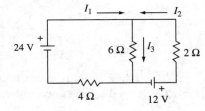

$$I_1 + I_2 = I_3 \qquad\qquad \text{(i)}$$

Substitute this value of I_3 in the two equations below.

$$24 - 6I_3 - 4I_1 = 0 \quad \rightarrow \quad 24 - 6(I_1 + I_2) - 4I_1 = 0 \qquad \text{(ii)}$$
$$12 - 2I_2 - 6I_3 = 0 \quad \rightarrow \quad 12 - 2I_2 - 6(I_1 + I_2) = 0 \qquad \text{(iii)}$$

Simplify (ii) and (iii). Solve (ii) for I_2 and substitute it in (iii) and solve for I_1: $I_1 = 2.73\,\text{A}$, $I_2 = -0.55\,\text{A}$, and $I_3 = 2.18\,\text{A}$. The negative sign for I_2 shows that I_2 flows opposite the direction assumed.

PROBLEM 25.6. Determine the current in each resistor and the voltage difference between points X and Y in the circuit shown here.

Solution

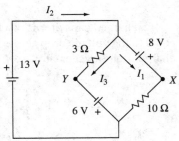

$$13 - 3I_3 + 6 = 0, \quad \text{so } I_3 = 6.33\,\text{A}$$
$$8 - 10I_1 - 6 + 3I_3 = 0, \quad \text{so } I_1 = 2.10\,\text{A}$$
$$I_2 = I_1 + I_3 = 8.43\,\text{A} \qquad V_{XY} = 6 + 10I_1 = 27\,\text{V}$$

25.3 RC Circuits

Charging and discharging of capacitors play an important role in many phenomena. In Fig. 25-5 a series RC circuit is shown. Initially the capacitor is uncharged, and when the switch is closed, charge begins to flow onto the capacitor plates. The potential difference between the battery terminals is equal to the sum of the drop across the capacitor plus the *IR* drop in the resistor.

$$\mathcal{E} = IR + \frac{q}{C} \tag{25.5}$$

Initially there is no voltage across the capacitor since no charge is initially on the plates. The current is large, and all of the voltage drop is across the resistor.

Initially the current through the resistor is large, but as the voltage on the capacitor builds up, the current decreases and finally falls to zero when the capacitor is completely charged. Equation 25.4 can be solved by substituting $I = dq/dt$.

$$R\frac{dq}{dt} = \mathcal{E} - \frac{q}{C} \quad \text{or} \quad \frac{dq}{q - C\mathcal{E}} = -\frac{1}{RC}dt$$

Integrate:

$$\int_0^q \frac{dq}{q - C\mathcal{E}} = -\int_0^t \frac{dt}{RC}$$

or

Thus

$$\ln(q - C\mathcal{E}) - \ln(-C\mathcal{E}) = -\frac{t}{RC}$$

$$q = C\mathcal{E}(1 - e^{-t/RC}) = Q(1 - e^{-t/RC}) \tag{25.6}$$

where $Q = C\mathcal{E}$ = maximum charge on capacitor. The current in the resistor is $I = dq/dt$.

$$I = \frac{\mathcal{E}}{R}e^{-t/RC} \tag{25.7}$$

At $t = 0$ the current is a maximum: $I = \mathcal{E}/R = I_0$. The current will have decreased to $(1/e)I_0$ after time $t = RC$. *RC* is the **time constant** τ. The time dependence of q and I is shown in Fig. 25-6.

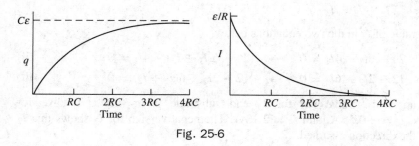

Fig. 25-6

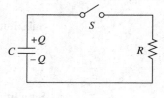

Fig. 25-7

If a charged capacitor is discharged through a resistor, as in Fig. 25-7, the voltage across the capacitor is equal to the voltage drop across the resistor, $IR = q/C$. Here the current measures the rate of *decrease* of the charge q on the capacitor, so $I = -(dq/dt)$. Integrating $dq/dt = -(q/RC)$ yields

$$\int_Q^q \frac{dq}{q} = -\int_0^t \frac{dt}{RC} \quad \text{or} \quad \ln\left(\frac{q}{Q}\right) = -\frac{t}{RC}$$

so

$$q = Qe^{-t/RC} \quad \text{and} \quad I = -\frac{dq}{dt} = I_0 e^{-t/RC} \tag{25.8}$$

Q is the initial charge on the capacitor, and $I_0 = Q/RC$.

PROBLEM 25.7.   A 6-μF capacitor is charged through a 5-kΩ resistor by a 500-V power supply. How long does it require for the capacitor to acquire 99 percent of its final charge?

Solution

$$\tau = C = (5000\,\Omega)(6 \times 10^{-6}\,\text{F}) = 0.030\,\text{s}$$

$$q = 0.99Q = Q(1 - e^{-t/\tau}),$$

so

$$e^{-t/\tau} = 0.01, \quad \frac{t}{\tau} = -\ln 0.01, \quad t = 0.14\,\text{s}$$

PROBLEM 25.8.   A miniature neon lamp consists of two electrodes embedded in a glass capsule filled with neon gas. When the voltage applied to the tube reaches 70 V, a gas discharge occurs (as evidenced by a flash of orange light). Before firing, the gas is an insulator, but when the discharge occurs, the ionized gas is a very good conductor (zero resistance). A very short time after the voltage across the lamp drops to zero, the ions recombine and again the gas is an insulator. A relaxation oscillator circuit that can be used to create a pulsating voltage to terminals X and Y is shown here. What is the pulsating frequency for this circuit if $V = 120\,\text{V}$, $R = 100\,\text{k}\Omega$, and $C = 2\,\mu\text{F}$?

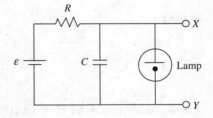

Solution

$$q = CV = C\mathcal{E}(1 - e^{-t/\tau})$$

$$70 = 120(1 - e^{-t/\tau}), \quad 120e^{-t/\tau} = 50, \quad -\frac{t}{\tau} = \ln\frac{50}{120} = -0.875$$

$$\tau = RC = (10^5\,\Omega)(2 \times 10^{-6}\,\text{F}) = 0.20\,\text{s}$$

$$t = (0.875)(0.205) = 0.175\,\text{s}, \quad f = \frac{1}{t} = 5.7/\text{s} = 5.7\,\text{Hz}$$

PROBLEM 25.9.   Show that when a capacitor is discharged through a resistor, the total energy stored in the capacitor is dissipated as heat in the resistor.

Solution   The total energy in the resistor:

$$E = \int_0^\infty I^2 R\,dt = RI_0^2 \int_0^\infty e^{-2t/RC}\,dt$$

$$E = RI_0^2 \frac{RC}{2}\left(-e^{-2t/RC}\right)_0^\infty = \frac{1}{2}R^2 C I_0^2$$

$$I_0 = \frac{Q}{RC}, \quad \text{so } E = \frac{1}{2}\frac{Q^2}{C} = \text{energy stored in the capacitor}$$

PROBLEM 25.10. A 6-μF capacitor is charged to 120 V. After 24 h the voltage has decreased to 116 V. What is the leakage resistance of the capacitor? What will be its voltage after 48 h?

Solution

$$q = Qe^{-t/RC}, \quad q = CV, \quad Q = C\mathcal{E}, \quad \text{so } V = \mathcal{E}e^{-t/RC}$$

$$-\frac{t}{RC} = \ln\left(\frac{V}{\varepsilon}\right), \quad t = -RC\ln\frac{V}{\mathcal{E}} = RC\ln\left(\frac{\mathcal{E}}{V}\right)$$

$$R = \frac{t}{C\ln(\mathcal{E}/V)} = \frac{(24)(3600\text{ s})}{(6\times10^{-6}\text{ F})\ln(120/116)} = 4.2\times10^{11}\,\Omega$$

$$V = \mathcal{E}e^{-t/RC} = 112\text{ V} \quad \text{after 48 h}$$

PROBLEM 25.11. A parallel plate capacitor is filled with dielectric material of dielectric constant κ and resistivity ρ. Determine the time constant in terms of ρ and κ and show that the time constant is independent of the dimensions of the capacitor plates. This result applies to all capacitors, not just to parallel plates.

Solution

$$C = \frac{\kappa\epsilon_0 A}{d}, \quad R = \rho\frac{d}{A}, \quad \text{so } \tau = RC = \epsilon_0\kappa\rho$$

25.4 Summary of Key Equations

EMF: $\quad \mathcal{E} = \int_A^B \mathbf{E}\cdot\mathbf{ds}$

Series resistors: $\quad R_s = R_1 + R_2 + R_3 + \cdots + R_n$

Parallel resistors: $\quad \frac{1}{R_p} = \frac{1}{R_1} + \frac{1}{R_2} + \frac{1}{R_3} + \cdots + \frac{1}{R_n}$

Two parallel resistors: $\quad R_p = \frac{R_1 R_2}{R_1 + R_2}$

Charging capacitor: $\quad q = C\mathcal{E}(1 - e^{-t/RC}), \quad I = \frac{\mathcal{E}}{R}e^{-t/RC}$

Discharging capacitor: $\quad q = Qe^{-t/RC}, \quad I = I_0 e^{-t/RC}$

Capacitative time constant: $\quad \tau = RC$

SUPPLEMENTARY PROBLEMS

25.12. An AA 1.5-V flashlight battery typically has an internal resistance of 0.30 Ω. What is its terminal voltage when it supplies 48 mA to a load? What power does it deliver to the load?

25.13. A bank of batteries with a terminal voltage of 50 V dissipates internally 20 W while delivering 1 A to a load. What is its EMF?

25.14. A 2-Ω resistor is placed in series with a 3-Ω resistor, and an 8-Ω resistor is placed in parallel with the two series resistors. What is the resultant resistance?

25.15. Five 4-Ω resistors are connected as shown here. What is the equivalent resistance?

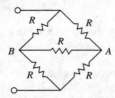

25.16. What current flows in the battery in the circuit here?

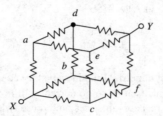

25.17. Twelve identical 1-Ω resistors are joined to form a cube. What is the resistance between terminals *X* and *Y*?

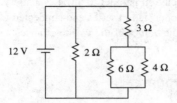

25.18. What is the resistance of the infinite network shown here?

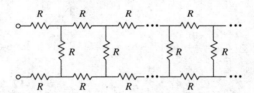

25.19. What is the value of $\mathcal{E}_1$ in this circuit?

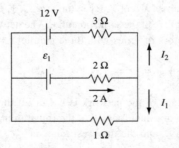

25.20. What is the potential difference between points *X* and *Y* in this circuit?

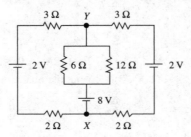

25.21. A camera flashlamp has a capacitor with a time constant of 2.2 s. If the capacitor charges through $200\,k\Omega$, what is the capacitance?

25.22. A 2-μF capacitor is charged to 8 V. It is then connected to a 12-V supply and given additional charge through a 10-MΩ resistor. How long will it take to reach a voltage of 10 V?

25.23. Resistors of 2 Ω, 3 Ω, and 4 Ω are connected in parallel. A 1-Ω resistor is connected to this combination. This array is connected to a 12-V battery. (a) What current flows in the 1-Ω resistor? (b) What current flows in the 2-Ω resistor?

25.24. Resistors R_1 and R_2 are connected in parallel to an 18-V battery, resulting in a current of 6 A through R_1. When these two resistors are connected in series and connected to the same battery, the potential difference across R_2 is 12 V. Determine the values of the two resistances.

SOLUTIONS TO SUPPLEMENTARY PROBLEMS

25.12. $V = \mathcal{E} - Ir = 1.5\,\text{V} - (0.048\,\text{A})(0.3\,\Omega) = 1.49\,\text{V}, \quad P = IV = (1.49\,\text{A})(0.048\,\text{V}) = 0.07\,\text{W}$

25.13. $P_{\text{intern}} = I(\mathcal{E} - 50\,\text{V}), \quad 20\,\text{W} = (1\,\text{A})(\mathcal{E} - 50\,\text{V}), \quad \mathcal{E} = 70\,\text{V}$

25.14. $R = \dfrac{(5)(8)}{5 + 8} = 3.08\,\Omega \quad \text{using } R_s = 2 + 3 = 5\,\Omega$

25.15. Redraw the circuit so it appears symmetric. Imagine a battery connected between the terminals. By symmetry the potential at A will be the same as at B, so no current will flow in the resistor between them. Remove it. Then the network is simply two resistors in series, and these two are in parallel with the other pair of series resistors. Thus

$$R = \frac{(8)(8)}{8 + 8} = 4\,\Omega$$

25.16. 6 Ω and 4 Ω in parallel, so $R_{46} = (4)(6)/(4 + 6) = 2.4\,\Omega$. R_{46} is in series with 3 Ω, so $R_s = 5.4\,\Omega$. R_s is in parallel with 2 Ω, so

$$R = \frac{(2)(5.4)}{(2 + 5.4)} = 1.46\,\Omega \quad \text{and} \quad I = \frac{\mathcal{E}}{R} = \frac{12}{1.46} = 8.2\,\text{A}$$

25.17. Imagine a battery connected across X and Y. By symmetry, a, b, and c are at the same potential. Join them with a wire. Similarly, d, e, and f are at the same potential. The network thus reduces to three parallel resistors in series with six parallel resistors in series with three parallel resistors, so

$$R = \frac{r}{3} + \frac{r}{6} + \frac{r}{3} = \frac{5}{6}r = \frac{5}{6\,\Omega} \quad (r = 1\,\Omega)$$

25.18. Suppose the equivalent resistance of the network is r. Cut off the right portion and replace it with r. The network then looks like the illustration. This is reducible since r is in parallel with R, and the final resistance is

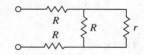

$$r = 2R + \frac{rR}{r + R}$$

Thus,

$$r^2 + rR = 2rR + 2R^2 + rR$$
$$r^2 - 2rR - 2R^2 = 0$$
$$r = (1 + \sqrt{3})R \quad \text{using the quadratic formula}$$

25.19. $I_1 + I_2 = 2, \quad 12 + (1)I_1 - 3I_2 = 0, \quad 12 - \mathcal{E}_1 - (2)(2) - 3I_2 = 0$

Solve: $I_1 = -1.5\,\text{A}, \quad I_2 = 3.5\,\text{A}, \quad \mathcal{E}_1 = -2.5\,\text{V}$

25.20. $6\,\Omega$ and $12\,\Omega$ in parallel:

$$R_p = \frac{(6)(12)}{6+12} = 4\,\Omega$$

By symmetry:

$$I_1 = I_2$$

so

$$8 - 4I_3 - 5I_1 - 2 = 0 \quad \text{and} \quad 2I_1 = I_3$$

Solve: $I_1 = 0.46\,\text{A}, \qquad I_3 = 0.92\,\text{A}$

$$V_{XY} = 8 - (0.92)(4) = 4.3\,\text{V}$$

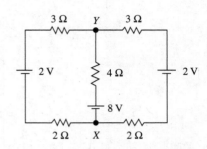

25.21. $\tau = RC = (200{,}000\,\Omega)(C) = 2.2\,\text{s}, \quad C = 11\,\mu\text{F}$

25.22. If initially uncharged, the time to reach 8 V is t_1, and the time to reach 10 V is t_2, where, using

$$q = CV, \quad V = \mathcal{E}(1 - e^{-t/\tau})$$
$$8 = 12(1 - e^{t_1/\tau}), \quad 10 = 12(1 - e^{-t_2/\tau})$$
$$\tau = RC = (10 \times 10^6\,\Omega)(2 \times 10^{-6}\,\text{F}) = 20\,\text{s}$$

Solve: $t_1 = 1.1\tau = 22\,\text{s}, \quad t_2 = 1.79\tau = 35.8\,\text{s}, \quad \Delta t = 13.8\,\text{s}$

25.23.

$$\frac{1}{R_P} = \frac{1}{2} + \frac{1}{3} + \frac{1}{4} = \frac{13}{12}$$

$$R_S = 1 + R_P = 1 + \frac{13}{12} = \frac{25}{13}\,\Omega$$

$$I_1 = \frac{\varepsilon}{R_S} = \frac{12}{25/13} = 6.24\,\text{A in 1-}\Omega\text{ resistor}$$

Potential drop across $1\,\Omega$ is $V_1 = I_1 R$

$$V_1 = (6.24\,\text{A})(1\,\Omega) = 6.24\,\text{V}$$

Drop across $2\,\Omega$ is thus $V_2 = 12 - 6.24 = 5.76\,\text{V}$

Current in $2\,\Omega$ is

$$I_2 = \frac{V_2}{R_2} = \frac{5.76\,\text{V}}{2} = 2.88\,\text{A}$$

25.24. In parallel 18 V is applied to R_1, so

$$I_1 = \frac{\varepsilon}{R_1}, \quad 6 = \frac{18}{R_1}, \quad R_1 = 3\,\Omega$$

In series, $\varepsilon = (R_1 + R_2)I$

$R_2 I = 12\,\text{V}$

$\varepsilon = 3I + R_2 I$

$18 = 3I + 12$

$I = 2\,\text{A}$

$18 = (3)(2) + 2R_2$

$R_2 = 6\,\Omega$

CHAPTER 26

Magnetic Fields

Magnetic effects have been known for thousands of years. Historically magnetism has been viewed as separate from, but related to, electricity. In fact, we now know from the theory of special relativity that magnetism and electricity are two different aspects of a single entity in nature, the electromagnetic field. One can make a good argument for learning about physics first from the theory of relativity and later treating classical newtonian mechanics as an approximation valid at low speeds. However, I will reluctantly continue to follow the traditional approach, and I will tell you about magnetism as if it is somehow intrinsically different from electricity, although this is only partially true.

26.1   The Magnetic Field

We can describe magnetic effects by means of a magnetic field, much as we did for electric fields. These magnetic fields exist all through space and are created by magnetic particles (like electrons and protons and iron atoms) and by moving electric charge. The magnetic field is strong where the magnetic field lines are close together. The tangent to a magnetic field line at a point is in the direction the north pole of a little bar magnet will point if placed at that point. We will encounter no single magnetic "charges" (called *magnetic monopoles*) analogous to electrons and protons, although their existence is theoretically possible (I think the Umpteenth Law of Nature is "In this universe, anything that is not prohibited is mandated"). Personally, I believe such particles exist, but not in this neighborhood of our galaxy. Magnetic fields do not exert a force on stationary electric charges, but they do push on moving particles. The magnetic force on a particle of charge q and velocity $\mathbf{v}$ in a magnetic field $\mathbf{B}$ (measured in **teslas**) is found experimentally to be

$$\boxed{\mathbf{F} = q(\mathbf{v} \times \mathbf{B}) \quad \text{where the magnitude of } F \text{ is } F = qvB \sin \theta} \tag{26.1}$$

Here θ is the smaller angle between $\mathbf{v}$ and $\mathbf{B}$. Use the **right-hand rule** to find the direction of $\mathbf{F}$, as follows: **Point your right fingers along v. Curl your fingers toward B. Your right thumb will point along F.** In finding forces, always treat every particle as if it has positive charge and follow the preceding rule. Then, if the particle is negative (like an electron), reverse the direction of $\mathbf{F}$. Here are some important features of the magnetic force:

1. The magnetic force acts only on moving electric charge, not on stationary charge.
2. The magnetic force is perpendicular to the plane of $\mathbf{B}$ and $\mathbf{v}$.
3. Since the magnetic force has no component parallel to $\mathbf{v}$, it can change the direction of motion but not the speed, and hence it does no work on the particle and does not change its kinetic energy.
4. The magnetic force is largest when the particle moves perpendicularly to the magnetic field, and it is zero when the particle moves parallel to the magnetic field.

An older unit for magnetic field is the gauss, where $10{,}000\,\text{G} = 1\,\text{T}$. At the Earth's surface, the Earth's magnetic field is about 0.5 G.

26.2   Motion of a Charged Particle in a Magnetic Field

Suppose that a particle of mass m and charge q moves perpendicular to a unifom magnetic field. Then $\theta = 90°$ and $F = qvB$. This sideways force will provide the centripetal force to make the particle move in a circle of radius r, where $mv^2/r = qvB$ or

$$r = \frac{mv}{qB} \quad \text{cyclotron radius} \tag{26.2}$$

This radius is called the **cyclotron radius** because of its application in the cyclotron, one of the early particle accelerators. The **cyclotron frequency** f_c is determined from

$$\omega_c = \frac{v}{r} = \frac{qB}{m}$$

where $f_c = \omega_c/2\pi$.

$$\omega_c = 2\pi f_c = \frac{qB}{m} \quad \text{cyclotron frequency} \tag{26.3}$$

PROBLEM 26.1. A proton (mass 1.67×10^{-27} kg) is observed to move along an arc of 32-cm radius when moving perpendicular to a magnetic field of 1.4 T. What is the cyclotron frequency and the momentum of the proton?

Solution

$$mv = qBr = (1.6 \times 10^{-19}\,\text{C})(1.4\,\text{T})(0.32\,\text{m}) = 7.17 \times 10^{-20}\,\text{kg} \cdot \text{m/s}$$

$$f_c = \frac{qB}{2\pi m} = \frac{(1.6 \times 10^{-19}\,\text{C})(1.4\,\text{T})}{(2\pi)(1.67 \times 10^{-27}\,\text{kg})} = 21\,\text{MHz}$$

PROBLEM 26.2. A certain type of velocity selector consists of a pair of parallel plates between which is established an electric field E. A beam of particles of mass m, charge q, and speed v is directed parallel to the plates in the region between them. A magnetic field **B** is applied perpendicular to both **E** and **v**. In Fig. 26-1, **B** is directed into the paper, as indicated by the symbol $\otimes$ (like the feathers on an arrow going away from you). A vector coming out of the paper is indicated by $\odot$ (like the point of an arrow coming at you). Determine an expression for the velocity of particles that are undeflected by this device.

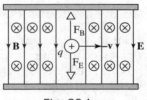

Fig. 26-1

Solution The magnetic force qvB balances the electric force qE, so $qvB = qE$ and $v = B/E$.

PROBLEM 26.3. A **mass spectrometer** is an instrument used to separate ions of slightly different masses. Often these are isotopes of an element, and as such they have very similar chemical properties. The construction of a mass spectrometer is shown in Fig. 26-2. Ions of charge $+q$ and mass m are accelerated through a potential difference V_0. The ions then move perpendicularly into a magnetic field B, where they are bent along a

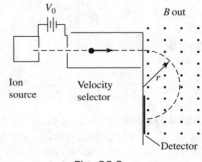

Fig. 26-2

semicircular path. They are detected a distance $d = 2r$ from the entry port. Determine the ion mass in terms of the known parameters.

Solution

$$\tfrac{1}{2}mv^2 = qV_0 \quad \text{and} \quad qvB = \frac{mv^2}{r}$$

Solve:

$$m = \frac{qr^2}{2V_0}B^2$$

B can be varied to cause different masses to hit the detector.

It is only the component of velocity perpendicular to the magnetic field that is changed by the magnetic force. Consequently, a particle moving at an angle other than 90° to the magnetic field will move in a helical path.

PROBLEM 26.4.   A velocity selector can be constructed using the following principle: Ions of charge q and mass m move upward starting at the origin with a speed v_0 at an angle θ to the z-axis. A magnetic field B is established along the z-axis. Determine the point where the ions first return to the z-axis.

Solution   The ions move in a helical path and will return to the z-axis after one period of the cyclotron frequency. During this time they will travel a distance $z = v_0 \cos \theta T$ along the axis.

$$fc = \frac{1}{2\pi}\frac{qB}{m} = \frac{1}{T} \quad \text{and} \quad z = \frac{2\pi m v_0 \cos \theta}{qB}$$

26.3   Magnetic Force on a Current-Carrying Wire

If the moving charges on which a magnetic field exerts a force are in a conductor, the force on the charges (that is, on the current) will be transferred to the wire when the electrons bump into the atoms of the wire. If n is the number of carriers per unit volume, each with charge q, and v_d is their drift velocity, then the force on a wire of length dL and cross-sectional area A will be $\mathbf{dF} = nAdL(\mathbf{v_d} \times \mathbf{B})$. From Eq. 24.2, $I = nAv_d$, so this may be written

$$\mathbf{dF} = I\mathbf{dL} \times \mathbf{B} \quad \text{force on current carrying wire} \tag{26.4}$$

$\mathbf{dL}$ points along the wire in the direction of current flow. If the wire is straight and in a uniform field, $F = BIL \sin \theta$.

PROBLEM 26.5.   A wire carrying 2.0 A lies along the x-axis. The current flows in the positive x direction. A magnetic field of 1.2 T is parallel to the xy plane and makes an angle of 30° with the x-axis (pointing into the first quadrant). What is the force on a segment of wire 0.40 m long?

Solution

$$\mathbf{F} = BIL \sin \theta\, \mathbf{k} = (1.2\,\text{T})(2\,\text{A})(0.40\,\text{m}) \sin 30°\, \mathbf{k} = 0.48\, \mathbf{k}N$$

PROBLEM 26.6.   A wire carrying 1.5 A lies on a horizontal surface in the xy plane. One end of the wire is at the origin and the other is at 3 m and 4 m. The wire follows an erratic path from one end to the other. A magnetic field of 0.15 T directed vertically downward is present. What magnetic force acts on the wire?

Solution

Break the path into tiny steps dx to the right and dy upward. The force on each little such segment is $\mathbf{dF} = BI\,dx\,\mathbf{j} + BI\,dy(-\mathbf{i})$. The total force is thus

$$\mathbf{F} = (0.15\,\text{T})(1.5\,\text{A})\left(\mathbf{j}\int_0^3 dx - \mathbf{i}\int_0^4 dy\right)$$

$$= 0.225(3\mathbf{j} - 4\mathbf{i})$$

The magnitude of the force is $F = (0.225)\sqrt{3^2 + (-4)^2} = 1.13N$

Note that the exact path of the wire doesn't matter, since when the wire zigzags back and forth, the forces cancel on the parts that backtrack. The total force is just what would result if the wire went in a straight line from (0, 0) to (3, 4).

PROBLEM 26.7.

A circular loop of radius R wire carries current I. A uniform magnetic field B is present perpendicular to the plane of the loop. What is the tension in the wire?

Solution

Look at the force on the top half of the loop. This force is balanced by the tension force on each end of the semicircle. Using the right-hand rule, I see that the magnetic force is directed radially outward. By symmetry the resultant force on the loop is directed in the z direction, where $F_z = F\cos\theta$.

$$F_z = \int_0^\pi BI\,dL\cos\theta, \quad \text{where } dL = R\,d\theta$$

$$= 2\int_0^{\pi/2} BIR\cos\theta\,d\theta = 2BIR(\sin\theta)_0^{\pi/2}$$

$$= 2BIR = 2T$$

so

$$T = BIR$$

26.4 Torque on a Current Loop

Consider a rectangular wire loop of sides a and b. Imagine it is positioned as shown in Fig. 26-3. The loop carries current I, and a uniform magnetic field B is present parallel to the y-axis. I have drawn in the force on each section of the wire, and you can see that the force on segment 2 cancels that on segment 4, since they

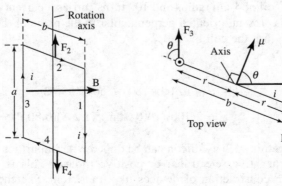

Fig. 26-3

are collinear and of equal magnitude. The forces on 1 and 3 cause a torque τ.

$$F = BaI, \qquad \tau = 2F\left(\frac{b}{2}\right)\sin\theta = BIba\sin\theta$$

But $ba = A$, the area of the loop, so

$$\boxed{\tau = BAI\sin\theta \quad \text{or} \quad \vec{\tau} = \vec{\mu} \times \mathbf{B} \quad \text{torque on a loop}} \tag{26.5}$$

Here I defined the **magnetic dipole moment** of the loop as $\vec{\mu} = I\mathbf{A}$, where A is the area of the loop. To find the direction of $\vec{\mu}$, curl your right fingers along the current, and your thumb will point along $\vec{\mu}$. $\vec{\mu}$ is perpendicular to the plane of the loop. For a circular loop, $\vec{\mu}$ is along the axis of the loop.

When a loop of current is placed in a magnetic field, it tries to line up with the plane of the loop perpendicular to the magnetic field. Another way of describing this is to say that the loop tries to enclose as many lines of B as possible, that is, tries to have maximum magnetic flux passing through the loop. You can also think of the magnetic moment $\vec{\mu}$ as if it is a small bar magnet. $\vec{\mu}$ tries to line up parallel to $\mathbf{B}$, just as a compass needle does. Notice that $\vec{\tau} = \vec{\mu} \times \mathbf{B}$ is analogous to the torque on an electric dipole moment in an electric field, $\vec{\tau} = \mathbf{p} \times \mathbf{E}$. It is tempting to imagine that a magnetic moment consists of two **magnetic poles**, a north pole analogous to a positive charge, and a south pole analogous to a negative charge, and this was done long ago. The point of the $\vec{\mu}$ vector is like a north magnetic pole, and the tail is like a south pole. You may still find this picture helpful for envisioning what is going on in magnetism. A magnetic field pushes a north pole in the direction of the field, just as an electric field pushes a positive charge in the direction of the field. However, when you cut an electric dipole in half, you get a positive charge and a negative charge. When you cut a small magnet in half, you get only two small magnets, no separate north and south poles, so in fact there are no north and south poles, in the sense that we have electric charges. In drawings, we show magnetic field lines as sprouting out of north poles and ending on south poles, just to help us envision the magnetic field.

PROBLEM 26.8. A circular coil of wire carries 50 mA of current. The coil has 50 turns and an area of $2.0\,\text{cm}^2$. A magnetic field of 0.300 T is present, oriented parallel to the plane of the coil. What torque acts on the coil?

Solution Fifty turns carrying 50 mA is equivalent to one turn carrying (50)(50 mA), so

$$\tau = NAIB\sin\theta = 50(2 \times 10^{-4}\,\text{m}^2)(50 \times 10^{-3}\,\text{A})(0.30\,\text{T})\sin 90° = 1.5 \times 10^{-4}\,\text{N}\cdot\text{m}$$

If one tries to rotate a magnetic moment in a magnetic field, a torque $-\mu B\sin\theta$ must be applied. The work done equals the gain in potential energy. Thus $U - U_0 = -\int_{\theta_0}^{\theta} \tau B\sin\theta\, d\theta = -\mu B(\cos\theta - \cos\theta_0)$. We choose U_0 such that $U = 0$ when $\theta = 90°$, so

$$\boxed{U = -\mu B\cos\theta = -\vec{\mu}\cdot\mathbf{B} \quad \text{potential energy of a magnetic dipole}} \tag{26.6}$$

PROBLEM 26.9. A circular coil of 4-cm radius and 100 turns carries a current of 1.2 A. A magnetic field of 0.80 T is present, oriented perpendicular to the plane of the coil. How much work is required to turn the coil by 180°?

Solution

$$W = -\mu B(\cos 180° - \cos 0°) = 2\mu B = 2NAIB$$

$$= 2(100)(\pi)(0.04\,\text{m})^2(1.2\,\text{A})(0.80\,\text{T}) = 0.97\,\text{J}$$

PROBLEM 26.10. Semiconductors such as silicon can be doped with impurities so that the charge carriers are either negative (electrons) or positive (holes). This is an important property to ascertain in construction of devices like transistors. Sketched in Fig. 26-4 is a setup to measure the **Hall effect**. Such a measurement can determine the sign and density of the carriers and, when calibrated, can be used to measure the strength of a magnetic field.

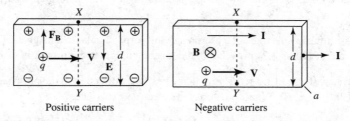

Fig. 26-4

Determine an expression in terms of given parameters for the Hall voltage measured between points X and Y in the arrangement shown.

Solution   The magnetic force bends the carriers up to the top of the sample until a sufficient electric field E builds up to cancel the magnetic force. When this happens, $qE = qv_dB$ and the voltage between X and Y is then $V_H = Ed = v_dBd$. The current is $I = nAqv_d$, where $A = ad$ and n is the carrier density. Thus

$$\frac{V_H}{d} = \frac{I}{nAq}B \quad \text{or} \quad V_H = \frac{1}{nq}\frac{IB}{a} \qquad (26.7)$$

$1/nq$ is the **Hall coefficient**. If the carriers are negative, and the current is to the right, as in the drawing, the carrier velocity is directed to the left, and again the magnetic force pushes the carriers upward. In this case V_H is negative, whereas V_H is positive for positive carriers. Thus measurement of the Hall coefficient determines the sign of the carriers and their density since almost always $q = e$.

26.5   Summary of Key Equations

Magnetic force:	$\mathbf{F} = q\mathbf{v} \times \mathbf{B}, \quad F = qvB\sin\theta$
Cyclotron radius:	$r_c = \dfrac{mv}{qB}$
Cyclotron frequency:	$f_c = \dfrac{1}{2\pi}\dfrac{qB}{m}$
Force on a conductor:	$\mathbf{dF} = I\,\mathbf{dL} \times \mathbf{B} \quad \text{and} \quad dF = I\,dLB\sin\theta$
Magnetic moment of a loop:	$\mu = NIA$
Torque on a loop:	$\vec{\tau} = \vec{\mu} \times \mathbf{B}, \qquad \tau = \mu B\sin\theta$
Potential energy of a magnetic dipole:	$U = -\vec{\mu} \cdot \mathbf{B} = -\mu B\cos\theta$
Hall effect:	$V_H = \dfrac{IB}{nqa}$

SUPPLEMENTARY PROBLEMS

26.11. An He^{2+} ion traveling 2×10^5 m/s moves perpendicular to a magnetic field of 0.75 T. What force does it experience?

26.12. In a photographic emulsion the track of a proton moving perpendicular to a magnetic field of 0.60 T is observed to be a circular arc of radius 1.2 cm. What is the kinetic energy of the proton in electronvolts?

26.13. An electron moves with speed 3.2×10^5 m/s in the positive x direction in the presence of a magnetic field $\mathbf{B} = 0.1\mathbf{i} + 0.3\mathbf{j} - 0.2\mathbf{k}$ (in teslas). What force does the electron experience?

26.14. A singly charged ion of lithium has a mass of 1.16×10^{-26} kg. It is accelerated through a voltage of 600 V and then enters a magnetic field of 0.60 T perpendicular to its velocity. What is the radius of the ion's path in the magnetic field?

26.15. A straight wire lies along the x-axis and carries a current of 2.0 A in the positive x direction. A uniform magnetic field of 0.08 T in the xy plane makes an angle of $60°$ with the wire. Determine the magnitude and direction of the magnetic force on a 1.5-m segment of the wire.

26.16. In a loudspeaker, a permanent magnet creates a magnetic field of 0.12 T directed radially outward from the z-axis. The voice coil of the speaker has 60 turns and a radius of 0.013 m and is positioned in the xy plane. What force acts on the coil when it carries a current of 1.5 A?

26.17. An electromagnetic rail gun can be constructed as follows: A conducting bar of mass m slides over two parallel horizontal conducting rails separated by a distance L. A power source causes current I to flow through the rails and the bar. A uniform vertical magnetic field B is maintained. If the bar is initially at rest, how fast will it be moving after it has moved a distance x? It has been suggested that this device could be used to project payloads into orbit around the Earth, or to transport ore from the surface of the Moon to a factory in space, or to induce nuclear fusion reactions through high-speed collisions.

26.18. A wire with mass per unit length of 0.04 kg/m carries 3 A horizontally to the east. What minimum magnetic field is required to support this wire against the force of gravity?

26.19. A square loop of side L and n turns carries a current I. One side of the square is along the z-axis, and current flows down in this side. The rest of the loop is in the positive xy quadrant, and the plane of the loop makes an angle $\phi < 90°$ with the x-axis. A magnetic field B is directed along the positive x-axis. What torque does the loop experience? When viewed from above, in what direction will the loop tend to rotate?

26.20. A length L of wire is formed into a rectangular loop. It carries current I. What should be its dimensions in order to maximize the torque on it when it is placed in a magnetic field?

26.21. In a Hall effect experiment, a sample 12-mm thick is used with a magnetic field of 1.6 T. When a current of 10 A is passed through the sample, a Hall voltage of 0.080 μV is observed. What is the carrier density, assuming $q = e$?

26.22. Uranium is found naturally, primarily in the form of the isotope ^{238}U (nuclear mass 3.95×10^{-25} kg). The isotope ^{235}U (mass 3.90×10^{-25} kg) is the form used in fission reactors to generate electricity (and to make bombs), and only a very small fraction of mined uranium is in this form. Separation of the lighter isotope from the heavier one is a major problem. Nowadays this is mainly done using centrifuges, but another method is to use a mass spectrometer arrangement. In one such device, singly charged uranium ions (carrying one proton charge) are accelerated through a potential difference of 10,000 V. They are then deflected by a magnetic field of 0.70 T and follow semicircular paths, at the ends of which the two beams are collected. (a) What is the radius of the path for each of the isotopes? (b) What is the separation of the two beams at the collection point?

26.23. Two parallel rails rest on a horizontal table. They are separated by a distance L. A rod of mass m carrying a current I slides on the rails. The rod is perpendicular to the rails and carries current I. A uniform vertical magnetic field B is present. The coefficient of friction between the rod and the rails is μ. What current will result in the rod moving with constant speed?

SOLUTIONS TO SUPPLEMENTARY PROBLEMS

26.11. $F = qvB \sin \theta = (2)(1.6 \times 10^{-19} \text{ C})(2 \times 10^5 \text{ m/s})(0.75 \text{ T}) \sin 90° = 4.8 \times 10^{-14} \text{ N}$

26.12. $r = \dfrac{mv}{qB}$

$$\text{KE} = \frac{1}{2}mv^2 = \frac{1}{2}\frac{r^2 e^2 B^2}{m}$$

$$= \frac{(0.5)(0.012 \text{ m})^2 (1.6 \times 10^{-19} \text{ C})^2 (0.60 \text{ T})^2}{(1.67 \times 10^{-27} \text{ kg})} = 3.97 \times 10^{-16} \text{ J}$$

$$= \frac{3.97 \times 10^{-16} \text{ J}}{1.6 \times 10^{-19} \text{ J/eV}} = 2480 \text{ eV}$$

26.13. $\mathbf{F} = q\mathbf{v} \times \mathbf{B} = -(1.6 \times 10^{-19} \text{ C})[\mathbf{i} \times (0.1\mathbf{i} + 0.3\mathbf{j} - 0.2\mathbf{k})] \text{ T}$

$$= -0.48 \times 10^{-19} \mathbf{k} - 0.32 \times 10^{-19} \mathbf{j}$$

$$= \sqrt{F_x^2 + F_y^2} = 5.8 \times 10^{-20} \text{ N}$$

26.14. $qV = \dfrac{1}{2}mv^2, \qquad r = \dfrac{mv}{qB} = \dfrac{1}{B}\sqrt{\dfrac{2mV}{q}}$

$$= \frac{1}{0.60 \text{ T}}\sqrt{\frac{(2)(1.16 \times 10^{-26} \text{ kg})(600v)}{(1.6 \times 10^{-19} \text{ C})}} = 0.016 \text{ m}$$

26.15. $F = BIL \sin \theta = (0.08 \text{ T})(2 \text{ A})(1.5 \text{ m}) \sin 60°$

$\qquad = 0.21 \text{ N in the } +z \text{ direction}$

26.16. $F = NBIL \sin \theta = (60)(0.12 \text{ T})(1.5 \text{ A})(2\pi)(0.013 \text{ m}) \sin 90°$

$\qquad = 0.88 \text{ N}$

26.17. $F = BIL, \qquad v^2 = v_0^2 + 2ax = 0 + 2\left(\dfrac{F}{m}\right)x, \qquad v = \sqrt{\dfrac{2BILx}{m}}$

26.18. $F = BIL \sin \theta = mg, \quad m = \lambda L, \qquad \theta = 90°, \qquad B = \dfrac{\lambda Lg}{IL} = \dfrac{(0.04 \text{ kg/m})(9.8 \text{ m/s}^2)}{3 \text{ A}} = 0.13 \text{ T}$

26.19. $\mu = nIA = nTL^2 \qquad \tau = \mu B \sin \theta = nIL^2 B \cos \theta \quad \text{since } \theta + \phi = 90°$

The loop will rotate counterclockwise.

26.20. $\tau = \mu B \sin \theta$, so τ is a maximum when $\mu = IA$ is a maximum, that is, when A is a maximum. Let $x =$ length of one side. Then

$$A = (x)\left(\frac{L}{2} - x\right), \qquad \frac{dA}{dx} = 0 = \left(\frac{L}{2} - x\right) - x, \quad x = \frac{L}{4}$$

A square loop gives maximum torque. (A circular loop gives even more torque for a given length of wire.)

26.21. $V_H = \dfrac{1}{nq}\dfrac{IB}{a},$

$$n = \frac{IB}{eV_H a} = \frac{(10 \text{ A})(1.6 \text{ T})}{(1.6 \times 10^{-19} \text{ C})(0.08 \times 10^{-6} \text{ V})(12 \times 10^{-3} \text{ m})}, \quad n = 1.04 \times 10^{29} \text{ m}^{-3}$$

26.22. $\frac{1}{2}mv^2 = eV$

$$v = \left[\frac{2eV}{m}\right]^{1/2}$$

$$evB = \frac{mv^2}{r}, \ r = \frac{mv}{eB} = \left[\frac{2mV}{e}\right]^{1/2}\frac{1}{B}$$

(a) $r_{235} = \left[\frac{(2)(3.90 \times 10^{-25}\,\text{kg})(10^4\,\text{V})}{1.6 \times 10^{-19}\,\text{C}}\right]^{1/2}\frac{1}{0.70\,\text{T}}$

$r_{235} = 0.315\,\text{m}$

$r_{238} = \left[\frac{(2)(3.95 \times 10^{-25}\,\text{kg})(10^4\,\text{V})}{1.6 \times 10^{-19}\,\text{C}}\right]^{1/2}\frac{1}{0.70\,\text{T}}$

$r_{238} = 0.317\,\text{m}$

(b) $2\Delta r = 2(r_{235} - r_{238}) = 0.004\,\text{m} = 0.04\,\text{cm}$

26.23. $F = BIL = \mu mg$

$$I = \mu\frac{mg}{BL}$$

CHAPTER 27

Sources of the Magnetic Field

Magnetic fields are created by moving electric charge (as when a current flows in a wire) or by electrons or protons (which act like little bar magnets) or by some magnetic atoms, such as iron. Some aspects of the magnetism of elements like iron can be imagined as due to the orbital motion of electrons circling the nucleus, and hence acting like little current loops, but even for iron the main magnetic effect results from the fact that electrons act somewhat like little spinning spheres of charge. Unfortunately, these models of atoms and electrons are wrong, even though we use them to help envision what is happening. As far as I can tell, no one understands exactly why an electron has a magnetic moment. Suffice to say that electrons and protons act like little bar magnets, and they have angular momentum as if they were little spinning spheres. The orbital motion of electrons in atoms is understood in terms of quantum mechanics, but it would take too much space to explain here. Consequently, I will focus on the magnetic fields originating in macroscopic currents. The magnetism due to magnetic material I will describe only briefly in phenomenological terms, since I don't understand it too well, and what I do know is very complicated.

27.1   Magnetic Fields due to Currents

Consider a current I flowing in a wire. Break the wire into little pieces of length ds. The magnetic field due to this little piece of current is then found experimentally to be

$$\mathbf{dB} = \frac{\mu_0}{4\pi} \frac{I\,\mathbf{ds} \times \hat{\mathbf{r}}}{r^2} \qquad \text{or} \qquad dB = \frac{\mu_0}{4\pi} \frac{I\,ds\sin\theta}{r^2} \qquad (27.1)$$

This is the **law of Biot–Savart**. Here r is the distance from the current element $I\,\mathbf{ds}$ to the field point P where we wish to find the magnetic field $\mathbf{B}$. $\hat{\mathbf{r}}$ is a unit vector pointing along $\mathbf{r}$. μ_0 is a constant of nature, the **permeability of free space**. $\mu_0 = 4\pi \times 10^{-7}\,\text{T} \cdot \text{m/A}$. The relationship between $\mathbf{ds}$, $\hat{\mathbf{r}}$, and $\mathbf{dB}$ is shown in Fig. 27-1. Remember that to determine the direction of $\mathbf{ds} \times \hat{\mathbf{r}}$, use your right hand, point your fingers along $\mathbf{ds}$, and curl them toward $\hat{\mathbf{r}}$. Your right thumb will point along $\mathbf{dB}$.

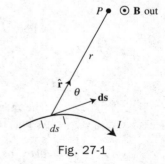

Fig. 27-1

To find the total magnetic field due to a conductor, we add up the contributions from each current element by integrating over the entire conductor. Thus

$$\mathbf{B} = \frac{\mu_0 I}{4\pi} \int \frac{\mathbf{ds} \times \hat{\mathbf{r}}}{r^2} \qquad (27.2)$$

The law of Biot–Savart was discovered experimentally, but it can be derived from Coulomb's law using the theory of special relativity.

PROBLEM 27.1.   Determine the magnetic field a distance R from a long straight wire carrying current I.

$$\boxed{B = \frac{\mu_0 I}{2\pi R}} \quad \text{long straight wire} \tag{27.3}$$

Solution

$$B = \frac{\mu_0 I}{4\pi} \int_{-\infty}^{\infty} \frac{dx \sin \theta}{r^2}$$

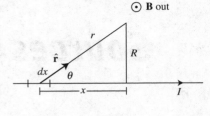

where

$$\sin \theta = \frac{R}{r} \quad \text{and} \quad r = \sqrt{x^2 + R^2}$$

$$B = \frac{\mu_0 I}{4\pi} \int_{-\infty}^{\infty} \frac{R\,dx}{(x^2 + R^2)^{3/2}} = \frac{\mu_0 IR}{4\pi} \left(\frac{x}{R^2 \sqrt{x^2 + R^2}} \right)_{-\infty}^{\infty}$$

In the top drawing (Fig. 27-2) **B** is directed out of the paper. The lines of B are concentric circles, with their spacing increasing as you move away from the wire. This example suggests another way of deducing the direction of the magnetic field B. **Point your right thumb along the current. Your fingers will curl around the current in the sense that the lines of B do**. We saw that the electrostatic field had sources and sinks (the positive and negative charges). E lines sprouted out of positive charges and ended on negative charges. The field lines looked like flow line for laminar flow in hydrodynamics. The magnetic field has no sources or sinks. The B lines end on themselves, and they look like flow lines for turbulent flow. A current is like a paddle wheel that stirs up little whirlpools in the B lines.

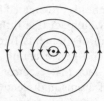

Fig. 27-2

PROBLEM 27.2.   Determine the magnetic field at the center of a circular loop of radius R carrying current I.

Solution

$$B = \frac{\mu_0 I}{4\pi} \int \frac{ds \, \sin \theta}{r^2} = \frac{\mu_0 I}{4\pi} \int_0^{2\pi} \frac{R\,d\theta}{R^2} = \frac{\mu_0 I}{4\pi R}(2\pi)$$

where $ds = R\,d\theta$ and $\theta = 90°$. Thus

$$\boxed{B = \frac{\mu_0 I}{2R}} \quad \text{circular loop} \tag{27.4}$$

The magnetic field of a small current loop (Fig. 27-3) is just like that of a small bar magnet, with B lines sprouting out of an imaginary north pole and looping back to end on an imaginary south pole. Thus the field of a small current-carrying loop is that of a magnetic dipole, with the same appearance as the field of an electric dipole.

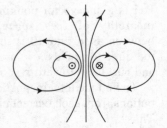

Fig. 27-3

PROBLEM 27.3.   A long straight wire carrying current I is bent 90° in a circular arc of radius R. What is the magnetic field at the center of the arc?

Solution   Each straight section is like one-half of an infinitely long straight wire, so the contribution from these two sections is just that found in Eq. 27.3. The contribution from the curved section is just one-fourth that of a full circle, Eq. 27.4. Thus at the

center

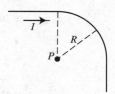

$$B = \frac{\mu_0 I}{2\pi R} + \frac{\mu_0 I}{8R} = 0.28 \frac{\mu_0 I}{R}$$

PROBLEM 27.4.   A long straight wire carrying current I has a semicircular "kink" in it of radius R. What is the magnetic field at the center of the semicircle?

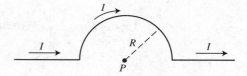

Solution   The straight sections contribute nothing to B, since for them $\theta = 0°$ and $dx \sin \theta = 0$. The contribution of the semicircle is just half that of a full circle, Eq. 27.4. Thus

$$B = \frac{\mu_0 I}{4R}$$

PROBLEM 27.5.   Two long parallel wires each carry current I in opposite directions. If they are separated by a distance d, what is the force per unit length on one of them?

Solution

$$F = BIL = \frac{\mu_0 I}{2\pi d} IL$$

so the force per unit length is

$$\frac{F}{L} = \frac{\mu_0 I^2}{2\pi d}$$

In general, oppositely directed currents always repel each other, and this means any closed circuit tends to expand and increase its area.

27.2   Ampere's Law

I have pointed out that steady currents act like "paddle wheels" that stir up circulation in the magnetic field, giving rise to B lines that look like the flow lines for whirlpools. A mathematical statement of this observation is **Ampere's law**:

$$\oint \mathbf{B} \cdot \mathbf{ds} = \mu_0 I \qquad\qquad (27.5)$$

This line integral is carried out around a closed path, and I is the current enclosed by the path. The integral itself is called the **circulation of B** around the path. Ampere's law is analogous to Gauss' law in that it is useful for calculating the magnetic field for highly symmetric current distributions, as illustrated by the following problems.

PROBLEM 27.6.   Calculate the magnetic field due to a long straight cylinder of radius R carrying current I_0 uniformly distributed over the cross-section of the conductor.

Solution   The trick in using Ampere's law is to identify a path along which $\mathbf{B} \cdot \mathbf{ds}$ is constant. Here we can see from the symmetry that B is constant along a circle of radius r concentric with the wire. For such a circle $\mathbf{B} \cdot \mathbf{ds} = B\,ds = Br\,d\phi$, so

$$\oint \mathbf{B} \cdot \mathbf{ds} = \int_{0}^{2\pi} Br\,d\phi = 2\pi rB = \mu_0 I \qquad \text{and} \qquad B = \frac{\mu_0 I}{2\pi r}$$

If $r \geq R$, then

$$I = I_0 \qquad \text{and} \qquad B = \frac{\mu_0 I_0}{2\pi r} \qquad r \geq R \qquad (27.6)$$

This is a much easier way to obtain the result of Eq. 27.3. For $r \leq R$,

$$\frac{I}{I_0} = \frac{\pi r^2}{\pi R^2}, \qquad \text{so } B = \frac{\mu_0 I_0 r}{2\pi R^2}, \quad r \leq R \qquad (27.7)$$

PROBLEM 27.7.   A very long solenoid has n turns per unit length and carries current I (Fig. 27-4). What is the magnetic field inside and outside the solenoid?

Solution   Using the right-hand rule one can see that the B lines go up through the interior of the solenoid. There will be a finite number of these lines inside the solenoid, and they come out the top of the solenoid and then loop back and enter the lower end. Since the solenoid is infinitely long, the density of the lines as they pass back through the xy plane is zero; that is, the magnetic field is zero outside an infinite solenoid. This is a pretty good approximation as long as the diameter of the solenoid is small compared to its length. One can arrive at this conclusion by applying Ampere's law to the rectangular loop 1 in Fig. 27-4. Here

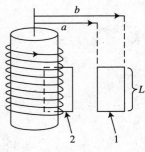

Fig. 27-4

$\oint \mathbf{B} \cdot \mathbf{ds} = [B(a) - B(b)]L = \mu_0 I = 0$, so $B(a) = B(b)$. This says B does not depend on the distance from the solenoid, but from the Biot–Savart law we know B goes to zero far away, so B must be zero everywhere outside the solenoid.

Now apply Ampere's law to loop 2. Since $B = 0$ outside the solenoid, there is no contribution to the integral from the path outside the solenoid. By symmetry, B is axial inside, so $\mathbf{B}$ is perpendicular to $\mathbf{ds}$ on the small sections of the loop, and thus $\mathbf{B} \cdot \mathbf{ds} = 0$ there. Along the inner section, B is constant, so the result is

$$\oint \mathbf{B} \cdot \mathbf{ds} = BL = \mu_0 nLI$$

where nL is the number of turns enclosed by the loop.

$$\boxed{B = \mu_0 nI \quad \text{long solenoid, } n \text{ turns per meter}} \qquad (27.8)$$

PROBLEM 27.8.   Determine the field inside a toroid of N turns carrying current I. A toroid (Fig. 27-5) is like a solenoid bent into a doughnut shape.

Solution   Apply Ampere's law to a circular path inside the toroid. By symmetry, B is tangential to this path and constant in magnitude along the path, so

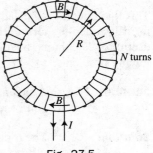

$$\oint \mathbf{B} \cdot \mathbf{ds} = B \oint ds = 2\pi RB = \mu_0 NI$$

Fig. 27-5

and

$$B = \frac{\mu_0 NI}{2\pi R} \quad \text{toroid} \tag{27.9}$$

If the integral path is outside the toroid, the current crossing the plane enclosed by the path is zero, so the field outside an ideal toroid is zero.

PROBLEM 27.9.   An infinite conducting sheet in the *xz* plane carries a uniform current density $\mathbf{J_s}$ in the *x* direction. Determine the magnetic field outside the sheet.

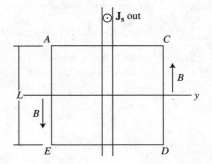

Solution   By use of the right-hand rule I see that **B** is directed as shown here. By symmetry, *B* is constant in magnitude. **B** and **ds** are perpendicular along *AC* and *DE*, so $\mathbf{B} \cdot \mathbf{ds} = 0$ there. Thus Ampere's law yields

$$\oint \mathbf{B} \cdot \mathbf{ds} = 2BL = \mu_0 J_s L \quad \text{or} \quad \boxed{B = \mu_0 \frac{J_s}{2}} \quad \text{infinite sheet} \tag{27.10}$$

27.3   Summary of Key Equations

Law of Biot–Savart: $$\mathbf{B} = \frac{\mu_0 I}{4\pi} \int \frac{\mathbf{ds} \times \hat{r}}{r^2}$$

Ampere's law: $$\oint \mathbf{B} \cdot \mathbf{ds} = \mu_0 I$$

Long solenoid: $B = \mu_0 nL$

Long straight wire: $B = \dfrac{\mu_0 I}{2\pi r}$

Toroid: $B = \dfrac{\mu_0 NI}{2\pi R}$

Center of circular loop: $B = \dfrac{\mu_0 I}{2R}$

Plane sheet: $B = \mu_0 \dfrac{J_s}{2}$

On axis of circular loop: $B = \dfrac{\mu_0 IR^2}{2(R^2 + z^2)^{3/2}}$

SUPPLEMENTARY PROBLEMS

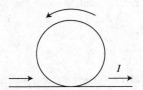

27.10.   What is the magnetic field on the axis of a circular loop of radius *R* carrying current *I* at a distance *z* from the center of the loop?

27.11.   A very long straight wire carrying current *I* has a circular loop in it of radius *R*. What is the magnetic field at the center of the loop?

27.12.   Determine the magnetic field at the center of a square of side 2*a* that carries current *I*.

27.13. Three long straight parallel wires each carry current I in the same direction. They are equidistant from each other with separation a. What force per unit length does one wire experience due to the other two?

27.14. A disk of radius R carries a uniform surface charge density σ. It rotates about its axis with angular velocity ω. What is the magnetic field at the center of the disk?

27.15. A long straight cylindrical conductor of radius R carries current with uniform current density J. Parallel to the axis of the cylinder is a cylindrical void whose axis is parallel to the axis of the conductor. The axis of the void is displaced a distance a from the axis of the conductor. What is the magnetic field within the void?

27.16. What magnetic field is produced by a very long solenoid with 150 turns per meter carrying a current of 20 A?

27.17. A long straight wire carrying current I_1 is placed in the plane of a rectangular loop carrying current I_2. What is the net force on the loop? Is it attracted or repelled by the wire?

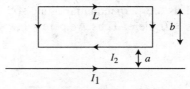

27.18. A long solenoid of radius R and n turns per meter carries current I_0. On the axis of the solenoid is a long straight wire carrying current I. What value of I will result in a magnetic field at the point $r = \frac{1}{2}R$ that is at 45° to the axis of the solenoid?

27.19. Show that the magnetic field at point P on the axis of a solenoid of finite length and n turns per meter and radius R carrying current I is $B = \frac{1}{2}\mu_0 nI(\cos \alpha_2 - \cos \alpha_1)$, where α_1 and α_2 are shown in the drawing.

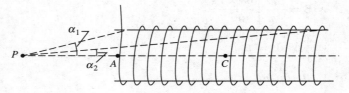

27.20. A long straight wire carries current I. It is bent at 90°. Half of the wire lies along the positive x-axis, and the other half lies along the positive y-axis. What is the magnetic field at the point $(-d, 0)$?

27.21. A wire carrying current I is bent into the shape of a square of side $2a$. Determine the magnetic field at the center of the square. Note that the integral of $(x^2 + a^2)^{-3/2}$ is $(x/a^2)(x^2 + a^2)^{-1/2}$.

SOLUTIONS TO SUPPLEMENTARY PROBLEMS

27.10. $B = \dfrac{\mu_0 I}{4\pi} \displaystyle\int_0^{2\pi} \dfrac{R\,d\phi}{r^2} \cos\theta$ $B = \dfrac{\mu_0 I R^2}{2(R^2 + z^2)^{3/2}}$

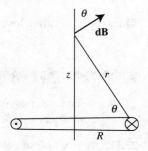

27.11. Superimpose the field of a straight wire, Eq. 27.3, with the field of a loop, Eq. 27.4.

$$B = \frac{\mu_0 I}{2\pi R} + \frac{\mu_0 I}{2R}$$

27.12. The field opposite the midpoint of a straight current segment of length $2a$ is given by the integral of Problem 27.1 with the limits of integration changed to $-a$ to a. There are four such segments in a square, so

$$B = 4\frac{\mu_0 I a}{4\pi}\left(\frac{2a}{a^2\sqrt{a^2 + a^2}}\right) = \frac{\sqrt{2}}{\pi}\frac{\mu_0 I}{a}$$

27.13. $B = 2\dfrac{\mu_0 I}{2\pi a}\cos 30° \qquad F = BIL = \dfrac{\sqrt{3}}{2\pi}\dfrac{\mu_0 I^2}{a}$

27.14. Consider annular rings of width dr. Each is like a current loop with current σdAf, where $dA = 2\pi r\,dr$ and $f = \omega/2\pi$. The field at the center of a loop is $\mu_0 I/2r$, so

$$B = \int_0^R \frac{\mu_0(r\sigma\omega\,dr)}{2r} = \frac{\mu_0\omega\sigma R}{2}$$

27.15. Consider a solid cylinder carrying current density J on top of which is superimposed a cylinder the size and location of the void, but carrying opposite current density. The field inside a cylinder with current flowing in the z direction is given by Eq. 27.7. The field is azimuthal in direction, so I can write it as

$$\mathbf{B}_+ = \frac{\mu_0 J}{2}(\mathbf{k}\times\mathbf{R}_+)$$

for the current in the solid cylinder and

$$\mathbf{B}_- = \frac{\mu_0 J}{2}(\mathbf{R}_-\times\mathbf{k})$$

for the current replacing the void.

$$\mathbf{B} = \mathbf{B}_+ + \mathbf{B}_- = \frac{\mu_0 J}{2}(\mathbf{k}\times\mathbf{R}_+ - \mathbf{R}_-)$$

But

$$\mathbf{a} + \mathbf{R}_- = \mathbf{R}_+, \quad \text{so } \mathbf{B} = \frac{\mu_0 J}{2}(\mathbf{k}\times\mathbf{a})$$

Amazingly, B is constant in the void!

27.16. $B = \mu_0 nI = (4\pi\times 10^{-7}\,\text{N/A}^2)(150/\text{m}^4)(20\,\text{A}) = 3.77\times 10^{-3}\,\text{T}$

27.17. The forces on the ends of the rectangle cancel, so

$$F = B_1 I_2 L - B_2 I_2 L = \frac{\mu_0 I_1 I_2 L}{2\pi}\left(\frac{1}{a} - \frac{1}{a+b}\right)$$

27.18. If the resultant field is at $45°$ to the axis, the field of the wire must have the same magnitude as the field of the solenoid, since they are perpendicular. Thus

$$\frac{\mu_0 I}{2\pi r} = \mu_0 nI_0 \qquad r = \frac{R}{2} \qquad I = \pi RnI_0$$

27.19. Consider the solenoid to be a series of coils, each carrying current $n\,dx\,I$. Integrate the result of Problem 27.10 to find the resultant field.

$$B = \frac{\mu_0 nI}{2}\int_{E_1}^{z_2} \frac{R^2\,dz}{(R^2 + z^2)^{3/2}}$$

Let

$$\sin\alpha = \frac{R}{r} \qquad r = \sqrt{z^2 + R^2}$$

$$\cos\alpha\,d\alpha = -\frac{R}{r^2}\frac{dr}{dz} = -\frac{R}{r^2}z\,dz \qquad \cos\alpha = \frac{z}{r}$$

$$B = \frac{\mu_0 nI}{2}\int_{\alpha_1}^{\alpha_2}(-\sin\alpha\,d\alpha) = \tfrac{1}{2}\mu_0 nI(\cos\alpha_2 - \cos\alpha_1)$$

27.20. The portion of the wire along the positive x-axis contributes nothing to the magnetic field at $(-d, 0)$, since $\mathbf{ds} \times \mathbf{r} = 0$ for $\mathbf{ds}$ along the x-axis.

For a wire extending from $-\infty$ to $+\infty$ along the y-axis, the field at $(-d, 0)$ would be $\frac{\mu_0 I}{2\pi d}$, Eq. 27.3. Here we have one half of such a wire, so

$$B = \frac{\mu_0 I}{4\pi d} \text{ at the point } (-d, 0)$$

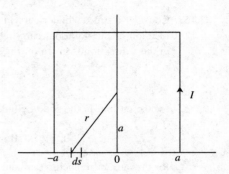

27.21. $B = 4\dfrac{\mu_0 I}{4\pi} \displaystyle\int_{-a}^{a} \dfrac{dx}{r^2} \sin\theta$

$\qquad = 4\dfrac{\mu_0 I}{4\pi} \displaystyle\int_{-a}^{a} \dfrac{dx}{(x^2 + a^2)} \left(\dfrac{a}{r}\right)$

$\qquad = 4\dfrac{\mu_0 I}{4\pi} \displaystyle\int_{-a}^{a} \dfrac{a\,dx}{(x^2 + a^2)^{3/2}}$

$\qquad = 4\dfrac{\mu_0 I a}{4\pi} \dfrac{x}{a(x^2 + a^2)^{1/2}} \Bigg|_{-a}^{a}$

$\qquad = 4\dfrac{\mu_0 I}{4\pi} \left(\dfrac{a}{a\sqrt{2a^2}} + \dfrac{a}{a\sqrt{2a^2}}\right)$

$\qquad = 4\dfrac{\mu_0 I}{4\pi a} \dfrac{2}{\sqrt{2}}$

$B = \sqrt{2}\dfrac{\mu_0 I}{\pi a}$

CHAPTER 28

Electromagnetic Induction and Inductance

In the early 1800s people knew that electric currents could cause magnetic fields, but it was not until Michael Faraday and others did some remarkable experiments that anyone understood how magnetic fields could cause electric currents. Their discoveries changed the course of human civilization profoundly. What we now call *electromagnetic induction* is the source of the electricity we use and is the basis for technology as diverse as electronics, computers, communications, and television.

28.1 Faraday's Law

I identified the electric flux through a surface as the number of E lines passing through the surface. I define **magnetic flux** in the same way. The magnetic flux Φ_B through a surface S is

$$\Phi_B = \int_S \mathbf{B} \cdot \mathbf{dA} \tag{28.1}$$

The magnetic flux through a surface is just the number of magnetic field lines passing through the surface. The flux through a small area dA is thus $d\Phi_B = \mathbf{B} \cdot \mathbf{dA} = B \, dA \cos\theta$. Suppose the surface is bounded by a closed loop. Faraday made the remarkable discovery that **an EMF is induced around a closed loop if the magnetic flux through the loop changes**. This is **Faraday's law**:

$$\mathcal{E} = -\frac{d\Phi_B}{dt} \tag{28.2}$$

$\mathcal{E}$ is measured in volts. If the loop is a conductor, a current will flow, but even for a path through empty space an EMF is created. This means that changing magnetic fields create electric fields in space, the lines of which are sort of like whirlpool flow lines. Previously we studied electrostatic fields, for which electric charge was a source or sink. The induced electric field caused by a changing magnetic flux is a different kind of electric field. It is not conservative, and no potential energy can be defined for it. These induced electric field lines have no beginning or end (that is, no sources or sinks), much like magnetic field lines.

There are different ways of changing the magnetic flux through a conducting loop, all of which cause a current to flow:

1. The loop can be moved from one place to another where the magnetic field has a different strength, thereby changing the flux through the loop.
2. The loop can be rotated, thereby changing the number of B lines passing through it.

3. The shape of the loop can be changed, thereby changing its area.
4. The magnetic field passing through the loop can be changed (perhaps by changing the current in a solenoid that is creating the magnetic field).

All of these mechanisms have important consequences and applications, and I will consider each in the following discussion.

It is important to note the significance of the minus sign in Faraday's law. This is meant to remind us that **any induced current resulting from an induced EMF is in a direction such that the flux due to it will oppose the change in flux that caused the induced EMF**. This is **Lenz's law**. In cases where no conductor is present, just imagine there is one in place in order to envision the direction of the induced EMF. The idea of Lenz's law is a consequence of conservation of energy, since if the induced effects did not oppose the changes that caused them, even an infinitesimal change in flux would grow and grow and produce infinite currents and voltages with no outside energy input, an impossible physical situation. If you try to pull a conductor out of a magnetic field, the forces on the induced current will oppose your pull. If you try to push a conductor into a magnetic field, the induced forces will hold it out. It is like trying to move something in viscous syrup. No matter what you do, Faraday opposes you. My version of Lenz's law is this: **Electromagnetic induction effects are like southern Idaho conservatives. They oppose all change.**

In Fig. 28-1 the left-hand coil carries a current and creates the magnetic field shown. When this coil moves to the right, it increases the flux in the right coil, and in so doing induces a current. This induced current flows as shown in order to create magnetic flux that will oppose the increase in flux due to the left coil. The two coils thus act like opposing bar magnets, as shown at the right. If one moved the left coil away from the other coil, the induced current would flow in the opposite direction.

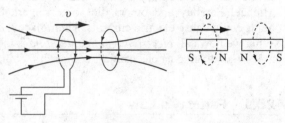

Fig. 28-1

PROBLEM 28.1.   A coil of 200 turns, 3-cm^2 area, and 24 Ω is placed perpendicular to the magnetic field of a large electromagnet. The field is reduced from 0.80 T to 0 in 20 ms. What current flows in the coil?

Solution
$$I = \frac{\mathcal{E}}{R} = \frac{1}{R}\frac{d\Phi_B}{dt} = \frac{NAB}{R\Delta t} = \frac{(200)(3 \times 10^{-4}\ \text{m}^2)(0.8\ \text{T})}{(24\ \Omega)(20 \times 10^{-3}\ \text{s})} = 0.10\ \text{A}$$

Note that the flux through N turns is N times the flux through one turn.

PROBLEM 28.2.   A magnetic field can be measured in the following way: A coil of 250 turns and area 1.80 cm^2 is placed in a permanent magnet so that maximum magnetic flux passes through the coil. It is connected to a galvanometer that measures total charge flow. When the coil is quickly jerked out of the magnet, a charge of 0.25 mC is observed to flow. The coil–galvanometer circuit has a resistance of 4 Ω. What is the magnetic field?

Solution
$$I = \frac{\Delta q}{\Delta t} = \frac{\mathcal{E}}{R} = \frac{\Delta \Phi_B}{R\Delta t} = \frac{NAB}{R\Delta t}, \quad B = \frac{R\Delta q}{NA} = \frac{(4\ \Omega)(0.25 \times 10^{-3}\ \text{C})}{(250)(1.8 \times 10^{-4}\ \text{cm}^2)} = 0.022\ \text{T}$$

PROBLEM 28.3.   A coil of N turns and area A is placed in a magnetic field B and rotated at constant angular velocity ω about a diameter perpendicular to the magnetic field. Derive an expression for the EMF induced in the coil.

Solution
$$\Phi_B = NAB \cos \theta = NAB \cos \omega t \qquad \mathcal{E} = -\frac{d\Phi_B}{dt} = NAB\omega \sin \omega t$$

This is the basis for an alternating current generator. By use of suitable contacts, direct current can also be generated.

PROBLEM 28.4. A rectangular loop of dimensions $a \times b$, resistance R, and mass m is oriented perpendicular to a horizontal uniform magnetic field. It is released from rest and falls so that part of the loop is outside the field, as shown here. What is the maximum terminal speed the loop achieves?

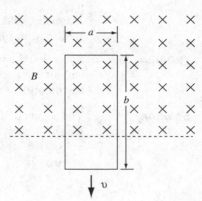

Solution When the magnetic force on the induced current is equal to the weight of the loop, the net force acting will be zero and the loop will have zero acceleration. The outward forces on the two side legs cancel, and the upward force on the top portion is BIa, where

$$I = \frac{\mathcal{E}}{R} = \frac{1}{R}\frac{d\Phi_B}{dt} = \frac{Ba}{R}\frac{dx}{dt} = \frac{Bav}{R}$$

Thus

$$\frac{B^2a^2v}{R} = mg \quad \text{and} \quad v = \frac{mgR}{B^2a^2}$$

Note that the induced current flows clockwise in an attempt to keep the flux within the loop from changing.

PROBLEM 28.5. A disk of resistivity ρ, radius a, and thickness b is placed with its plane perpendicular to a time-varying magnetic field, $B = B_0 \sin \omega t$. Circular currents called **eddy currents** are induced, and energy is dissipated in the disk. Calculate the rate of power dissipation. This thermal energy loss presents a problem in devices such as transformers. The effect finds constructive application in the metal detectors used in airports or for finding buried coins and in induction furnaces.

Solution Consider a narrow annular ring of width dr and radius r. The flux through this ring is $\Phi = \pi r^2 B_0 \sin \omega t$, and the EMF induced in the ring is $\mathcal{E} = -d\Phi/dt = \pi r^2 \omega B_0 \cos \omega t$. The resistance of the ring is $\rho(2\pi/b\,dr)$, so the total power dissipated in the disk is

$$P = \int \frac{\mathcal{E}^2}{R} = \int_0^a \frac{(\pi r^2 \omega B_0 \cos \omega t)^2 b\,dr}{2\pi r\rho} = \frac{1}{8\rho}\pi ba^4 \omega^2 B_0^2 \cos^2 \omega t$$

The power dissipation depends on the fourth power of the radius of the disk, and this dependence is found for other shapes also, so laminating transformers' cores with thin sheets of metal helps reduce power loss significantly. The average of $\cos^2 \omega t$ over one period is $\frac{1}{2}$, so

$$P_{\text{av}} = \frac{1}{16\rho}\pi ba^4 \omega^2 B_0^2$$

A changing magnetic field in space creates whirlpool-like electric fields, and this is true whether or not a conductor is present. Such fields are used in the **betatron** (Fig. 28-2) to accelerate electrons to high speed. An evacuated "doughnut" is placed in the field of a large electromagnet, and the electrons travel in circular orbits. The magnetic field is varied sinusoidally, and for one-fourth of the cycle electrons are accelerated by the induced electric field. Thus the magnetic field serves both to turn the electrons in circular orbits and to accelerate them. In order to accomplish this, the magnetic field must be weaker at the orbit than in the interior of the circular path, as indicated in Fig. 28-2.

PROBLEM 28.6.   Show that the magnetic field at the orbit in a betatron must be half the average magnetic field over the area enclosed by the orbit.

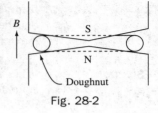

Fig. 28-2

Solution   The magnetic force provides the needed centripetal force, so $qvB = mv^2/r$. The induced EMF is $\mathcal{E} = \int \mathbf{E} \cdot \mathbf{ds} = 2\pi r E = d\Phi_B/dt$. By Newton's second law, the force on the electron is

$$F = \frac{d(mv)}{dt} = qE = \frac{1}{2\pi r}\frac{d\Phi_B}{dt}$$

Since $v = 0$ at $t = 0$,

$$mv = \frac{1}{2\pi r}\Phi_B = qrB$$

But $\Phi_B = \pi r^2 B_{av}$, so $B = \frac{1}{2}B_{av}$.

28.2   Motional EMF

An interesting example of an induced EMF occurs when a conductor moves through a magnetic field. In Fig. 28-3 a metal bar of length l is oriented perpendicular to a uniform magnetic field and moved at constant velocity perpendicular to its length and to the field. A positive charge carrier will experience a force $q\mathbf{v} \times \mathbf{B}$ directed toward the top of the bar. Positive charge will build up at the top end of the bar and set up a back electric field E, such that in the steady state $qE = qvB$. This means an EMF will develop between the ends of the bar, $\mathcal{E} = \int E \, dy = El = Bv$. We could also arrive at this result using Faraday's law. Consider an imaginary rectangular "loop," indicated by the dashed line. The bar forms one side of the loop. In time dt the bar moves a distance dt to the right, increasing the area of the loop by $v \, dt \, l$. This increases the flux through the loop by $d\Phi_B = Bvl \, dt$, so $\mathcal{E} = d\Phi_B/dt = Bvl$. This kind of EMF is called a **motional EMF**.

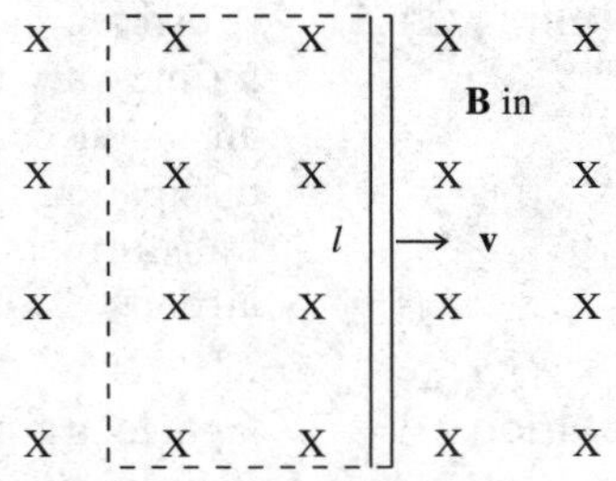

Fig. 28-3

$$\boxed{\mathcal{E} = Bvl \quad \text{motional EMF}} \tag{28.3}$$

PROBLEM 28.7.   It has been suggested that birds might use the EMF induced between their wing tips by the Earth's magnetic field as a means of helping them navigate during migration. What EMF would be induced for a Canada goose with a wingspread of 1.5 m flying 10 m/s in a region where the vertical component of the Earth's field is 2×10^{-5} T?

Solution $\mathcal{E} = Bvl = (2 \times 10^{-5}\,\text{T})(10\,\text{m/s})(1.5\,\text{m}) = 3 \times 10^{-4}\,\text{V} = 0.3\,\text{mV}$. This is probably too small for the birds to detect, since cellular voltages are typically 70 mV. However, for a 747 jet liner with a wingspan of 60 m and a speed of 900 km/h, the effect is appreciable.

PROBLEM 28.8. Faraday invented an ingenious device called a **homopolar generator**, or **Faraday disk** (Fig. 28-4). A copper disk of radius r is mounted with its axis parallel to a uniform magnetic field B. The disk rotates at angular frequency ω. Electrical contact with conducting brushes is made at points A and C, on the perimeter and axis of the disk. What EMF is generated between terminals A and C?

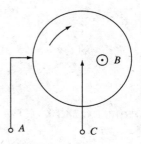

Solution Imagine a closed loop consisting of the connections shown, plus a line segment connecting A and C, plus

Fig. 28-4

a radial piece of the disk from the center to a point on the circumference, plus an arc along the circumference. In time dt this radial piece sweeps out a small triangle of area $\frac{1}{2}r(r\,d\theta) = \frac{1}{2}r^2\omega\,dt$. This increases the flux through the loop by $d\Phi_B = B\,dA = \frac{1}{2}Br^2\omega\,dt$. Thus the EMF induced between A and C is

$$\mathcal{E} = \frac{d\Phi_B}{dt} = \frac{1}{2}Br^2\omega$$

28.3 Inductance

Suppose you close a switch and cause current to flow through a circuit that includes a battery or other voltage source. This current will create a magnetic field enclosed by the circuit, so this means magnetic flux has been created through the circuit. But according to Faraday's law, this increase in flux will induce an EMF that tries to oppose the buildup of flux; that is, the induced EMF (called a **back EMF**) will oppose the voltage due to the battery. This occurs in all circuits, but the effect is very strong for circuits that contain a coil or solenoid with many turns (for example, motors, transformers, generators, and so on). If current I creates magnetic flux Φ_B in a circuit element, we define the **self-inductance L**, in **henries**, of the circuit element to be

$$L = \frac{\Phi_B}{I} \quad \text{self-inductance} \tag{28.4}$$

Circuit elements with significant self-inductance are called **inductors**, and they find many important applications. Since $\Phi_B = LI$, the induced back EMF $\mathcal{E} = -(d\Phi_B/dt)$ is

$$\mathcal{E} = -L\frac{dI}{dt} \quad \text{back EMF} \tag{28.5}$$

PROBLEM 28.9. What is the self-inductance of a long solenoid of radius a with n turns per meter?

Solution The field inside a solenoid is $\mu_0 nI$, so the flux through one turn is $\pi a^2 B = \pi r^2 \mu_0 nI$. In 1 m there are n turns, so the flux through these n turns is $\Phi_B = n(\pi r^2 \mu_0 nI)$ and $L = \Phi_B/I = \pi\mu_0 n^2 a^2$.

It can be shown that two inductors L_1 and L_2 connected in series are equivalent to a single inductor $L = L_1 + L_2$ and that two in parallel are equivalent to $L = L_1 L_2/(L_1 + L_2)$ (inductors combine as resistors do).

Suppose coils in two different circuits are near each other. Then current I_1 in coil 1 can create flux Φ_{B21} in coil 2 and induce an EMF in it. I define the *mutual inductance* of coil 2 with respect to coil 1 as

$$M_{21} = \frac{\Phi_{B21}}{I_1}$$

Thus $\Phi_{B21} = M_{21}I_1$, and the EMF induced in coil 2 is

$$\mathcal{E}_2 = \frac{d\Phi_{B21}}{dt} = -M_{21}\frac{dI_1}{dt}$$

It is possible to show that $M_{21} = M_{12} = M$, so

$$\mathcal{E}_2 = -M\frac{dI_1}{dt} \quad \text{and} \quad \mathcal{E}_1 = -M\frac{dI_2}{dt} \qquad (28.6)$$

M and L are both measured in henries.

PROBLEM 28.10. Two circular coils of radii a and b ($a \gg b$) are concentric and lie in the same plane. The larger coil has N_1 turns, and the smaller one has N_2 turns. What is their mutual inductance?

Solution The magnetic field at the center of the large coil is $B_1 = \mu_0 N_1 I_1/2a$. Thus the flux through the small coil is approximately $\Phi_{B21} = N_2 \pi b^2 B_1$, so $M = \Phi_{B21}/I_1 = \mu_0 N_1 N_2 \pi (b^2/2a)$.

28.4 Energy Storage in a Magnetic Field

Consider the circuit in Fig. 28-5. When the switch is closed, the battery tries to push current through the resistor and the inductor. The induced back EMF in the inductor acts like another battery connected with opposite polarity to the actual battery. Thus $\mathcal{E}_0 - \mathcal{E}_{\text{back}} = IR$ or

$$\mathcal{E}_0 - L\frac{dI}{dt} = IR \qquad (28.7)$$

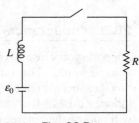

Rearrange this equation and multiply by I, $\mathcal{E}_0 I = I^2 R + LI(dI/dt)$. I interpret this equation as follows: $\mathcal{E}_0 I$ is the rate at which the battery is doing work. $I^2 R$ is the rate at which heat is being generated in the resistor. $LI(dI/dt)$ is the rate at which magnetic energy is being stored in the inductor. The last expression can be integrated to yield the stored magnetic energy.

Fig. 28-5

$$U_B = \int_0^t LI\frac{dI}{dt}\, dt = \int_0^I LI\, dI$$

or

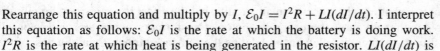

$$U_B = \frac{1}{2}LI^2 \quad \text{stored magnetic energy} \qquad (28.8)$$

This is analogous to the electric energy stored in a capacitor:

$$U_E = \frac{1}{2}\frac{Q^2}{C}$$

PROBLEM 28.11. Determine the energy stored per unit length and the energy stored per unit volume in a very long solenoid of radius a with n turns per meter carrying current I?

Solution From Problem 28.9, $L = \mu_0 \pi n^2 a^2$. Thus $U_B = \frac{1}{2}\mu_0 \pi n^2 a^2 I^2$. The volume per unit length is πa^2, so the *magnetic energy density* u_B is $u_B = \frac{1}{2}\mu_0 n^2 I^2$ (in joules per cubic meter). In terms of the magnetic field $B = \mu_0 nI$, this may be written

$$u_B = \frac{1}{2}\frac{B^2}{\mu_0} \quad \text{magnetic energy density} \qquad (28.9)$$

Equation 28.9 proves to be valid for all magnetic fields, not just that of a solenoid. Energy is stored in a magnetic field analogous to the way it is stored in an electric field with density $\frac{1}{2}\epsilon_0 E^2$.

PROBLEM 28.12.   How much energy is stored in an air core solenoid of length 10 cm and diameter 1.2 cm if it has 200 turns and carries 1.20 A?

Solution   Approximate the solenoid as being long, so $L = \mu_0 \pi n^2 a^2$ (see Problem 28.9).

$$U_B = \frac{1}{2}\mu_0 \pi n^2 a^2 I^2 = (0.5)(4\pi \times 10^{-7}\,\text{N/A}^2)(\pi)\left(\frac{200}{0.10\,\text{m}}\right)^2 (0.006\,\text{m})^2 (1.20\,\text{A})^2$$

$$= 4.1 \times 10^{-4}\,\text{J}$$

28.5   Magnetic Materials

In some materials, such as iron, the atoms behave like little current loops or magnetic dipoles. They act something like little bar magnets, and sometimes they interact so that they spontaneously align themselves and give rise to a "magnetized" material. Such substances are said to be **ferromagnetic**, and they are used to make the permanent magnets with which you are familiar. If a ferromagnet is heated above a certain critical temperature (the Curie temperature), thermal agitation causes the little magnets to become disordered, and the material is no longer ferromagnetic. When the magnets no longer align spontaneously, they are said to be **paramagnetic**. Paramagnetism, as well as ferromagnetism, proves to be of great practical importance. For example, the field in an air core solenoid due to current in the windings is typically very weak. However, when the solenoid is filled with paramagnetic iron, the weak field due to the current is sufficient to cause the iron dipoles to become mostly aligned, and they in turn produce a very large magnetic field. Consequently, in devices such as electromagnets or inductors, iron or some other "permeable" material is inserted to enhance the magnetic effects. A measure of the degree of magnetization obtained is the **relative permeability** μ_r. In calculations of magnetic effects, for example, inductance or resulting magnetic field, it is usually sufficient to replace μ_0 in our previous equations with $\mu = \mu_r \mu_0$. μ_r is dimensionless.

28.6   RLC Circuits

Consider a circuit in which a resistor R and an inductor L are connected in series to a battery (Fig. 28-5). When the switch is closed, the battery tries to push current through the circuit, but the large back EMF $-L(dI/dt)$ opposes the battery, so at first not much current flows. As the rate of change of the current decreases, the back EMF also decreases, until finally there is no back EMF and a steady current $I = \mathcal{E}_0/R$ flows. The circuit is described by $\mathcal{E}_0 - L(dI/dt) = IR$. The solution is

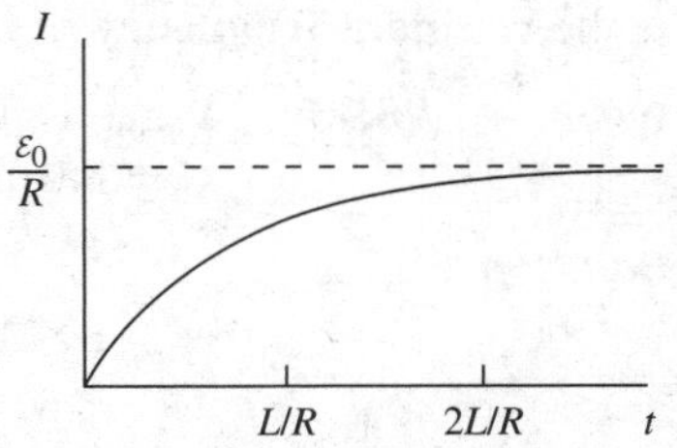

$$\boxed{I(t) = \frac{\mathcal{E}_0}{R}(1 - e^{-t/\tau}) \quad RLC \text{ circuit}} \qquad (28.10)$$

Fig. 28-6 shows $I(t)$. The *time constant* of the circuit is τ.

$$\boxed{\tau = \frac{L}{R}} \qquad (28.11)$$

Once a current is flowing through an inductor, it is hard to stop it because if you try to reduce the flux in an inductor, a back EMF will try to keep the current flowing. In Fig. 28-7 a steady current flows through the resistor and inductor after a long time with the battery connected. If now the switch is thrown downward, the induced EMF in the inductor will cause current to continue flowing through

Fig. 28-6

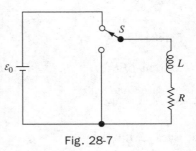

Fig. 28-7

R for a little while. In this case $-L(dI/dt) = IR$, so

$$I(t) = \frac{\mathcal{E}_0}{R} e^{-t/\tau} \qquad (28.12)$$

Incidentally, if after a long time you tried suddenly to open the switch in Fig. 28-5, a large voltage could be induced, and this could cause a spark across the contacts of the switch. In circuits with huge inductances (for example, with big motors or electromagnets), you simply cannot break the circuit quickly. A huge voltage can be generated, and this will spark across the switch. One has to be very careful working with such equipment because if a wire breaks, a very high voltage may be generated and electrocute you. In a car, voltages as high as 50,000 V are generated for the spark plugs by breaking a circuit carrying current through an ignition coil.

Interesting effects occur in circuits containing both inductance and capacitance. In Fig. 28-8 the capacitor initially carries charge Q_0. When the switch is closed, the circuit is described by

$$-L\frac{dI}{dt} = \frac{Q}{C} \qquad \text{or} \qquad L\frac{d^2Q}{dt^2} + \frac{Q}{C} = 0 \qquad (28.13)$$

where I used $I = dQ/dt$. Thus, writing $I_0 = \omega Q_0$,

Fig. 28-8

$$\boxed{Q(t) = Q_0 \cos \omega t \quad \text{and} \quad I(t) = \omega Q_0 \sin \omega t, \quad \text{where } \omega = \frac{1}{\sqrt{LC}}} \qquad (28.14)$$

The current oscillates periodically, and the stored energy is

$$U = \frac{1}{2}\frac{Q^2}{C} + \frac{1}{2}LI^2 = \text{constant}$$

Thus

$$f = \frac{\omega}{2\pi}$$

is the **resonant frequency** of the circuit.

PROBLEM 28.13.   What is the resonant frequency of the circuit of Fig. 28-8 if $L = 2.60\,\text{mH}$ and $C = 8.0\,\text{pF}$?

Solution

$$f = \frac{1}{2\pi}\frac{1}{\sqrt{LC}} = (2\pi)^{-1}(2.6 \times 10^{-3}\,\text{H})^{-1/2}(8 \times 10^{-12}\,\text{F})^{-1/2} = 1.1\,\text{MHz}$$

28.7 Summary of Key Equations

Magnetic flux: $\Phi_B = \displaystyle\int_S \mathbf{B} \cdot \mathbf{dA}$

Faraday's law: $\mathcal{E} = -\dfrac{d\Phi_B}{dt}$

Motional EMF: $\mathcal{E} = Blv$

Self-inductance: $L = \dfrac{\Phi_B}{I}$

Back EMF: $\mathcal{E} = -L\dfrac{dI}{dt}$

Stored magnetic energy: $U = \dfrac{1}{2}LI^2$

Magnetic energy density: $\quad u = \dfrac{1}{2\mu_0}B^2$

RL circuit: $\qquad\qquad\qquad I = \dfrac{\mathcal{E}_0}{R}(1 - e^{-t/\tau})$ current buildup

$\qquad\qquad\qquad\qquad I = I_0 e^{-t/\tau}$ current decay

Inductive time constant: $\quad \tau = \dfrac{L}{R}$

LC resonant frequency: $\quad \omega = \dfrac{1}{\sqrt{LC}}$

SUPPLEMENTARY PROBLEMS

28.14. An airplane propeller 2.0 m from tip to tip rotates at 18,000 rev/min. If the plane is flying due north at a point where the horizontal component of the Earth's magnetic field is 1.2×10^{-5} T, what voltage is generated between the tips of the propeller?

28.15. Two long parallel conducting rails are separated by a distance d. At one end they are joined by a resistance R. A conducting bar of length d is made to slide at constant velocity v along the rails. Both the bar and the rails have negligible resistance. (a) What current flows in the circuit? (b) How much power is required to move the bar? (c) How does the power dissipated in the resistance compare with the power required to move the bar?

28.16. A wire of negligible resistance is bent into a rectangle of sides a and b. A uniform magnetic field is present perpendicular to the plane of the loop. A bar of resistance R is placed between points X and Y on the loop and moved to the right at constant speed v. What current flows in the bar? In what direction does current flow?

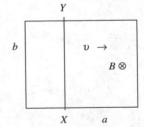

28.17. A long, thin solenoid has 750 turns per meter, and its current is increased at the rate of 60 A/s. What is the induced electric field within the solenoid at a point 5 mm from the axis and midway between the ends?

28.18. Determine the self-inductance of a toroid with N turns and of square cross-section $a \times a$ and inner radius R.

28.19. Determine the self-inductance per unit length of a long straight coaxial cable of inner radius a and outer radius b.

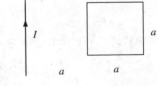

28.20. A long straight wire is positioned in the plane of a conducting square loop of side a and resistance R. The wire is parallel to the closest side of the square and distance a from it. What is the average power dissipated in the loop when the current in the wire is $I = I_0 \sin \omega t$?

28.21. It has been suggested that surplus power from a generating plant might be stored in the magnetic field of a giant toroid. If the magnetic field in the toroid were 12 T (very large), what volume would be needed to store 10^6 kWh of energy? If the toroid were doughnut-shaped, with inner radius R and outer radius $2R$, about how large would R have to be?

28.22. The axis of a coil of 250 turns and area 0.002 m² is aligned at 40° to a uniform magnetic field. The magnetic field is decreased from 0.08 to 0.02 T in 0.020 s. What is the average EMF generated in the coil?

28.23. A coil of 10 turns is wrapped tightly around a long solenoid of radius 2 cm and 200 turns per meter. What is the mutual inductance of the system?

28.24. An amplitude modulation (AM) radio receiver uses an *LC* resonant circuit whose frequency matches the frequency of incoming radio waves. If an inductance of $16\,\mu H$ is used, what range of variable capacitance is required in order to tune over the AM band from 500 to 1700 kHz?

28.25. A sensitive detection instrument in a space vehicle has a resistance of $180\,\Omega$ and is designed to operate with a current of 42 mA. However, it is necessary that the current rise to no more than 10 percent of this operating value within the first $50\,\mu s$ after the voltage is applied. To achieve this, an inductor is connected in series with the device. (a) What voltage is required? (b) What minimum inductance is required? (c) What is the time constant of the circuit?

28.26. A coil of 20 turns and radius 2 cm is wrapped around a long solenoid of radius 0.8 cm and 1200 turns/m. The solenoid is connected to an AC generator, resulting in a current of $6.00 \sin 360\,t$ in the solenoid. What is the induced EMF in the coil?

28.27. A rectangular loop of wire of mass *m*, resistance *R* and side *a* falls with two of its sides horizontal. It enters a region where a uniform horizontal field *B* is present. The plane of the loop is perpendicular to the magnetic field. While the upper edge of the loop is still outside the magnetic field, the flux through the loop is changing and an induced EMF causes a current to flow in the loop. The magnetic force on this current causes a "drag" force on the loop, similar to that encountered by an object falling through a viscous fluid. As a result, the loop reaches a constant terminal speed. (You can observe this impressive effect by dropping an aluminum pie tin through the field of a large magnet.) Calculate the terminal speed of the loop.

SOLUTIONS TO SUPPLEMENTARY PROBLEMS

28.14. Each half of the propeller sweeps out area πr^2 each revolution, so $\mathcal{E} = 2B\pi r^2 f$:

$$\mathcal{E} = 2(1.20 \times 10^{-5}\,\text{T})(\pi)(1.0\,\text{m})^2 \left(\frac{18,000}{60}/\text{s}\right) = 0.003\,\text{V}$$

28.15. $\mathcal{E} = Bvd$:

(a) $I = \dfrac{\mathcal{E}}{R} = \dfrac{Bvd}{R}$,

(b) $P_1 = Fv = (BId)v = \dfrac{B^2v^2d^2}{R}$, and

(c) $P_2 = I^2R = \left(\dfrac{Bvd}{R}\right)^2 = P_1$

28.16. Current flows downward in the bar. Both loops contribute induced EMF, so

$$I = \frac{2Bbv}{R}$$

28.17. Consider the flux through a circle of radius *r*:

$$\mathcal{E} = \frac{d\Phi_B}{dt} = \pi r^2 \frac{dB}{dt} = \pi r^2 \mu_0 n \frac{dI}{dt} = 2\pi r E$$

$$E = \frac{1}{2}\mu_0 rn \frac{dI}{dt} = (0.5)(4\pi \times 10^{-7}\,\text{N/A}^2)(5 \times 10^{-3}\,\text{m})(750/\text{m})(60\,\text{A/s})$$

$$= 1.4 \times 10^{-3}\,\text{V/m}$$

28.18. From Ampere's law, $2\pi rB = \mu_0 NI$:

$$\Phi_B = N \int B\,dA = N \int_{R}^{R+a} \left(\frac{\mu_0 NI}{2\pi r}\right)(a\,dr) = \frac{\mu_0 N^2 Ia}{2\pi}\left[\frac{1}{R^2} - \frac{1}{(R+a)^2}\right]$$

$$L = \frac{\Phi_B}{I} = \frac{\mu_0 N^2 a}{2\pi}\left[\frac{1}{R^2} - \frac{1}{(R+a)^2}\right]$$

28.19. From Ampere's law: $2\pi rB = \mu_0 I$. For the unit length:

$$L = \frac{\Phi_B}{I} = \frac{1}{I}\int_a^b \left(\frac{\mu_0 I}{2\pi r}\right)(dr) = \frac{\mu_0}{2\pi}\ln\left(\frac{b}{a}\right)$$

28.20. The field of the wire is

$$B = \frac{\mu_0 I}{2\pi r}$$

The flux in the loop is

$$\Phi_B = \int_a^{2a}\left(\frac{\mu_0 I}{2\pi r}\right)a\,dr = \frac{\mu_0 Ir}{2\pi}\ln 2$$

The EMF in the loop is

$$\mathcal{E} = \frac{d\Phi_B}{dt} = \frac{\mu_0 a\,\ln 2}{2\pi}\frac{dI}{dt}$$

The power in the loop is

$$P = \frac{\mathcal{E}^2}{R} = \frac{1}{R}\left(\frac{\mu_0 a\,\ln 2}{2\pi}\right)^2 (I_0\omega\cos\omega t)^2$$

$$(\cos^2\omega t)_{ave} = \frac{1}{T}\int_0^T\cos^2\omega t\,dt,\quad T = \frac{2\pi}{\omega},\quad \cos^2\omega t = \frac{1-\cos^2\omega t}{2}$$

so

$$(\cos^2\omega t)_{ave} = \tfrac{1}{2}$$

$$P_{ave} = \frac{\mu_0 a\,\ln 2(I_0^2\omega^2)}{8\pi^2 R}$$

$$= \frac{1}{2R}\left[\frac{(\mu_0 a\,\ln 2)(I_0\omega)}{2\pi}\right]^2$$

28.21. $U = \left(\dfrac{B^2}{2\mu_0}\right)(\text{volume}) \simeq \left(\dfrac{B^2}{2\mu_0}\right)(2\pi)\left(\dfrac{3}{2}R\right)\left(\dfrac{\pi R^2}{4}\right)\quad U \simeq \dfrac{3}{8}\pi^2 R^3 B^2$

$R = \left(\dfrac{8U}{3\pi^2 B^2}\right)^{1/3} \simeq \left[\dfrac{8\times 10^6\times 3.6\times 10^6\,\text{J}}{3\pi^2(12\,\text{T})^2}\right]^{1/3} = 1890\,\text{m}\quad$ Too big!

28.22. $\mathcal{E} = \dfrac{\Delta\Phi_B}{\Delta t} = \dfrac{NA\,\Delta B}{\Delta t}\cos\theta = \dfrac{(250)(0.002\,\text{m}^2)(0.06\,\text{T}(\cos 40°)}{0.02\,\text{s}}\quad \mathcal{E} = 1.15\,\text{V}$

28.23. $M = \dfrac{\Phi_B}{I_1} = \dfrac{N\mu_0 nI_1\pi r^2}{I_1} = (10)(4\pi\times 10^{-7}\,\text{N/A}^2)(200/\text{m})(\pi)(0.02\,\text{m})^2$

$\qquad = 3.2\times 10^{-6}\,\text{H}$

28.24. $f = \dfrac{1}{2\pi\sqrt{LC}}$

so C varies from $5.5\times 10^{-10}\,\text{F}$ to $6.3\times 10^{-9}\,\text{F}$.

28.25. (a) $\mathcal{E} = IR = (42 \times 10^{-3}\,\text{A})(180\,\Omega) = 7.56\,\text{V}.$

(b) $I = I_0(1 - e^{-t/\tau}) = 0.1 I_0$ and $e^{-t/\tau} = 0.9.$

$$\ln\left(-\frac{t}{\tau}\right) = \ln(0.9), \quad \tau = 2.46t = \frac{L}{R} = 2.46t = 2.46(50 \times 10^{-6}\,\text{s})$$

$$L = 0.022\,\text{H}$$

(c) $\tau = \dfrac{L}{R} = 0.122\,\text{ms}$

28.26. $\varepsilon = \dfrac{\Delta\phi}{\Delta t} = \mu_0 n \dfrac{dI}{dt} NA, \quad I = I_0 \sin(\omega t), \quad \dfrac{dI}{dt} = \omega I_0 \cos(\omega t)$

$= \mu_0 n I_0 \omega \cos(\omega t) N \pi r^2$

$\varepsilon = \left(4\pi \times 10^{-7}\,\dfrac{\text{T}\cdot\text{m}}{\text{A}}\right)(1200/\text{m})(6\,\text{A})[360 \cos(360t)\,\text{A}](20\pi)(0.008\,\text{m})^2$

$\varepsilon = 0.013 \cos(360t)\,\text{V}$

28.27. When the magnetic force balances the downward gravity force, the net force is zero and the acceleration is zero, resulting in a constant velocity.

$$F = BIa = mg$$

$$I = \frac{\varepsilon}{R} = \frac{1}{R}B\frac{dA}{dt}, \quad dA = a\,dy$$

$$= \frac{1}{R}Ba\frac{dy}{dt} = \frac{1}{R}Bav$$

Thus,

$$B\left(\frac{Bav}{R}\right)a = mg$$

$$v = \frac{mgR}{B^2 a^2}$$

Alternating Current Circuits

Currents that vary in time, particularly those that vary sinusoidally, have very many useful applications. Most important, they make it possible to use induced voltages and currents based on Faraday's law. I refer to quantities that vary sinusoidally as *alternating current* (ac) even though this is not quite grammatically correct.

29.1 Transformers

Consider an **ideal transformer** consisting of two coils, one with N_1 turns (the primary) and one with N_2 turns (the secondary). Assume all of the magnetic flux from one coil goes through the other. This can be accomplished by wrapping one coil on top of the other or, as is done more commonly, by wrapping both coils on an iron core (Fig. 29-1). In the latter case, the magnetic field lines tend to stay completely within the iron. The lines between the coils indicate that the transformer is filled with iron.

Suppose an ac voltage V_1 is applied to the primary. Since I assume no resistance in the coils, the applied voltage will equal the induced back EMF. If the flux through one coil is Φ_B, then

$$V_1 = -N_1 \frac{d\Phi_B}{dt}$$

Similarly, the voltage across the secondary coil will be

$$V_2 = -N_2 \frac{d_B}{dt}$$

Divide these equations and obtain

$$\boxed{\frac{V_1}{V_2} = \frac{N_1}{N_2}} \quad \text{ideal transformer} \tag{29.1}$$

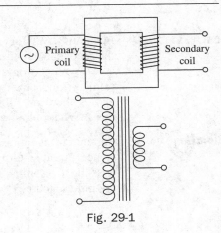

Fig. 29-1

If the transformer is ideal and has no losses, then the power into the primary is equal to the power out of the secondary, so $P = I_1 V_1 = I_2 V_2$. Thus

$$\boxed{\frac{I_1}{I_2} = \frac{N_2}{N_1}} \quad \text{ideal transformer} \tag{29.2}$$

A transformer may be used to step up or step down a voltage or a current.

PROBLEM 29.1.   A transformer used to operate a neon sign steps 120 Vac up to 6000 V. The primary has 150 turns. How many turns does the secondary have?

Solution

$$N_2 = \frac{V_2}{V_1} N_1 = \frac{6000\,\text{V}}{120\,\text{V}}(150) = 7500 \text{ turns}$$

PROBLEM 29.2   A filament transformer is required to provide 32 A to a 0.2-Ω load. The primary is operated off a 120-Vac line. What turns ratio is required for the transformer?

Solution   The secondary voltage required is $V_2 = I_2 R = (32\,\text{A})(0.2\,\Omega) = 6.4\,\text{V}$. Thus

$$\frac{N_2}{N_1} = \frac{V_2}{V_1} = \frac{6.4\,\text{V}}{120\,\text{V}} = 0.053$$

PROBLEM 29.3.   In an **autotransformer** a single coil is used for both the primary and the secondary. In the coil shown here there are three leads attached to a coil wrapped on an iron core. Any two leads may be used as the primary, and any two may be used as the secondary. Between A and B there are 300 turns, and between B and C there are 900 turns. What output voltages are possible for a primary input voltage of 120 Vac?

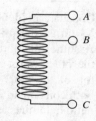

Solution   Possible turns ratios are 300 : 900, 300 : 1200, 900 : 1200, and the reciprocals of these. Hence possible turns ratios are 1 : 3, 1 : 4, 3 : 4, 3 : 1, 4 : 1, and 4 : 3. 1 : 1 is also possible (this might be done to provide dc isolation between two circuits). Hence possible secondary voltages are 30, 40, 90, 120, 160, 360, and 480 Vac.

29.2   Single Elements in ac Circuits

In Fig. 29-2a an ac generator applies a voltage $v = V_{max} \sin \omega t$ to a resistor, resulting in current

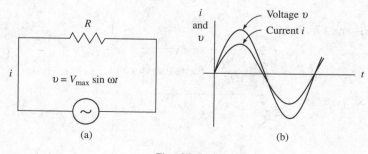

Fig. 29-2

$$i = \frac{1}{R} V_{max} \sin \omega t \qquad (29.3)$$

As shown in Fig. 29-2b, current and voltage are in phase in a resistor (that is, they are both a maximum at the same time).

Now consider the circuit of Fig. 29-3. Here an ac voltage is applied to a capacitor, so $v = Q/C$:

$$i = \frac{dQ}{dt} = C\frac{dv}{dt} = \omega C V_{max} \cos \omega t$$

$$i = \frac{1}{X_C} V_{max} \cos \omega t, \quad \text{where } X_C = \frac{1}{\omega C} \qquad (29.4)$$

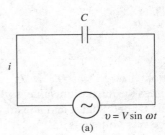

X_C is the **capacitive reactance**, in ohms. It relates current and voltage for a capacitor, just as resistance R does for a resistor. Note that no charge moves between the plates of a capacitor, so no actual conduction current flows. However, it is useful to imagine a kind of ac current, called *displacement current*, flowing between the plates. The displacement current between the plates is equal to the conduction current flowing onto the plates. This idea lets us envision the current as being continuous throughout an ac circuit, just as for dc circuits. Observe that the voltage across the capacitor varies as $\sin \omega t$, whereas the current flowing to the capacitor varies as $\cos \omega t$. I say the **current leads the voltage by 90° in a capacitor**. This is shown in Fig. 29-3b. The peak of the current occurs at an earlier time than the nearest peak of the voltage. (Be careful. This drawing is tricky. Since the peak of the voltage is to the right of the peak of the current, you might think it is the one leading. Not so. The voltage peak occurs at a later time than the current peak.)

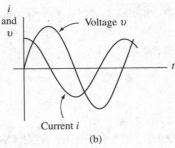

Fig. 29-3

As a final case, consider the circuit of Fig. 29-4. Here an ac voltage is applied to an inductor, and $v - L(di/dt) = 0$:

$$\frac{di}{dt} = \frac{1}{L} V_{max} \sin \omega t$$

Integrate and obtain

$$i = -\frac{V_{max}}{\omega L} \cos \omega t$$

or

$$i = -\frac{V_{max}}{X_L} \cos \omega t, \quad \text{where } X_L = \omega L \qquad (29.5)$$

X_L is the *inductive reactance* in ohms. As shown in Fig. 29-4b, the *current in an inductor lags the applied voltage by 90°*. Here is a mnemonic to help you remember these important phase relationships:

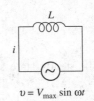

$$\mathcal{E} \text{L I} \quad \text{the} \quad \text{I C} \, \mathcal{E} \quad \text{man}$$

This means EMF ($\mathcal{E}$) leads current (I) by 90° in an inductor (L), and current (I) leads EMF ($\mathcal{E}$) by 90° in a capacitor (C). You should note that **no power is dissipated in an ideal capacitor or inductor. Power is dissipated only in resistors**. If current $i = I_{max} \sin \omega t$ flows through a resistance R, the instantaneous

Fig. 29-4

power dissipation is $P(t) = i^2R = I_{max}^2 R \sin^2 \omega t$. Of more interest is the average power dissipated over the period of one cycle, $T = 2\pi/\omega$.

$$P_{av} = \frac{1}{T} \int_0^T I_{max}^2 R \sin^2 \omega t \, dt$$

$$= \frac{I_{max}^2 R}{T} \int_0^T \left(\frac{1 - \cos 2\omega t}{2}\right) dt = \frac{I_{max}^2 R}{T} \left(\frac{1}{2}t - \frac{\sin 2\omega t}{4\omega}\right)\Bigg|_0^T = \frac{1}{2}I_{max}^2 R$$

I define the *root mean square* (rms) value of the current as

$$I_{rms} = \frac{1}{\sqrt{2}} I_{max} = 0.707 I_{max}$$

Thus

$$\boxed{P_{av} = I_{rms}^2 R} \tag{29.6}$$

Note that $I_{rms}^2 = (i^2)_{av} = \frac{1}{2}I_{max}^2$. The rms value of any quantity that varies as $\sin \omega t$ or $\cos \omega t$ is always $1/\sqrt{2}$ times the maximum value, so $V_{rms} = 0.707 V_{max}$. Domestic electricity is provided at a frequency of 60 Hz in the United States, and a residential outlet provides 120 Vac. This means the voltage is alternating (ac) and the rms voltage is 120 V.

PROBLEM 29.4.   At a frequency of 100 kHz, what is the reactance of a 2-μF capacitor? Of a 6-mH inductor?

Solution

$$X_C = \frac{1}{\omega C} = \frac{1}{2\pi(10^5 \text{ Hz})(2 \times 10^{-6} \text{ F})} = 0.80 \, \Omega$$

$$X_L = \omega L = 2\pi(10^5 \text{ Hz})(6 \times 10^{-3} \text{ H}) = 3770 \, \Omega$$

29.3   The Series *RLC* Circuit and Phasors

In Fig. 29-5 is a series *RLC* combination to which is applied a voltage $v = V \sin \omega t$ (for brevity, I dropped the subscript "max" on the amplitude of the voltage). The circuit equation is

$$v - v_L = v_R + v_C$$

or

$$V \sin \omega t = L\frac{di}{dt} + Ri + \frac{q}{C} \tag{29.7}$$

The steady-state solution for $i(t)$ is

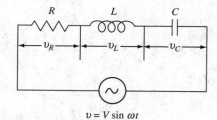

$v = V \sin \omega t$

Fig. 29-5

$$\boxed{i(t) = \frac{V}{\sqrt{R^2 + (X_L - X_C)^2}} \sin(\omega t - \phi)} \tag{29.8}$$

The **phase angle** ϕ is given by

$$\boxed{\tan \phi = \frac{X_L - X_C}{R}} \tag{29.9}$$

ϕ is the phase difference between the voltage applied to the circuit and the resulting current. It can have any value between $\pi/2$ and $-\pi/2$ rad (90° to −90°).

The **impedance Z** of the circuit, in ohms, is defined as

$$Z = \sqrt{R^2 + (X_L - X_C)^2} \qquad (29.10)$$

The amplitude of the current I is related to the amplitude of the applied voltage V by

$$V = IZ \qquad (29.11)$$

This is analogous to $V = IR$ in dc circuits, and Z is analogous to resistance R.

We can avoid the complications of having to deal with confusing trig functions with different phases by means of an ingenious representation using **phasors**. A phasor is a rotating vector used to help us picture what is going on in a circuit. Observe that the current is the same in every element of the circuit. I will use the current phasor I as a reference for measuring the phase of the various voltage drops in the circuit. In Fig. 29-6a start by drawing an arrow representing the current phasor so that it is pointing to the right. This phasor acts as a vector. $\mathbf{V}_R$ is in phase with the current, so I draw it as a phasor pointing in the same direction as **I**. In an inductor, the current lags the voltage in phase by 90°, so I draw the phasor $\mathbf{V}_L$ 90° ahead of the current phasor **I**. In a capacitor the current leads the voltage by 90°, so the phasor $\mathbf{V}_C$ is 90° behind the current phasor.

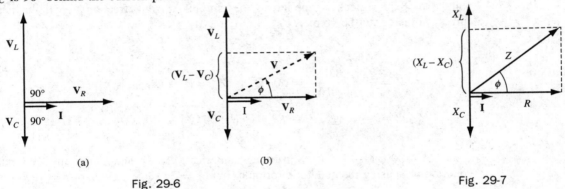

Fig. 29-6   Fig. 29-7

The applied voltage **V** must equal the vector sum of the voltages across the three elements, so $\mathbf{V} = \mathbf{V}_R + \mathbf{V}_L + \mathbf{V}_C$. This is illustrated in Fig. 29-6b. Here I first add $\mathbf{V}_L$ and $\mathbf{V}_C$, and then I add the result vectorially to $\mathbf{V}_R$. In the example I have drawn here, $V_L = IX_L$ is larger than $V_C = IX_C$, so the circuit is inductive and the current is seen to lag the voltage by the phase angle ϕ.

Equation 29.10 shows that Z, R, and $(X_L - X_C)$ are proportional to the sides of a right triangle. This is also apparent from Fig. 29-6b, since each voltage phasor is proportional to the current and $V_R = IR$, $V_L = IX_L$, $V_C = IX_C$, and $V = IZ$. The relationships are shown in an **impedance diagram** (Fig. 29-7).

Now imagine that the phasor diagram, Fig. 29-6b, is rotated counterclockwise with constant angular velocity ω. The projections of the phasors on the vertical axis represent the instantaneous values of i and v, as shown in the right-hand part of Fig. 29-8. Here I show the voltages across each element. The current varies as v_R.

PROBLEM 29.5.   A series *RLC* circuit has $R = 150\,\Omega$, $L = 0.25\,\text{H}$, and $C = 16.0\ \mu\text{F}$. It is driven by a 60-Hz voltage of peak value 170 V. Determine the current, the maximum voltage across each element, and the phase angle of the circuit.

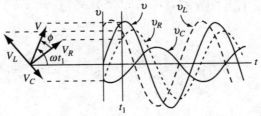

Fig. 29-8

Solution
$$X_L = \omega L = 2\pi f L = 2\pi(60\,\text{Hz})(0.25\,\text{H}) = 94\,\Omega$$

$$X_C = \frac{1}{\omega C} = \frac{1}{2\pi C} = [2\pi(60\,\text{Hz})(16 \times 10^{-6}\,\text{F})] = 166\,\Omega$$

$$Z = \sqrt{R^2 + (X_L - X_C)^2} = \sqrt{(150\,\Omega)^2 + (94\,\Omega - 166\,\Omega)^2} = 166\,\Omega$$

$$\tan\phi = \frac{X_L - X_C}{R} = \frac{94\,\Omega - 166\,\Omega}{150\,\Omega} = -0.48 \quad \phi = -25.6°$$

The negative phase angle indicates that the voltage lags the current; that is, the circuit is capacitive overall.

$$I = \frac{V}{Z} = \frac{170\,\text{V}}{166\,\Omega} = 1.02\,\text{A} \qquad V_R = IR = (1.02\,\text{A})(150\,\Omega) = 154\,\text{V}$$

$$V_L = IX_L = (1.02\,\text{A})(94\,\Omega) = 95.9\,\text{V} \qquad V_C = IX_C = (1.02\,\text{A})(166\,\Omega) = 170\,\text{V}$$

PROBLEM 29.6 A series *RLC* circuit has $R = 400\,\Omega$, $L = 0.50\,\text{mH}$, and $C = 0.04\,\mu\text{F}$. An alternating voltage of 8.00-V peak is applied to the circuit. At what frequency will the peak current in the circuit be 16 mA?

Solution
$$Z = \frac{V}{I} = \frac{8.0\,\text{V}}{0.016\,\text{A}} = 500\,\Omega \qquad Z^2 = R^2 + (X_L - X_C)^2$$

$$\omega L - \frac{1}{\omega C} = \pm\sqrt{Z^2 - R^2} = \pm\sqrt{(500\,\Omega^2) - (400\,\Omega)^2} = \pm 300\,\Omega$$

$\omega^2 LC \mp 300\omega C - 1 = 0$. Substitute numerical values and solve the quadratic equation in ω using the quadratic formula. Find $f = 2\pi/\omega = 12\,\text{kHz}$ or $f = 107\,\text{kHz}$.

29.4 Power in ac Circuits

The instantaneous power in a series *RLC* circuit is $P = iv = I_{max}\sin(\omega t - \phi)V_{max}\sin\omega t$. Of more interest is the average power. To calculate P_{av}, expand $\sin(\omega t - \phi)$ and integrate over one period. Thus $\sin(\omega t - \phi) = \sin\omega t\cos\phi - \cos\omega t\sin\phi$, so

$$P_{av} = I_{max}V_{max}\frac{1}{T}\left(\cos\phi\int_0^T\sin^2\omega t\,dt - \sin\phi\int_0^T\cos\omega t\sin\omega t\,dt\right)$$

$$= I_{max}V_{max}\left[\cos\phi\int_0^T\left(\frac{1 - \cos 2\omega t}{2T}\right)dt - \sin\phi\frac{\sin^2\omega t}{2T}\bigg|_0^{2\pi/\omega}\right]$$

$$= I_{max}V_{max}\left[\cos\phi\left(\frac{t}{2T} - \frac{\sin 2\omega t}{4\omega T}\right)\bigg|_0^T\right]$$

where

$$\sin^2\omega t\bigg|_0^{2\pi/\omega} = 0 \quad\text{and}\quad \sin 2\omega t\bigg|_0^{2\pi/\omega} = 0, \quad\text{so } P_{av} = \tfrac{1}{2}I_{max}V_{max}\cos\phi$$

Since

$$I_{rms} = \frac{1}{\sqrt{2}} I_{max} \quad \text{and} \quad V_{rms} = \frac{1}{\sqrt{2}} V_{max}$$

$$\boxed{P_{av} = I_{rms} V_{rms} \cos \phi} \tag{29.12}$$

$\cos \phi$ is the **power factor**. From Fig. 29-7, I see that

$$\cos \phi = \frac{R}{Z} \quad \text{and} \quad Z = \frac{V_{max}}{I_{max}}$$

so

$$\cos \phi = \frac{RI_{max}}{V_{max}} = \frac{RI_{rms}}{V_{rms}}$$

Thus

$$\boxed{P_{av} = I_{rms}^2 R} \tag{29.13}$$

Power is dissipated only in resistors, not in capacitors or inductors.

29.5   Resonance in ac Circuits

When an alternating voltage is applied to an *RLC* circuit, the current that flows is strongly dependent on the frequency of the voltage. It is this response feature of such circuits that makes them so useful. Since $I = V/Z$, the current will be largest when the impedance Z is smallest. Since $Z = \sqrt{R^2 + (X_L - X_C)^2}$, this requires $X_L = X_C$ or $\omega_0 L = 1/\omega_0 C$. This condition is called **resonance**, and ω_0 is the **resonant frequency** in the series *RLC* circuit.

$$\boxed{X_L = X_C \quad \omega_0 = \frac{1}{\sqrt{LC}}} \quad \text{resonance} \tag{29.14}$$

For a fixed applied voltage, the variation in the current in the circuit as a function of frequency is shown in Fig. 29-9. With more losses (bigger resistance) in the circuit, the response is broader. The power curve as a function of frequency looks much the same as the current curve. If $\Delta \omega$ is the frequency width between the half-power points (these are the frequencies where the power is half its maximum), one can define a **quality factor** Q_0 as $Q_0 = \omega_0/\Delta \omega$. Further, it is possible to show that $Q_0 = \omega_0 L/R$.

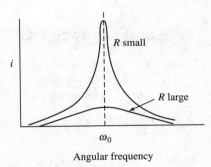

Fig. 29-9

At resonance the current and applied voltage are in phase. At high frequencies the inductive reactance $(X_L + \omega L)$ predominates, and the current lags the voltage. At low frequencies the capacitive reactance dominates $(X_C = 1/\omega C)$ and the current leads the voltage.

PROBLEM 29.7.   An alternating voltage of amplitude 60 V is applied to a series *RLC* circuit where $R = 24\,\Omega$, $L = 25\,mH$, and $C = 0.10\,\mu F$. Determine the resonant frequency, the value of Q_0, the current at resonance, the frequency width between the half-power points, and the maximum voltage across the inductor at resonance.

Solution

$$\omega_0^2 = \frac{1}{LC} = [(0.025\,\text{H})(0.10 \times 10^{-6}\,\text{F})]^{-1} \quad \omega_0 = 2 \times 10^4/\text{s}$$

$$Q_0 = \frac{\omega_0 L}{R} = \frac{(2 \times 10^4/\text{s})(0.025\,\text{H})}{24\,\Omega} = 20.8 \quad \Delta\omega = \frac{\omega_0}{Q_0} = 962/\text{s}$$

At resonance, $Z = R$, so

$$I_{max} = \frac{V_{max}}{R} = \frac{60\text{ V}}{24\text{ }\Omega} = 2.5\text{ A}$$

Thus

$$V_L + IX_L = \omega LI = (2 \times 10^4/\text{s})(0.025\text{ H})(2.5\text{ A}) = 1250\text{ V}$$

Observe that the voltage across the inductor or the capacitor at resonance can be much larger than the applied voltage.

29.6   Summary of Key Equations

Transformer: $\dfrac{V_1}{V_2} = \dfrac{N_1}{N_2}$ $\dfrac{I_1}{I_2} = \dfrac{N_2}{N_1}$

Series *RLC* impedance: $Z = \sqrt{R^2 + (X_L - X_C)^2}$ $V = IZ$

Reactance: $X_L = \omega L$ $X_C = \dfrac{1}{\omega C}$

Root mean square quantities: $I_{rms} = \dfrac{1}{\sqrt{2}} I_{max}$

Power: $P = I_{rms}^2$ $R = I_{rms} V_{rms} \cos\phi$

Power factor: $\cos\phi = \dfrac{R}{Z}$

Phase angle: $\tan\phi = \dfrac{X_L - X_C}{R}$

Series resonance: $X_L = X_C$ $\omega_0^2 = \dfrac{1}{LC}$

SUPPLEMENTARY PROBLEMS

29.8. Appliances in Europe are designed to operate on 240 Vac. Suppose you try to operate a German coffeemaker on a 120-Vac outlet. If it was intended to provide 720 W to its 5-Ω heating coil, what current will it now deliver? What power will it deliver?

29.9. For safety, lights around a residential swimming pool are often operated on a 6-Vac line instead of the normal 120-Vac line. What turns ratio is required in the step-down transformer used for this purpose?

29.10. What is the reactance of a 2-mH inductor at 10 kHz? At 60 Hz? What is the reactance of a 2-μF capacitor at 10 kHz? At 60 Hz?

29.11. The FM radio band ranges from 88 to 108 MHz. In a certain receiver a variable capacitor is used in a resonant circuit with an inductor. The capacitor can be varied from 10.9 to 16.4 pF. What value of inductance is needed to cover the entire FM band?

29.12. A series *RLC* circuit has a resistance of 120 Ω and an impedance of 400 Ω. What is the power factor, $\cos\phi$? If the rms current is 180 mA, how much power is dissipated in the circuit?

29.13. In a certain *RLC* series circuit, the maximum voltages across the resistor, inductor, and capacitor are 24 V, 180 V, and 120 V, respectively. What is the phase angle between the current and the voltage? Does the current lag or lead the voltage?

29.14. A series RLC circuit has $R = 500\,\Omega$, $L = 4.0\,H$, and $C = 7.0\,\mu F$. It is driven by a peak voltage of 36 V at 60 Hz. At the instant that the applied voltage is a maximum, what is the voltage across each circuit element?

29.15. An inductor with a reactance of $30\,\Omega$ and appreciable resistance uses 12 W of power when it carries an rms current of 0.60 A. What is the impedance of the inductor?

29.16. A resistor R, inductor L, and capacitor C are all connected in parallel to an ac source of EMF. What is the maximum current delivered by the source?

29.17. A series RLC circuit has elements with the values R_1, L_1, and C_1 and resonant frequency ω_0. A second series RLC circuit has elements R_2, L_2, and C_2 and the same resonant frequency ω_0. Show that if all six of these elements are connected together in series, the resonant frequency will again be ω_0.

29.18. What is the half-power bandwidth for a series RLC circuit with $L = 2\,mH$, $C = 8\,\mu F$, and $Q_0 = 50$?

29.19. A series RLC circuit has the parameters $R = 100\,\Omega$, an inductance of 90 mH, and a capacitance of $8\,\mu F$. What is the power factor for this circuit when it is driven by an ac generator with $\omega = 800/s$?

29.20. A factory that uses machinery involving large motors can be modeled as an electrical load consisting of a resistance R in series with an inductance L (because of the windings in the motors). Such an arrangement results in a power factor that reduces the actual power delivered compared to the case where no inductance is present. To overcome this problem, capacitors are placed in series with the resistive load to increase the power factor. To see how this works, consider a series RL circuit with $R = 40\,\Omega$ and $L = 50\,mH$. A voltage with $\omega = 360\,Hz$ is applied. (a) What is the power factor? (b) What capacitance must be added in series to result in a power factor of one?

SOLUTIONS TO SUPPLEMENTARY PROBLEMS

29.8. Initially $I_2^2 R = 720\,W$, and $R = 5\,\Omega$, so $I_2 = 12\,A$, and $V_2 = I_2 R = 60\,V$. Thus

$$\frac{V_2}{V_1} = \frac{240\,V}{60\,V} = \frac{1}{4}$$

With

$$V_1' = 120\,V \quad V_2' = 30\,V \quad P' = \frac{V_2'^2}{R} = 180\,W \quad I_2' = 6\,A$$

29.9. $\dfrac{N_2}{N_1} = \dfrac{120\,V}{6\,V} = 20$

29.10. $X_L = \omega L = 2\pi f L \quad X_L = 26\,\Omega$ at 10 kHz $\quad X_L = 0.75\,\Omega$ at 60 Hz

$$X_C = \frac{1}{\omega C} \quad X_C = 8.0\,\Omega \text{ at } 10\,kHz \quad X_C = 1300\,\Omega \text{ at } 60\,Hz$$

29.11. $\omega^2 = \dfrac{1}{LC}$ and $L = \dfrac{1}{C\omega^2}$, so $L = 2 \times 10^{-7}\,H$

29.12. $\cos\phi = \dfrac{R}{Z} = \dfrac{120\,\Omega}{400\,\Omega} = 0.30 \qquad P = I_{rms}^2 R = (0.18\,A)^2 (120\,\Omega) = 3.9\,W$

29.13. $\tan\phi = \dfrac{V_L - V_C}{V_R} = \dfrac{180\,V - 120\,V}{24\,V} = 2.5, \quad \phi = 68°$
The current lags the voltage.

29.14.
$$X_L = \omega L = 1500\,\Omega \qquad X_C = \frac{1}{\omega C} = 380\,\Omega$$

$$\tan\phi = \frac{X_L - X_C}{R} = \frac{1500\,\Omega - 380\,\Omega}{500\,\Omega} = 2.24 \quad \phi = 66°$$

$$Z = \sqrt{R^2 + (X_L - X_C)^2} = 1230\,\Omega \quad I_{max} = \frac{V_{max}}{Z} = 0.029\,A$$

$$V_{R\,max} = RI_{max} = 14.5\,V \quad V_{L\,max} = X_L I_{max} = 44\,V \quad V_{C\,max} = X_C I_{max} = 11.2\,V$$

When V is a maximum, $\quad V_R = V_{R\,max}\cos\phi = 60\,V \quad V_L = V_{L\,max}\sin\phi = 40\,V$

$$V_C = -V_{C\,max}\sin\phi = -10\,V$$

29.15. $P_{av} = I_{rms}^2 R$, so $R = \dfrac{12\,W}{(0.6\,A)^2} = 33.3\,\Omega$

$$Z = \sqrt{R^2 + X_L^2} = \sqrt{(33.3\,\Omega)^2 + (30\,\Omega)^2} = 45\,\Omega$$

29.16. The applied voltage is

$$V\sin\omega t = V_R = V_L = V_C \quad I_R = \frac{V\sin\omega t}{R}$$

$$V_L = -L\frac{dI_L}{dt} = V\sin\omega t, \quad \text{so } I_L = \frac{V}{\omega L}\cos\omega t \quad I_L = \frac{V}{\omega L}\sin\left(\omega t - \frac{\pi}{2}\right)$$

$$V_C = \frac{1}{C}q = V\sin\omega t \quad \frac{1}{C}\frac{dq}{dt} = \frac{1}{C}I_C = \omega V\cos\omega t \quad I_C = \omega C V\sin\left(\omega t + \frac{\pi}{2}\right)$$

$$I_C = \frac{V}{X_C}\sin\left(\omega t + \frac{\pi}{2}\right)$$

Represent the currents with a phasor diagram as shown here. The total current is thus obtained by vector addition:

$$I = \sqrt{\left(\frac{1}{R}\right)^2 + \left(\frac{1}{X_L} - \frac{1}{X_C}\right)^2}$$

$$\tan\phi = \frac{1/X_L - 1/X_C}{1/R}$$

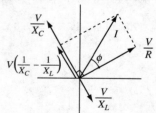

29.17. $\omega_0^2 = \dfrac{1}{L_1 C_1} = \dfrac{1}{L_2 C_2}$

For the combined circuit, $L = L_1 + L_2$ and

$$C = \frac{C_1 C_2}{C_1 + C_2}$$

The capacitors are in series.

$$\omega^2 = \frac{1}{(L_1 + L_2)(C_1 C_2 / C_1 + C_2)} = \frac{C_1 + C_2}{L_1 C_1 C_2 + C_1 L_2 C_2} = \frac{C_1 + C_2}{1/\omega_0^2 C_2 + 1/\omega_0^2 C_1} = \omega_0^2$$

29.18. $\omega_0 = \dfrac{1}{\sqrt{LC}} = \dfrac{1}{\sqrt{(2\times 10^{-3}\,H)(8\times 10^{-6}\,F)}} = 7906 \qquad Q = \dfrac{\omega_0}{\Delta\omega}, \quad \Delta\omega = \dfrac{\omega_0}{Q} = 158/s$

29.19. $x_c = \dfrac{1}{wc} = \dfrac{1}{(800)(8 \times 10^{-6})} = 156\,\Omega$

$x_L = wL = (800)(90 \times 10^{-3}) = 72\,\Omega$

$R = 100\,\Omega$

$z = \left[R^2 + (x_L - x_c)^2\right]^{1/2} = \left[100^2 + (72 - 156)^2\right]^{1/2}$

$z = 131\,\Omega$

Power factor $\cos\phi = \dfrac{R}{z} = \dfrac{100\,\Omega}{131\,\Omega} = 0.77$

29.20. $x_L = wL(360)(0.05) = 18\,\Omega$

$R = 40\,\Omega,$

$z = \sqrt{R^2 + x_L^2} = (40^2 + 18^2) = 44\,\Omega$

$\cos\phi' = \dfrac{R'}{z} = \dfrac{40}{44} = 0.91$

For a power factor of one require $X_L = X_c$

$\dfrac{1}{wc} = wL$

$c = \dfrac{1}{w^2 L} = \dfrac{1}{(360)^2(0.05)} = 154\,\mu\text{F}$

CHAPTER 30

Electromagnetic Waves

One cannot imagine a more fascinating aspect of nature than electromagnetic (EM) waves. They are everywhere. The wonderful colors of a rainbow or of a peacock's feathers, the TV signals that circle the world, the x rays the dentist uses, and the deadly gamma rays from radioactive sources are all part of this magnificent panoply that shrouds our universe. These are all electromagnetic waves, and although they are not fully understood even today, we do know a lot about them.

30.1   Maxwell's Equations and the Wave Equation

In the 1860s James Clerk Maxwell made one of the all-time greatest human intellectual accomplishments. First, he recognized on mathematical grounds alone that Ampere's law as we have used it could not be correct. He modified it by adding to I a term $\epsilon_0(d\Phi_E/dt)$. This is called a **displacement current**, because it has the dimensions of current. I think this is a poor name, but we're stuck with it. The laws of electricity and magnetism in free space then took the form of the integrals shown on the left in Table 30-1. Further, Maxwell was able to write these relations as differential equations, as shown in the right column of Table 30-1. Here $\nabla = (\partial/\partial x)\mathbf{i} + (\partial/\partial y)\mathbf{j} + (\partial/\partial z)\mathbf{k}$, ρ is the volume charge density, and $\mathbf{J}$ is the current density.

TABLE 30-1   Maxwell's Equations

$\oint \mathbf{E} \cdot d\mathbf{A} = \dfrac{Q}{\epsilon_0}$	$\nabla \cdot \mathbf{E} = \dfrac{\rho}{\epsilon_0}$	(30.1)
$\oint \mathbf{B} \cdot d\mathbf{A} = 0$	$\nabla \cdot \mathbf{B} = 0$	(30.2)
$\oint \mathbf{E} \cdot d\mathbf{s} = -\dfrac{d\Phi_B}{dt}$	$\nabla \times \mathbf{E} = -\dfrac{\partial \mathbf{B}}{\partial t}$	(30.3)
$\oint \mathbf{B} \cdot d\mathbf{s} = \mu_0 I + \epsilon_0 \mu_0 \dfrac{d\Phi_E}{dt}$	$\nabla \times \mathbf{B} = \mu_0 \mathbf{J} + \epsilon_0 \mu_0 \dfrac{\partial \mathbf{E}}{\partial t}$	(30.4)

The top equation tells us that charges are the sources and sinks of the electrostatic field. The second equation states that there are no "magnetic charges," that is, sources or sinks of the magnetic field. The third equation shows that changing magnetic fields stir up "whirlpool-like" electric fields. The last equation says that magnetic fields are stirred up by paddle wheel-like conduction currents or changing electric fields.

Next, Maxwell made a brilliant discovery, using mathematics (do your homework and you may end up rich and famous, too). He combined Eqs. 30.3 and 30.4 to obtain the **electromagnetic wave equations** for E and B.

$$\frac{\partial^2 E}{\partial x^2} = \mu_0 \epsilon_0 \frac{\partial^2 E}{\partial t^2} \quad \text{and} \quad \frac{\partial^2 B}{\partial x^2} = \mu_0 \epsilon_0 \frac{\partial^2 B}{\partial t^2} \tag{30.5}$$

I have written these for the simple case of a plane wave moving in the x direction. This is called a *wave equation* because the solutions are of the form

$$E = E_{max} \cos(kx - \omega t) \quad \text{and} \quad B = B_{max} \cos(kx - \omega t) \tag{30.6}$$

Here $\omega = 2\pi f$ and $k = 2\pi/\lambda$, where f is the frequency and λ the wavelength. Further, Maxwell showed that the speed of the wave is $v = \omega/k = 1/\sqrt{\mu_0 \epsilon_0}$. If you substitute numerical values for ϵ_0 and μ_0, you obtain the remarkable result

$$\lambda f = v = c = \frac{1}{\sqrt{\mu_0 \epsilon_0}} = 3.00 \times 10^8 \, \text{m/s} \tag{30.7}$$

Maxwell immediately recognized this value as the speed of light, and so he had deduced what light is: a traveling electro-magnetic wave. Who would ever have guessed that from rubbing cat's fur on amber or playing around with magnetic rocks we would finally figure out what light is? And radio waves and infrared and ultraviolet and x rays, too.

Notice that all electromagnetic waves travel at the same speed in vacuum. Since $\lambda f = c$, waves of shorter wavelength have higher frequency, as shown in Fig. 30-1. For historical reasons different portions of the EM spectrum are given different names, although they are all the same kind of thing. Visible light constitutes a narrow range of the spectrum, from wavelengths of about 400–800 nm.

One can combine Maxwell's equations to show that

$$\frac{E_{max}}{B_{max}} = \frac{E}{B} = c \tag{30.8}$$

In a plane EM wave traveling in the x direction, **E** and **B** are perpendicular to each other and to the velocity vector **v**, as shown in Fig. 30-2. E and B are in phase.

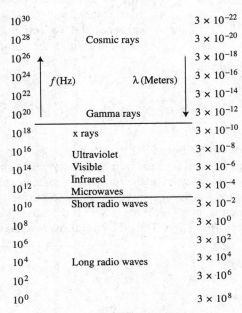

Fig. 30-1

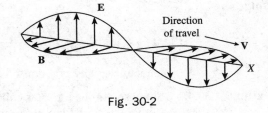

Fig. 30-2

PROBLEM 30.1. Radio station KRPL broad-casts an AM signal at a carrier frequency of 1400 kHz. What is the wavelength of this radio wave?

Solution

$$\lambda = \frac{c}{f} = \frac{3 \times 10^8 \, \text{m/s}}{1400 \times 10^3 /\text{s}^{-1}} = 214 \, \text{m}$$

30.2 Energy and Radiation Pressure

We have seen that electric and magnetic fields store energy, and so as an EM wave propagates, it carries energy. The rate of energy flow, in watts per square meter, can be shown to be

$$\mathbf{S} = \frac{1}{\mu_0} \mathbf{E} \times \mathbf{B} \tag{30.9}$$

S is the **Poynting vector**. (It "points" in the direction the energy is going.) In a plane wave, $\mathbf{E} \perp \mathbf{B}$, and $E = cB$, so

$$S = \frac{1}{\mu_0}EB = \frac{1}{c\mu_0}E^2 = \frac{c}{\mu_0}B^2 \qquad (30.10)$$

PROBLEM 30.2. Consideration has been given to the use of powerful lasers as weapons to shoot down missiles or aircraft. Such laser beams are limited by the fact that an electric field that is too large will cause electrical breakdown in the air, causing the air to become a plasma and reflective for the laser beam. Assuming air breaks down at about 3×10^6 V/m, what is the intensity of the strongest laser beam one can use?

Solution

$$S = \frac{E^2}{c\mu_0} = \frac{(3 \times 10^6 \text{ V/m})^2}{(3 \times 10^8 \text{ m/s})(4\pi \times 10^{-7} \text{ N/A}^2)} = 2.4 \times 10^{10} \text{ W/m}^2$$

Maxwell showed that an electromagnetic wave carries forward momentum $p = U/c$, where U is the energy transported forward in time t. If the radiation is absorbed by a surface, the momentum drops to zero and a force is exerted on the surface in a manner similar to what was found with Newton's second law: $F = dp/dt$. Since S is the energy per second incident on 1 m^2, the force per unit area (pressure) on a perfectly absorbing (that is, black) surface is S/c. If the wave is reflected back, the change in momentum is twice as great, and then $P = 2S/c$. Thus the **radiation pressure** of an electromagnetic wave is

$$P = \frac{S}{c} \quad \text{black surface}, \qquad P = 2\frac{S}{c} \quad \text{reflecting surface} \qquad (30.11)$$

PROBLEM 30.3. Sunlight strikes the Earth's atmosphere with an intensity of about 1000 W/m^2. What pressure would the sunlight exert on a black surface?

Solution

$$P = \frac{S}{c} = \frac{(1000 \text{ W/m}^2)}{(3 \times 10^8 \text{ m/s})} = 3.3 \times 10^{-6} \text{ N/m}^2 = 3 \times 10^{-11} \text{ atm}$$

You can see why you don't usually notice this pressure!

PROBLEM 30.4. Suppose you had a solar collector 2×4 m on your roof, and you were able to collect sunlight for 8 h/day with an average intensity of 500 W/m^2. You might try to save this energy to heat your house at night when it is cold. Suppose you stored the energy by heating water by 20°C. How many liters could you heat in this way?

Solution The total energy collected is

$$U = S\,At = (500 \text{ W/m}^2)(8 \text{ m}^2)(8 \text{ h})(3600 \text{ s/h}) = 1.15 \times 10^8 \text{ J}$$

One liter of water is 1 kg, and to heat 1 kg by 20°C requires energy $U_1 = (4200 \text{ J/kg°C})(20°C) = 84,000 \text{ J}$. Thus one could heat N liters, $N = U/U_1 = (1.15 \times 10^8 \text{ J})/(8.4 \times 10^4 \text{ J}) = 1370 \text{ L}$.

30.3 Polarization

The plane defined by the electric field vector and the velocity vector (the xy plane in Fig. 30-2) is called the **plane of polarization**. An EM wave in which **E** always oscillates in the same plane is **plane polarized**. Light from most sources, for example, an incandescent lamp or the Sun, is **unpolarized**. This means it comes in short bursts, with each piece being randomly polar-ized. There are ways, however, of producing polarized light, for example, by reflection or by passing the wave through some polarizing material (for example, Polaroid) that trans-mits only waves with the electric field vibrating in a certain selected direction, the polarization axis. In Fig. 30-3 an unpo-larized light wave with intensity I_0 W/m^2 is incident on a polarizing sheet P_1 which has its axis vertical. Half of the light is absorbed, and it leaves the sheet polarized vertically with intensity $\frac{1}{2}I_0$. When it strikes the second sheet P_2, whose axis is rotated through angle θ from vertical, only

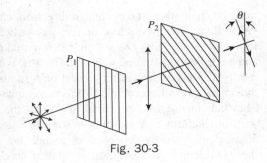

Fig. 30-3

the component of **E** parallel to the polarizing axis, $E\cos\theta$, is transmitted. Since intensity is proportional to E^2, the final transmitted intensity is $\frac{1}{2}I_0\cos^2\theta$. In general, if a polarized wave of intensity I_0 is incident on a polarizer whose axis makes angle θ with the plane of polarization, the transmitted intensity is

$$I = I_0 \cos^2\theta \quad \text{law of Malus} \tag{30.12}$$

PROBLEM 30.5. Unpolarized light of intensity I_0 is incident on two polarizing sheets whose axes are perpendicular. What is the intensity of the transmitted light? What will be the transmitted intensity if a third sheet is placed between the first two with its axis at $45°$ to theirs?

Solution With two sheets, after one sheet, $I_1 = \frac{1}{2}I_0$. After the second sheet, $I_2 = I_1\cos^2 90° = 0$. In the second case, $I_1 = \frac{1}{2}I_0$, $I_2 = I_1\cos^2 45°$, $I_3 = I_2\cos^2 45° = I_1\cos^4 45°$, and $I_3 = (\frac{1}{2}I_0)(0.707)^4 = 0.125I_0$.

PROBLEM 30.6. Suppose you want to rotate the plane of polarization of a beam of polarized light by $90°$, but you do not want the final intensity to be less than 50 percent of the initial intensity. What is the minimum number of polarizing sheets you must use?

Solution The final sheet must be oriented at $90°$ to the incident plane of polarization. If you insert one additional sheet at $45°$ (which means the light goes through two sheets, each with $\theta = 45°$), the transmitted intensity will be $I_0(\cos^2 45°)^2 = 0.25I_0$. For three sheets, each rotated by $30°$, $I = I_0(\cos^2 30°)^3 = 0.42I_0$. For four sheets, each rotated by $90°/4 = 22.5°$, $I = I_0(\cos^2 22.5°)^4 = 0.53I_0$. Thus four sheets are required.

30.4 Reflection and Refraction of Light

From now on I will focus on the most important of electromagnetic waves, those we can see, called **light**. When a wave strikes a new material, it can be reflected, transmitted, or absorbed. Here I will consider only substances that reflect or transmit the waves. Both processes can occur; that is, some light is reflected and some is transmitted, as when light strikes clear glass. If we consider plane waves, we can imagine a **wave front** (a plane) over which the **E** vector is constant. This is a plane of constant phase, and it moves along with the wave. Rather than draw wave fronts, it is more convenient to draw rays. A *ray* is a line drawn per-pendicular to a wave front. It is directed along the velocity vector of the wave. Most people have seen a laser beam, and this is a good way to imagine a light ray to behave. If light is reflected from a surface, it is found that it bounces off at the same angle it came in at. This is the **law of reflection**.

$$\theta_1 = \theta_1' \quad \text{reflection} \tag{30.13}$$

Caution: Be certain you use the angles shown in Fig. 30-4. The **angle of incidence** θ_1 is between the ray and the normal to the surface (that is, the line N perpendicular to the surface). The **angle of reflection** θ_1' is between the reflected ray and the normal N.

In vacuum all electromagnetic waves travel at the speed c. However, in matter they slow down and travel at some lesser speed v. The ratio of speeds is the **index of refraction**, $n = c/v$. When light goes from one material to another, it can change direction, an effect called **refraction**. The frequency of the light always stays the same, but since $v = f\lambda$, slowing down makes λ smaller, and speeding up makes λ larger. To see why this bends the light, imagine a marching band on a football field. The musicians are all marching to the same beat (frequency). They advance across the field at 45° to the 50-yard line with instructions that as soon as someone steps across the 50-yard line, she is supposed to reduce her stride length (her wavelength) to one-tenth of what it was. You can see what will happen. As they cross the 50-yard line the rows will swing in the direction of the first person to slow

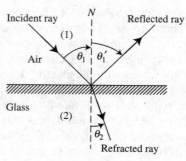

Fig. 30-4

down. Once everyone is across the 50, they will march on straight ahead at reduced speed but having deviated from their original direction. The angle of refraction θ_2 is related to the angle of incidence θ_1 by **Snell's law**:

$$n_1 \sin \theta_1 = n_2 \sin \theta_2 \quad \text{refraction} \tag{30.14}$$

 Observe that when light goes from a low-density medium like air (small index n_1) to a more dense medium like glass (larger index n_2), a ray is bent toward the normal, as in Fig. 30-4. When going in the opposite direction, it is bent away from the normal. When incident at 90° (straight on), it is not deviated at all.

   The incident ray, the reflected ray, the refracted (transmitted) ray, and the normal to the surface all lie in one plane.

PROBLEM 30.7. Light strikes a plate of glass at an angle of incidence of 60°. The index of refraction of the glass is 1.50 and it is 3 mm thick. What is the lateral displacement of the light ray after it has traversed the glass plate? (For air, $n \simeq 1.00$.)

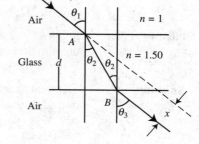

Solution $n_1 \sin \theta_1 = n_2 \sin \theta_2$ and $n_2 \sin \theta_2 = n_3 \sin \theta_3$, so $\theta_1 = \theta_3$, since $n_1 = n_3$.

   Let distance $AB = r$; then $d/r = \cos \theta_2$:

$$x = r \sin (\theta_1 - \theta_2) = \frac{d(\sin \theta_1 - \theta_2)}{\cos \theta_2}$$

$$1.00 \sin 60° = 1.50 \sin \theta_2, \quad \theta_2 = 35.3°, \quad x = \frac{(3 \text{ mm}) \sin (60° - 35.3°)}{\cos 35.3°} = 1.54 \text{ mm}$$

PROBLEM 30.8. When I was a kid I did a lot of skin diving, and it seemed to me that when I looked down at a fish or a shell from out of the water, it was a lot bigger than it was when I brought it up. I decided that the water magnified the image of the fish. In fact, the water doesn't magnify the image, but it does make it appear closer, and then your brain decides it must be a bigger fish (things up close look bigger). Suppose you look almost straight down at an object a distance d below the surface of the water ($n = 1.33$). How deep does it appear to be?

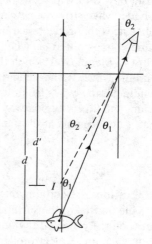

Solution Two rays of light leaving the object are shown here. One goes straight up and is not bent, and the other enters your eye after being refracted. Observers seeing these rays would believe they emanated from point I, where they believe the fish to be. From the drawing, $\tan \theta_2 = x/d'$ and $\tan \theta_1 = x/d$. For small angles, $\tan \theta \simeq \sin \theta \simeq \theta$ and $n \sin \theta_1 = (1) \sin \theta_2$. So

$$\frac{\tan \theta_1}{\tan \theta_2} \simeq \frac{xd'}{xd} \simeq \frac{\sin \theta_1}{\sin \theta_2} = \frac{1}{n} \quad \text{and} \quad d' = \frac{1}{n}d = \frac{1}{1.33}d$$

PROBLEM 30.9. Light strikes an equilateral glass prism ($n = 1.42$) and travels through it parallel to one side. What is the angle of incidence? Through what angle is the ray deviated after it leaves the prism?

Solution

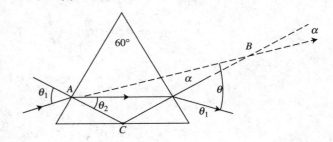

From the drawing I see that $\theta_2 = 30°$, so $(1) \sin \theta_1 = 1.42 \sin 30°$. Thus $\theta_1 = 44.8°$. By symmetry the ray is refracted by the same angle at the second surface, so the total angle of deviation is $\theta = \theta_1 - \alpha$. By geometry I see that $\angle ACB = 120°$, so for triangle ABC, $120° + \theta_1 + \alpha = 180°$. Since $\theta_1 = 44.8°$, $\alpha = 15.2°$, and the deviation is $\theta = \theta_1 - \alpha = 44.8° - 15.2° = 29.6°$.

One can also derive the law of refraction from *Fermat's principle*, which states that **light will travel from point A to point B along the path that requires the least time**. In Fig. 30-5 light travels from one medium to another, with a plane interface between the two. The total time of travel is

$$t = \frac{\sqrt{a^2 + x^2}}{v_1} + \frac{\sqrt{b^2 + (d - x)^2}}{v_2}$$

For minimum time,

Fig. 30-5

$$\frac{dt}{dx} = 0, \quad \text{so} \quad \frac{x}{v_1\sqrt{a^2 + x^2}} - \frac{(d - x)}{v_2\sqrt{b^2 + (d - x)^2}} = 0$$

But

$$\sin \theta_1 = \frac{x}{\sqrt{a^2 + x^2}}, \quad \sin \theta_2 = \frac{(d - xx)}{\sqrt{b^2 + (d - x)^2}}, \quad v_1 = \frac{c}{n_1}, \quad v_2 = \frac{c}{n_2}$$

Thus

$$n_1 \sin \theta_1 = n_2 \sin \theta_2 \quad \text{Snell's law}$$

30.5 Total Internal Reflection

If light travels from a medium with high index of refraction into one with smaller index of refraction, Snell's law shows that as θ_1 is increased, it will reach a finite value θ_c at which $\theta_2 = 90°$, as shown in Fig. 30-6. For angles of incidence less than θ_c, some of the light is reflected and some is transmitted (refracted). However, for an angle of incidence $\theta_1 \geq \theta_c$, *all* of the light is reflected. This **total internal reflection** requires $n_1 \sin \theta_c = n_2 \sin 90°$, so the **critical angle** θ_c is determined by

$$\boxed{\sin \theta_c = \frac{n_2}{n_1} \quad \text{critical angle}} \qquad (30.15)$$

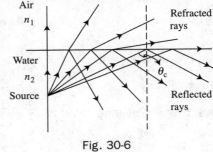

Fig. 30-6

PROBLEM 30.10.   Total internal reflection is used to keep light rays within a glass fiber, and this forms the basis for fiber-optics technology. Suppose a light ray in air strikes the end of a fiber ($n_2 = 1.31$) at point A in the drawing here. What is the maximum value of θ_1 that will ensure that the ray undergoes total reflection at point B and hence remains within the fiber?

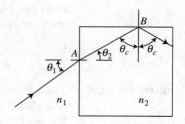

Solution $n_1 \sin \theta_1 = n_2 \sin \theta_2$ and $\theta_2 = 90° - \theta_c$

Thus

$$\cos \theta_2 = \sin \theta_c = \frac{n_1}{n_2} \quad \text{(Note: } n_1 \text{ and } n_2 \text{ switched from Eq. 30.15.)}$$

$$\sin \theta_1 = \frac{n_2}{n_1} \sin \theta_2 = \frac{n_2}{n_1} \sqrt{1 - \cos^2 \theta_2} = \sqrt{\left(\frac{n_2}{n_1}\right)^2 - 1}$$

$$= \sqrt{\left(\frac{1.31}{1}\right)^2 - 1} = 0.716, \quad \theta_1 = 45.7°$$

PROBLEM 30.11.   If you lie on the bottom of a swimming pool and look up, you can see the entire outside world by looking up from the vertex of a cone of half angle θ. For water, $n = 1.33$. What is θ?

Solution $\theta = 90° - \theta_c$ $\sin \theta_c = \frac{1}{n} = \frac{1}{1.33}$ $\theta_c = 48.8°$ $\theta = 41.2°$

PROBLEM 30.12.   A large beaker contains water ($n_3 = 1.33$) to a depth $d_3 = 10\,\text{cm}$, on top of which floats a layer of benzene ($n_2 = 1.50$) of thickness $d_2 = 6\,\text{cm}$. A coin rests on the bottom of the beaker. How deep does it appear to be when you look nearly straight down at it?

Solution   From Problem 30.8, an observer in the benzene would see the coin at a depth d_2' below the benzene–water interface, where $d_2' = (n_2/n_3)d_3$. An observer in the air will thus see the coin at a depth

$$d_2' = \frac{n_1}{n_2}\left(d_2 + d_2'\right) = \frac{n_1}{n_2}\left(d_2 + \frac{n_2}{n_3}d_3\right) = \frac{1}{1.50}(6\,\text{cm}) + \frac{1}{1.33}(10\,\text{cm}) = 11.5\,\text{cm}$$

30.6   Summary of Key Equations

Maxwell's equations:

$$\oint \mathbf{E} \cdot d\mathbf{A} = \frac{Q}{\epsilon_0} \qquad\qquad \nabla \cdot \mathbf{E} = \frac{\rho}{\epsilon_0}$$

$$\oint \mathbf{B} \cdot d\mathbf{A} = 0 \qquad\qquad \nabla \cdot \mathbf{B} = 0$$

$$\oint \mathbf{E} \cdot ds = -\frac{d\Phi_B}{dt} \qquad\qquad \nabla \times \mathbf{E} = -\frac{\partial \mathbf{B}}{\partial t}$$

$$\oint \mathbf{B} \cdot ds = \mu_0 I + \epsilon_0 \mu_0 \frac{d\Phi_E}{dt} \quad \nabla \times \mathbf{B} = \mu_0 \mathbf{J} + \epsilon_0 \mu_0 \frac{\partial \mathbf{E}}{\partial t}$$

The EM wave equation: $\dfrac{\partial^2 E}{\partial x^2} = \mu_0 \epsilon_0 \dfrac{\partial^2 E}{\partial t^2}$ and $\dfrac{\partial^2 B}{\partial x^2} = \mu_0 \epsilon_0 \dfrac{\partial^2 B}{\partial t^2}$

Wave velocity: $v = f\,\lambda$

Power flow (Poynting vector): $\mathbf{S} = \dfrac{1}{\mu_0}\mathbf{E} \times \mathbf{B}$ $E = cB$ $S = \dfrac{1}{c\mu_0}E^2$

Polarized light transmission: $I = I_0 \cos^2 \theta$

Law of reflection: $\theta_1 = \theta_1'$

Law of refraction (Snell's law): $n_1 \sin \theta_1 = n_2 \sin \theta_2$

Total internal reflection: $\sin \theta_c = \dfrac{n_2}{n_1}$

Polarization by reflection (Brewster's angle): $\tan \theta_p = \dfrac{n_2}{n_1}$

SUPPLEMENTARY PROBLEMS

30.13. The helium-neon lasers usually used in classrooms typically provide an intensity of about $1\,\text{m/W/mm}^2$. What are the rms electric and magnetic fields for such a beam?

30.14. Type 2 comets have tails composed of very fine dust particles. The tail of such a comet points away from the Sun because the radiation pressure of sunlight is appreciable. In fact, sunlight radiation pressure has pushed all very small dust particles out of our solar system. Calculate the ratio of the gravitational and radiation forces on a small black particle of density $5000\,\text{kg/m}^3$. For what size will the radiation force exceed the gravitational force? The Sun radiates $3.8 \times 10^{26}\,\text{W}$, and its mass is $2 \times 10^{30}\,\text{kg}$. The universal gravitation constant is $6.7 \times 10^{-11}\,\text{N} \cdot \text{m}^2/\text{kg}^2$.

30.15. Suppose a 100-W light bulb emits 20 W of light uniformly in all directions. What are the rms values of the electric and magnetic fields at a distance of 30 cm from the lamp?

30.16. It has been suggested that space vehicles might be launched or propelled by shining an intense laser beam on them. To gain a feeling for what sort of beam power is required, estimate the mass of the largest aluminum sphere a 10-kW laser could levitate (that is, support against the force of gravity). Assume the beam strikes a small area of the sphere and is totally reflected.

30.17. On Earth we don't usually notice radiation pressure, but in some places in the universe it plays an important role. In the interior of a star the radiation can be so intense that radiation pressure is an important factor in determining the structure of the star. Approximately what electric field would be required to give a pressure of 1 atm ($10^5\,\text{N/m}^2$) on a perfect absorber?

30.18. Two polarizers have their axes at $40°$ to each other. Unpolarized light of intensity I_0 is incident on them. What is the transmitted intensity?

30.19. Unpolarized light is incident on four polarizing sheets, for each of which the axis is rotated by an angle θ from that of adjacent sheets. Find θ if 25 percent of the incident intensity is transmitted by the sheets.

30.20. By looking through Polaroid sunglasses, you can tell if the light you are seeing is polarized because the transmission axis of the glasses is vertical. Rotate the glasses and polarized light will appear alternately bright and dark. Light can be totally or partially polarized by reflection, with the reflected light having its plane of polarization parallel to the reflecting surface. Light in material of index n_1 reflected from material with index n_2 is totally polarized if incident at *Brewster's angle* θ_p, where $\tan \theta_p = n_2/n_1$. Determine the angle between the reflected ray and the refracted ray when light is incident at θ_p.

30.21. The index of refraction of diamond is 2.42. How fast does light travel in diamond?

30.22. If you've ever done any fly fishing in Idaho for eastern brook trout, you probably know that a fish lying in a quiet pool gets pretty nervous if you try to sneak up on him. You don't want him to see you, but he can sort of see around the edge of the bank because of refraction of light. To see how this works, suppose the fish is 1.00 m deep and a distance of 0.80 m from the edge of a straight bank (assume the water goes right to the top of the bank). (a) If you are 1.8 m tall, how close to the edge of the bank could you come without being seen by the fish if light from your head to the fish were not refracted? (b) How close could you come if the water refracts the light ($n = 1.33$)?

30.23. A surfer floats on quiet water ($n = 1.33$) waiting for a big wave. A shark swims toward him at a depth of 4 m. How close (horizontally) can the shark approach before the surfer can spot him? The surfer is lying on his board with his eyes close to water level.

30.24. Total internal reflection in a 45°–90°–45° prism is used in prism binoculars to "fold" the light's path length and make for a more compact instrument. The prisms and lenses must be carefully aligned in binoculars, as illustrated by this problem. If light is incident on a short side of the prism, what is the maximum value of θ_1 for which total reflection will occur at point P if the index of refraction of the glass is $n = 1.50$?

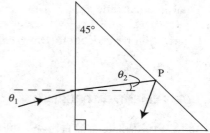

30.25. Suppose that in Problem 30.24 the angle of incidence is 45°. By how much will the incident ray have been deviated when it leaves the prism at point P?

30.26. A cube of plastic is positioned with its faces in the xy, yz and xz planes. A beam of light in the xy plane strikes the plastic on its xz plane face. The beam then passes out of the cube through the yz face, having been deflected by 15.8° from its original direction. What is the index of refraction of the plastic?

30.27. In a shoe store a small mirror is placed on the floor so that a customer can see the shoes she is trying on. In order for her to do this, the mirror is tipped back slightly from vertical. If the customer stands 60 cm from the mirror and her eyes are 160 cm above the floor, at what angle should the mirror be tipped?

SOLUTIONS TO SUPPLEMENTARY PROBLEMS

30.13. $S = \dfrac{E^2}{c\mu_0}$, $\quad E = \sqrt{(3 \times 10^8 \text{ m/s})(4\pi \times 10^{-7} \text{ N/A}^2)\left(\dfrac{10^{-3} \text{ W}}{10^{-6} \text{ m}^2}\right)}$

$\qquad E = 600 \text{ V/m}, \qquad B = \dfrac{E}{c} = 2 \times 10^{-6} \text{ T}$

30.14. The particle looks like a disk of area πa^2. Sunlight spreads uniformly over a sphere of radius r, so

$$S = \frac{P}{4\pi r^2}, \quad F_R = 2\left(\frac{S}{c}\right)\frac{\pi a^2}{r^2} = \frac{Pa^2}{scr^2}$$

The gravity force:

$$F_G = G\frac{M_5 m}{r^2} = G\frac{M_5(4\pi/3)\rho a^3}{r^2}$$

These forces are equal when

$$\frac{4\pi G M_5 \rho a^3}{3r^2} = \frac{Pa^2}{2cr^2}, \quad \text{so } a = \frac{3}{8\pi}\frac{P}{\rho M_5 Gc}$$

$$a = \frac{(3)(3.8 \times 10^{-26} \text{ W/m}^2)}{8\pi(5000 \text{ kg/m}^3)(2 \times 10^{30} \text{ kg})(6.7 \times 10^{-11} \text{ N} \cdot \text{m}^2/\text{kg}^2)(3 \times 10^8 \text{ m/s})}$$

$$= 0.2 \times 10^{-6} \text{ m}$$

Particles smaller than this experience a radiation force pushing them away from the Sun that is greater than the attractive gravity force. They are thus pushed out of the solar system.

30.15. $S = \dfrac{E^2}{c\mu_0} = \dfrac{20 \text{ W}}{4\pi r^2}$, $\quad r = 0.30 \text{ m}$, $\quad$ so $E = 82 \text{ V/m}$, $\quad B = 2.7 \times 10^{-7} \text{ T}$

30.16. $\text{Pressure} = 2\dfrac{S}{c}, \qquad F = (\text{pressure})(\text{area}) = \dfrac{2SA}{c} = \dfrac{2P}{c}$

$$F = mg, \quad m = \frac{2P}{gc} = \frac{(2)(10^4\,\text{W})}{(9.8\,\text{m/s}^2)(3 \times 10^8\,\text{m/s})} = 6.8 \times 10^{-6}\,\text{kg}$$

30.17. $P = \dfrac{S}{c} = \dfrac{E^2}{c^2 \mu_0}, \quad E = c\sqrt{\mu_0 P}$

$$E = (3 \times 10^8\,\text{m/s})\sqrt{(4\pi \times 10^{-7}\,\text{N/A}^2)(1 \times 10^5\,\text{N/m}^2)} = 1/1 \times 10^8\,\text{V/m}$$

30.18. $I = (\frac{1}{2}I_0)(\cos^2 \theta) = \frac{1}{2}I_0 \cos^2 40° = 0.29I_0$

30.19. $I = \frac{1}{2}I_0(\cos^2 \theta)^3, \qquad 0.25I_0 = (0.5I_0)(\cos^6 \theta), \quad \theta = 27°$

30.20. $\theta_p = \theta_1 = \theta_1' \quad \text{Angle of reflection} \qquad \tan \theta_p = \dfrac{\sin \theta_1}{\cos \theta_1} = \dfrac{n_2}{n_1} \qquad \text{and} \qquad n_1 \sin \theta_1 = n_2 \sin \theta_2$

Thus

$$\sin \theta_1 = \frac{n_2}{n_1}\cos \theta_1 = \frac{n_2}{n_1}\sin \theta_2, \quad \cos \theta_1 = \sin \theta_2$$

so

$$\theta_1 + \theta_2 = 90°$$

and the reflected ray is thus perpendicular to the refracted ray.

30.21. $v = \dfrac{c}{n} = \dfrac{3 \times 10^8\,\text{m/s}}{2.42} = 1.24 \times 10^8\,\text{m/s}$

30.22. (a) $\dfrac{x}{1.8\,\text{m}} = \dfrac{0.8\,\text{m}}{1.0\,\text{m}}, \quad x = 1.44\,\text{m}$

(b) $\dfrac{x}{1.8\,\text{m}} = \tan \theta, \quad \dfrac{0.8\,\text{m}}{1.0\,\text{m}} = \tan \theta, \quad \theta_2 = 38.7°, \qquad (1) \sin \theta_1 = 1.33 \sin \theta_2$

$\theta_1 = 56.2°, \qquad x = 1.8 \tan 56.2° = 2.69\,\text{m}$

30.23. At closest approach a light ray from the shark travels almost parallel to the water surface, so

$$\theta_2 = 90°, \qquad n_1 \sin \theta_1 = n_2 \sin \theta_2 = (1)(1) = 1$$

$$\sin \theta_1 = \frac{1}{n_1} = \frac{1}{1.33}, \quad \theta_1 = 48.8°, \quad \frac{x}{d} = \tan \theta_1, \quad \text{so } x = 4 \tan 48.8° = 4.6\,\text{m}$$

30.24. Light is incident at P at θ_c:

$$\sin \theta_c = \frac{1}{n_2} = \frac{1}{1.50}, \quad \theta_c = 41.8°$$

Draw the normal line at P. It intercepts the dashed line in the drawing at 135°, and so for the triangle for which θ_2 and θ_c are interior angles, $135° + \theta_2 = 180°$:

$$\theta_2 = 3.2°, \qquad n_1 \sin \theta_1 = n_2 \sin \theta_2, \qquad \sin \theta_1 = \left(\frac{1.50}{1.0}\right)\sin 3.2°, \quad \theta_1 = 4.8°$$

30.25. $\theta_1 = 45°,$ $(1)\sin 45° = (1.50)\sin \theta_2,$ $\theta_2 = 28.1°$

At P, the incident angle is

$$\theta_3 = \theta_2 = 28.1°, \quad \text{so } (1.50)\sin \theta_3 = (1.0)\sin \theta_4, \quad \theta_4 = 45°$$

The normal at P is parallel to the incident ray, so the incident ray is deviated by 45°.

30.26. $n_1 \sin \theta_1 = n \sin \theta_2$

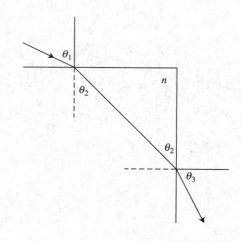

$n \sin (90 - \theta_2) = n_1 \sin \theta_3$

$n \cos \theta_2 = n_1 \sin \theta_3$

$n\left[1 - \sin^2 \theta_2\right]^{1/2} = n_1 \sin \theta_3$

$n\left[1 - \dfrac{\sin^2 \theta_1}{n^2}\right]^{1/2} = n_1 \sin \theta_3$

Square both sides,

$n^2 - \sin^2 \theta_1 = n_1^2 \sin^2 \theta_3$

$n^2 = \sin^2 \theta_1 + n_1^2 \sin^2 \theta_3$

Given $n_1 = 1$, $\theta_1 = 60°$, $\theta_3 = 75.8°$

$n^2 = \sin^2 60° + \sin^2 75.8°$

$n = 1.30$

30.27. Use the Law of Sines, $\dfrac{h}{\sin \theta} = \dfrac{r + d}{\sin (90° + \theta)}$

$r = \left(h^2 + d^2\right)^{1/2}$

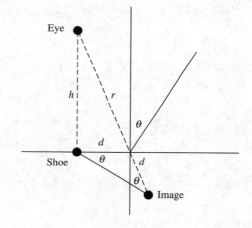

$\sin (90° + \theta) = \cos \theta$

$r = \left(160^2 + 60^2\right)^{\frac{1}{2}} = 171 \text{ cm}$

$\dfrac{h}{\sin \theta} = \dfrac{r + d}{\cos \theta}$

$\dfrac{\sin \theta}{\cos \theta} = \tan \theta = \dfrac{h}{r + d}$

$\tan \theta = \dfrac{160}{171 + 60}$

$\theta = 35°$

Mirror is tipped 35° from vertical

Mirrors and Lenses

Light provides the most important sensory input to our brains, and as a result, light plays a major role in our existence. More and more technology uses light for control and communication, and mirrors and lenses are vital for manipulating light.

31.1 Plane Mirrors

A few objects, such as a lamp or the Sun, are seen by the light they emit. However, most objects are seen in reflected light. Every point from which light is reflected can be considered a kind of secondary source of light. In Fig. 31-1 I've drawn an object O in front of a mirror. I've shown rays coming only from the top, but they come from all parts. The rays reflect off the mirror according to the law of reflection, with $\theta_1 = \theta_1'$. If you were to look toward the reflected rays, they would all appear to come from point I, the image of point O. When light appears to come from an image but does not actually do so, the image is a **virtual image. An image formed behind a mirror is always a virtual image.** From the drawing you can see that for a plane mirror, the image size is equal to the object size, $h = h'$, and the object distance is equal to the image distance, $d_0 = d_i$. Note that the object does not have to be directly in front of the mirror in order for an image to be seen. In Fig. 31-1c you could imagine removing the upper half of the mirror and the image would still be seen. However, to see the image, you do have to position your eye correctly. Imagine that the piece of mirror you are looking at is like a window in a house. To see something inside, you have to be lined up in the right place. The same is true for a mirror.

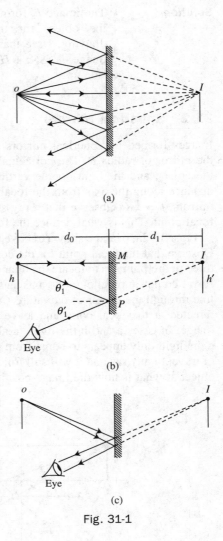

Fig. 31-1

PROBLEM 31.1.   How tall does a wall mirror have to be in order for you to see all of yourself in the mirror? Does it matter how far you stand from the mirror?

Solution   If you stand a distance d from the mirror and your height is h, then from the drawing I see that the length L of mirror required is given by the law of similar triangles:

$$\frac{L}{h'} = \frac{d}{2d}, \quad \text{so } L = \tfrac{1}{2}h' = \tfrac{1}{2}h \quad \text{independent of } d$$

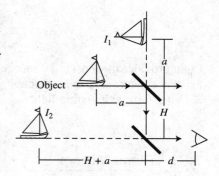

PROBLEM 31.2.   A periscope consists of two parallel plane mirrors, positioned as shown here. Let $a =$ distance from the object viewed to the first mirror, $H =$ separation of the mirrors, and $d =$ distance of your eye from the lower mirror. How far from your eye is the image you see? Is it right side up or upside down?

Solution   The image I_1 formed in the upper mirror acts as the object for the lower mirror. From the drawing, I see that the distance from the eye to image I_2 is $a + H + d$, and the image is right side up.

31.2 Spherical Mirrors

Images formed by spherical mirrors are illustrated in Fig. 31-2. Point C is the center of curvature of the mirror of radius R. The horizontal line through C is the **axis** of the mirror. The dark vertical line is the object, and the light double vertical line is the image. Point F is the **focal point** of the mirror. The distance along the axis from the focal point to the mirror is the **focal length** of the mirror. For spherical mirrors, $f = \tfrac{1}{2}R$. Observe that a ray striking the mirror parallel to the axis is reflected back through the focal point. This is what defines the focal point.

Figs. 31-2a, b, and c show **concave mirrors** (they are "caved in"). Figs. 31-2d and e are convex mirrors. Observe that the focal point for the convex mirror is to the right of the reflecting surface. I have drawn in three **principal rays** to locate the images (only two are needed, but you should know how to draw all three). Ray 1 comes in parallel to the axis and leaves along a line through the focal point F. Ray 2 goes in along a line through the center of curvature C and reflects back along the same line. Ray 3 comes in along a line directed at the focal point and leaves parallel to the axis. The reflected rays appear to come from the image. In cases a and b the light "really" passes through the image (these are **real images**). In c, d, and e the light only appears to come from the image. These are **virtual images**. You can reverse the direction of travel of any ray and it will still follow the same path (the principle of reciprocity). Thus if you place the object at what is now the image position, the image will form at the present object position.

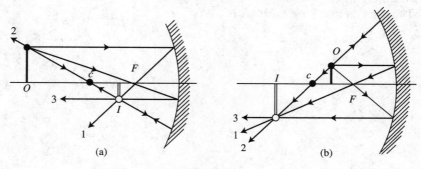

Fig. 31-2

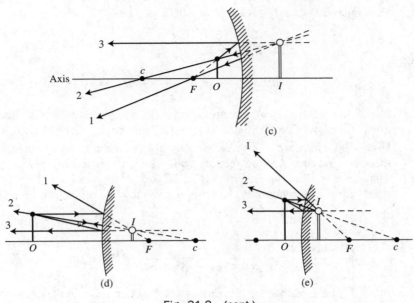

Fig. 31-2   (cont.)

Always draw ray diagrams for mirrors and lenses with the incident light moving from left to right. Using geometry, you can deduce the following equation for locating more exactly the position and size of the image.

$$\frac{1}{d_0} + \frac{1}{d_i} = \frac{1}{f} \quad \text{mirror equation}$$

(31.1)

Here

$d_0 =$ distance from the object to the mirror
$d_i =$ distance from image to the mirror
$f =$ focal length of the mirror $= \frac{1}{2}R$

d_0 and d_i are positive if the object or image is in front of the mirror, negative if behind. For a concave mirror, $f > 0$, and for a convex mirror, $f > 0$. The **magnification** is defined as

$$M = -\frac{\text{image size}}{\text{object size}} - \frac{d_i}{d_0} \quad \text{magnification}$$

(31.2)

If $M > 0$, the image is upright. If $M < 0$, the image is upside down.

PROBLEM 31.3.   You've probably noticed on the right-hand side mirror of your car the warning: CAUTION: Objects are closer than they appear. This warning appears because a convex mirror is used there (unlike the plane mirror on the left side). It forms a diminished image, like that in Fig. 31-2d. If you thought you were looking in a plane mirror (as your brain automatically assumes), the object would be much farther away than it actually is. The advantage of the convex mirror is that it provides a very wide field of view. Such mirrors are also used in stores so that a cashier can watch out for shoplifters. Suppose that in the mirror of Fig. 31-2d you see the image of a truck 2 m tall in a mirror with radius 30 cm. If the truck is 20 m behind the mirror, how large will the image be? If you saw this same image in a plane mirror, how far away would your brain think the truck is?

Solution   Here $d_0 = 20\,\text{m}$, and $f = -R/2 = -0.15\,\text{m}$. Thus

$$\frac{1}{20\,\text{m}} + \frac{1}{d_i} = \frac{1}{-0.15\,\text{m}} \quad \text{and} \quad d_1 = -0.15\,\text{m}$$

$$h' = -\frac{d_i}{d_0}h = -\frac{-0.15\,\text{m}}{20\,\text{m}}(2\,\text{m}) = 0.015\,\text{m}$$

If you saw such a small image in a flat mirror, you would think the distance to the truck was about $(2\,\text{m}/0.15\,\text{m})(20\,\text{m}) = 267\,\text{m}$ away !!! CAREFUL!

PROBLEM 31.4.   High school kids are always worried about zits. When I was an adolescent, I had one of those magnifying shaving mirrors with which I perused my physiognomy diligently. The setup is like that in Fig. 31-2c. If you place your face 15 cm from the mirror, what radius of curvature is required to provide a magnification of 1.33?

Solution $$d_0 = 0.15\,\text{m} \quad \text{and} \quad M = -\frac{d_i}{d_0} = 1.33, \quad \text{so } d_i = -1.33\,d_0 = -0.20\,\text{m}$$

$$\frac{1}{d_0} + \frac{1}{d_i} = \frac{1}{f} \quad \frac{1}{f} = \frac{1}{0.15\,\text{m}} + \frac{1}{-0.20\,\text{m}} \quad f = 0.60\,\text{m} \quad R = 2f = 1.20\,\text{m}$$

With such a large radius, the mirror looks almost flat, and you may never have noticed that it is spherical. As you can see, mirrors can magnify.

PROBLEM 31.5.   An amateur astronomer wants to build a telescope using a concave mirror to collect light from distant objects. She builds a spherical dish with a radius of curvature 4.00 m. She has a second small concave mirror of focal length 0.200 m which she proposes to place facing the big mirror on the axis of the big mirror just outside its focal point. She makes a hole in the large mirror on its axis. Her purpose is to make it possible to view the image by placing her eye (or a camera) to the right of the big mirror. It would be awkward if she put her head to the left of the big mirror where she would block the incoming light. She would like the final image formed 10 cm to the right of the big mirror. Where should she place the small mirror (see Fig. 31-2c)?

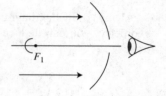

Solution   Since the stellar objects are very far away, $d_0 = \infty$ for the big mirror, so for it $1/\infty + 1/d_i' = 1/2\,\text{m}$ (since $f = \frac{1}{2}R = 2\,\text{m}$). Thus $d_i' = 2\,\text{m}$. If the object distance for the small mirror is x, the image distance must be $d_i = 2.10\,\text{m} + x$. Thus for the small mirror,

$$\frac{1}{0.20} = \frac{1}{2.10 + x} + \frac{1}{x} \quad (2.1 + x)x = 0.2x + 0.2(2.10 + x)$$

$$x^2 + 2.1x - 0.2x - 0.2x - 0.42 = 0, \quad x^2 + 1.7x - 0.42 = 0$$

$$x = \frac{-1.70 \pm \sqrt{(1.70)^2 + 4(0.42)}}{2}$$

so

$$x = 0.22\,\text{m}$$

Place the small mirror 2.22 m from the big one.

31.3 Thin Lenses

Lenses form images by refraction. They bend light much as a prism does (Fig. 31-3). By using spherical surfaces, parallel light rays may be brought to a fairly sharp focus at the focal point. In fact, the rays have to be very near the axis to get a sharp focus, but I draw them as shown for clarity. The upper lens here is a **converging lens** that bends the rays together. Any lens thicker on the axis than on the edge is a converging lens. The lower is a **diverging lens**. A lens thicker on the edge than on the axis is a diverging lens in which parallel incident rays diverge away from the focal point. The distance from the center of the lens to the focal point is the **focal length**. If light were to come from the right, it would be focused at a focal point to the left of the lens. Thus the lens has two focal points, one on each side, each a distance f from the midpoint (assuming the same outside medium on both sides).

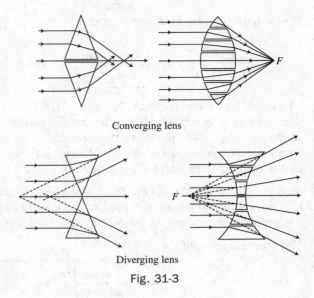

Converging lens

Diverging lens

Fig. 31-3

One can locate images for lenses by ray tracing, as shown in Fig. 31-4. Three principal rays are shown. The light is refracted at both surfaces of the lens, but for simplicity I do not show this. Ray 1 travels parallel to the axis and then leaves the lens along a line through the focal point. Ray 2 goes straight through the center of the lens without deviation. The center of the lens is like a flat plate of glass, and it displaces the ray slightly laterally but does not deviate it. Ray 3 goes through the focal point and then leaves the lens parallel to the axis. The point from which the rays appear to emanate is the top of the image. Diverging lenses only form virtual images. Converging lenses can form real or virtual images, as shown in Fig. 31-4. Only two principal rays are needed to locate an image, but you should know how to draw all three of the rays shown.

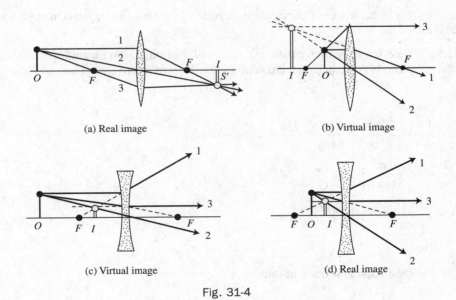

(a) Real image

(b) Virtual image

(c) Virtual image

(d) Real image

Fig. 31-4

It is possible to obtain an equation to determine image size and location. In Fig. 31-5, triangles $AA'O$ and $CC'O$ are similar triangles, so $h/h' = d_0/d_i$. Also, triangles BOF and $CC'F$ are similar, so $h/h' = f/(d_i - f)$. Equate these expressions.

$$\frac{d_0}{d_i} = \frac{f}{d_i - f} \qquad \text{or} \qquad d_0 d_i - fd_0 = fd_i$$

Divide both sides by fd_0d_i and rearrange.

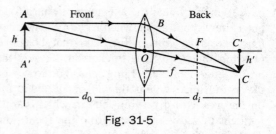

Fig. 31-5

$$\boxed{\frac{1}{d_0} + \frac{1}{d_i} = \frac{1}{f}} \quad \text{lens equation} \qquad (31.3)$$

This has the same form as the mirror equation (Eq. 31.1). It describes all of the arrangements shown in Fig. 31-4, provided that the following sign conventions are observed. The "front" of the lens is the side from which the light comes.

d_0 is positive if the object is in front of the lens, negative if in back of the lens.
d_i is positive if the image is in back of the lens, negative if in front of the lens.
f is positive for converging lenses, negative for diverging lenses.

The **magnification** of the lens is

$$\boxed{M = -\frac{\text{image size}}{\text{object size}} = -\frac{h'}{h} = -\frac{d_i}{d_0}} \quad \text{lens magnification} \qquad (31.4)$$

We can derive an expression for the focal length of a thin lens in terms of the radii of curvature of the surfaces. If the curvatures of the front and back surfaces are R_1 and R_2, respectively, and the lens is in air, then

$$\boxed{\frac{1}{f} = (n-1)\left(\frac{1}{R_1} - \frac{1}{R_2}\right)} \quad \text{lens maker's formula} \qquad (31.5)$$

R_1 and R_2 are positive if the center of curvature is in back of the lens and negative if it is in front. n is the index of refraction of the lens.

PROBLEM 31.6.   Determine the image distance, nature of the image, and magnification when an object is placed (a) 5 cm, (b) 10 cm, and (c) 20 cm from a converging lens of focal length 10 cm.

Solution

(a) $\dfrac{1}{f} = \dfrac{1}{d_0} + \dfrac{1}{d_i}$ $\qquad \dfrac{1}{10\,\text{cm}} = \dfrac{1}{5\,\text{cm}} + \dfrac{1}{d_i}$ $\qquad d_i = -10\,\text{cm}$

$M = -\dfrac{d_i}{d_0} = -\dfrac{10\,\text{cm}}{5\,\text{cm}} = -2$

Since $d_i < 0$, the image is virtual.

(b) $\dfrac{1}{10\,\text{cm}} = \dfrac{1}{10\,\text{cm}} + \dfrac{1}{d_i}$

so $d_i = \infty$ and the image is formed at infinity.

(c) $\dfrac{1}{10\,\text{cm}} = \dfrac{1}{20\,\text{cm}} + \dfrac{1}{d_i}$ $\qquad d_i = 20\,\text{cm}$ $\qquad M = -1$

The image is real ($d_i > 0$).

PROBLEM 31.7.   Combinations of lenses are often used. The image of the first lens serves as the object for the second lens. Suppose two thin coaxial lenses with focal lengths $f_1 = 25\,\text{cm}$ and $f_2 = 10\,\text{cm}$ are separated by 8 cm. An insect is positioned 6 cm from lens 1 (on the opposite side from lens 2). Where is its final image?

Solution

The image position for the first lens is d_{1i}, where $1/25\,\text{cm} = 1/6\,\text{cm} + 1/d_{1i}$. $d_{1i} = -7.9\,\text{cm}$ (virtual image). Thus the object distance for the second lens is $d_{2o} = 8\,\text{cm} + 7.9\,\text{cm} = 15.9\,\text{cm}$.

$$\frac{1}{10\,\text{cm}} = \frac{1}{15.9\,\text{cm}} + \frac{1}{d_i} \quad d_i = 27.0\,\text{cm}$$

The final image (real) is thus 27 cm from the second lens and 41 cm from the bug.

PROBLEM 31.8.

A diverging lens in conjunction with a converging lens can be used to make a "beam expander" for parallel rays, as in a laser beam. If you have a diverging lens of focal length f_1 and a converging lens of focal length f_2, by what factor will a beam be expanded?

Solution

Place the lenses so that their focal points coincide, with the incoming beam striking the diverging lens first. From the drawing I see that $R/f_2 = r/f_1$, so $R/r = f_2/f_1$.

PROBLEM 31.9.

Show that the lens equation may be written $f^2 = xx'$, where x and x' are the object and image distances measured from the focal points. This is the *newtonian* form of the lens equation.

Solution

$d_0 = f + x$ and $d_i = f + x' \dfrac{1}{f+x} + \dfrac{1}{f+x'} = \dfrac{1}{f}$. Multiply by $f(f+x)(f+x')$.

$$f(f+x') + f(f+x) = (f+x)(f+x')$$

Simplify:

$$f^2 = xx'$$

31.4   Optical Instruments

You have no doubt noticed that you can see an object such as a tree more clearly when it is up close to you. This is not because the tree is bigger when it is closer, but rather because when it is closer, the rays of light from it subtend a larger angle at your eye. If an object of height h subtends an angle θ_0 at your eye, then for small angles $\theta_0 \simeq h/d$. If you now look through an optical instrument at the object and find the image subtends an angle θ, the angular magnification is defined as

$$\boxed{m = \frac{\theta}{\theta_0}} \quad \text{angular magnification} \tag{31.6}$$

The average young adult cannot comfortably focus on objects closer than about 25 cm (the near point). Thus an object of height h at the near point will subtend an angle of $\theta_0 = h/25\,\text{cm}$ when it is as close as can be seen clearly. A relaxed eye easily focuses on an image at infinity, however, so with a **simple magnifier** (a converging lens) the object is placed at the focal point. This results in an image at infinity because

$$\frac{1}{f} = \frac{1}{d_0} + \frac{1}{d_i} = \frac{1}{f} + \frac{1}{d_i}, \quad \text{so} \quad \frac{1}{d_i} = 0 \quad \text{and} \quad d_i = \infty$$

An object of height h at the focal point a distance f from the lens will subtend an angle θ, where $\tan\theta \simeq \theta = h/f$. Thus the angular magnification of the simple magnifier is

$$m = \frac{\theta}{\theta_0} = \frac{h/f}{h/25}$$

or

$$\boxed{m = \frac{25\,\text{cm}}{f}} \quad \text{simple magnifier} \tag{31.7}$$

If the image is formed at the near point instead of at infinity, the maximum magnification can be obtained. In this case $d_i = -25$ cm (neglecting the distance from the eye to the magnifier) and $(1/-25) + 1/d_0 = 1/f$. Thus

$$d_0 = \frac{25f}{f+25} \qquad \text{and} \qquad \theta = \frac{h}{d_0} = \frac{h(f+25)}{25f}$$

$$m = \frac{\theta}{\theta_0} = \frac{h(f+25)}{25f}\left(\frac{25}{h}\right) = 1 + \frac{25}{f}$$

PROBLEM 31.10. A jeweler examines a diamond with a loupe (a simple magnifier) with a focal length of 8.0 cm. What magnification is obtained if the gem is positioned so that its image is at the normal near point, 25 cm? How far from the lens must the diamond be held?

Solution

$$m = 1 + \frac{25}{f} = 1 + \frac{25}{8} = 4.1 \qquad \frac{1}{d_0} + \frac{1}{-25} = \frac{1}{8}, \quad d_0 = 6.1 \text{ cm}$$

The diamond should be held 6.1 cm from the lens. The angular magnification will be 4.1.

A **refracting telescope** is used for viewing very distant objects (for example, stars or ships on the horizon). Incoming light first strikes an **objective lens** that forms an inverted image near its focal point (since the object distance is infinite). This image serves as the object for the eyepiece lens, and a virtual final image is then formed at infinity and viewed by the eye. Thus the focal points of the two lenses coincide, and their separation is then $f_1 + f_2$ (Fig. 31-6). If the intermediate image height is h, $\tan \theta_1 \simeq \theta_1 = h/f_1$ and $\tan \theta_2 \simeq \theta_2 = h/f_2$; then

$$\boxed{m = \frac{\theta_2}{\theta_1} = \frac{\text{focal length of objective}}{\text{focal length of eyepiece}} = \frac{f_1}{f_2} \qquad \text{telescope}} \qquad (31.8)$$

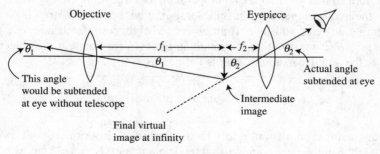

Fig. 31-6

PROBLEM 31.11. The refracting telescope at the Yerkes Observatory in Wisconsin has an objective lens focal length of 20 m and an eyepiece focal length of 2.5 cm. The diameter of the objective is 40 in (made large in order to gather more light, not for greater magnification). It is difficult to make such large lenses because of many problems, so most modern large telescopes use reflecting mirrors. What magnification of the Moon can the Yerkes telescope provide?

Solution

$$m = \frac{f_1}{f_2} = \frac{20 \text{ m}}{0.025 \text{ m}} = 800$$

A **compound microscope** is used for looking at small objects up close. An objective and an eyepiece are used (Fig. 31-7). The object of height h to be viewed is placed just outside the focal point of the objective at an object distance $d_0 \simeq f_0$. The first image is formed a distance d_i from the objective, resulting in magnification $M = -(d_i/d_0) = -(d_i/d_0)$ (see Eq. 31.4). The eyepiece is essentially a simple magnifier used to view the second image. It provides magnification $25/f_e$ (see Eq. 31.7, with f_e in centimeters). The resultant magnification of the microscope is the product of the magnification of the objective and the magnification of the eyepiece:

$$M = \frac{h_2}{h} = \frac{25d_i}{f_0 f_e} \quad \text{compound microscope} \tag{31.9}$$

Here f_0 = objective focal length, f_e = eyepiece focal length, and d_i = distance from objective lens to first intermediate image. Typically $f_0, f_e \ll d_i$, so d_i is roughly the length of the instrument (about 18 cm).

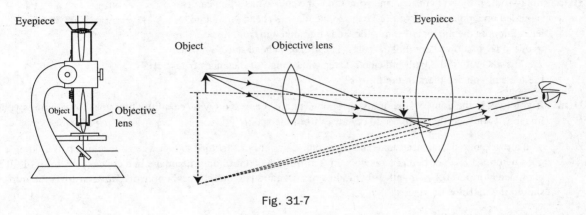

Fig. 31-7

31.5 Summary of Key Equations

Mirror or lens equation: $\dfrac{1}{f} = \dfrac{1}{d_0} + \dfrac{1}{d_i}$

Mirror or lens magnification: $M = \dfrac{h_{\text{image}}}{h_{\text{object}}} = -\dfrac{d_i}{d_0}$

Lens maker's formula: $\dfrac{1}{f} = (n-1)\left(\dfrac{1}{R_1} - \dfrac{1}{R_2}\right)$

Angular magnification: $m = \dfrac{\theta}{\theta_0}$

Simple magnifier: $m = \dfrac{25\,\text{cm}}{f}$

Telescope: $m = \dfrac{\theta_2}{\theta_1} = \dfrac{\text{objective focal length}}{\text{eyepiece focal length}}$

Microscope: $M = \dfrac{h_2}{h} = \dfrac{25d_i}{f_0 f_e}, \quad d_i \simeq \text{tube length}$

SUPPLEMENTARY PROBLEMS

31.12. A rectangular room of dimensions $a \times b$ has mirrors on three walls. A horizontal laser beam is directed through a hole in the unsilvered wall (of length a) a distance x from one end. It is reflected off each of the other walls once. It then leaves the room through the hole it entered. At what angle to the first wall should the beam be directed?

31.13. A light beam traveling in the xy plane strikes a plane mirror in the xz plane. When the mirror rotates about the z-axis through an angle ϕ, through what angle does the light beam rotate?

31.14. A person sits in a chair 3 m from a wall on which hangs a plane mirror. Three meters directly behind him stands a woman 1.6 m tall. What minimum height of plane mirror on the wall will enable him to see the woman's full height?

31.15. I play some hilly golf courses, and sometimes the course has a plane mirror mounted on a pole so that you (U) can see the players ahead on a blind tee shot. Suppose the mirror is 1.0 m wide, and you stand 4 m from it and 1.0 m to one side, as shown here (not to scale). The group G ahead starts at the tee T and walks along the path indicated. Over what range of distances from the tee can you see them in the mirror?

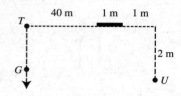

31.16. If you look in the convex side of a soup spoon (radius 3 cm), you will see a little tiny you. How large is your image if your face (height h) is 20 cm away from the spoon?

31.17. You are asked to design a servomechanism control system in a factory so that when a machine part 0.60 m tall goes by on a conveyor belt at a distance of 2 m from the control unit, its image in a concave mirror will fill a photodetector port 0.50 cm tall. What radius of curvature is required for the mirror? How far from the mirror should the detector be placed?

31.18. A short rod of length L is placed on the axis of a spherical mirror a distance d_0 from the mirror, where $L \ll d_0$. Determine the length L' of the image. The ratio L'/L is the *longitudinal magnification* of the mirror. How is this magnification related to the lateral magnification, Eq. 31.2?

31.19. Sketched here is the arrangement of mirrors used for a Cassagrain reflector astronomical telescope. A small spherical mirror is placed 2.5 m from a large concave spherical mirror of radius of curvature 6.0 m. A hole is made on the axis of the large mirror, and the design provides for the image of a distant object to be formed 0.80 m behind the large mirror, where it can conveniently be observed or photographed.

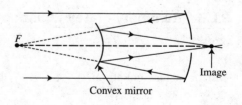

What radius of curvature is required for the small mirror, and is it concave or convex?

31.20. When fitting contact lenses or preparing for eye surgery, it is important to measure the radius of curvature of the cornea. This can be done with a keratometer, a device in which an illuminated object is placed a known distance from the eye and the size of the virtual image formed is observed and measured with a small telescope. Suppose that for an object distance of 120 mm a magnification of 0.040 is obtained. What is the radius of curvature of the cornea?

31.21. My single-lens reflex camera has two interchangeable lenses, one of focal length 55 mm and one of focal length 200 mm. A person 1.70 m tall stands 10 m away from the camera. What is the size of her image on the film for each of the two lenses? Which of these lenses do you believe is called a "telephoto lens"?

31.22. Two thin converging lenses of focal lengths f_1 and f_2 are coaxial and placed very close together. What is the effective focal length of the combination?

31.23. The *power* of a lens is defined as $P = 1/f$, where f is in meters and P is in *diopters* (optometrists use these units when they prescribe glasses). What is the power of two lenses in close contact? Do you see why this definition of lens power is reasonable?

31.24. In a common physics lab experiment a lighted source is placed a fixed distance D from a screen. A converging lens of focal length f is placed between the source and the screen. Determine the separation between the two possible images that can be formed, and calculate the ratio of their sizes.

31.25. An object is placed 60 cm in front of a diverging lens of focal length -15 cm. A converging lens of focal length 20 cm is placed 10 cm behind the first lens. Where is the final image located, and what is the overall magnification?

31.26. What is the angular magnification of a simple magnifier of focal length 5 cm if it forms an image 25 cm from the eye?

31.27. Blood cells are viewed in a microscope whose eyepiece has a focal length of 3.0 cm and an objective lens with a focal length of 0.40 cm. The distance between the eyepiece and the objective is 18 cm. If a blood cell subtends an angle of 2×10^{-5} rad when viewed with the naked eye, what angle does it subtend when viewed through the microscope?

31.28. A telescope's objective lens has a focal length of 50 cm, and the eyepiece has a focal length of 2 cm. How far apart should the lenses be to form an image at infinity? What then is the magnification? What should be the lens separation if the image is formed at the near point, 25 cm?

31.29. An object is placed 20.0 cm from a converging lens of focal length 50 mm. If the object moves toward the lens at a speed of 12 cm/s, at what speed does the image move away from the lens?

31.30. The focal length on a camera with a zoom lens can vary from 35 mm to 120 mm. You photograph a very distant object 2.0 m tall, first using the 35 mm focal length and then using the 120 mm focal length. What is the approximate ratio of the sizes of the two images?

SOLUTIONS TO SUPPLEMENTARY PROBLEMS

31.12. Draw the ray diagram here. For the ray to emerge from the hole, $\tan \theta = y/x$, where $x/a = y/b$ so $\tan \theta = b/a$. Observe that whenever a ray strikes two perpendicular mirrors, the reflected ray is parallel to the incident ray, no matter what the angle of the angle of incidence. This is also true for three perpendicular mirrors, and this is the basis of the corner cube reflector, a mirror that always sends a ray straight back. When people first went to the corner cube reflector was left there from which laser beams could be reflected. This has important uses, including measuring exactly the distance to the Moon.

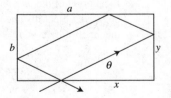

31.13. If the angle of incidence increases by angle α, the angle of reflection also increases by α. Thus the angle between the incident beam and the reflected beam increases by 2α when the incident beam rotates by angle α.

31.14. The image of the woman is 6 m behind the mirror, or 9 m from the person in the chair. Draw rays from the sitting person's eyes to the top and bottom of the image. Use the law of similar triangles with h the height of the mirror, $h/3\,\text{m} = 1.6\,\text{m}/9\,\text{m}$ and $h = 0.53\,\text{m}$.

31.15. Draw the ray diagram and use the law of similar triangles.

$$\frac{x_2}{41\,\text{m}} = \frac{2\,\text{m}}{1\,\text{m}}, \quad x_2 = 82\,\text{m} \qquad \frac{x_1}{40\,\text{m}} = \frac{2\,\text{m}}{1\,\text{m} + 1\,\text{m}}, \quad x_1 = 40\,\text{m}$$

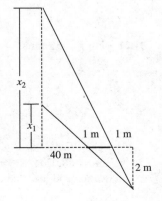

31.16. $f = \frac{1}{2}R = 1.5\,\text{cm}$ $\dfrac{1}{f} = \dfrac{1}{d_0} + \dfrac{1}{d_i}$ $\dfrac{1}{1.5\,\text{cm}} = \dfrac{1}{20\,\text{cm}} + \dfrac{1}{d_i}$

$$d_i = 1.6\,\text{cm} \qquad M = -\frac{d_i}{d_0} = -\frac{-1.6\,\text{cm}}{20\,\text{cm}} = 0.08, \quad \text{so } h' = 0.08\,h \simeq \frac{1}{13}\,h$$

31.17. Use a concave mirror:

$$M = -\frac{d_i}{d_0} = -\frac{0.005\,\text{m}}{0.60\,\text{m}}$$

$$d_i = -Md_0 = -\left(\frac{0.005\,\text{m}}{0.60\,\text{m}}\right)(2.0\,\text{m}) = 0.017\,\text{m}$$

$$\frac{1}{f} = \frac{1}{0.017\,\text{m}} + \frac{1}{2.0\,\text{m}}, \quad f = \tfrac{1}{2}R = 0.0165\,\text{m}, \quad R = 0.033\,\text{m}$$

Place the detector 1.7 cm from mirror.

31.18. Let $d_0 = x$ and $d_i = y$. Then

$$\frac{1}{f} = \frac{1}{x} + \frac{1}{y}$$

Take the derivative:

$$0 = -\frac{1}{x^2} - \frac{1}{y^2}\frac{dy}{dx}$$

But for a short rod, $L \simeq dx$ and $L' \simeq dy$, so

$$M_L = \frac{L'}{L} \simeq \frac{dy}{dx} = -\left(\frac{y}{x}\right)^2 = -\left(\frac{d_i}{d_0}\right)^2 = -M^2$$

31.19. For the large mirror, $f_1 = 1/2R_1 = 3.0\,\text{m}$. The object viewed (a star) is far away, so its image is formed at the focal point of the large mirror. The object distance for the small mirror is thus $d_0 = 2.5\,\text{m} - 3\,\text{m} = -0.50\,\text{m}$. A negative object distance means the object point is *behind* the mirror (a virtual object). The image distance for the small mirror is $d_i = -(2.5\,\text{m} + 0.8\,\text{m}) = -3.3\,\text{m}$. $d_i < 0$ because the image is in *front* of the mirror. Thus

$$\frac{1}{f_2} = \frac{1}{-0.5\,\text{m}} + \frac{1}{-3.3\,\text{m}} \qquad f_2 = \tfrac{1}{2}R_2 = -0.434\,\text{m} \qquad R_2 = 0.87\,\text{m}$$

31.20. $M = -\dfrac{d_i}{d_0}, \quad 0.040 = -\dfrac{d_i}{120\,\text{mm}}, \quad d_i = -4.8\,\text{mm}, \qquad \dfrac{1}{f} = \dfrac{1}{d_i} + \dfrac{1}{d_0} = \dfrac{1}{-4.8\,\text{mm}} + \dfrac{1}{120\,\text{mm}}$

$$f = -5.0\,\text{mm}, \qquad R = 2f, \quad R = -10.0\,\text{mm} \quad \text{convex cornea}$$

31.21. For $f = 55\,\text{mm}$,

$$\frac{1}{f} = \frac{1}{d_0} + \frac{1}{d_i}, \qquad \frac{1}{55\,\text{mm}} = \frac{1}{10^4\,\text{mm}} + \frac{1}{d_i}, \qquad d_i = 55.3\,\text{mm}$$

$$M = -\frac{d_i}{d_0} = \frac{55.3\,\text{mm}}{10^4\,\text{mm}} = -5.53 \times 10^{-3}$$

The image size $h' = Mh$ $(-5.53 \times 10^{-3})(1.70\,\text{mm} = -9.4\,\text{mm})$. For $f = 200\,\text{mm}$, find: $d_i = 204\,\text{mm}$, $M = -0.0204$, and $h' = -35\,\text{mm}$ (inverted image). A long focal length lens in a camera gives more magnification and is called a *telephoto lens*.

31.22. For the first lens, $1/f_1 = (1/d_0) + (1/d_i)$. The image distance for the first lens is the object distance for the second lens, but for it the object distance is negative (the "object" is behind the second lens). Thus

$$\frac{1}{f_2} = -\frac{1}{d_i} + \frac{1}{d_i'} = \frac{1}{d_0} - \frac{1}{f_1} + \frac{1}{d_i'}, \quad \text{so} \quad \frac{1}{f_1} + \frac{1}{f_2} = \frac{1}{d_0} + \frac{1}{d_i'}$$

Thus the effective focal length f is given by

$$\frac{1}{f} = \frac{1}{f_1} + \frac{1}{f_2} \quad \text{or} \quad f = \frac{f_1 f_2}{f_1 + f_2}$$

31.23. From Problem 30.21, $1/f = 1/f_1 + 1/f_2$, so $P = P_1 + P_2$. A short focal length means more "power," that is, more magnification, as seen in Eq. 31.7 for the simple magnifier.

31.24. Given that

$$D = d_0 + d_i \quad \frac{1}{f} = \frac{1}{d_0} + \frac{1}{D - d_0}, \quad \text{then } d_0(D - d_0) = f(D - d_0) + f d_0 \quad \text{and} \quad d_0^2 - D d_0 + fD = 0$$

$$d_0 = \frac{D \pm \sqrt{D^2 - 4fD}}{2}, \quad \text{so } d_{0+} - d_{0-} = \sqrt{D^2 - 4fD}$$

Let

$$\delta = \sqrt{D^2 - 4fD}, \quad h_+' = \frac{D + \delta}{D - \delta} h \quad \text{and} \quad h_-' = \frac{D - \delta}{D + \delta} h$$

Thus

$$\frac{h_+'}{h_-'} = \left(\frac{D + \sqrt{D^2 - 4fD}}{D - \sqrt{D^2 - 4fD}} \right)^2$$

31.25. Use the lens formula twice:

$$\frac{1}{60\,\text{cm}} + \frac{1}{d_{i1}} = \frac{1}{-15\,\text{cm}}, \quad d_{i1} = -12\,\text{cm}$$

The image of the first lens is the object for the second lens:

$$\frac{1}{22\,\text{cm}} + \frac{1}{d_{i2}} = \frac{1}{20\,\text{cm}}, \quad d_{i2} = 220\,\text{cm}$$

Thus the final image is 220 cm to the right of lens 2:

$$M = M_1 M_2 = \left(-\frac{d_{i1}}{d_{01}} \right) \left(-\frac{d_{i2}}{d_{02}} \right) = \left(-\frac{-12}{60} \right) \left(-\frac{220}{22} \right) = -2 \quad \text{inverted}$$

31.26. From Eq. 31.7,

$$m = \frac{25}{f} = \frac{25}{5} = 5$$

31.27. From Eq. 31.9,

$$M = \frac{h_2}{h} = \frac{\theta_2}{\theta} = \frac{25d_i}{f_e f_0} = \frac{25(18\,\text{cm})}{(3\,\text{cm})(0.40\,\text{cm})} = 375$$

I assume d_i is approximately equal to the length of the microscope tube. Thus, $\theta_2 = 375\,\theta = 375(2 \times 10^{-5}\,\text{rad}) = 7.5 \times 10^{-3}\,\text{rad}$.

31.28. From Fig. 31-6. I see that the lens spacing is approximately $L = f_0 + f_e = 50\,\text{cm} + 2\,\text{cm} = 52\,\text{cm}$. The magnification (Eq. 31.8) is

$$m = \frac{f_0}{f_e} = \frac{50\,\text{cm}}{2\,\text{cm}} = 25$$

If the final image is formed at the near point, 25 cm, then $d_{ie} = -25\,\text{cm}$ and

$$\frac{1}{f_e} = \frac{1}{d_{0e}} + \frac{1}{d_{ie}} \quad \text{or} \quad \frac{1}{2\,\text{cm}} = \frac{1}{d_{0e}} + \frac{1}{-25\,\text{cm}}, \quad d_{0e} = 1.85\,\text{cm}$$

In this case, the lens separation should be $L = f_0 + d_{ie} = 50\,\text{cm} + 1.85\,\text{cm} = 51.85\,\text{cm}$.

31.29.

$$\frac{1}{d_0} + \frac{1}{d_i} = \frac{1}{f}$$

$$-\frac{1}{d_0^2}\frac{dd_0}{dt} - \frac{1}{d_1^2}\frac{dd_1}{dt} = 0, \quad v_0 = \frac{dd_0}{dt}, \quad v_i = \frac{dd_i}{dt}$$

$$-\frac{1}{d_0^2}v_0 - \frac{1}{d_i^2}v_i$$

$$\frac{1}{20} + \frac{1}{d_i} = \frac{1}{5}, \quad d_i = 6.7\,\text{cm}$$

$$v_i = -\left(\frac{6.7}{20}\right)^2(-12\,\text{cm/s})$$

$$v_i = 1.3\,\text{cm/s}$$

31.30. From Eq. 31.4, $\dfrac{h'}{h_0} = \dfrac{d_i}{d_0}$

$$\frac{1}{d_0} + \frac{1}{d_i} = \frac{1}{f}$$

$$\frac{1}{d_i} = \frac{1}{f} - \frac{1}{d_0}$$

If $d_0 \gg f$, $d_i \simeq f$

The two image sizes are

$$h_1 \simeq \frac{d_{i1}h_0}{d_0}, \quad h_2 \simeq \frac{d_{i2}h_0}{d_0}$$

So $\dfrac{h_1}{h_2} = \dfrac{d_{i1}}{d_{i2}} \simeq \dfrac{f_1}{f_2}$

$$\frac{h_1}{h_2} \simeq \frac{120\,\text{mm}}{35\,\text{mm}} = 3.4$$

The long focal length lens produces a larger image.

CHAPTER 32

Interference

Imagine a light wave that is like a very long sine wave. If you start at a crest and move an integral number of wavelengths, you will come to another crest. Such a wave is **coherent**. Laser beams have this property, and that is perhaps their most important characteristic. Sunlight, on the other hand, consists of lots of short segments of a sine wave, and the phase of one segment is not related to another. Two coherent light waves of the same wavelength may **interfere**. This means they can reinforce each other and produce a larger amplitude (**constructive interference**), or they may cancel each other out (**destructive interference**). In Fig. 32-1, the top two waves on the left are **in phase**. They add together to produce the lower curve there. The top two waves in the center are **out of phase** (more exactly, 180° out of phase). They cancel, as indicated by the straight line at the bottom. The waves on the right have some phase difference between 0° and 180°, and they result in a wave whose amplitude is reduced and whose phase is shifted.

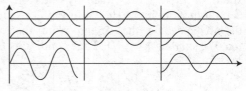

Fig. 32-1

32.1 Double Slit Interference

In 1801 Thomas Young performed a brilliant experiment that established the wave nature of light. He shined light on a narrow slit which then acted like a point source. In Fig. 32-2 the circles emanating from S_0 represent the wave fronts of the light (crests of the wave). The light then struck a second screen containing two slits. Each of these slits acted as a secondary source emitting coherent light. Two sets of wave fronts emanate from these slits. Where the crests coincide, the light is reinforced, and where a crest falls on top of a trough, the light cancels. As a result light directed along the black dots is bright, and light directed along the small circles cancels. On the screen one sees a pattern of bright and dark bands.

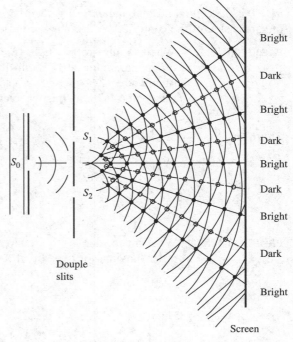

Fig. 32-2

343

Let $L =$ path length from a slit to the screen. If the difference in path lengths is $\Delta L = \lambda, 2\lambda, 3\lambda, \ldots, n\lambda$, the waves will interfere constructively and a bright band will appear. If $\Delta L = 1/2\lambda, 3/2\lambda, 5/2\lambda, \ldots, n/2\lambda$, the waves will cancel and a dark band will appear. I assume that the distance D to the screen is large compared to the slit spacing d, so the rays go nearly parallel toward the screen. From Fig. 32-3 you can see that $\Delta L = d \sin \theta$. Thus the positions of bright and dark bands on either side of the central maximum are given by

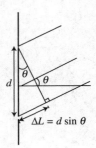

$$d \, \sin \theta = m\lambda, \quad m = 0, 1, 2, \ldots, n \quad \text{bright} \qquad (32.1)$$

$$d \, \sin \theta = \left(m + \tfrac{1}{2}\lambda\right), \quad m = 0, 1, 2, \ldots, n \quad \text{dark} \qquad (32.2)$$

One can work out the intensity for any θ (see Section 32.3 below) with the result that

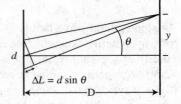

$$\boxed{I = 4I_0 \cos^2\left(\frac{\pi d \, \sin \theta}{\lambda}\right)} \qquad (32.3)$$

Here I_0 is the intensity from one of the slits alone. This is valid for small θ. In fact, the intensity falls off as one goes away from the central maximum, as shown in Fig. 32-4. In practice $D \gg y$, where y is the distance on the screen from the central axis to a band, so $\tan \theta \simeq \sin \theta \simeq \theta = y/D$. Thus conditions for bright or dark bands are

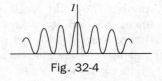

$$m = 0, 1, 2, \ldots, n, \qquad y_m = \frac{\lambda D}{d} m \qquad \text{bright} \qquad (32.4)$$

Fig. 32-4

$$y_m = \frac{\lambda D}{d}\left(m + \tfrac{1}{2}\right) \qquad \text{dark} \qquad (32.5)$$

$m = 0$ is the central bright band, $m = 1$ is the first-order bright band (one on either side of the center), $m = 2$ is the second-order, and so on.

PROBLEM 32.1.   When sodium yellow light ($\lambda = 589 \, \text{nm}$) is used in a double slit experiment, the first-order maximum is 0.035 cm from the central maximum. When another light source of unknown wavelength is used, the first-order maximum occurs at 0.032 cm from the center. What is this wavelength?

Solution
$$y_1 = \frac{\lambda_1 D}{d}(1) \qquad \text{and} \qquad y_2 = \frac{\lambda_2 D}{d}(1)$$

Divide:

$$\frac{\lambda_2}{\lambda_1} = \frac{y_2}{y_1} \qquad \lambda_2 = \left(\frac{0.032 \, \text{cm}}{0.035 \, \text{cm}}\right)(589 \, \text{nm}) = 539 \, \text{nm}$$

PROBLEM 32.2.   What slit spacing is required to give a separation of 2 cm between the second- and third-order maxima for two slits if $\lambda = 550 \, \text{nm}$ and $D = 1.50 \, \text{m}$?

Solution
$$y_3 = \frac{\lambda D}{d}(3) \qquad \text{and} \qquad y_2 = \frac{\lambda D}{d}(2)$$

$$y_3 - y_2 = \frac{\lambda D}{d}(3 - 2) \qquad d = \frac{\lambda D}{\Delta y}$$

$$d = \frac{(550 \, \text{nm})(1.5 \, \text{m})}{0.02 \, \text{m}} = 0.041 \, \text{mm}$$

32.2 Multiple Slit Interference and Phasors

Consider the two waves radiating from the two slits of Fig. 32-3. Let the electric field of the upper be $E_1 = E_0 \sin \omega t$ and that of the lower ray be $E_2 = E_0 \sin(\omega t + \phi)$. The difference in phase ϕ is due to the difference in path length ΔL. $\phi = 2\pi(\Delta L/\lambda)$. I wish to add these two electric fields together to obtain the resultant field E. I represent each electric field by a phasor that rotates with angular frequency ω, as shown in Fig. 32-5a. The vector sum of $\mathbf{E}_1$ and $\mathbf{E}_2$ is the resultant field $\mathbf{E}$. There will be a phase angle β between $\mathbf{E}$ and $\mathbf{E}_1$, and from the isosceles triangle in Fig. 32-5b I see that $2\beta = \phi$. Thus from the triangle, $E = 2E_0 \cos \beta = 2E_0 \cos \frac{1}{2}\phi$. The intensity of a wave is proportional to the square of its amplitude. If I is the intensity of the resultant wave, and I_0 is the intensity of one of the individual waves, then

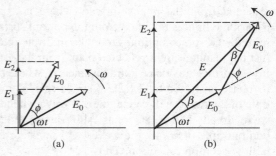

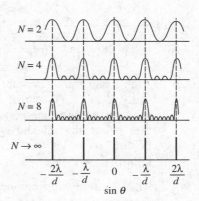

(a)   (b)

Fig. 32-5

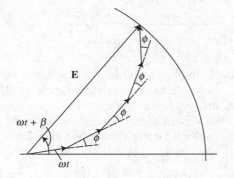

$$\frac{I}{I_0} = \frac{E^2}{E_0^2} = \frac{1}{E_0^2}\left(2E_0 \cos \tfrac{1}{2}\phi\right)^2$$

$$I = 4I_0 \cos^2 \frac{1}{2}\phi, \quad \phi = 2\pi\frac{\Delta L}{\lambda} = 2\pi\frac{d \sin \theta}{\lambda}$$

$$= 4I_0 \cos^2\left(\frac{\pi d \sin \theta}{\lambda}\right)$$

This is Eq. 32.4.

One can apply this method of phasor addition to any number of slits, as in Fig. 32-6. As the number of slits is increased, the principal maxima become sharper, and secondary maxima appear between them. In Fig. 32-7 the pattern due to N slits is shown. As $N \to \infty$, the lines become very sharp. The principal maxima occur when the difference in path length satisfies $\Delta L = m\lambda$ and m = integer.

32.3 Interference in Thin Films

Almost all interference effects result from splitting a light beam and allowing one beam to travel a different path than the other, resulting in a phase shift between them when they are recombined. Interference effects in thin films are a striking example of this. First note the following important points:

- When light traveling in a medium of index of refraction n_1 is reflected from a medium of index n_2, it undergoes a 180° phase shift if $n_2 > n_1$, and it undergoes no phase shift if $n_1 > n_2$.

• When light of wavelength λ travels in a medium whose index of refraction is n, its wavelength is $\lambda_n = v/f = c/nf = \lambda/n$.

In Fig. 32-8, a light ray is incident normally on a thin soap film in air. (For clarity I drew the rays at a slight angle to the normal.) Part of the light is reflected from the upper surface, where it experiences a 180° phase shift. Part of the light is reflected from the bottom surface. Some of the light is transmitted, and some continues to make multiple internal reflections. I will look at just the resultant light reflected back from the top. The light reflected from the lower surface travels an extra distance $2t$, where t = film thickness. This means its phase is shifted by $2\pi(2t/\lambda_n) = (4n\pi t/\lambda)$ rad. If this phase shift is a multiple of 2π rad, it will cancel the top wave when they interfere (because the top wave already had a 180° phase shift due to reflection). If the phase shift is an odd multiple of π rad, the waves will interfere constructively and bright light will be seen on reflection. Thus

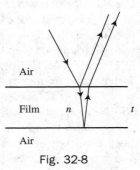

Air

Film n t

Air

Fig. 32-8

$$2nt = \left(m + \tfrac{1}{2}\right)\lambda \quad (m = 0, 1, 2, \ldots, n) \quad \text{maximum reflection} \qquad (32.6)$$

$$2nt = m\lambda \quad (m = 0, 1, 2, \ldots, n) \quad \text{minimum reflection} \qquad (32.7)$$

PROBLEM 32.3. After a rain you've probably noticed brightly colored films of gasoline or oil floating on water puddles in the street. The white sunlight falling on the film contains all colors, but some wavelengths (colors) are more strongly reflected than others in various parts of the film where the thickness varies. What minimum thickness of gasoline ($n = 1.40$) floating on water ($n = 1.33$) will result in strong destructive interference for blue light ($\lambda = 470$ nm)?

Solution The light reflected from the top surface undergoes a 180° phase shift. The light reflected from the water does not undergo a reflection phase shift since $n_{\text{gas}} > n_{\text{water}}$. Thus

$$2nt = \lambda \quad \text{and} \quad t = \frac{470\,\text{nm}}{2(1.4)} = 168\,\text{nm}$$

PROBLEM 32.4. Optical lenses ($n = 1.40$) in cameras are coated with a thin film of magnesium fluoride ($n = 1.38$) to reduce reflections. What minimum thickness is required to give minimum reflection at $\lambda = 550$ nm?

Solution Waves reflected from the air–MgF$_2$ surface and from the MgF$_2$–glass surface both experience a 180° phase shift due to reflection from a medium of greater index, so minimum reflection will occur if

$$t = \frac{1}{2}\lambda n = \frac{1}{2n}\lambda = \frac{550\,\text{nm}}{2(1.38)}$$

$$= 199\,\text{nm}$$

Monochromatic light reflected from an **air wedge** formed between two pieces of glass shows interference effects, as indicated in Fig. 32-9. Light is directed down perpendicular to the top plate, and some of it is reflected from the top surface of the

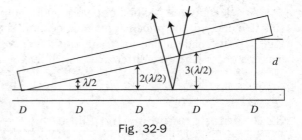

$\lambda/2$ $2(\lambda/2)$ $3(\lambda/2)$ d

D D D D D

Fig. 32-9

wedge, undergoing no reflection phase shift. Some light continues on and is reflected from the bottom surface of the wedge at the air–glass interface. This wave undergoes a 180° reflection phase shift. It continues back into the air above the wedge where it interferes with the first reflection. From above one sees a series of bright and dark D bands. The waves will cancel if the travel distance across the wedge is zero or a multiple of $\tfrac{1}{2}\lambda$.

PROBLEM 32.5. Observation of the interference pattern in a thin air wedge provides a technique for measuring very small thicknesses. Suppose a thin plastic sheet of thickness d is placed between two flat pieces of glass, as in Fig. 32-9. The width of the glass is $\omega = 6.00$ cm. Light of wavelength $\lambda = 550$ nm is incident from above, and a series of bright and dark bands are observed. The separation between the dark bands is measured to be $x = 0.48$ mm. What is the thickness of the plastic sheet?

Solution The wedge thickness must increase by $\frac{1}{2}\lambda$ between successive dark bands, so if θ is the angle of the wedge, then $\tan\theta = d/\omega = 0.5\lambda/x$,

Thus

$$d = \frac{(0.5)(550 \times 10^{-9}\,\text{m})(60\,\text{mm})}{0.48\,\text{mm}} = 0.034\,\text{mm}$$

A variation of the air wedge experiment is used in grinding lenses. A convex lens is placed on an optical flat surface and illuminated from above with monochromatic light (Fig. 32-10). Viewed from above, a number of concentric dark rings are observed (**Newton's rings**), with a central dark spot. Any distortion from a pure circle indicates an irregularity in the lens.

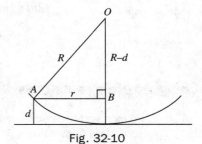

Fig. 32-10

PROBLEM 32.6. Using light with $\lambda = 580$ nm, the radius of the 16th dark ring is observed to be 8.0 mm. What is the radius of curvature of the lens?

Solution The gap width increases by $\frac{1}{2}\lambda$ from one dark ring to the next, so in Fig. 32-10, $d = (16)(\frac{1}{2}\lambda) = 8\lambda$. From the drawing I see that

$$R^2 = r^2 + (R - d)^2 = r^2 + R^2 - 2Rd + d^2$$

$d^2 \ll r^2$ so neglect it. Thus

$$2Rd = r^2 \quad \text{and} \quad R = \frac{r^2}{2d} = \frac{(8 \times 10^{-3}\,\text{m})^2}{2(8)(580 \times 10^{-9}\,\text{m})} = 6.90\,\text{m}$$

32.4 The Michelson Interferometer

An interferometer is an instrument used to measure extremely small displacements. The **Michelson interferometer** is shown in Fig. 32-11. Monochromatic light from a source strikes a partially silvered mirror P. Part of the light, ray A, is transmitted and continues on to strike mirror M_A. It is reflected back, and part of the ray is reflected down to the detector. Ray B is reflected up to hit mirror B. It is reflected back, and part goes through the mirror M to the detector. If the paths for the two rays are the same, they will interfere constructively. If now mirror B is moved back, there will be a path difference. If M_B is moved back $\frac{1}{4}\lambda$, the path length will change by $\frac{1}{2}\lambda$ (because the beam goes up and back) and the rays will interfere destructively. Thus as M_B is moved, the field of view will go alternately bright and dark. By counting the number of changes, we can measure very accurately how far M_B moves. Plate CP is inserted as a compensating plate to equalize the distances traveled in glass by the two beams.

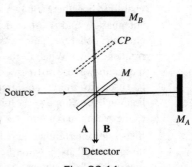

Fig. 32-11

PROBLEM 32.7.   Light of 587.5 nm is used in a Michelson interferometer. When the movable mirror is moved, 1780 changes from dark to dark patterns are observed. How far was the mirror moved?

Solution
$$x = N\frac{\lambda}{2} = (1780)\left(\frac{587.5\,\text{nm}}{2}\right) = 5.229 \times 10^{-4}\,\text{m}$$

32.5   Summary of Key Equations

Double slit interference: $d \sin\theta = m\lambda, \quad m = 0, 1, 2, \ldots \qquad$ bright

$\qquad\qquad\qquad\qquad\qquad d \sin\theta = \left(m + \frac{1}{2}\right)\lambda, \quad m = 0, 1, 2, \ldots \qquad$ dark

Thin film in air: $\qquad 2nt = \left(m + \frac{1}{2}\right)\lambda, \quad m = 0, 1, 2, \ldots \qquad$ bright

$\qquad\qquad\qquad\qquad 2nt = m\lambda, \quad m = 0, 1, 2, \ldots \qquad$ dark

SUPPLEMENTARY PROBLEMS

32.8.   Two narrow slits are separated by 0.10 mm and are illuminated with light of wavelength 550 nm. A screen is placed 2.2 m from the slits. What is the lateral separation of the first- and second-order maxima on the screen?

32.9.   What minimum thickness of soap film ($n = 1.33$ nm) in air will show very little reflected blue light ($\lambda = 480$ nm) and strongly reflected red light ($\lambda = 640$ nm)?

32.10.   In Fig. 32-9 a nylon glass fiber of diameter 0.081 mm is placed between the plates, and the plates are illuminated from above with light of wavelength 550 nm. If the width of the plates is 4 cm, how many dark bands would you see across the width of the plates?

32.11.   Light of wavelength 610 nm is incident at 45° on a soap film ($n = 1.33$). What is the minimum film thickness for which the reflected light will appear bright?

32.12.   In the Lloyd's mirror experiment shown here, light striking the screen directly from the source interferes with light reflected from the mirror. Here $L \gg x, d$. For what values of x will a bright spot appear on the screen?

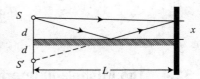

32.13.   Interference effects can be used to provide an aircraft guidance system for use in low visibility weather. Shown here are two antennas A_1 and A_2 spaced 50 m apart and emitting coherent radio waves of 33-MHz frequency. (a) What is the wavelength of the radio waves? (b) How can this system be used to guide an aircraft to a safe landing? (c) Suppose the plane "locks on" to the wrong maximum, for example, to the

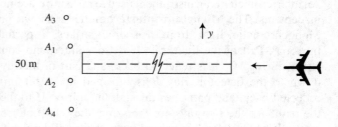

first-order interference maximum. If the aircraft is 2400 m from the antennas, how far horizontally in the y direction will its position be in error? (d) Explain how the pilot can detect having locked on to the wrong maximum if the antenna radiates two frequencies simultaneously, such as 33 and 35 MHz, both of which can be detected independently in the aircraft. (e) Why is it important that the frequencies not be in the ratio of two integers, for example, 3 : 4? (f) What would be the effect of adding two more antennas A_3 and A_4, each spaced 50 m from the adjacent antenna?

32.14. A Michelson interferometer can be used to make a careful measurement of the index of refraction of a gas. This might be useful for detecting pollutants in the atmosphere or for investigating the atmosphere of Venus with a space probe. An evacuated tube is placed in one arm of the interferometer, and the path lengths are adjusted so that the field of view is dark. Now gas is slowly allowed to enter the evacuated tube. Its presence changes the optical path length, and N changes from dark to dark are observed (called N *fringes*). Derive an expression for the index of refraction of the gas in terms of the vacuum wavelength λ of the light, the length of the tube L, and N.

32.15. A Michelson interferometer is adapted to measure the displacement of an ear drum when sound impinges on it. If green light of wavelength 552 nm is used and 94 fringes (dark to dark bands) pass by the detector, how far did the mirror move?

32.16. A sheet of plastic of index of refraction 1.50 is placed over one slit in a double slit interference experiment. When the slits are then illuminated with light of wavelength 640 nm, the center of the screen appears dark, rather than bright. What is the minimum thickness of the plastic sheet that will produce this effect?

32.17 A double slit experiment can be used to determine the wavelength of a light beam. In such an experiment, monochromatic light is incident on two slits separated by 0.025 mm. The screen on which the interference pattern is viewed is 1.25 m from the slits. The second order bright fringe is 5.2 cm from the central bright fringe. What is the wavelength of the light used?

SOLUTIONS TO SUPPLEMENTARY PROBLEMS

32.8. From Eq. 32.5,

$$y_2 - y_1 = \frac{\lambda D}{d}(m_2 - m_1) = \frac{(550\,\text{nm})(2.2\,\text{m})}{0.1 \times 10^{-3}\,\text{m}} = 1.1\,\text{mm}$$

32.9. From Eqs 32.7 and 32.8,

$$2nt = \left(m_1 + \tfrac{1}{2}\right)\lambda_1 \quad \text{dark}$$

$$2nt = m_2\lambda_2 \quad \text{bright}$$

$$\left(m_1 + \tfrac{1}{2}\right)(640\,\text{nm}) = m_2(480\,\text{nm}), \qquad 4\left(m_1 + \tfrac{1}{2}\right) = 3m_2, \qquad 4m_1 + 2 = 3m_2$$

By trial and error,

$$m_1 = 1, \, m_2 = 3, \qquad t = \frac{m_2\lambda_2}{2n} = \frac{(3)(480\,\text{nm})}{2(1.33)} = 540\,\text{nm}$$

32.10. Between each band the gap width increases by $1/2\lambda$, so if $\omega =$ band separation, then

$$\frac{0.5\lambda}{\omega} = \frac{0.081\,\text{mm}}{40\,\text{mm}}$$

Number of bands: $N = 40\,\text{mm}/\omega = 0.081\,\text{mm}/0.5(550\,\text{nm}) = 295$ bands. Note that the result is independent of 4-cm width.

32.11. $n_1 \sin\theta_1 = n_2 \sin\theta_2,$ (1) $\sin 45° = 1.33 \sin\theta_2,$ $\theta_2 = 32.1°$

The path length in the film is

$$2d\cos\theta_2 = \frac{\lambda_n}{4} = \frac{\lambda}{4n} \qquad d = \frac{\lambda}{8n\cos\theta_2} = \frac{610\,\text{nm}}{8(1.33)(\cos 32.1°)} = 68\,\text{nm}$$

32.12. The reflected ray acts as if it came from a source at S'. This results in the equivalent of double slit interference, with d replaced by $2d$. Also, the reflected ray undergoes a 180° phase shift on reflection, so the position of a bright spot is that described by Eq. 32.6:

$$y_m = \frac{\lambda L}{2d}\left(m + \tfrac{1}{2}\right)$$

32.13. (a) $\lambda = \dfrac{c}{f} = \dfrac{3 \times 10^8 \, \text{m/s}}{33 \times 10^6 / \text{s}} = 9.1 \, \text{m}$

(b) The antennas produce a double slit interference pattern. The plane should "lock on" to the central maxima. This is done by flying in a direction to keep the received signal strong.

(c) $d \sin \theta = \lambda$ $\sin \theta = \dfrac{9.1 \, \text{m}}{50 \, \text{m}} = 0.18$ $y \simeq D \sin \theta = (2400 \, \text{m})(0.18) = 440 \, \text{m}$

(d) Both signals are a maximum when locked to the central maxima. However, if the pilot is locked to the first-order maxima of one signal the other receiver will not be locked to a maxima so to stay on course the pilot must keep *both* receivers locked to a maximum. Then she knows she is headed for the runway.

(e) If the frequencies were in the ratio of, say, 2 : 3, the third maxima from the shorter wavelength could fall on top of the second maximum of the longer wavelength and mislead the pilot into thinking she was locked on two central maxima.

(f) Four antennas act like four slits and produce a sharper and more intense pattern of interference maxima, so the flight path will be more sharply defined (see Fig. 32-7).

32.14. Initially the number of wavelengths in the tube length was $N_1 = L/\lambda$. With the gas in, $N_2 = L/\lambda_n = nL/\lambda$. Thus the number of fringes passing the viewer is $N = N_2 - N_1 = (n - 1)(L/\lambda)$.

32.15. The mirror moved $\frac{1}{2}\lambda$ for each fringe, so $x = \frac{1}{2}(94)(552 \, \text{nm}) = 2.59 \times 10^{-5} \, \text{m}$.

32.16. For sheet thickness t the number of wavelengths in the sheet is $t/(\lambda/n) = n(t/\lambda)$.

Here λ = wavelength in vacuum and λ/n is the wavelength in the plastic.

The number of wavelengths in air in distance t is t/λ.

To cause destructive interference at the center of the screen we require

$$n\frac{t}{\lambda} - \frac{t}{\lambda} = \frac{1}{2}, \quad \text{so } t = \frac{\lambda}{2(n-1)} = \frac{640 \, \text{nm}}{2(1.5-1)} = 640 \, \text{nm}$$

32.17. $d \sin \theta = m\lambda$

$\sin \theta; \dfrac{y}{D}$

$\dfrac{dy_m}{D} = m\lambda$

$\lambda = \dfrac{dy_m}{Dm} = \dfrac{(0.025 \times 10^{-3} \, \text{m})(0.052 \, \text{m})}{(1.25 \, \text{m})(2)}$

$\lambda = 520 \, \text{nm}$

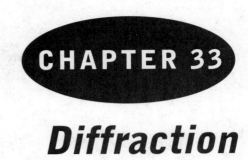

Diffraction

Diffraction and interference are the same phenomenon, and it is unfortunate that two different terms are used. Both involve the interaction (interference) of two waves. These effects occur for all waves, not just light. The term *diffraction* is used when light passes through a single opening or next to an opaque object and one part of the wave front interferes with another part of the same wavefront. Because of the wave nature of light and the spreading out of secondary waves, the shadow of something like a razor blade is not sharply defined. Instead, the edge of the shadow is a series of closely spaced light and dark regions. These are very closely spaced and hard to see with the naked eye. In a similar fashion, if light is shined through a hole onto a screen, the result is not a single, well-defined spot of light but rather a series of light and dark rings. All of these effects have important consequences. In the following, I will consider cases where the incident light rays are parallel (Fraunhofer diffraction). In practice, a lens is used to focus these rays on a screen. The case of nonparallel rays (Fresnel diffraction) is more complicated, but it shows the same qualitative features.

33.1 Single Slit Diffraction

Consider monochromatic plane waves incident on a narrow slit of width b (Fig. 33-1). In Fig. 33-1a I have drawn many little arrows to suggest that the wavefront can be imagined to consist of many coherent secondary sources. Rays going straight ahead are all in phase and when focused on a screen, will give a bright band. Now consider Fig. 33-1b. Consider the two rays C and B. Treat these as two light sources a distance $\frac{1}{2}b$ apart. They give an interference pattern much like that due to double slits; that is, when $(\frac{1}{2}b)\sin\theta = \frac{1}{2}\lambda$, these two waves cancel, as shown. Now consider another pair of waves just below C and B. They, too, will cancel. Keep moving down the slit in this fashion and you will see that each ray can be paired off with another ray that will cancel it. Hence the condition for the first minimum is $b\sin\theta = \lambda$. We can continue this process by dividing the slit into 4 zones, as in Fig. 33-1c. At the angle shown there, the path difference between adjacent rays is $\frac{1}{2}\lambda$, and again as we move up across the slit, the rays cancel in pairs. In this case the effective "slit spacing" is $\frac{1}{4}b$, so $(\frac{1}{4}b)\sin\theta = \frac{1}{2}\lambda$ or $b\sin\lambda = 2\lambda$. We can next divide the slit width into 6 zones, then 8 zones, and so on. The result in general is

$$b\sin\theta = m\lambda, \quad m = 1, 2, 3, \ldots, n \quad \text{single slit minima} \tag{33.1}$$

Do not confuse this with the condition for *maxima* in the double slit experiment, even though the equations appear similar.

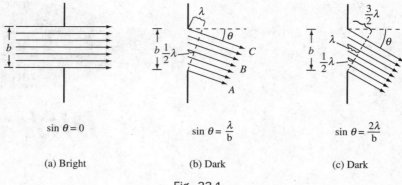

$$\sin \theta = 0 \qquad\qquad \sin \theta = \frac{\lambda}{b} \qquad\qquad \sin \theta = \frac{2\lambda}{b}$$

(a) Bright   (b) Dark   (c) Dark

Fig. 33-1

I can work out the intensity $I(\theta)$ using phasors. I want to find the resultant electric field $E(\theta)$ at point P on the screen. Imagine the slit divided into many narrow zones, each of width Δy (Fig. 33-2). Treat each as a secondary source of light contributing electric field amplitude ΔE to the field at P. The difference in path length between adjacent zones is $\Delta L = \Delta y \sin \theta$, so the difference in phase between adjacent zones is

$$\Delta \phi = \frac{2\pi \Delta y \sin \theta}{\lambda}$$

Now add up all of the incremental $\Delta \mathbf{E}$'s to obtain the field $\mathbf{E}(\theta)$ at point P. Do this vectorially by placing the vectors together head-tail-head-tail (Fig. 33-3). Each small vector is rotated from the previous one by $\Delta \phi$, as in Fig. 32-6. In the limit $\Delta y \to 0$, the total phase shift is

$$\phi = \sum (\Delta \phi)_i = 2\pi \frac{\sin \theta}{\lambda} \sum \Delta y_i = 2\pi \sin \theta \frac{b}{\lambda}$$

As $\Delta y \to 0$, the line of small vectors becomes an arc of length E_0, where $E_0 = R\phi$. E_θ is one side of an isosceles triangle, so $\frac{1}{2} E_\theta = R \sin \frac{1}{2}\phi$. Note that

$$R = \frac{E_0}{\phi}, \quad \text{so } E_\theta = 2E_0 \left[\frac{\sin(\phi/2)}{\phi} \right]$$

When $\theta = 0$, $\Delta \phi = 0$ and all of the small vectors are parallel (Fig. 33-2b), giving a large resultant (the central maximum) of magnitude E_0. It is convenient to let $\beta = \frac{1}{2}\phi$. The intensity is proportional to E^2, so

$$\boxed{I = I_0 \left(\frac{\sin \beta}{\beta} \right)^2, \quad \text{where } \beta = \pi \frac{b}{\lambda} \sin \theta \quad \text{single slit diffraction}} \qquad (33.2)$$

Here $b = $ slit width, and $I_0 \propto E_0^2$.

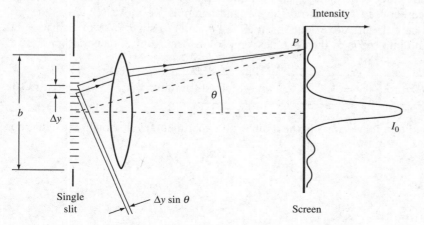

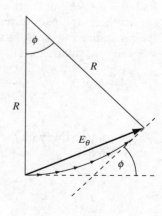

Fig. 33-2   Fig. 33-3

Fig. 33-4 is a qualitative sketch of the intensity (the central maximum is actually much larger). The first minimum occurs when all of the small phasors add to zero; that is, they form a closed circle, as in Fig. 33-5b. The first secondary maximum occurs when they go around $1\frac{1}{2}$ times (Fig. 33-5c). We can envision in this way how all of the maxima and minima occur. The minima occur when $\beta = m\pi$, where $m = 1, 2, 3, \ldots, n$ in Eq. 33.2. In terms of θ,

$$b\sin\theta = m\lambda \qquad m = 1, 2, 3, \ldots, n \qquad \text{single slit minima} \qquad (33.3)$$

This is the same result obtained in Eq. 33.1. Observe that a narrower slit gives a more spread out diffraction pattern, as does a longer wavelength.

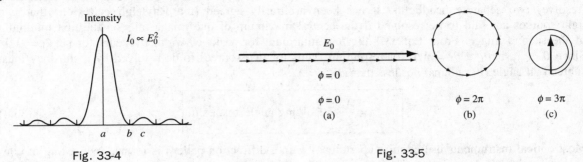

Fig. 33-4   Fig. 33-5

PROBLEM 33.1. In a lecture demonstration a helium–neon laser ($\lambda = 632.8$ nm) illuminates a slit of width 0.12 mm. The diffraction pattern is observed on a screen 12 m from the slit. What is the width of the central maximum (between the first minima on either side)?

Solution For $m = 1$,

$$\sin\theta \simeq \theta = \frac{\lambda}{b} = \frac{(632.8 \text{ nm})}{0.12 \times 10^{-3} \text{ m}} = 5.27 \times 10^{-3} \text{ rad}$$

$$y = D\theta = (12 \text{ m})(5.27 \times 10^{-3} \text{ m}) = 0.063 \text{ m}, \quad \text{so } \Delta y = 2y = 0.126 \text{ m}$$

PROBLEM 33.2. Determine approximately the ratio of the intensity of each of the first two secondary maxima to that of the intensity of the central maximum for a single slit diffraction pattern.

Solution The secondary maxima occur approximately halfway in between the minima; that is,

$$\beta = \frac{3\pi}{2} \quad \text{and} \quad \frac{5\pi}{2}$$

Thus

$$\frac{I_1}{I_0} = \left(\frac{\sin\beta}{\beta}\right)^2 = \left[\frac{\sin(3\pi/2)}{(3\pi/2)}\right]^2 = 0.045$$

$$\frac{I_2}{I_0} = \left[\frac{\sin(5\pi/2)}{(5\pi/2)}\right]^2 = 0.016$$

33.2   Resolution and Diffraction

Consider two distant incoherent light sources that illuminate a single slit as in Fig. 33-6. According to geometric optics, we would expect to see two distinct bright spots on the screen at points P_1 and P_2. However, because of the wave nature of light, we see instead a diffraction pattern consisting of a series of bright and dark bands. If the angle α is sufficiently large, the two central maxima will be sufficiently separated so that they are clearly two bright bands. However, as α decreases, the patterns move together, as shown in Fig. 33-7. At some small value of α, we will no longer be able to distinguish two distinct images. You have probably noticed this effect when viewing the headlights of a distant car. When it is far away (small α), you see only a single light. As the car approaches, you gradually can distinguish (resolve) two separate headlights. It has been arbitrarily agreed (the **Rayleigh criterion**) that two light sources are said to be resolved if the central maximum of one image falls on the first minimum of the second image. From Eq. 33.1 the first minimum for a slit of width b occurs for $\sin \theta_1 = \lambda/b$. Since $\theta \ll 1$, $\sin \theta_1 \simeq \theta_1 = \lambda/b$ for resolution. Thus for two sources to be resolved by a slit, they must subtend an angle (in radians) not less than $\theta_{min} = \lambda/b$.

$$\theta_{min} = \frac{\lambda}{b} \quad \text{resolving angle for a slit} \tag{33.4}$$

Most optical instruments use circular apertures. Again a diffraction pattern is formed, consisting of concentric bright and dark circles. The intensity varies with θ approximately as it does for a slit. It can be shown that for a circular aperture of diameter D, the minimum angle of resolution is

$$\theta_{min} = 1.22 \frac{\lambda}{D} \quad \text{resolving angle for circular aperture} \tag{33.5}$$

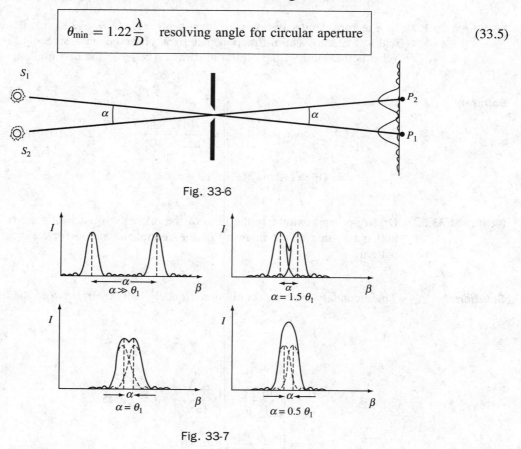

Fig. 33-6

Fig. 33-7

PROBLEM 33.3.   The Earth–Moon distance is 3.84×10^8 m. What is the minimum separation of two objects on the Moon if they are to be barely resolved by a telescope on Earth with a 60-cm-diameter objective lens? Assume $\lambda = 550$ nm.

Solution   If x = separation of the objects, they subtend an angle $\alpha = x/r = \theta_{min} = 1.22\ \lambda/D$. Thus

$$x = \frac{(1.22)(3.84 \times 10^8\ \text{m})(550\ \text{nm})}{0.60\ \text{m}} = 429\ \text{m}$$

PROBLEM 33.4.   It is rumored that a U-2 spy plane flying at an altitude of 30 km has cameras that can resolve an object as small as a man (2 m). Using visible light (500 nm), what diameter lens would be required?

Solution

$$\theta_{min} = 1.22\frac{\lambda}{D} = \frac{x}{r} \qquad D = \frac{(1.22)(3 \times 10^4\ \text{m})(500\ \text{nm})}{2\ \text{m}} = 9.2\ \text{mm}$$

33.3   The Diffraction Grating

A *diffraction grating* (actually an interference grating, but misnamed) consists of thousands of equally spaced narrow slits (Fig. 33-8). Diffraction gratings are very useful for analyzing light and for obtaining pure wavelengths. When monochromatic light is incident on a grating and then projected onto a screen, an interference pattern is observed analogous to that obtained in the double slit experiment. As for the double slit, strong constructive interference will occur when the path length difference between adjacent slits is a multiple of the wavelength. If d is the slit spacing,

$$\boxed{d\sin\theta = m\lambda, \quad m = 0, 1, 2, 3, \ldots, n \quad \text{diffraction grating maxima}} \tag{33.6}$$

m is called the **order** of the diffraction grating maximum.

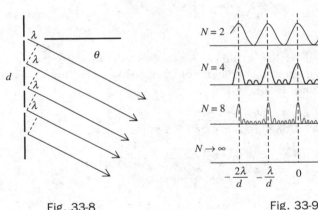

Fig. 33-8   Fig. 33-9

Note that as the number of slits in the grating increases, the intensity pattern becomes increasingly sharp (Fig. 33-9). One can see qualitatively why this is so. Recall that in the case of two slits, to reach the first minimum at θ_1, you must increase θ until the path difference is $\frac{1}{2}\lambda$. But suppose instead you have 100 slits. Consider the rays from the first slit and the fifty-first slit. Since these are 50 times as far apart as two adjacent slits, you only have to increase θ by $1/50\theta_1$ in order to make the path difference between these two slits $\frac{1}{2}\lambda$. Continue in this way, pairing slits 2 and 52, 3 and 53, and so on. Thus the minimum is reached very quickly as the number of slits is increased. Here I assumed each slit was narrow enough so that single slit diffraction did not change the pattern appreciably. For wider slits, we must multiply the many slit pattern by the single slit pattern to obtain the resultant intensity.

PROBLEM 33.5.   A certain grating gives a second-order line for a helium−neon laser ($\lambda = 632.8\ \text{nm}$) at $35.48°$. What is the grating spacing?

Solution

$$d = \frac{n\lambda}{\sin\theta} = \frac{2(632.8 \times 10^{-9}\ \text{m})}{\sin 35.48°} = 2.18 \times 10^{-6}\ \text{m}$$

PROBLEM 33.6.   A mercury arc emits green light at 546.1 nm. What is the angular separation between the first- and second-order lines with a diffraction grating with 6000 slits per centimeter?

Solution $d \sin \theta_1 = \lambda$ and $d \sin \theta_2 = 2\lambda$,

$$\sin \theta_1 = \frac{546.1 \times 10^{-9} \text{ m}}{6000^{-1} \text{ cm}} = 0.3277 \qquad \theta_1 = 19.1°$$

Similarly,

$$\sin \theta_2 = 0.6553 \quad \theta_2 = 40.9° \qquad \theta_2 - \theta_1 = 21.8°$$

   Spectroscopy (the measurement of wavelengths) has been one of the most powerful tools of science, and it depends on being able to distinguish between two spectral lines whose wavelengths, λ_1 and λ_2, are very close together. Define the **resolving power** of a device as

$$R = \frac{\lambda}{|\lambda_2 - \lambda_1|} = \frac{\lambda}{\Delta\lambda}$$

$\lambda_1 \simeq \lambda_2$, so it doesn't matter which value is in the numerator. In order for two lines to be resolved, it is required that the principal maximum of one not be closer than the first minimum of the other (the Rayleigh criterion). Thus in Fig. 33-10, I require that $\theta_1 = \Delta\theta$. If the total number of lines in the grating is N, and the mth order is observed, I can show that the resolving power of the grating is

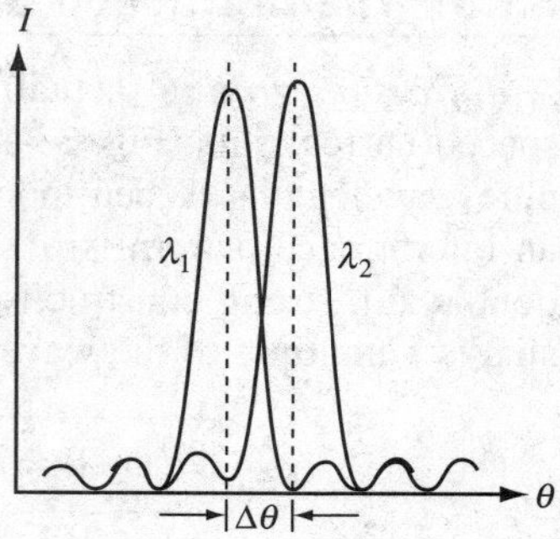

$$\boxed{R = \frac{\lambda}{\Delta\lambda} = mN \quad \text{resolving power of a grating}} \qquad (33.7)$$

Fig. 33-10

PROBLEM 33.7.   Resolution of the yellow sodium *D* lines at 588.995 and 589.592 nm played an important role in learning about the structure of atoms. How many lines must a grating have to resolve these lines in first order?

Solution $R = mN = \dfrac{\lambda}{\Delta\lambda} = \dfrac{589 \text{ nm}}{|588.995 \text{ nm} - 589.592 \text{ nm}|} = 987$

                  $m = 1$,   so $N \approx 1000$ lines

This is not difficult to achieve, and it is easy to resolve these lines.

33.4 Summary of Key Equations

Single slit minimum: $b \sin \theta = m\lambda \quad m = 0, 1, 2, 3, \ldots, n$

Single slit intensity: $I = I_0 \left(\dfrac{\sin \beta}{\beta}\right)^2$,   where $\beta = \pi \dfrac{b}{\lambda} \sin \theta$

Angular resolution: $\theta_{min} = \dfrac{b}{\lambda}$ slit

                  $\theta_{min} = 1.22 \dfrac{\lambda}{D}$ circular hole

Diffraction grating: $d \sin \theta = m\lambda$, $m = 0, 1, 2, \ldots$ maxima

Grating resolving power: $R = \dfrac{\lambda}{\Delta\lambda} = mN$, N lines total, mth order

SUPPLEMENTARY PROBLEMS

33.8. What is the intensity ratio for the third secondary maximum and the central maximum for a single slit diffraction pattern?

33.9. When 450-nm light passes through a slit 0.080 mm wide, it forms a diffraction pattern on a screen 1.20 m from the slit. What is the distance on the screen from the central maximum to the first minimum in the diffraction pattern?

33.10. Birds of prey have keen eyesight. If the pupil of an eagle has a diameter of 10 mm, at what altitude could it fly and still see clearly a mouse 7 cm long? Assume $\lambda = 550\,\text{nm}$.

33.11. One of the world's largest radiotelescopes is at Arecibo, Puerto Rico. It has a 305-m-diameter reflecting "dish" that fits into a natural mountain basin. What is the angular resolution of this telescope when it reflects 10-cm microwaves? The Mt. Palomar reflecting telescope has a diameter of 5.08 m. What is its angular resolution for $\lambda = 550\,\text{nm}$?

33.12. Some surveillance satellites have cameras with lenses 35 cm in diameter. What is the angular resolution for such a camera if $\lambda = 550\,\text{nm}$? If the satellite is at an altitude of 160 km, what is the minimum spacing between two objects on the ground that the camera can barely resolve?

33.13. Driving across the Mojave Desert at night you can see a long way. If you see the headlights of an oncoming car in the distance, they look like a single spot of light. However, as the car nears, it reaches a point where you can make out two headlamps. If the headlamps are separated by 1.6 m and your pupil diameter is 5 mm, how far away is the car when you can barely resolve the lights? Assume $\lambda = 550\,\text{nm}$.

33.14. A screen is placed 1.30 m away from a diffraction grating illuminated with light of wavelength 614 nm. If the second- and third-order spectra are to be separated by 1.00 cm on the screen, how many lines per centimeter are needed for the grating?

33.15. Three and only three lines on either side of the central maximum can be seen when a grating is illuminated with light ($\lambda = 520\,\text{nm}$). What is the maximum number of lines per centimeter for the grating?

33.16. Monochromatic light is incident on a diffraction grating at angle α to the normal to the grating. At what angles θ will maxima of the diffraction pattern be observed?

33.17. The **dispersion** of a diffraction grating, defined as $d\theta/d\lambda$, is a measure of how much the grating spreads out the diffraction pattern. Obtain an expression for the dispersion of a grating when $d \gg \lambda$.

33.18. Spacing of atoms in crystals is of the order of 0.1 nm, and this is comparable to the wavelength of x rays. Consequently, it is possible to observe three-dimensional diffraction by crystals, analogous to what is observed with a diffraction grating and light. This has proved extremely useful in determining crystal structure. In the sketch here, x rays are scattered from adjacent planes in the crystal separated by a distance d. What is the condition for constructive interference in the scattered beam?

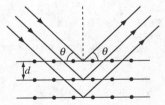

33.19. Astronomers have observed planets orbiting a distant star. The star is believed to be about $4.0 \times 10^{17}\,\text{m}$ from the Earth. If a planet is $1.3 \times 10^{11}\,\text{m}$ from this star, what diameter aperture must a telescope have in order to resolve the star–planet pair, using visible light of wavelength 560 nm?

33.20. At what angle does the first-order maximum occur for green light of wavelength 550 nm incident on a grating with 12,000 lines/cm?

SOLUTIONS TO SUPPLEMENTARY PROBLEMS

33.8. The third secondary maximum occurs approximately halfway between the fourth and fifth minima, that is, at

$$\beta = \frac{7\pi}{2}$$

From Eq. 33.2,

$$I = I_0\left(\frac{\sin\beta}{\beta}\right)^2, \quad \frac{I}{I_0} = \left(\frac{2}{7\pi}\right)^2 = 0.0083$$

33.9. From Eq. 33.1, $b\sin\theta = m\lambda$, $m = 1$, and

$$y \simeq L\sin\theta = \frac{L\lambda}{b} = \frac{(1.20\,\text{m})(450 \times 10^{-9}\,\text{m})}{(8.0 \times 10^{-5}\,\text{m})} = 6.8\,\text{mm}$$

33.10. $\theta_{\min} = 1.22\dfrac{\lambda}{D}, \quad x = r\theta_{\min}, \quad r = \dfrac{xD}{1.22\lambda} = \dfrac{(0.07\,\text{m})(0.010\,\text{m})}{1.22(550 \times 10^{-9}\,\text{m})} = 1000\,\text{m}$

33.11. $\theta_{\min} = 1.22\dfrac{\lambda}{D} = 1.22\left(\dfrac{0.10\,\text{m}}{305\,\text{m}}\right) = 0.0004\,\text{rad}$ (Arecibo)

$$= 1.22\left(\frac{550 \times 10^{-9}\,\text{m}}{5.08\,\text{m}}\right) = 1.32 \times 10^{-7}\,\text{rad} \quad \text{(Palomar)}$$

33.12. $\theta_{\min} = 1.22\dfrac{\lambda}{D} = \dfrac{550\,\text{mm}}{35\,\text{cm}} = 1.6 \times 10^{-6}\,\text{rad}$

$$x = r\theta_{\min} = (160 \times 10^3\,\text{m})(1.6 \times 10^{-6}\,\text{rad}) = 0.26\,\text{m}$$

33.13. $\theta_{\min} = 1.22\dfrac{\lambda}{D} = \dfrac{x}{r}, \quad r = \dfrac{xD}{1.22\lambda} = \dfrac{(1.6\,\text{m})(5 \times 10^{-3}\,\text{m})}{1.22(550 \times 10^{-9}\,\text{m})} = 11.9\,\text{km}$

33.14. $d\sin\theta = m\lambda, \quad y \simeq L\sin\theta = L\dfrac{m\lambda}{d}, \quad y_3 - y_2 = L\dfrac{\lambda}{d}(m_3 - m_2)$

$$N = \frac{1}{d} = \frac{y_3 - y_2}{L\lambda(m_3 - m_2)} = \frac{0.01\,\text{m}}{(1.3\,\text{m})(614 \times 10^{-9}\,\text{m})(3 - 2)} = 12{,}500/\text{m} = 125 \text{ lines per centimeter}$$

33.15. $d\sin\theta = m\lambda$

For the maximum order $\theta = 90°$, so $d = m\lambda = 3(520\,\text{nm}) = 1.56 \times 10^{-6}\,\text{m}$.

$$N = \frac{1}{d} = 6.41 \times 10^5/\text{m} = 6410 \text{ lines per centimeter}$$

33.16. Consider a ray that is distance $m\lambda$ behind the ray through an adjacent slit. Before hitting the slit, it gained a distance $d\sin\alpha$ on the adjacent ray, so the net path difference between the two rays is $d\sin\theta - d\sin\alpha$. Thus for constructive interference, $d\sin\theta - d\sin\alpha = m\lambda$, and $\lambda = $ integer.

33.17. $d\sin\theta = m\lambda, \quad \cos\theta\dfrac{d\theta}{d\lambda} = \dfrac{m}{d}, \quad \cos\theta = \sqrt{1 - \sin^2\theta} = \sqrt{1 - \left(\dfrac{m\lambda}{d}\right)^2} \simeq 1 - \dfrac{1}{2}\left(\dfrac{m\lambda}{d}\right)^2$

$$\frac{d\theta}{d\lambda} = \frac{m}{d\cos\theta} \simeq \frac{m}{d[1 - 1/2(m\lambda/d)^2]} \simeq \frac{m}{d}\left(1 - \frac{1}{2}\frac{m^2\lambda^2}{d^2}\right) \simeq \frac{m}{d}$$

33.18. From the drawing I see that the path difference between rays a and b is $2d \sin \pi$. Thus for constructive interference $2d \sin \theta = m\lambda$, $m =$ integer. This looks like the equation for the maxima from the diffraction grating, but notice that θ is defined differently here, and an added factor of 2 is present.

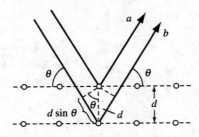

33.19. $\theta_{\min} = 1.22 \dfrac{\lambda}{D}; \dfrac{y}{L}$

$$D: \frac{1.22\lambda L}{y} = \frac{(1.22)(560 \times 10^{-9}\,\text{m})(4.0 \times 10^{17}\,\text{m})}{1.3 \times 10^{11}\,\text{m}}$$

$D = 2.1\,\text{m}$ Aperture diameter

33.20. $d \sin \theta = m\lambda$, $m = 1$

$$\sin \theta = \frac{m\lambda}{d} = \frac{(1)(550 \times 10^{-9}\,\text{m})}{(1.2 \times 10^4 \times 10^2)^{-1}\,\text{m}}$$

$\theta = 41.2°$

CHAPTER 34

Special Relativity

We live in a strange and wonderful universe. Few of its characteristics are more striking and puzzling than the behavior described by Einstein's theory of special relativity. The results seem to go against all common sense, but every aspect of the theory has been confirmed by experiment. There are lessons to be learned here, not only in physics but for life in general. Our understanding is constrained and limited by experience.

34.1   The Basic Postulates

When he developed his theory in 1905, Einstein had set out to understand the electric and magnetic fields of moving objects and observers. He was led to make the following postulates:

The laws of physics are the same in all inertial reference frames.

Light travels at the same speed in vacuum in all inertial reference frames, independent of the velocity of the source or of the observer.

An *inertial frame* is one in which a free body, subject to no forces, travels in a straight line at constant velocity (that is, Newton's first law holds). Any reference frame moving with constant velocity with respect to an inertial frame is also an inertial frame. Einstein's first postulate seems quite reasonable. If in a train moving at constant velocity, you drop a pen, you see it fall straight down. The fact that the train is moving with respect to the ground makes no difference. This makes sense. After all, you could think of the train as stationary and the ground moving by underneath. Clearly the pen doesn't know what the ground is doing.

The second postulate makes no sense at all to me. Suppose you stand still and someone shines a light in your direction. You measure the light's speed as c. Now suppose you run toward the light. I would think you would see the light coming at you at a speed greater than c. Not so. It still approaches at the speed c. Consider another example. A person driving by in a car turns on a light at the instant she passes you. You see the light radiate out in a sphere with you at the center. The person in the car keeps going, but she also sees the light radiate out in a sphere centered on her. How can two people at different positions both be at the center of the same sphere? Don't ask me. That is the way it is, but none of us have first-hand awareness of such phenomena. It's crazy, but true.

34.2   Simultaneity

Two events are simultaneous if an observer stationed midway between them sees them happen at the same time. Until 1900 people thought that simultaneity was absolute; that is, either two events were simultaneous or they were not. Einstein showed this not to be true with the following "gedankenexperiment." A railroad flatcar is moving at constant velocity. A trainman in the car stands midway between the ends of the car. A bystander is on the ground next to the track. At the instant the trainman and the bystander are right next to each other, two bolts of lightning strike the ends of the car and are reflected off. The bystander on the ground sees both lightning bolts at the same time and thus concludes they struck the car simultaneously. He observes the car move forward in the instant after the lightning strike, so that light from the front of the car hits the trainman before the light from the rear does. The trainman, however, has a different interpretation of what happened. The light from the front hits him first (he agrees on that), but he concludes that since he is midway between the ends of the car, lightning must have struck the front of the car first. He concludes the lightning strikes were not simultaneous. Thus **the simultaneity of events is relative** to the motion of the observers (hence the term, theory of relativity). Time measurements are related to simultanity, so time is also relative. Further, the same is true of length measurement for moving objects. For example, if you want to measure the length of a train from locomotive to caboose, you must note the position of the front and back of the train simultaneously. If you mark on the ground the position of the front and then 2 s later mark the position of the back, you will not get the same measurement of the length that a person riding with the train will obtain.

34.3   The Lorentz Transformation Equations

I wish to describe how two observers, one moving with constant velocity with respect to the other, will describe an event. In Fig. 34-1 are two inertial reference frames. S is stationary (the lab frame), and S' is moving with speed v in the x direction. When their origins coincide, clocks there are set to zero ($t = 0$, $t' = 0$). I recognize that time is not absolute, so I use different symbols for time in the two coordinate systems. According to newtonian mechanics, $t' = t$ and $x' = x - vt$ for an object positioned distance x from the S origin and x' from the S' origin. Using this reasoning, observers in the two frames could see light moving at different speeds, and this

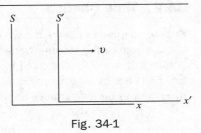

Fig. 34-1

is inconsistent with Einstein's second postulate. Einstein showed that the correct relationship between (x, y, z, t) and (x', y', z', t') is given by the **Lorentz transformation equations**:

$$y' = y \qquad z' = z \qquad x' = \gamma(x - vt) \qquad x = \gamma(x' + vt')$$

$$t' = \gamma\left(t - \frac{v}{c^2}x\right) \qquad t = \gamma\left(t' + \frac{v}{c^2}x'\right)$$

where

$$\gamma = \frac{1}{\sqrt{1 - v^2/c^2}} \tag{34.1}$$

The equations for x and t are obtained from x' and t' by replacing v by $-v$. Note that as $v/c \to 0$, these approach the classical equations. Only the coordinate in the direction of motion is affected.

PROBLEM 34.1.   Suppose that a space ship (S') moves past the Earth (S) at speed $0.6c$. Clocks on the Earth and on the ship are synchronized at that instant ($t = 0$, $t' = 0$). The space ship continues on to a star 12 ly away. (One *light year* is the distance light travels in vacuum in 1.00 year; 1 ly $= 9.46 \times 10^{15}$ m.) Clocks at the star are synchronized with those on Earth. (a) According to an observer on Earth, how far did the ship travel and how long did it take? (b) According to an observer on the ship, how far did he travel and how long did it take?

Solution   Consider the arrival of the ship at the star to be an "event" with coordinates (x, t) in frame S and (x', t') in frame S'. (a) For an Earth-based observer, $x = 12$ ly and $t = x/v = (12\,\text{ly})/0.6c = 20$ y. (b) For an observer in the space ship, $x' = 0$ (he is still at his origin when he arrives at the star). Check this with Eq. 34.1.

$$x' = \gamma(x - vt) = 1.25[12\,\mathrm{ly} - (0.6c)(20\,\mathrm{y})] = 0,$$

where

$$\gamma = \frac{1}{\sqrt{1 - (0.6)^2}} = 1.25$$

The time for the space ship observer is

$$t' = \gamma\left(t - \frac{vx}{c^2}\right) = (1.25)\left[20\,\mathrm{y} - \frac{(0.6c)(12\,\mathrm{ly})}{c^2}\right] = 16\,\mathrm{y}$$

Thus the observer on Earth sees 20 years pass, whereas the traveler sees only 16 years pass. The Earth-based observer always believes that **moving clocks run slow**.

34.4 Time Dilation

Suppose a number of synchronized clocks are placed along the x-axis in the fixed frame S. A clock is placed at the origin of the moving frame S'. As the moving clock passes point x_1, it reads time t_1', and as it passes point x_2, it reads t_2'. The fixed clocks read corresponding times t_1 and t_2. We can relate these times using Eq. 34.1, noting that the moving clock is always at $x' = 0$. Thus $t_1 = \gamma t_1'$ and $t_2 = \gamma t_2'$. The time intervals measured in the two frames are thus $\Delta t = t_2 - t_1 = \gamma(t_2' - t_1') = \gamma\Delta t_0$. Thus

$$\Delta t = \gamma\,\Delta t_0$$

or

$$\boxed{\Delta t = \frac{1}{\sqrt{1 - v^2/c^2}}\,\Delta t_0 \quad \text{time dilation}} \tag{34.2}$$

Since $\gamma > 1$, $\Delta t > \Delta t_0$. We say that **moving clocks run slow**. This has been confirmed by flying accurate clocks around the world. A stationary clock at home records a longer elapsed time than does the flying clock. Time measured with a clock that stays at one place in its reference frame (like the clock in S' above) measures **proper time**. If you observe your wristwatch while you are walking around, you are measuring proper time intervals.

PROBLEM 34.2. A space traveler in a ship moving at $0.99c$ travels for what he believes is 2 y. How long did the trip take according to an observer on Earth?

Solution
$$\Delta t = \frac{1}{\sqrt{1 - v^2/c^2}}\,\Delta t_0 = \frac{1}{\sqrt{1 - v^2/c^2}}(2\,\mathrm{y}) = 14\,\mathrm{y}$$

If you go fast enough, you can go a long way without getting very much older.

34.5 Length Contraction

Suppose a rod of length L_0 is at rest along the x'-axis in the moving frame S'. Its ends are at x_1' and x_2' and $L_0 = x_2' - x_1'$. Since the rod is at rest in S', L_0 is the **proper length** of the rod. An observer in the fixed frame S measures the length of the rod, noting the positions x_1 and x_2 simultaneously at time t. He determines the length to be $L = x_2 - x_1$. Using Eq. 34.1, $x'_2 - x'_1 = \gamma(x_2 - vt) - \gamma(x_1 - vt) = \gamma(x_2 - x_1)$ or $L_0 = \gamma L$.

$$\boxed{L = \frac{1}{\gamma}L_0 = \sqrt{1 - \frac{v^2}{c^2}}\,L_0 \quad \text{length contraction}} \tag{34.3}$$

Here $L_0 =$ length of object at rest.

PROBLEM 34.3.   A space ship 150 m long flies by a stationary observer at a speed 0.5c. How long does it appear to the observer?

Solution
$$L = \sqrt{1 - \frac{v^2}{c^2}}\,L_0 = \sqrt{1 - (0.5)^2}(150\,\text{m}) = 130\,\text{m}$$

34.6   Relativistic Velocity Transformation

Suppose an object is moving with velocity components (u_x, u_y, u_z) in the rest frame S. In the moving frame S', it will have velocity components (u'_x, u'_y, u'_z). To see how these are related, take differentials of Eq. 34.1. $dx = \gamma(dx' + v\,dy')$ and $dt = \gamma[dt' + (v/c^2)dx']$. Divide these and factor out dt' from the numerator and the denominator.

$$u_x = \frac{dx}{dt} = \frac{\gamma(dx' + v\,dt')}{\gamma(dt' + v/c^2\,dx')} = \frac{dt'(dx'/dt' + v)}{dt'(1 + v/c^2\,dx'/dt')} = \frac{v'_x + v}{1 + vu'_x/c^2}$$

In a similar fashion I obtain u_y and u_z, where $u_y = dy/dt$ and $u_z = dz/dt$. The result is

$$u_x = \frac{u'_x + v}{1 + v'_x v/c^2} \qquad v'_x = \frac{u_x - v}{1 - u_x v/c^2} \tag{34.4}$$

$$u_y = \frac{u'_y}{\gamma(1 + u'_x v/c^2)} \qquad u'_y = \frac{u_y}{\gamma(1 - u_x v/c^2)} \qquad \gamma = \frac{1}{\sqrt{1 - v^2/c^2}} \tag{34.5}$$

$$u_z = \frac{u'_z}{\gamma(1 + u'_x v/c^2)} \qquad u'_z = \frac{u_z}{\gamma(1 - u_x v/c^2)} \tag{34.6}$$

PROBLEM 34.4.   A space ship moving toward you at speed 0.5c shines a light at you. At what speed do you see the light approaching?

Solution   In the moving frame (the space ship), the light speed is $u' = c$ and the moving frame is coming at you at speed v. In the rest frame (you), the light speed is u:

$$u = \frac{u' + v}{1 + vu/c^2} = \frac{c + 0.5c}{1 + 0.5c^2/c^2} = c$$

Light travels at speed c in all frames.

PROBLEM 34.5.   An Earth-based observer sees rocket A moving at 0.70c directly toward rocket B, which is moving toward A at 0.80c. How fast does rocket A see rocket B approaching?

Solution   In newtonian mechanics, A would see B approaching at speed $v = 0.7c + 0.8c = 1.5c$. This is wrong. Suppose rocket A is the moving reference frame (A is at rest in this frame, and it is moving with speed $v = 0.7c$ toward Earth). Then the velocity of B in the Earth rest frame is $u = -0.8c$. From Eq. 34.4, the velocity u' of B as seen from A is

$$u' = \frac{u_x - v}{1 - vu_x/c^2} = \frac{-0.8c - (0.7c)}{1 - (0.7c)(-0.8c)/c^2} = 0.96c$$

You can see that no object is ever observed to move faster than c, the speed of light.

PROBLEM 34.6.   In the laboratory (frame S) a particle moves with velocity $\mathbf{u} = 0.3c\mathbf{i} + 0.4c\mathbf{j}$. The moving frame S' in Fig. 34-1 has speed $v = 0.5c$. Determine for both frames the speed of the particle and the angle between its velocity and the x-axis.

Solution   In the lab,

$$u = \sqrt{u_x^2 + u_y^2} = \sqrt{(0.3c)^2 + (0.4c)^2} = 0.5c$$

and

$$\tan \theta = \frac{u_y}{u_x} = \frac{0.4c}{0.3c} = 1.33 \quad \theta = 53°$$

In S',

$$u_x' = \frac{u_x - v}{1 - u_x v/c^2} = \frac{0.3c - 0.5c}{1 - (0.3c)(0.5c)c^2} = 0.24c$$

$$\gamma = \frac{1}{\sqrt{1 - v^2/c^2}} = \frac{1}{\sqrt{1 - (0.5c/c)^2}} = 1.15$$

$$u_y' = \frac{u_y}{\gamma(1 - u_x v/c^2)} = \frac{0.4c}{1.15[1 - (0.3c)(0.5c)/c^2]} = 0.41c$$

Thus

$$u' = \sqrt{u_x'^2 + u_y'^2} = \sqrt{(0.24c)^2 + (0.41c)^2} = 0.48c$$

$$\tan \theta' = \frac{0.41c}{0.24c} \quad \theta' = 60°$$

34.7   Relativistic Momentum and Force

We saw that in a collision between two particles, linear momentum is conserved. This proves not to be true if momentum is defined as $p = mv$. The relativistically correct definition of momentum of a particle of mass m and velocity $\mathbf{u}$ is

$$\boxed{\mathbf{p} = \gamma m \mathbf{u} = \frac{1}{\sqrt{1 - u^2/c^2}} m\mathbf{u}} \tag{34.7}$$

As $v/c \to 0$, $\mathbf{p} \to m\mathbf{u}$, the classical value.

Observe that as $u \to c$, $p \to \infty$. The equation relating force and rate of change of momentum is still valid:

$$\boxed{\mathbf{F} = \frac{d\mathbf{p}}{dt}} \tag{34.8}$$

PROBLEM 34.7.   An electron of mass m is subject to a constant force F. What is its velocity as a function of time if the electron is initially at rest?

Solution   Nonrelativistically,

$$u = at = \frac{F}{m}t$$

and the velocity increases without limit. Relativistically,

$$\frac{dp}{dt} = F \qquad \int_0^p dp = \int_0^t F\,dt, \quad \text{so } p = Ft$$

Substitute Eq. 34.7:

$$\frac{mu}{\sqrt{1 - u^2/c^2}} = Ft$$

Solve for u:

$$u = \frac{Ft}{m\sqrt{1 + (Ft/mc)^2}} \qquad\qquad (34.9)$$

Note that $u \to c$ as $t \to \infty$. At low speeds, when you push on a particle, it gains speed. However, as the speed gets higher and higher, the acceleration decreases. It is as if the particle gets more and more massive. You can never make its speed exceed (or reach) c, the speed of light. Rather than call c the speed of light, it would be more accurate to call it something like the "cosmic speed limit." Light happens to move at that speed, but objects with mass don't.

For short times where

$$\left(\frac{Ft}{mc}\right)^2 \ll 1 \qquad u = t$$

in agreement with the classical result.

PROBLEM 34.8. A rocket experiences a force equal to its weight on Earth (mg) for a period of 3 y (9.5×10^7 s). What speed does it reach according to classical mechanics? According to relativistic mechanics?

Solution From Eq. 34.9,

$$u = \frac{Ft}{m\sqrt{1 + (Ft/mc)^2}} = \frac{mgt}{m\sqrt{1 + (mgt/mc)^2}}$$

$$= \frac{(9.8\,\text{m/s}^2)(9.5 \times 10^7\,\text{s})}{\sqrt{1 + [(9.8\,\text{m/s}^2)(9.5 \times 10^7\,\text{s})/3 \times 10^8\,\text{m/s}]^2}}$$

$$= 2.8 \times 10^8\,\text{m/s}$$

Classically: $u = at = \frac{F}{m}t = gt = (9.8\,\text{m/s}^2)(9.5 \times 10^7\,\text{s}) = 9.3 \times 10^8\,\text{m/s}$

The classical result is more than three times the speed of light, and clearly wrong.

PROBLEM 34.9. What is the momentum of a particle of mass m with speed $0.8c$ in terms of the classical value, mu?

Solution $\gamma = \dfrac{1}{\sqrt{1 - u^2/c^2}} = \dfrac{1}{\sqrt{1 - (0.8c)^2/c^2}} = 1.67 \qquad p = \gamma mu = 1.67\,mu$

34.8   Relativistic Energy

The kinetic energy of a particle is defined as equal to the work required to accelerate the particle from rest to a final speed v. Thus $KE = \int_0^x F\,dx$. But $F = dp/dt$ and $dx = v\,dt$, so

$$KE = \int_0^t \left(\frac{dp}{dt}\right)(v\,dt) = \int_0^p v\,dp$$

From Eq. 34.7,

$$p = \frac{mv}{\sqrt{1 - v^2/c^2}}$$

Solve for v and find

$$v = \frac{(p/m)}{\sqrt{1 + (p/mc)^2}}, \quad \text{so } KE = \int_0^p \frac{(p/m)}{\sqrt{1 + (p/mc)^2}}\,dp = mc^2\left\{\left[1 + \left(\frac{p}{mc}\right)^2\right]^{1/2} - 1\right\}$$

Use Eq. 34.7 again to express KE in terms of v. Thus

$$\boxed{KE = \frac{mc^2}{\sqrt{1 - v^2/c^2}} - mc^2 = (\gamma - 1)mc^2} \tag{34.10}$$

Note that for $v/c \ll 1$,

$$\sqrt{1 - \frac{v^2}{c^2}} \simeq 1 - \frac{v^2}{2c^2} \quad \text{and} \quad \frac{1}{1 - v^2/2c^2} \simeq 1 + \frac{v^2}{2c^2}$$

Thus for $v/c \ll 1$,

$$KE \simeq mc^2\left(1 + \frac{v^2}{2c^2}\right) - mc^2 = \frac{1}{2}mv^2$$

the classical result.

The term mc^2 in the kinetic energy is independent of the velocity. It is called the *rest energy*, $E_0 = mc^2$, of the particle. We identify the total energy E as $E = \text{kinetic energy} + \text{rest energy} = (\gamma - 1)mc^2 + mc^2$. Thus

$$\boxed{E = \gamma mc^2 = \frac{mc^2}{\sqrt{1 - v^2/c^2}}} \quad \text{total relativistic energy} \tag{34.11}$$

When a particle is at rest ($\gamma = 1$), $E = mc^2$. This is Einstein's famous equation. This suggests that mass and energy are equivalent, and indeed this is the case. In nuclear reactions mass is converted to other forms of energy, such as light and heat. Hence we speak of the mass-energy equivalence. Because c^2 is a very large number, a small amount of mass corresponds to a large amount of energy.

It is useful to express E in terms of p, and the result of doing this is

$$\boxed{E^2 = p^2c^2 + (mc^2)^2} \tag{34.12}$$

Fig. 34-2 shows a mnemonic for remembering Eq. 34.12. Think of a right triangle with E the hypotenuse. Apply Pythagorean theorem. You can remember the other two sides by remembering that they have the dimensions of energy (like $\frac{1}{2}mv^2$).

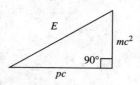

Fig. 34-2

PROBLEM 34.10. Derive Eq. 34.12.

Solution
$$E = \gamma mc^2 \quad \text{and} \quad p = \gamma mv$$

$$E^2 - p^2c^2 = \gamma^2(mc^2)^2 - (\gamma mvc)^2 = (mc^2)^2\left(\frac{1}{1 - v^2/c^2}\right)\left(1 - \frac{v^2}{c^2}\right) = m^2c^4$$

so

$$E^2 = p^2c^2 + (mc^2)^2$$

PROBLEM 34.11. At what speed is the kinetic energy of a particle equal to its rest energy?

Solution
$$E = \gamma mc^2 = KE + mc^2 = mc^2 + mc^2 = 2mc^2, \quad \text{so} \quad \gamma = 2 \quad \text{and} \quad \gamma^2 = 4$$

Thus

$$\frac{1}{1 - v^2/c^2} = 4 \quad \text{and} \quad 1 - \frac{v^2}{c^2} = 0.25, \quad \text{so} \quad v = 0.87c$$

There are three particles with zero mass: the photon (a particle of light), the neutrino, and the graviton (not yet observed). These particles all travel at the speed of light and have only kinetic energy and no rest energy. Particles with mass all travel at speeds less than the speed of light.

34.9 The Doppler Effect for Light

Anyone who has heard a speeding ambulance race by with siren screaming is familiar with the Doppler effect for sound. The pitch of the approaching siren is raised, and as the ambulance passes and moves away, there is a noticeable drop in the frequency. A similar effect occurs with electromagnetic waves. If a source of light is approaching you (or you are approaching it), the frequency is raised and the wavelength shortened, and if the source is receding, the frequency is lowered and the wavelength is increased. The formula for light differs from that for sound, however, and it turns out to be

$$f = \sqrt{\frac{1 - v/c}{1 + v/c}}\, f_0 \quad \text{receding source;} \quad f = \sqrt{\frac{1 + v/c}{1 - v/c}}\, f_0 \quad \text{approaching source} \qquad (34.13)$$

Here f_0 is the frequency emitted by a stationary source. Astronomers use this effect to estimate the speed of galaxies moving away from us. Light that is normally yellow, say, is shifted to longer wavelengths, for example, to the red end of the spectrum. The effect is called the **red shift**, and its observation plays an important role in cosmology.

PROBLEM 34.12. A police radar uses the Doppler effect to measure the speed of cars. A microwave signal of 8×10^9 Hz is transmitted and reflected back from the car. The signal experiences two Doppler shifts, one when it hits the car and another when it is reflected back from a moving reflector. The received signal is made to beat against the transmitted frequency, and the beat frequency is calibrated in terms of vehicle speed. What beat frequency would be detected from a car approaching at 108 km/h?

Solution $v = 108$ km/h $= 30$ m/s. The frequency incident on the car is

$$f_1 = \sqrt{\frac{1 + v/c}{1 - v/c}}\, f_0$$

The frequency reflected is

$$f_2 = \sqrt{\frac{1 + v/c}{1 - v/c}} f_1 = \frac{1 + v/c}{1 - v/c} f_0 \simeq \left(1 + \frac{v}{c}\right)\left(1 + \frac{v}{c}\right)f_0 \simeq \left(1 + 2\frac{v}{c}\right)f_0$$

The beat frequency is

$$\Delta f = f_2 - f_0 = 2\frac{v}{c}f_0 = 2\left(\frac{30 \text{ m/s}}{3 \times 10^8 \text{ m/s}}\right)(8 \times 10^9 \text{ Hz}) = 1600 \text{ Hz}$$

34.10   Summary of Key Equations

Lorentz transformation: $x' = \gamma(x - vt)$ $x = \gamma(x' + vt')$ $\gamma = \dfrac{1}{\sqrt{1 - v^2/c^2}}$

$$t' = \gamma\left(t - \frac{v}{c^2}x\right) \qquad t = \gamma\left(t' + \frac{v}{c^2}x'\right)$$

$$y' = y \quad z' = z$$

Time dilation: $\Delta t = \gamma \Delta t_0$

Length contraction: $L = \dfrac{1}{\gamma}L_0$

Velocity transformation: $u_x = \dfrac{u'_x + v}{1 + vu'_x/c^2}$ $u_y = \dfrac{u'_y}{\gamma(1 + vu_x/c^2)}$ $u_z = \dfrac{u'_z}{\gamma(1 + vu_x/c^2)}$

$$u'_x = \frac{u_x - v}{1 - vu_x/c^2} \qquad u'_y = \frac{u_y}{\gamma(1 - vu_y/c^2)} \qquad u'_z = \frac{u_z}{\gamma(1 - vu_x/c^2)}$$

Momentum: $\mathbf{P} = \gamma m\mathbf{u}$

Force: $\mathbf{F} = \dfrac{d\mathbf{p}}{dt}$

Kinetic energy: $KE = (\gamma - I)mc^2$

Energy: $E = \gamma mc^2$ $E^2 = (pc)^2 + (mc^2)^2$

Doppler effect: $f = \sqrt{\dfrac{c - v}{c + v}} f_0$ receding

$$f = \sqrt{\frac{c + v}{c - v}} f_0 \quad \text{approaching}$$

SUPPLEMENTARY PROBLEMS

34.13. Positive pions are unstable, living an average time of 2.6×10^{-8} s in their own rest frame. If such a particle is moving in the laboratory at $0.982c$, what lifetime is observed in the laboratory? How far does a pion move in the laboratory in this time?

34.14. A space ship of length 200 m (at rest) passes by a satellite in 8 μs. What is its speed relative to the satellite?

34.15. Relativistic effects are not usually noticeable in everyday life. For example, suppose a jet fighter has a length of 12 m on the ground. When flying by at 900 km/h (250 m/s), by how much would its length appear to be shortened?

34.16. A radioactive atom is moving in the *x* direction in the laboratory at speed $0.3c$. It emits an electron having speed $0.8c$ in the rest frame of the atom. What will be the velocity of the electron in the laboratory when it is ejected in (a) the *x* direction, (b) the $-x$ direction, and (c) the *y* direction in the atom's rest frame?

34.17. A light beam moves with speed *c* in a direction making an angle θ with the *x*-axis, so $u_x = \cos \theta$ and $u_y = c \sin \theta$. Determine the velocity of the light in moving frame S' in Fig. 34-1. What angle does the light make with the x'-axis if $\theta = 45°$ and $v = 0.6c$?

34.18. An electron (rest mass 9.11×10^{-31} kg) is accelerated to an energy of 30×10^9 eV (30 GeV). What is its kinetic energy? Its momentum? Its speed?

34.19. A Klingon spy ship races away from a stationary space habitat at a speed of $0.3c$. The starship *Ironsides* follows it at a speed of $0.2c$ with respect to the spy ship. To an observer in the habitat, with what speed does the *Ironsides* overtake the spy ship?

34.20. What velocity must a particle of rest mass *m* and velocity *v* have if its momentum is 10 mv?

34.21. Show that relativistically force and acceleration are not necessarily parallel.

34.22. When a ^{235}U nucleus undergoes fission in a power plant it releases about 200 MeV of energy. When 1 kg of coal is burned, about 3.0×10^7 J of energy is released. How many kilograms of coal would be needed to provide as much energy as 1 kg of ^{235}U if each nuclei underwent fission? The molar mass of ^{235}U is 235 g. What fraction of the uranium mass is converted to energy in the fission process?

34.23. In observing a galaxy in the Virgo cluster, astronomers utilize a spectral line associated with calcium atoms. They find a shift in wavelength from 396.85 nm for atoms in the laboratory to 398.4 nm in the light from the receding galaxy (this is called the *cosmological red shift* because the shift is to longer wavelength, that is, toward the red end of the visible spectrum). Based on this information, estimate how fast the galaxy is moving away from us. Assuming the galaxy always moved at the same speed, estimate how long ago the galaxy was at the same place as our galaxy if the distance to the galaxy is determined to be 78×10^6 ly. Consistent measurements like this on many different galaxies have led us to conclude that at one time about 10 to 20 billion years ago, all of the mass in the universe was at one point and experienced a "Big Bang" that sent it hurtling outward. It seems that every galaxy is receding from every other galaxy. At this time, the best estimate of this "age of the universe" is about 12 billion years.

34.24. My pick-up truck does not get very good mileage, but I can drive about 400 miles on a 20 gal tank of gas. One gallon of gas yields about 1.1×10^8 J of energy. Suppose that all of the mass in a bean (mass 300 mg) could be converted into thermal energy. How far could I drive on this energy?

34.25. A proton is moving at speed $0.95c$. Calculate its kinetic energy using (a) the incorrect classical expression $KE = mv^2/2$ and (b) the correct relativistic expression $KE = (\gamma - 1)mc^2$.

SOLUTIONS TO SUPPLEMENTARY PROBLEMS

34.13. $t = \dfrac{t_0}{\sqrt{1 - v^2/c^2}} = \dfrac{2.6 \times 10^{-8}\,\text{s}}{\sqrt{1 - (0.982c/c)^2}} = 13.8 \times 10^{-8}\,\text{s}$

$x = vt = (0.982)(3 \times 10^8\,\text{m/s})(13.8 \times 10^{-8}\,\text{s}) = 40.6\,\text{m}$

34.14. An observer on the satellite sees the length of the spaceship to be

$$L = \frac{1}{\gamma} L_0$$

He measures the passage time to be

$$t = \frac{L}{v} = \frac{1}{v}\sqrt{1 - \frac{v^2}{c^2}}\, L_0$$

Solve: $\quad \dfrac{1}{v^2} - \dfrac{1}{c^2} = \left(\dfrac{t}{L}\right)^2 \qquad \dfrac{1}{v^2} = \left(\dfrac{1}{3\times 10^8\,\text{m/s}}\right)^2 - \left(\dfrac{8\times 10^{-6}\,\text{s}}{200\,\text{m}}\right)^2$

Thus $v = 2.49 \times 10^7$ m/s $= 0.083c$.

34.15. $\quad \Delta L = L_0 - \sqrt{1 - \dfrac{v^2}{c^2}}\, L_0 = \left[1 - \sqrt{1 - \left(\dfrac{250\,\text{m/s}}{3\times 10^8\,\text{m/s}}\right)^2}\,\right](12\,\text{m}) = 5\times 10^{-6}\,\text{m}$

34.16. (a) $\quad u_x = \dfrac{u_x' + v}{1 + vu_x'/c^2} = \dfrac{0.8c + 0.3c}{1 + (0.8c)(0.3c)/c^2} = \dfrac{1.1c}{1.24} = 0.89c$

(b) $\quad u_x = \dfrac{-0.8c + 0.3c}{1 + (-0.8c)(0.3c)/c^2} = -0.66c$

(c) $\quad \gamma = \dfrac{1}{\sqrt{1 - v^2/c^2}} = \dfrac{1}{\sqrt{1 - (0.3)^2}} = 1.05 \qquad u_y = \dfrac{u_y'}{\gamma(1 + vu_x'/c^2)} = \dfrac{0.8c}{1.05(1 + 0)} = 0.76c$

$u_x = \dfrac{u_x' + v}{1 + vu_x'/c^2} = \dfrac{0 + 0.3c}{1 + 0} = 0.3c \qquad u = \sqrt{u_x^2 + u_y^2} = \sqrt{(0.3c)^2 + (0.76c)^2} = 0.82c$

$\tan\theta = \dfrac{u_y}{u_x} = \dfrac{0.76c}{0.3c} = 2.53 \quad \theta = 68.5° \text{ to the } x\text{-axis}$

34.17. $\quad u_x' = \dfrac{u_x - v}{1 - vu_x/c^2} = \dfrac{c\cos\theta - v}{1 - cv\cos\theta/c^2} = \dfrac{c\cos\theta - v}{1 - v/c\cos\theta}$

$u_y' = \sqrt{1 - \dfrac{v^c}{c^2}}\,\dfrac{u_y}{1 - vu_x/c^2} = \sqrt{1 - \dfrac{v^2}{c^2}}\,\dfrac{c\sin\theta}{1 - v/c\cos\theta}$

$u'^2 = u_x'^2 + u_y'^2 = \left(\dfrac{c\cos\theta - v}{1 - v/c\cos\theta}\right)^2 + \left(1 - \dfrac{v^2}{c^2}\right)\left(\dfrac{c\sin\theta}{1 - v/c\cos\theta}\right)^2$

$\qquad\quad = \dfrac{c^2\cos^2\theta - 2cv\cos\theta + v^2 + c^2\sin^2\theta - v^2\sin^2\theta}{(1 - v/c\cos\theta)^2}$

$u'^2 = \dfrac{c^2 - 2cv\cos\theta + v^2(1 - \sin^2\theta)}{(c - v\cos\theta)^2} \qquad c^2 = \dfrac{v^2 - 2cv\cos\theta + v^2\cos^2\theta}{(c - v\cos\theta)^2} \qquad c^2 = \dfrac{(c - v\cos\theta)^2}{(c - v\cos\theta)^2} \qquad c^2 = c^2$

Thus $u' = c$, as expected. In the moving frame, the light makes angle θ' with the x'-axis, where

$$\tan\theta = \dfrac{u_y'}{u_x'} = \sqrt{1 - \dfrac{v^2}{c^2}}\,\dfrac{c\sin\theta}{c\cos\theta - v} = \sqrt{1 - (0.6)^2}\left(\dfrac{c\sin 45°}{c\cos 45° - 0.6c}\right) = 5.28, \quad \theta' = 79°$$

34.18. $mc^2 = (9.11 \times 10^{-31}\,\text{kg})(3 \times 10^8)^2 = 8.20 \times 10^{-14}\,\text{J} = (8.20 \times 10^{-14}\,\text{J})(1.6 \times 10^{-19}\,\text{J/eV})^{-1}$

$mc^2 = 5.12 \times 10^5\,\text{eV} = 0.51\,\text{MeV}$

$$\text{KE} = E - mc^2 = E\left(1 - \frac{mc^2}{E}\right) = E\left(1 - \frac{0.51\,\text{MeV}}{30 \times 10^3\,\text{MeV}}\right) = 0.99998\,E$$

$$pc = \sqrt{E^2 - (mc^2)^2} \simeq E, \quad \text{since } \frac{mc^2}{E} \ll 1,$$

$$p = \frac{(30 \times 10^9\,\text{eV})(1.6 \times 10^{-19}\,\text{J/eV})}{(3 \times 10^8\,\text{m/s})} \qquad p = 1.6 \times 10^{-17}\,\text{kg}$$

$$E = \frac{mc^2}{\sqrt{1 - v^2/c^2}}, \quad 1 - \frac{v^2}{c^2} = \left(\frac{mc^2}{E}\right)^2, \quad \frac{v^2}{c^2} = 1 - \left(\frac{mc^2}{E}\right)^2, \quad \frac{v}{c} = \sqrt{1 - \left(\frac{mc^2}{E}\right)^2} \simeq 1 - \frac{1}{2}\left(\frac{mc^2}{E}\right)^2$$

$$\frac{v}{c} = 1 - \frac{1}{2}\left(\frac{0.51\,\text{MeV}}{3 \times 10^4\,\text{MeV}}\right)^2 = 1 - 1.45 \times 10^{-10}$$

Thus $v = (1 - 1.45 \times 10^{-10})c$, so $v \simeq c$

34.19. Let the spy ship be the moving frame S' and the platform be the fixed frame. Then $v = 0.3c$ and $u' = 0.20c$.

$$u = \frac{u'_x + v}{1 + vu'_x/c^2} = \frac{0.2c + 0.3c}{1 + (0.3c)(0.2c)/c^2} = 0.47c$$

The habitat observer sees the *Ironsides* moving at $0.43c$ and the spy ship moving at $0.3c$, so the *Ironsides* is closing on the spy ship at $0.43c - 0.30c = 0.13c$.

34.20. $p = \gamma mv = 10\,mv$, so $\gamma^2 = 100 = \dfrac{1}{1 - v^2/c^2}$, $v = 0.995c$

34.21. $\mathbf{F} = \dfrac{d\mathbf{p}}{dt} = \dfrac{d(\gamma m\mathbf{u})}{dt} = \gamma m\dfrac{d\mathbf{u}}{dt} + m\mathbf{u}\dfrac{d\gamma}{dt} = \gamma m\dfrac{d\mathbf{u}}{dt} + m\mathbf{u}\left[\dfrac{u}{c^2(1 - u^2/c^2)^{3/2}}\right]\dfrac{du}{dt}$

$$= \gamma m\dfrac{d\mathbf{u}}{dt} + \gamma^3 m\left(\frac{u}{c^2}\right)\dfrac{du}{dt}\mathbf{u}$$

$\mathbf{u}$ and $d\mathbf{u}/dt$ are not necessarily parallel, so $\mathbf{F}$ need not be parallel to $d\mathbf{u}/dt$.

34.22. One kilogram of ^{235}U contains N nuclei:

$$N = \left(\frac{1000\,\text{g}}{235\,\text{g}}\right)(6 \times 10^{23}) = 2.6 \times 10^{24}$$

The total energy released by 1 kg is thus

$$E = (2.6 \times 10^{24})(200\,\text{MeV})(1.6 \times 10^{-13}\,\text{J/MeV}) = 8.3 \times 10^{13}\,\text{J}$$

To provide this energy requires mass M_c of coal:

$$M_c = \frac{8.3 \times 10^{13}\,\text{J}}{3 \times 10^7\,\text{J/kg}} = 2.8 \times 10^6\,\text{kg of coal}$$

$$\frac{\Delta m}{m} = \frac{E/c^2}{m} = \frac{E}{mc^2} = \left(\frac{0.235\,\text{kg}}{6 \times 10^{23}}\right)^{-1}\left(\frac{1}{3 \times 10^8\,\text{m/s}}\right)^{-2}(200\,\text{MeV})(1.6 \times 10^{-13}\,\text{J/MeV}) = 9.1 \times 10^{-4}$$

34.23. $f = \dfrac{c}{\lambda}$,   so $\dfrac{c}{\lambda} = \sqrt{\dfrac{c-v}{c+v}}\,\dfrac{c}{\lambda_0}$, $\dfrac{1}{\lambda^2} = \dfrac{c-v}{\lambda_0^2(c+v)}$, $\lambda_0^2(c+v) = \lambda^2(c-v)$

$\dfrac{v}{c} = \dfrac{\lambda^2 - \lambda_0^2}{\lambda^2 + \lambda_0^2} \simeq \dfrac{(\lambda - \lambda_0)(\lambda + \lambda_0)}{2\lambda_0^2}$, $\dfrac{v}{c} \simeq \dfrac{2\lambda_0(\lambda - \lambda_0)}{2\lambda_0^2} = \dfrac{\lambda - \lambda_0}{\lambda_0} = \dfrac{398.4\,\text{nm} - 396.85\,\text{nm}}{396.85\,\text{nm}} = 0.0039$

$v = 0.0039c$, $t = \dfrac{d}{v} = \dfrac{76 \times 10^6\,\text{ly}}{0.0039c} = 19.5 \times 10^9\,\text{y}$

34.24. To travel 1 mi requires energy

$$U_1 = \frac{(20)(1.1 \times 10^8\,\text{J})}{400\,\text{mi}} = 5.5 \times 10^6\,\text{J/mi}$$

One bean could yield energy

$$U = mc^2 = (0.30 \times 10^{-3}\,\text{kg})(3 \times 10^8\,\text{m/s})^2$$
$$U = 2.7 \times 10^{13}\,\text{J}$$

The distance this would drive my truck is

$$d = \frac{2.7 \times 10^{13}\,\text{J}}{5.5 \times 10^6\,\text{J/mi}} = 4.9 \times 10^6\,\text{mi}$$

A little mass is worth a lot of energy!

34.25. (a) $\text{KE} = \tfrac{1}{2}mv^2 = \tfrac{1}{2}(1.67 \times 10^{-27}\,\text{kg})(0.95 \times 3.0 \times 10^8\,\text{m/s})^2$

$\text{KE} = 6.78 \times 10^{-11}\,\text{J}$

(b) $\text{KE} = (\gamma - 1)mc^2$

$\gamma = \dfrac{1}{(1 - (v^2/c^2))^{\frac{1}{2}}} = \dfrac{1}{(1 - 0.95^2)^{\frac{1}{2}}} = 3.20$

$\text{KE} = (3.20 - 1)(1.67 \times 10^{-27}\,\text{kg})(3.0 \times 10^8\,\text{m/s})^2$

$\text{KE} = 33 \times 10^{-11}\,\text{J}$

For such a high velocity there is a big difference between the relativistic result and the classical result.

CHAPTER 35

Atoms and Photons

The history of the development of modern physics in the twentieth century reads like a detective story. The theories of relativity and quantum mechanics changed radically our view of the universe. Their impact on our lives has been tremendous.

35.1 Atoms and Photons

The story begins around 1900 when attempts were made to understand the radiation emitted by hot blackbodies. It was well known that when an object is heated, it radiates energy. If a horseshoe is held in a flame, it first radiates dull red light, and as it gets hotter and hotter, the intensity increases and the light becomes brighter red and then yellow. This behavior is shown in Fig. 35-1. Classical theory predicted that the intensity of radiation should continue to increase indefinitely at shorter wavelengths, but instead the curves start turning downward around 400 nm (the "ultraviolet catastrophe"). Max Planck was able to reproduce the experimental curves with a theory based on the radical assumption that radiation of frequency f and wavelength λ is emitted only in discrete packets, which he called **photons**. He postulated that the energy of a photon is

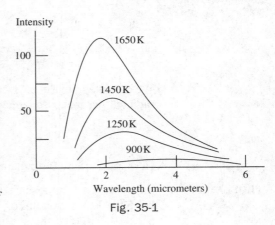

Fig. 35-1

$$\boxed{E = hf = \frac{hc}{\lambda} \quad \text{photon energy}} \tag{35.1}$$

Here h is **Planck's constant**: $h = 6.63 \times 10^{-34} \, \text{J} \cdot \text{s}$. On this basis he calculated the intensity of blackbody radiation to be

$$\boxed{I(\lambda, T) = \frac{2\pi c^2 h}{\lambda^5} \frac{1}{e^{hc/k_B T\lambda} - 1} \quad \text{blackbody radiation}} \tag{35.2}$$

k_B is Boltzmann's constant. Planck envisioned a collection of radiators whose energy was $E = nhf = n(hc/\lambda)$, where $n =$ integer. When the system emits a photon, it drops down one energy level, and when it absorbs a photon, it moves up one energy level. Such discrete energy levels are **quantized**, and the packets of radiation are called *photons* or **quanta**.

PROBLEM 35.1. Calculate the energy in joules and electronvolts for the following photon wavelengths: 1×10^{-10} m (x ray), 200 nm (ultraviolet), 500 nm (yellow light near the peak of the Sun's spectrum), 10 μm (infrared radiated by your body), and 200 m (AM radio station).

Solution Substituting numerical values in $E = hc/\lambda$ and using $1\,\mathrm{eV} = 1.6 \times 10^{-19}$ J, $E = 2.0 \times 10^{-15}\,\mathrm{J} = 12.4\,\mathrm{keV}$ (x rays), $E = 9.9 \times 10^{-19}\,\mathrm{J} = 6.2\,\mathrm{eV}$ (ultraviolet) $E = 4.0 \times 10^{-19}$ J $= 2.5\,\mathrm{eV}$ (yellow light), $E = 9.9 \times 10^{-28}\,\mathrm{J} = 6.2 \times 10^{-9}\,\mathrm{eV}$ (radio waves)

Note that it usually takes a few electronvolts to break chemical bonds, which is why UV light is much more likely to cause skin cancer and tanning than is visible light. X rays and gamma rays are even more dangerous.

From Eq. 35.1 we can deduce the following useful relation for blackbody radiation:

$$\lambda_{max}T = \text{constant} = 0.0029\ \mathrm{m \cdot K} \quad \text{Wien's displacement law} \tag{35.3}$$

T is the absolute temperature, and λ_{max} is the wavelength at which the radiation intensity is a maximum.

PROBLEM 35.2. The peak of the Sun's spectrum is at 484 nm. Estimate the temperature of the Sun's surface. Human skin is at about 33°C (306 K). At what wavelength is the peak of its radiation, assuming it is a blackbody?

Solution

$$T_s = \frac{0.0029\ \mathrm{m \cdot K}}{484 \times 10^{-9}\ \mathrm{m}} \simeq 6000\ \mathrm{K}$$

$$\lambda_{max} = \frac{0.0029\ \mathrm{m \cdot K}}{306\ \mathrm{K}} = 9.5 \times 10^{-6}\ \mathrm{m} = 9.5\ \mathrm{\mu m} \quad \text{infrared}$$

We can integrate Eq. 35.1 to obtain the total energy radiated by a blackbody. The energy radiated per meter squared per second is

$$S = \sigma T^4 \quad \text{Stefan-Boltzmann law} \tag{35.4}$$

Here the Stefan–Boltzmann constant is $\sigma = 5.67 \times 10^{-8}\ \mathrm{W/m^2 \cdot K^4}$.

PROBLEM 35.3. Suppose that from a star's radiation spectrum its surface temperature is determined to be 5400 K. Astronomical observation determines its distance from Earth to be 5.2×10^{18} m, and the starlight reaching us has intensity $1.4 \times 10^{-4}\ \mathrm{W/m^2}$. From this data, estimate the size of the star.

Solution The surface area of a star of radius r is $4\pi r^2$, so the total power radiated is $P = 4\pi r^2 S$, where $S = \sigma T^4$. By the time the light reaches us, it is spread over a sphere of radius R, so $P = 4\pi R^2 S_e$. Equating these, $4\pi r^2 \sigma T^4 = 4\pi R^2 S_e$:

$$r^2 = \frac{R^2 S_e}{\sigma T^4} = \frac{(5.2 \times 10^{18}\ \mathrm{m})^2 (1.4 \times 10^{-8}\ \mathrm{W/m^2})}{(0.0029\ \mathrm{W/m^2 \cdot K^4})(5400\ \mathrm{K})^4}$$

so

$$r = 3.9 \times 10^8\ \mathrm{m}$$

about half as big as the Sun.

35.2 The Photoelectric Effect

When light of an appropriate wavelength is shined on certain metals, electrons are emitted. This can give rise to a current in an arrangement such as that in Fig. 35-2. Two electrodes, an anode *A* and a cathode *c*, are enclosed in an evacuated glass envelope. When the cathode is illuminated, a photocurrent will flow in the ammeter if the battery is connected so that the anode is positive. When the battery polarity is reversed, and a reverse voltage, with the anode negative, is applied until the current flow stops. This stopping voltage is then measured as a function of the wavelength of the incident light. For wavelengths longer than a certain critical value, no current flows, no matter how intense the light. In 1905 Einstein explained what is happening by using Planck's idea of photons. Einstein suggested that a light beam consists of a stream of particles (photons) each carrying energy hf. A certain amount of energy ϕ, the **work function** for the metal, is required to break an electron loose. One and only one photon knocks loose one electron, so hundreds of red photons may fail to eject an electron, whereas a single blue photon (shorter wavelength, more energetic) may free an electron. If the kinetic energy of the emitted electrons is KE,

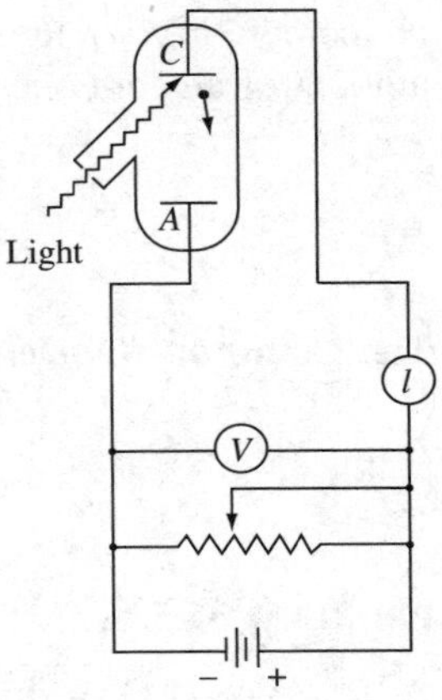

$$hf = \frac{hc}{\lambda} = KE + \phi \qquad (35.5)$$

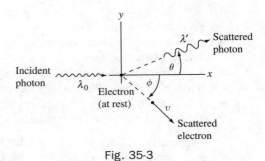

Fig. 35-2

The kinetic energy is determined by applying a stopping voltage until no current flows, at which time $eV = KE$.

The importance of this experiment is that it definitely established the particle nature of light. In general, when light is emitted or absorbed, it acts like a particle. When it is traveling, it acts like a wave. This behavior is called the **duality of light**. Depending on the circumstances, light can act like either a particle or a wave.

PROBLEM 35.4. Light of wavelength 200 nm is incident on a cadmium surface. A stopping voltage of 2.15 eV is required to reduce the photocurrent to zero. What is the work function of cadmium?

Solution The energy of an incoming photon is

$$E = \frac{hc}{\lambda} = \frac{(6.63 \times 10^{-34}\,\text{J}\cdot\text{s})(3 \times 10^{8}\,\text{m/s})}{(200 \times 10^{-9}\,\text{m})(1.6 \times 10^{-19}\,\text{eV/J})} = 6.22\,\text{eV}$$

Thus

$$\phi = E - KE = 6.22\,\text{eV} - 2.15\,\text{eV} = 4.07\,\text{eV}$$

35.3 The Compton Effect

Further confirmation of the particle nature of light came in 1923 from experiments by Arthur Compton on scattering of x rays. He noted that photons act like particles with energy $E = hf$ and momentum $p = hc/\lambda$. One can see that the latter must be true, because $E = (1 - v^2/c^2)^{-1/2}mc^2$, and $v = c$ for photons, so $m = 0$ if E is finite. Thus **all particles that travel at speed c have zero rest mass**. Since $E^2 = p^2c^2 + (mc^2)^2 = p^2c^2 + 0$, $pc = E = hf$ and $p = hf/c = h/\lambda$. In summary, photons of all frequencies travel at speed c in vacuum and never slower, and they have zero rest mass, energy $E = hf = hc/\lambda$, and momentum $p = hf/c = h/\lambda$.

$$p = \frac{hf}{c} = \frac{h}{\lambda} \quad \text{photon momentum} \qquad (35.6)$$

Compton recognized that when a photon is scattered from an electron at rest, the process is like an elastic collision in which a photon with energy hc/λ_1 and momentum h/λ_1 is scattered from a stationary mass at rest, resulting in a scattered photon with energy hc/λ_2 and momentum h/λ_2 and an electron with momentum γmv and energy γmc^2. The process is illustrated in Fig. 35-3. Energy and momentum are conserved before and after the collision, resulting in

Fig. 35-3

an increase in λ for the scattered x ray. The shift in wavelength depends on the scattering angle. With a fair amount of algebra one can show that

$$\lambda_2 - \lambda_1 = \Delta\lambda = \frac{h}{m_e c}(1 - \cos\theta) \tag{35.7}$$

The **Compton wavelength** is defined as

$$\lambda_c = \frac{h}{m_e c} = 0.0024\,\text{nm} \tag{35.8}$$

PROBLEM 35.5.   A 5.5-MeV gamma ray is scattered at $60°$ from an electron. What is the energy in megaelectronvolts of the scattered photon?

Solution $\lambda_1 = \dfrac{hc}{E_1} = \dfrac{(6.63 \times 10^{-34}\,\text{J}\cdot\text{s})(3 \times 10^8\,\text{m/s})}{(5.5 \times 10^6\,\text{eV})(1.6 \times 10^{-19}\,\text{J/eV})} = 2.26 \times 10^{-13}\,\text{m}$

$\lambda_2 = \lambda_1 + \dfrac{h}{m_e c}(1 - \cos\theta) = 2.26 \times 10^{-13}\,\text{m} + \dfrac{(6.63 \times 10^{-34}\,\text{J}\cdot\text{s})(1 - \cos 60°)}{(9.11 \times 10^{-31}\,\text{kg})(3 \times 10^8\,\text{m/s})}$

$\qquad = 2.26 \times 10^{-13}\,\text{m} + 1.21 \times 10^{-12}\,\text{m} = 1.44 \times 10^{-12}\,\text{m}$

$E_2 = \dfrac{hc}{\lambda_2} = \dfrac{(6.63 \times 10^{-34}\,\text{J}\cdot\text{s})(3 \times 10^8\,\text{m/s})}{(1.44 \times 10^{-12}\,\text{m})(1.6 \times 10^{-19}\,\text{J/eV})} = 0.87\,\text{MeV}$

35.4   Atomic Spectra and Bohr's Model of the Atom

All material at a given temperature emits thermal electromagnetic radiation over a continuous range of wavelengths, with the intensity distribution described reasonably well by the blackbody radiation curves. However, when a low-pressure gas undergoes an electric discharge (as in a neon lamp), projecting the emitted light through a diffraction grating or a prism shows a series of sharp bright lines of pure colors. Further, this emission spectrum is characteristic of each element, and no two elements have the same spectrum. One can also shine a continuous rainbow spectrum of light on a container of gas of a given element and observe in the transmitted light a continuous spectrum on which are superimposed discrete dark lines. These dark absorption lines occur at the same wavelengths as the emission lines. Observation of absorption spectra is useful for determining the composition of a gas. It is through this method that we are able to determine the composition of stars. During the 1800s scientists worked out empirical formulas to enable them to determine what wavelengths would be emitted by hydrogen, but there was no theoretical basis for their work until 1913, when Niels Bohr put forth the brilliant ideas that were to form the basis for modern quantum theory. Bohr knew that an atom is electrically neutral and consists of a heavy, positively charged nucleus and a number of light electrons carrying negative electric charge. He first considered the hydrogen atom, for which the nucleus is a single proton of charge $+e$ accompanied by one electron of charge $-e$. Bohr made the following assumptions:

- The electron moves in orbits around the proton, with the electric coulomb force providing the needed centripetal force.
- The orbits in which the electron moves are stable; that is, in these orbits the electron does not radiate and its energy is constant.
- Radiation is emitted or absorbed only when the electron jumps from one orbit to another, and the frequency of the radiation depends only on the energies of the electron in the initial and final orbits.
- The allowed orbits are those for which the orbital angular momentum mvr is an integer multiple of $\hbar = h/2\pi$; that is, $mvr = n\hbar$, $n = 1, 2, 3, \ldots, n$.

The last three postulates were very radical, and the only justification for them is that they work; that is, Bohr was able to calculate the observed spectrum of hydrogen.

Since the proton is much heavier than the electron ($m_p \simeq 1840m_e$), assume the proton mass is infinite and that the proton remains fixed while the electron circles it in a circular orbit. Then

$$\frac{mv^2}{r} = k\frac{e^2}{r^2}, \quad k = \frac{1}{4\pi\epsilon_0}$$

The energy $E = \text{KE} + \text{PE}$:

$$E = \tfrac{1}{2}mv^2 - k\frac{e^2}{r} = \tfrac{1}{2}k\frac{e^2}{r} - k\frac{e^2}{r} = -\tfrac{1}{2}k\frac{e^2}{r}$$

Combine this with $mvr = n\hbar$. Then

$$E_n = \frac{ke^2}{2a_0}\left(\frac{1}{n^2}\right) = -\frac{13.6\,\text{eV}}{n^2}$$

$$r_n = n^2 a_0, \quad n = 1, 2, 3, \ldots, n$$

$$\text{Bohr radius } a_0 = \frac{\hbar^2}{kme^2} = 0.0529\,\text{nm} \tag{35.8}$$

Each energy value is called an **energy level**, and some are shown in Fig. 35-4, an **energy level diagram**. The lowest state, $n = 1$, is the **ground state**. The first **excited state** is $n = 2$. $n = \infty$ is a free electron, with $E = 0$, at rest infinitely far from the nucleus. The bound states all have negative energies. A photon is emitted when an electron makes a transition from a higher level to a lower level. When an electron jumps from a lower level to a higher level, a photon is absorbed. Groups of spectral lines are given names, for example, the Lyman series and the Balmer series. The frequency of a spectral line is given by

$$\Delta E = |E_f - E_i| = hf = \frac{hc}{\lambda} \tag{35.9}$$

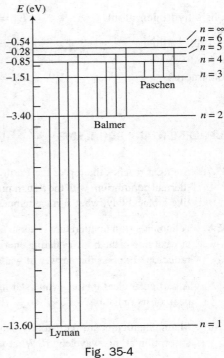

Fig. 35-4

PROBLEM 35.6. What is the wavelength emitted for the first Balmer line, when a hydrogen atom makes a transition from the $n = 3$ to the $n = 2$ state?

Solution

$$\Delta E = -13.6\,\text{eV}\left(\frac{1}{3^2} - \frac{1}{2^2}\right) = 1.89\,\text{eV}$$

$$\lambda = \frac{hc}{\Delta E} = \frac{(6.63 \times 10^{-34}\,\text{J}\cdot\text{s})(3 \times 10^8\,\text{m/s})}{(1.89\,\text{eV})(1.6 \times 10^{-19}\,\text{J/eV})} = 658\,\text{nm}$$

The Bohr model of hydrogen was successful in explaining many features of the spectrum. Most important, it introduced the concept of quantized energy levels and transitions between levels associated with absorption and emission of photons. One should not, however, take too literally the picture of little point mass electrons circling the nucleus like planets around the Sun. For example, in the ground state of hydrogen the orbital angular momentum is zero, contrary to what is predicted here. Electrons have more of a will-o'-the-wisp nature, and they are as much wave as particle, as we shall see in the next chapter.

35.5   Summary of Key Equations

Photon energy: $E = hf = \dfrac{hc}{\lambda}$

Photon momentum: $p = \dfrac{h}{\lambda}$

Planck's blackbody radiation law: $I(\lambda, T) = \dfrac{2\pi c^2 h}{\lambda^5} \dfrac{1}{e^{hc/k_a \lambda} - 1}$

Wien's displacement law: $\lambda_{max} T = 0.0029 \text{ m} \cdot \text{K} = \text{constant}$

Stefan–Boltzmann law: $S = \sigma T^4$

Photoelectric effect: $hf = \text{KE} + \phi$

Compton scattering: $\lambda_2 - \lambda_1 = \lambda_c(1 - \cos\theta)$

Compton wavelength: $\lambda_c = \dfrac{h}{m_e c} = 0.0024 \text{ nm}$

Bohr's hydrogen atom: $E_n = -\dfrac{13.6 \text{ eV}}{n^2}, \quad n = 1, 2, 3, \ldots, n$

nth orbit radius: $r_n = n^2 a_0$

Bohr radius: $a_0 = \dfrac{\hbar^2}{kmc^2} = 0.0529 \text{ nm}, \quad \hbar = \dfrac{h}{2\pi}$

SUPPLEMENTARY PROBLEMS

35.7.   Sunlight reaches the top of the Earth's atmosphere with an intensity of about 1400 W/m^2. If the Earth is in thermal equilibrium with the Sun, it must radiate infrared energy at this same rate. Assuming the Earth radiates like a blackbody, what surface temperature for the Earth would you estimate?

35.8.   A human's skin temperature is about $33°\text{C}$, and total skin area is about 1.4 m^2. Assuming blackbody radiation, at what rate is the body radiating energy? If the environment is at $23°\text{C}$, what is the net rate of heat loss due to radiation? Express the answer in watts and in kilocalories per hour.

35.9.   A particular class 0 blue–white star has a surface temperature of about $5 \times 10^4 \text{ K}$. At what wavelength is the peak of its radiation?

35.10. Mammalian cells are particularly subject to radiation damage at 260 nm (UV) because DNA has a strong absorption at this wavelength. What is the energy in electronvolts of these photons?

35.11. Aluminum has a work function of 4.08 eV. What is the maximum wavelength that will eject electrons from aluminum?

35.12. The work function of sodium is 2.28 eV. What is the speed of an electron emitted from sodium by a photon with $\lambda = 500 \text{ nm}$?

35.13. An x ray with $\lambda = 0.025 \text{ nm}$ experiences a fractional change in wavelength $\Delta\lambda/\lambda = 0.05$ in Compton scattering. At what angle is it scattered?

35.14. Derive an expression for the kinetic energy of an electron that has experienced Compton scattering by an incident photon of frequency f_1 and wavelength λ_1. Express the result in terms of h, f_1, λ_1, and $\Delta\lambda = \lambda_2 - \lambda_1$.

35.15. Calculate the frequency of revolution for an electron in the ground state of hydrogen. For comparison, find the frequency emitted when an electron falls from the $n = 2$ state to the ground state.

35.16. What is the maximum wavelength of a photon that will ionize a hydrogen atom in the ground state. To "ionize" means to free the electron from the atom.

35.17. A hydrogen atom in the ground state absorbs a photon of wavelength 88.8 nm. What is the kinetic energy of the electron after the absorption?

35.18. A certain metal has a work function of 2.6 eV. (a) What is the shortest wavelength of electromagnetic radiation that will eject electrons from this material? (b) If light of wavelength 420 nm is used, what will be the speed of the emitted electrons?

35.19. About 2% of the energy used by a 100-W incandescent light bulb appears as light energy. Approximately how many photons are emitted per second by such a lamp, assuming an average wavelength of 540 nm?

SOLUTIONS TO SUPPLEMENTARY PROBLEMS

35.7. $S = \sigma T^4, \quad T^4 = \dfrac{S}{\sigma} = \dfrac{1400 \text{ W/m}^2}{5.67 \times 10^{-8} \text{ W/m}^2 \cdot \text{K}^4} \quad T = 396 \text{ K} = 1213°\text{C}$

The Earth's surface temperature is less than this because much of the incident sunlight is reflected without being absorbed.

35.8. $P_h = SA = A\sigma T^4 = (1.4 \text{ m}^2)(5.67 \times 10^{-8} \text{ W/m}^2 \text{ K}^4)(306 \text{ K})^4 = 700 \text{ W}$

$$P_e = P_h \left(\frac{T_\le}{T_h}\right)^4 = (700 \text{ W})\left(\frac{296 \text{ K}}{306 \text{ K}}\right)^4 = 613 \text{ W}$$

Net heat loss: $P = P_h - P_e = 87 \text{ W} = 70 \text{ kcal/h}$

35.9. $\lambda_{\max} T = 0.0029 \text{ m} \cdot \text{K}, \quad \lambda_{\max} = \dfrac{0.0029 \text{ K} \cdot \text{m}}{5 \times 10^4 \text{ K}} = 58 \text{ nm}$

35.10. $E = \dfrac{hc}{\lambda} = \dfrac{(6.63 \times 10^{-34} \text{ J} \cdot \text{s})(3 \times 10^8 \text{ m/s})}{(260 \times 10^{-9} \text{ m})(1.6 \times 10^{-19} \text{ J/eV})} = 4.78 \text{ eV}$

35.11. $\dfrac{hc}{\lambda} = \text{KE} + \phi = 0 + \phi, \quad \lambda = \dfrac{hc}{\phi} = \dfrac{(6.63 \times 10^{-34} \text{ J} \cdot \text{s})(3 \times 10^8 \text{ m/s})}{(4.08 \text{ eV})(1.6 \times 10^{-19} \text{ J/eV})} = 304 \text{ nm}$

35.12. $hf = \dfrac{hc}{\lambda} = \text{KE} + \phi = \dfrac{1}{2}mv^2, \quad v^2 = \dfrac{2}{m}\left(\dfrac{hc}{\lambda} - \phi\right)$

$$v^2 = \left(\frac{2}{9.11 \times 10^{-31} \text{ kg}}\right)\left[\frac{(6.63 \times 10^{-34} \text{ J} \cdot \text{s})(3 \times 10^8 \text{ m/s})}{500 \times 10^{-9} \text{ m}} - (2.28 \text{ eV})(1.6 \times 10^{-19} \text{ J/eV})\right]$$

$v = 2.69 \times 10^5 \text{ m/s}$

35.13. $\Delta\lambda = \dfrac{h}{m_e c}(1 - \cos\theta) = \lambda_c(1 - \cos\theta), \quad \cos\theta = \dfrac{\lambda_c}{\lambda} - \dfrac{\Delta\lambda}{\lambda} = \dfrac{0.0024 \text{ nm}}{0.025 \text{ nm}} - 0.05 = 0.046$

and

$$\theta = 87.4°$$

35.14. $\text{KE} = hf_1 - hf_2 = \dfrac{hc}{\lambda_1} - \dfrac{hc}{\lambda_2} = \dfrac{hc}{\lambda_1 \lambda_2}(\lambda_2 - \lambda_1) = \dfrac{hf_1 \lambda_1}{\lambda_1 \lambda_2}\Delta\lambda = \dfrac{hf_1}{\lambda_2}\Delta\lambda = hf_1\left(\dfrac{\Delta\lambda}{\lambda_1 + \Delta\lambda}\right)$

35.15. From derivation of Eq. 35.8,

$$\text{KE} = \frac{1}{2}mr\omega^2 = \frac{1}{2}k\frac{e^2}{r} = \frac{13.6\,\text{eV}}{n^2}, \qquad n=1, \quad r=a_0 = 0.0529\,\text{nm}$$

$$\omega^2 = \frac{2(13.6\,\text{eV})(1.6\times10^{-19}\,\text{J/eV})}{(1)^1(9.11\times10^{-31}\,\text{kg})(0.0529\times10^{-9}\,\text{m})}, \quad \omega = 2\pi f$$

$$f = 4.78\times10^{10}\,\text{Hz} \quad \text{frequency of revolution}$$

For

$$n=3 \text{ to } n=2, \quad hf_{32} = \Delta E = (13.6\,\text{eV})\left(\frac{1}{2^2} - \frac{1}{3^2}\right) = 1.89\,\text{eV}$$

$$f_{32} = \frac{(1.89\,\text{eV})(1.6\times10^{-19}\,\text{J/eV})}{(6.63\times10^{-34}\,\text{J}\cdot\text{s})} = 4.56\times10^{14}\,\text{Hz}$$

$\lambda_{32} = c/f_{32} = 658\,\text{nm}$ (red). The photon frequency is about 10^4 times larger than the rotation frequency.

35.16. $\Delta E = E_\infty - E_1 = -\dfrac{13.6\,\text{eV}}{\infty^2} + \dfrac{13.6\,\text{eV}}{1^2} = 13.6\,\text{eV}$ $\qquad \Delta E = hf = \dfrac{hc}{\lambda} \qquad \lambda = \dfrac{hc}{\Delta E}$

$$\lambda = \frac{(6.63\times10^{-34}\,\text{J}\cdot\text{s})(3\times10^8\,\text{m/s})}{(13.6\,\text{eV})(1.6\times10^{-19}\,\text{J/eV})} = 91.4\,\text{nm}$$

35.17. The energy of the photon is

$$E = hf = \frac{hc}{\lambda} = \frac{(6.63\times10^{-34}\,\text{J}\cdot\text{s})(3\times10^8\,\text{m/s})}{(88.8\times10^{-9}\,\text{m})(1.6\times10^{-19}\,\text{J/eV})} = 14.0\,\text{eV}$$

The ionization energy is

$$E_\infty - E_1 = -\frac{13.6\,\text{eV}}{\infty^2} + \frac{13.6\,\text{eV}}{1^2} = 13.6\,\text{eV}$$

Thus the KE of the electron is

$$\text{KE} = 14.0\,\text{eV} - 13.6\,\text{eV} = 0.60\,\text{eV}$$

35.18. (a) $hf = \text{KE} + \varphi = 0 + \varphi$

$$f = \frac{c}{\lambda}, \quad \lambda = \frac{hc}{\varphi} = \frac{(6.63\times10^{-34}\,\text{J}\cdot\text{s})(3.0\times10^8\,\text{m/s})}{(2.6)(1.6\times10^{-19}\,\text{J})}$$

$$\lambda = 478\,\text{nm}$$

(b) $\text{KE} = hf - \varphi = \dfrac{hc}{\lambda} - \varphi$

$$\text{KE} = \frac{1}{2}mv^2, \quad v = \left[\frac{2}{m}\left(\frac{hc}{\lambda} - \varphi\right)\right]^{\frac{1}{2}}$$

$$v = \left\{\left[\frac{2}{9.11\times10^{-31}\,\text{kg}}\right]\left[\frac{(6.63\times10^{-34}\,\text{J}\cdot\text{s})(3.0\times10^8\,\text{m/s})}{420\times10^{-9}\,\text{m}} - (2.6)(1.6\times10^{-19}\,\text{J})\right]\right\}^{\frac{1}{2}}$$

$$v = 3.56\times10^5\,\text{m/s}$$

35.19. If n photons are emitted per second,

$$nhf = eP$$

$$f = \frac{c}{\lambda}, \text{ so } n = \frac{eP\lambda}{hc}, \quad e = \text{efficiency}$$

$$n = \frac{(0.02)(100\,\text{W})(540 \times 10^{-9}\,\text{m})}{(6.63 \times 10^{-34}\,\text{J} \cdot \text{s})(3.0 \times 10^8\,\text{m/s})}$$

$$n = 5.4 \times 10^{18}/\text{s}^4$$

CHAPTER 36

Quantum Mechanics

Bohr's theory was fairly successful in describing the spectrum of hydrogen, but it did not explain the varying intensity of different spectral lines, nor did it adequately describe the spectrum of heavier atoms. Further, it did not explain the fact that under high resolution, some lines were found to be doublets. Sommerfeld showed that including elliptical orbits in Bohr's theory helped explain how the periodic table of elements was determined, but there was still no clear basis for some of Bohr's assumptions. More radical ideas were called for. The early 1920s was the time of another great leap forward in our understanding of nature. The theory that evolved is called **quantum mechanics**.

36.1   de Broglie Waves

In 1924 Louis de Broglie, a French nobleman, reasoned as follows: It is known that light exhibits both wave properties (interference, diffraction, refraction) and particle properties (photoelectric effect, Compton scattering). One of the salient features of nature is its symmetry, so is it not reasonable that just as a wave like light sometimes acts like a particle, should not a particle like an electron act like a wave? de Broglie knew that the momentum of a photon is $p = E/c = hf/c = h/\lambda$, or $\lambda = h/p$, so he made the bold proposal that an electron has wave properties with wavelength λ:

$$\boxed{\lambda = \frac{h}{p} \quad \text{de Broglie wavelength}} \tag{36.1}$$

Here the momentum is

$$p = \frac{1}{\sqrt{1 - v^2/c^2}} mv \simeq mv \quad \text{if } v \ll c$$

   This relation holds true for all particles, from electrons to baseballs or locomotives. However, for large objects, λ is so small that the wave properties are not usually observable. For electrons, however, de Broglie's hypothesis was quickly confirmed.

PROBLEM 36.1.   What is the energy of an electron with $\lambda = 550$ nm (like green light)?

Solution

$$E = \frac{1}{2} mv^2 = \frac{p^2}{2m} = \frac{h^2}{2m\lambda^2} = \frac{(6.63 \times 10^{-34}\,\text{J} \cdot \text{s})^2}{2(9.11 \times 10^{-31}\,\text{kg})(550 \times 10^{-9}\,\text{m})^2}$$

$$= 7.98 \times 10^{-25}\,\text{J} = 5.0 \times 10^{-6}\,\text{eV}$$

PROBLEM 36.2. Determine the de Broglie wavelength for the following: (a) a moving golf ball (mass 0.05 kg, $v = 40\,\text{m/s}$), (b) an orbiting electron in the ground state of hydrogen (13.6 eV), and (c) an electron accelerated through 100 kV in an electron microscope.

Solution

(a) $\lambda = \dfrac{h}{mv} = \dfrac{6.63 \times 10^{-34}\,\text{J} \cdot \text{s}}{(0.05\,\text{kg})(40\,\text{m/s})} = 3.3 \times 10^{-34}\,\text{m}$

(b) $\text{KE} = \dfrac{1}{2}mv^2 = \dfrac{p^2}{2m} = \dfrac{h^2}{2m\lambda^2}$

$\lambda = \dfrac{h}{\sqrt{2m\text{KE}}} = \dfrac{6.63 \times 10^{-34}\,\text{J} \cdot \text{s}}{\sqrt{2(9.11 \times 10^{-31}\,\text{kg})(13.6\,\text{eV})(1.6 \times 10^{-19}\,\text{J/eV})}} = 0.33\,\text{nm}$

(c) $\lambda = 3.88 \times 10^{-12}\,\text{m}$

36.2 Electron Diffraction

Davisson and Germer scattered electrons from nickel crystals, and they found that the intensity of the scattered electrons was consistent with scattering from a diffraction grating whose slit spacing corresponded to the lattice spacing for the nickel crystal. Similar results were obtained for other crystals. Further, when a beam of electrons was directed at two parallel slits, the famous result of Young for the interference of light was reproduced. The electron detector counts at the screen showed a series of bright and dark bands, indicating where the electrons landed. Note that even though the electrons act like waves when they travel through the slits, showing interference, when they strike the detector they do so as individual particles, each making a spot, say, on a photographic film. The pattern observed for two slits is shown in Fig. 36-1. It is the same pattern as that obtained for light. As more and more electrons hit the screen, the light and dark bands become more pronounced.

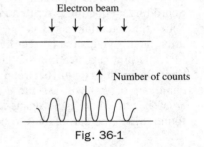

Fig. 36-1

If in the setup of Fig. 36-1 one of the slits is covered, the pattern of electrons hitting the screen changes dramatically. Now one obtains the single slit diffraction pattern. A puzzling question is this. How does an electron going through one slit "know" if the other slit is open or not. Something fishy is going on. The best explanation people have come up with is to say that the "wave" associated with the electron is a probability wave. This wave moves through space with the electron, and its nature is influenced by the physical surroundings. The wave is different depending on whether one or two slits are open. The wave sort of "guides" where the electron will be found. Where the wave has a large amplitude, the electron is likely to be found. Where the wave is small, the electron is not likely to be. It is interference in this wave as it goes through the slits that gives rise to the observed patterns on the screen.

36.3 The Schrödinger Equation

If $y(x, t)$ is the transverse displacement of a vibrating string, $y(x, t)$ is determined by the wave equation:

$$\frac{\partial^2 y}{\partial t^2} = v^2 \frac{\partial^2 y}{\partial x^2} \tag{36.2}$$

We have seen that standing waves can be set up in a string, as is the case for a violin string. The form of a standing wave is $y(x, t) = \psi(x) \sin \omega t$. Substitute this in the wave equation above. Then

$$-\omega^2 \psi(x) = v^2 \frac{d^2\psi}{dx^2} \quad \text{or} \quad \frac{d^2\psi}{dx^2} + \frac{\omega^2}{v^2}\psi = 0$$

But

$$\frac{\omega}{v} = \frac{2\pi f}{v} = \frac{2\pi}{\lambda} \quad \text{and} \quad \lambda = \frac{h}{p}, \quad \text{so} \quad \frac{\omega}{v} = \frac{p}{\hbar}, \quad \text{where } \hbar = \frac{h}{2\pi}$$

If the potential energy of the particle is $U(x)$, the total energy is

$$E = \frac{p^2}{2m} + U(x)$$

so

$$p^2 = 2m[E - U(x)]$$

Thus the wave equation in one dimension takes the form

$$\boxed{\frac{d^2\psi}{dx^2} + \frac{2m}{\hbar^2}[E - U(x)]\psi(x) \quad \text{time-independent Schrödinger equation}} \tag{36.3}$$

This is not a rigorous derivation, but it gives us a clue as to how one might arrive at a relation like Eq. 36.3. There $\psi(x)$ is the **wave function** for a state in which the energy E is constant in time. Such states are called **stationary states**. We have seen that for a vibrating violin string, fixed at both ends, standing waves can occur when the length of the string L is a multiple of $\frac{1}{2}\lambda$, resulting in only certain definite vibration frequencies (the harmonics). In similar fashion, when suitable boundary conditions are applied to the solutions of the Schrödinger equation, we find solutions only for certain discrete values of the energy E. When the Schrödinger equation is written in three dimensions, its solution is generally difficult to obtain analytically. When the potential energy is that for an electron in a hydrogen atom, $U(r) = -k(e^2/r)$, we can find the solution, and it yields the correct hydrogen atom energy levels.

I interpret the meaning of the wave function as follows: $|\psi(x)|^2\,dx$ **is the probability of finding a particle in the region between x and $x + dx$.** (Note: $|\psi|^2 = \psi\psi*$ if ψ is complex. $\psi*$ is the complex conjugate of ψ. I will assume ψ is real in the following.) Thus $\psi^2(x)$ is the **probability density**. Where $\psi^2(x)$ is zero, there is no chance of finding the particle. Where $\psi^2(x)$ is large, the particle is likely to be found. Since the particle has to be found somewhere, the sum of all of the probabilities is 1.

$$\boxed{\int_{-\infty}^{\infty} \psi^2(x)\,dx = 1} \tag{36.4}$$

When Eq. 36.4 is satisfied, the wave function ψ is said to be *normalized*. The average value (called the *expectation value*) of any physical property $f(x)$ can be calculated if ψ is known. Thus the average value of $f(x)$ in a state with wave function $\psi(x)$ is

$$\boxed{\langle f(x)\rangle = \int_{-\infty}^{\infty} f(x)\psi^2\,dx} \tag{36.5}$$

For example, the average value of the position is found by letting $f(x) = x$. The above ideas can be extended to deal with three-dimensional problems. Also, a time-dependent Schrödinger equation can be obtained to describe time-varying wave functions. There are only a few potential energy functions for which the Schrödinger equation can be solved exactly in three dimensions. Sophisticated approximation methods have been developed to deal with complicated problems, such as determining the states of heavy atoms. One can get better insight into these ideas by considering a specific example, as in the next section.

36.4   A Particle in a Box

Consider a particle of mass m confined in a square potential well, that is, in a box, of width L. For $0 \leq x \leq L$, the potential energy $U(x) = 0$. Outside the well the potential energy is infinite, so there is no chance of finding the particle there. Thus the wave function must satisfy the boundary conditions $\psi(0) = \psi(L) = 0$. The Schrödinger wave equation is thus

$$\frac{d^2\psi}{dx^2} + \frac{\hbar^2}{2m}(E - 0)\psi = 0 \quad \text{or} \quad \frac{d^2\psi}{dx^2} + k^2\psi = 0$$

where

$$k^2 = \frac{2mE}{\hbar^2}$$

The solution is $\psi(x) = A \sin kx + B \cos kx$. Since $\psi(0) = 0 = 0 + B$, $B = 0$. Thus $\psi(x) = A \sin kx$. Since $\psi(L) = A \sin kL = 0$, $kL = n\pi$, where $n = 1, 2, 3, \ldots$. The solution for the wave function is

$$\psi(x) = A \sin\left(\frac{n\pi x}{L}\right), \quad n = 1, 2, 3, \ldots \tag{36.6}$$

The wavelength of these standing waves is $\lambda = 2L/n$, and the momentum of the particle is $p = h/\lambda = n(h/2L)$. This means that the energy is quantized because

$$E = \frac{1}{2}mv^2 = \frac{(mv)^2}{2m} = \frac{p^2}{2m}$$

Thus

$$E_n = \frac{h^2}{8\,mL^2}n^2, \quad n = 1, 2, 3, \ldots \tag{36.7}$$

We see that the boundary conditions have resulted in a set of quantized energy levels, analogous to what was found for the hydrogen atom. Here the energy varies as n^2, whereas in hydrogen $E \propto 1/n^2$. The integer n is a **quantum number** with which a state is labeled. The energy levels for the first three states are shown in Fig. 36-2. The first three wave functions for a particle in a box are shown in Fig. 36-3a, and the probability density ψ^2 for each of these states is shown in Fig. 36-3b.

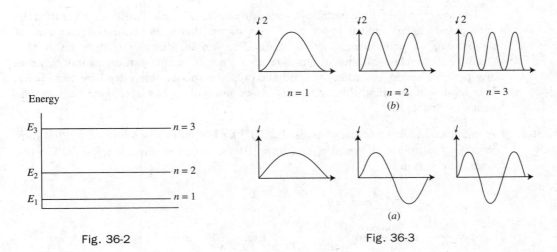

Fig. 36-2

Fig. 36-3

The wave function of Eq. 36.6 is normalized by the condition $\int_0^L \psi^2 \, dx = 1$:

$$\int\limits_0^L A^2 \sin^2\left(\frac{n\pi x}{L}\right) dx = 1 \qquad A^2 \int\limits_0^L \frac{1}{2}\left[1 - \cos\left(\frac{2n\pi x}{L}\right)\right] dx = \frac{1}{2}LA^2 = 1, \quad \text{so } A = \sqrt{\frac{2}{L}}$$

$$\boxed{\psi(x) = \sqrt{\frac{2}{L}} \sin\left(\frac{n\pi x}{L}\right), \quad n = 1, 2, 3, \ldots \quad \text{normalized wave function}} \qquad (36.8)$$

Notice that classically the particle is equally likely to be found anywhere in the well, but this is not the case here.

PROBLEM 36.3. When the particle is in its ground state ($n = 1$), what is the probability of finding it in the left quarter of the well, that is, between $x = 0$ and $x = \frac{1}{4}L$? What is the probability of finding the particle between $x = \frac{1}{4}L$ and $x = \frac{3}{4}L$?

Solution

$$P = \int\limits_0^{L/4} \psi^2 \, dx = \frac{2}{L} \int\limits_0^{L/4} \sin^2\left(\frac{\pi x}{L}\right) dx = \frac{2}{L} \int\limits_0^{L/4} \frac{1}{2}\left[1 - \cos\left(\frac{2\pi x}{L}\right)\right] dx$$

$$= \frac{1}{L}\left[x - \frac{L}{2\pi}\sin\left(\frac{2\pi x}{L}\right)\right]\Bigg|_0^{L/4} = \frac{1}{L}\left(\frac{L}{4} - \frac{L}{2\pi}\right) = 0.09$$

whereas we expect 0.25 classically. By symmetry the probability of finding the particle between $x = \frac{3}{4}L$ and $x = L$ is also 0.09, so the probability of finding it between $x = \frac{1}{4}$ and $x = \frac{3}{4}L$ is $P' = 1 - 2(0.09) = 0.82$.

PROBLEM 36.4. An electron is confined in an infinite square well. In the ground state its energy is 1 eV. What is the width of the well? How much energy is required to excite the electron from the ground state to the second excited state ($n = 3$)?

Solution From Eq. 36.7,

$$E_n = \frac{n^2 h^2}{8mL^2} \qquad L^2 = \frac{n^2 h^2}{8mE_n} = \frac{(1)(6.63 \times 10^{-34}\,\text{J}\cdot\text{s})^2}{8(9.11 \times 10^{-31}\,\text{kg})(1\,\text{eV})(1.6 \times 10^{-19}\,\text{J/eV})}$$

$$L = 0.614\,\text{nm} \qquad E = 1\,\text{eV} \qquad E_3 = E_1 n^2 = 9E_1 = 9\,\text{eV}$$

The lowest energy for a particle in an infinite well is $E_1 = h^2/8mL^2$. This is called the **zero-point energy**, and it is not zero. This means that even at the absolute zero of temperature, a particle confined to a finite region of space can never be at rest, contrary to the classical idea of absolute zero. Further, such a particle cannot travel at any possible speed, but rather it can have only the speeds determined by $\frac{1}{2}mv^2 = E_n$. It is hard to understand this intuitively, since none of us has ever seen such strange behavior firsthand.

PROBLEM 36.5. A microscopic dust particle of mass 2×10^{-8} kg is confined to a box of width 1 mm. What is the minimum speed it can have? If the speed of the particle is 0.1 mm/s, what state is it in?

Solution

$$\frac{1}{2}mv^2 = E_1 = \frac{h^2}{8mL^2}$$

$$v = \frac{h}{2mL} = \frac{6.63 \times 10^{-34}\,\text{J}\cdot\text{s}}{2(2 \times 10^{-8}\,\text{kg})(0.001)} = 1.66 \times 10^{-23}\,\text{m/s}$$

You can see that for all practical purposes, even a particle as small as a speck of dust is at rest in its ground state. In the *n*th state, $1/2mv^2 = h^2n^2/8mL^2$:

$$n = \frac{2mLv}{h} = \frac{2(2 \times 10^{-8}\,\text{kg})(10^{-3}\,\text{m})(10^{-4}\,\text{m/s})}{6.63 \times 10^{-34}\,\text{J}\cdot\text{s}}, \quad n = 6 \times 10^{18}$$

At such a high quantum number, transitions from a state $n+1$ to a state n are not observable on a macroscopic scale. Further, the wave function goes through so many oscillations in the width of the well that the probability density is essentially constant across the well, in agreement with what we expect classically. This is an example of Bohr's **correspondence principle**, which states that quantum mechanical behavior must agree with classical theory in the limit of very high quantum numbers.

36.5   A Particle in a Finite Well and Tunneling

Consider now a particle in a potential well of finite depth U between $x = 0$ and $x = L$. The potential energy is zero at the bottom of the well. Within the well (Region II) the Schrödinger equation is $(d^2\psi/dt^2) + k^2\psi = 0$, where $k^2 = 2mE/\hbar$. As in Section 36.4, the solution is $\psi_{\text{II}} = C \sin kx$. However, we no longer assert that $\psi = 0$ at $x = 0$ and $x = L$. In Regions I and III, outside the well, $U > E$, so Schrödinger's equation becomes

$$\frac{d^2\psi}{dx^2} = \frac{2m(U - E)}{\hbar^2}\psi = K^2\psi \tag{36.9}$$

Here $K^2 > 0$, so the solution for ψ is of the form $\psi = Ae^{Kx} + Be^{-Kx}$ outside the well. We require $\psi \to 0$ as $x \to \pm\infty$, so $\psi_I + Ae^{Kx}(x < 0)$ and $\psi_{III} = Be^{-Kx}(x > L)$. The constants A, B, and C are determined by applying the boundary conditions, $\psi_I = \psi_{II}$ and $d\psi_I/dx = d\psi_{II}/dx$ at $x = 0$, and $d\psi_{II}/dx = d\psi_{III}/dx$ and $\psi_{II} = \psi_{III}$ at $x = L$. The first three wave functions are shown in Fig. 36-4. The most interesting feature of these curves is that they show a finite probability of finding the particle *outside* the well, in a region that would be forbidden classically. This has many important practical consequences, for example, for constructing electronic devices.

A related phenomenon occurs when a free particle encounters a potential barrier of height $U > E$. In Fig. 36-5 a particle moving to the right encounters a barrier. Classically we expect it to be reflected, but when Schrödinger's equation is solved, we find an exponential decay for the wave function within the forbidden region. If the barrier is not too thick, the wave function may not decay completely to zero on the other side. This means there is a finite probability that the particle will tunnel through the barrier, and magically appear on the other side. This effect occurs in the tunnel diode, in superconducting Josephson junctions, and in the emission of alpha particles in radioactive decay.

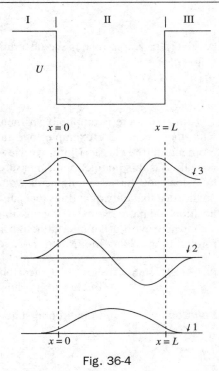

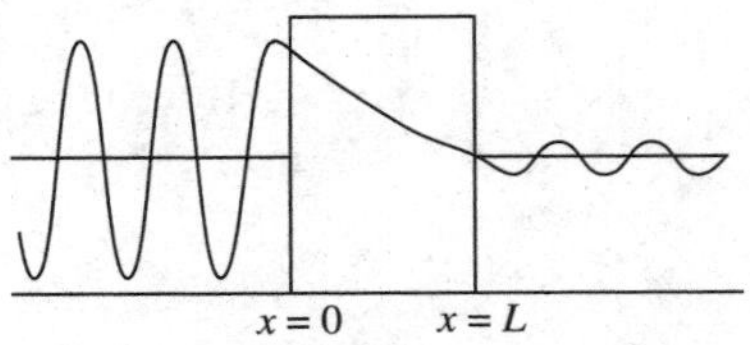

Fig. 36-4

Fig. 36-5

36.6   The Heisenberg Uncertainty Principle

The wave function of a free electron is a sine function spread throughout all space. It has a well-defined wavelength and hence a well-defined momentum, $p = h/\lambda$. We say that the uncertainty in the momentum

of the electron is zero, but the uncertainty in its position is infinite, since there is equal probability of its being anywhere. If
the electron is not free but rather is localized in some region
(as in the particle in a box problem), the wave function takes
the form of a **wave packet**. In Fig. 36-6 I have shown a pure
sine wave, representing a particle that is not localized, and a
wave packet, representing a particle that is located somewhere
within the range shown. The wave packet may be thought to be
made up of a superposition of many waves of different wavelengths. Thus the uncertainty in the momentum increases when
the uncertainty in the position decreases. The **Heisenberg uncertainty principle** (which can be derived

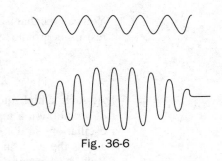

Fig. 36-6

from some fundamental considerations) states that there is a lower limit to the uncertainties in position
Δx and momentum Δp_x.

$$\Delta x \, \Delta p_x > \hbar \quad \text{uncertainty principle} \tag{36.10}$$

A related principle relates the uncertainties in the energy ΔE and the time a system is in a given energy
state Δt:

$$\Delta E \, \Delta t > \hbar \tag{36.11}$$

Although there is ample experimental evidence to support these principles, it is not completely clear
what they mean. For example, one interpretation is to assert that a particle has a definite position and
momentum (the classical intuitive view) and that measurement disturbs it by some unknown amount. Supposedly it is this interaction of measurement that causes the uncertainty. Another view is to say that a particle does not "have" simultaneous values of its position and its momentum. This line of reasoning is based
on the idea that only quantities that can be measured have physical reality, and simultaneous exact values of
position and momentum are not possible. Thus it makes no sense to refer to an electron as if it definitely "is"
somewhere with a definite, although unknown, momentum. I lean to the latter view, but it certainly is
foreign to everyday life. Much more remains to be learned.

PROBLEM 36.6.   Show that the Bohr model of hydrogen, with the electrons moving in well-defined
circular orbits, cannot be correct in view of the uncertainty principle.

Solution   In the ground state of hydrogen, the energy in the Bohr model was

$$\text{KE} = \frac{ke^2}{2\tau} = -E = 13.6 \, \text{eV} \qquad \text{KE} = \frac{p^2}{2m}$$

$$p = \sqrt{2m \, \text{KE}}$$

$$= \sqrt{2(9.11 \times 10^{-31} \, \text{kg})(13.6 \, \text{eV})(1.6 \times 10^{-19} \, \text{J/eV})} = 2 \times 10^{-24} \, \text{kg} \cdot \text{m/s}$$

If we take the uncertainty in the momentum to be of the same order as the momentum
itself, then

$$\Delta p \simeq 5 \times 10^{-24} \, \text{kg} \cdot \text{m/s} \quad \text{and} \quad \Delta r > \frac{\hbar}{\Delta p} = \frac{1 \times 10^{-34} \, \text{J} \cdot \text{s}}{5 \times 10^{-24} \, \text{kg} \cdot \text{m/s}} = 0.02 \, \text{nm}$$

But the Bohr radius is about 0.05 nm, so the uncertainty in the position of the electron is
comparable to the Bohr radius. Thus the picture of an electron traveling a sharply
defined orbit is incorrect.

PROBLEM 36.7.   In a laser it is necessary to excite an electron to an excited state where it stays for a time
of perhaps 10^{-8} s (its lifetime) before making a transition to a lower energy level and
thereby emitting a photon. Estimate the frequency spread (the "line width") of the
laser light emitted. What is the fractional line width $\Delta f/f$ if the wavelength emitted
is 540 nm?

Solution
$$\Delta E = h\Delta f, \qquad \Delta E \Delta t = h\Delta f \Delta t \approx \hbar, \quad \text{so } \Delta f \approx \frac{1}{2\pi\Delta t} \approx 1.6 \times 10^7 \text{ Hz}$$

$$\frac{\Delta f}{f} = \frac{\lambda \Delta f}{c} = \frac{(540 \times 10^{-9}\text{ m})(1.6 \times 10^7 \text{ Hz})}{(3 \times 10^8 \text{ m/s})} = 2.9 \times 10^{-8}$$

Doppler shifts associated with thermal motion and collisions broaden spectral lines in excess of this "natural" line width, but this is an important limiting value that determines just how sharp a spectral line can be.

Recall that the ground state energy for a particle in a box is $E = h^2/8mL^2$ (Eq. 36.7). This means that if we reduce L (that is, reduce Δx), the ground state energy will increase. Since $E = p^2/2m$, this means momentum p also increases, and this means Δp will increase, consistent with the uncertainty principle.

36.7 Spin Angular Momentum

It is possible to solve the Schrödinger equation for the hydrogen atom and calculate the energy levels and the spectrum. This could also be done for atoms such as sodium. However, when this was done in the 1920s, it was found that some spectral lines were in fact closely spaced doublets, an effect not predicted by the theory. For example, when the sodium line at 589.3 nm was examined with very high resolution, it was found to be composed of two lines, at 589.0 and 589.6 nm. It was possible to explain this effect by postulating a new property of the electron, **spin angular momentum**. It is as if the electron is a little spinning ball of charge. It has spin angular momentum and a magnetic moment as well. It thus acts like a small bar magnet. If the component of the angular momentum is measured in some direction, say, along the z-axis, the result of the measurement is always $+\frac{1}{2}\hbar$ or $-\frac{1}{2}\hbar$. I won't go through all of the mathematics to show you how this funny behavior comes about; suffice it to say that one of the labels that identifies a state of an electron is its **spin quantum number**, $s = \frac{1}{2}$. The **magnitude of the spin angular momentum** S in terms of the spin quantum number s is

$$S = \sqrt{s(s+1)}\hbar = \sqrt{\frac{1}{2}\left(\frac{1}{2}+1\right)}\;\hbar = \frac{\sqrt{3}}{2}\hbar \qquad (36.12)$$

Classically one would expect the magnitude of S to be simply $s\hbar$. The curious expression here results in a straightforward way from mathematical calculations, but it has no simple intuitive explanation.

The z component of the spin angular momentum is

$$S_z = m_s\hbar, \quad \text{where } m_s = +\tfrac{1}{2} \quad \text{or} \quad m_s = -\tfrac{1}{2} \qquad (36.13)$$

m_s is the **spin magnetic quantum number**. The spin magnetic moment of the electron $\vec{\mu}_s$ is related to the spin angular momentum **S** by

$$\vec{\mu}_s = -\frac{e}{m}\mathbf{S} \qquad (36.14)$$

Since $S_z = \pm\frac{1}{2}\hbar$, $\mu_z = \pm e\hbar/2m = \pm\mu_B$. μ_B is the **Bohr magneton**. $\mu_B = 9.27 \times 10^{-27}$ J/T. If an electron is placed in a magnetic field in the z direction, it will have its magnetic moment aligned either parallel to the magnetic field ($m_s = -\frac{1}{2}$) or antiparallel to the magnetic field ($m_s = +\frac{1}{2}$). Since the energy of a magnetic moment μ in a magnetic field B is $U = -\vec{\mu} \cdot \mathbf{B}$, an electron in a magnetic field will be in one of two states with energies $+\mu_B B_z$ ($m_s = +\frac{1}{2}$) and $-\mu_B B_z$ ($m_s = -\frac{1}{2}$). These are sometimes referred to as states with "spin up" or with "spin down." This splitting of energy levels is called the **Zeeman effect**.

In a nonuniform magnetic field a magnetic moment will experience a force. In the Stern–Gerlach experiment, a beam of neutral silver atoms, each with a single valence electron, was directed through a nonuniform magnetic field. The beam was observed to split into two distinct components, and this was an early confirmation of the spatial quantization of spin angular momentum. The result is in sharp contrast to what one expects classically, where magnetic moments can be aligned at any angle to a magnetic field.

Although it is common to use the term "spin angular momentum," do not take this term literally. Electrons, as far as we know, are point particles. They definitely are not little spinning spheres. Further, the property called "spin angular momentum" is better called "intrinsic angular momentum." It does not change in magnitude, and when measured, it has only one of two possible values ("up" or "down"). The concept is encountered only in the realm of quantum physics, not with macroscopic systems. In one of the great all-time achievements in theoretical physics, in 1928, P. A. M. Dirac was able to show that a relativistic treatment of quantum mechanics predicts not only electron spin but also the existence of an antielectron, the positron. That is not to say that this "explains" electron spin in any intuitive way, but it does remove some of the arbitrary nature of the model.

PROBLEM 36.8.   Electron spin resonance experiments involve placing electrons in a magnetic field and then inducing transitions between the two Zeeman levels by irradiating the system with photons. It is common to use a magnetic field of about 0.30 T. What wavelength photon would induce a transition from the spin-down to the spin-up state in this case?

Solution
$$\Delta E = \mu_B B - (-\mu_B B) = 2\mu_B B = hf = \frac{hc}{\lambda}$$

$$\lambda = \frac{hc}{2\mu_a B} = \frac{(6.63 \times 10^{-34}\,\text{J} \cdot \text{s})(3 \times 10^8\,\text{m/s})}{2(9.27 \times 10^{-24}\,\text{J/T})(0.3\,\text{T})} = 3.6\,\text{cm} \quad \text{in the microwave X band}$$

36.8   The Quantum Theory of Hydrogen

The Schrödinger wave equation in three dimensions in spherical coordinates appears rather formidable:

$$-\frac{\hbar^2}{2m}\left[\frac{1}{r^2}\frac{\partial}{\partial r}\left(r^2\frac{\partial \psi}{\partial r}\right) + \frac{1}{r^2 \sin\theta}\frac{\partial}{\partial\theta}\left(\sin\theta\frac{\partial\psi}{\partial\theta}\right) + \frac{1}{r^2 \sin^2\theta}\frac{\partial^2\psi}{\partial\phi^2}\right] + U(r)\psi = E\psi \qquad (36.15)$$

However, the equation can be solved for the hydrogen atom potential energy function, $U = -k(e^2/r)$. I will not pursue the mathematics here (too lengthy), but I will point out some salient features. The solution is of the form $(\psi r, \theta, \phi) = R(r)\Theta(\theta)\Phi(\phi)$. I require that the solutions be single-valued and that $\psi \to 0$ as $r \to \infty$. The resulting "standing wave" or "stationary" solutions represent the possible allowed quantum states of the atom. Each spatial variable (r, θ, and ϕ) gives rise to a quantum number. We can think of the spin variable as a sort of "fourth dimension" that gives rise to another quantum number, making a total of four in all. This spin variable is a funny one in that it is a discrete variable and has only two values, $+\frac{1}{2}$ and $-\frac{1}{2}$. Thus each state is labeled by a set of four quantum numbers that result naturally from the mathematics, as follows:

1.   The radial function $R(r)$ exists only for integral values 1, 2, 3, ..., of the **principal quantum number *n***. *n* is called the "principal quantum number" because the energy depends principally on it, with the result that

$$E_n = -\left(\frac{ke^2}{2a_0}\right)\frac{1}{n^2}, \quad n = 1, 2, 3, \ldots \qquad (36.16)$$

This is the same result obtained with the Bohr model.

2.   The polar function $\Theta(\theta)$ gives rise to quantization of the **orbital angular momentum** of the electron as it moves around the nucleus. One finds that the magnitude of the orbital angular momentum is

$$L = \sqrt{\ell(\ell+1)}\hbar, \quad \ell = 0, 1, 2, \ldots, (n-1) \qquad (36.17)$$

Here ℓ is the **orbital quantum number** and *n* is the principal quantum number. Note that this does not agree with the incorrect Bohr assumption $L = n\hbar$.

3.   The solution for $\Phi(\phi)$ gives rise to the quantum number m_ℓ, the **magnetic quantum number**. The value of m_ℓ determines the *z* component of the orbital angular momentum.

$$L_z = m_\ell \hbar, \quad m_\ell = 0, \pm 1, \pm 2, \ldots, \pm \ell \qquad (36.18)$$

Associated with the orbital angular momentum of the electron is an orbital magnetic dipole moment μ_ℓ.

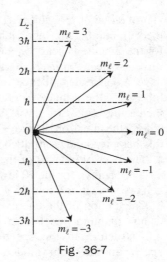

$$\mu_\ell = -\left(\frac{e}{2m}\right)\mathbf{L} \quad \text{and} \quad (\mu_\ell)_z = -m_\ell\left(\frac{e\hbar}{2m}\right) = -m_\ell\mu_B \qquad (36.19)$$

Fig. 36-7 shows a pictorial representation of the relation between the orbital angular momentum vector $\mathbf{L}$ and the allowed values of L_z for the case $\ell = 3$. Note that the length of L is $\sqrt{\ell(\ell+1)}\hbar$.

4. Electron spin gives rise to the **spin magnetic quantum number** m_s. This is related to the z component of the spin angular momentum S_z.

$$S = \sqrt{s(s+1)}, \quad \text{where } s = \tfrac{1}{2}$$
$$S_z = m_s\hbar, \quad \text{where } m_{ss} = \pm\tfrac{1}{2} \qquad (36.20)$$

Fig. 36-7

The above restrictions on the quantum numbers can be summarized as follows:

> The principal quantum number $\quad n = 1, 2, 3, \ldots$
> The orbital quantum number $\quad \ell = 0, 1, 2, \ldots, (n-1)$.
> The magnetic quantum number $\quad m_\ell = 0, \pm1, \pm2, \ldots, \pm\ell$.
> The spin quantum number $\quad m_s = \pm\tfrac{1}{2}$.

(36.21)

A somewhat arcane scheme is used to label hydrogenic wave functions (based on some historical reasons). States with a given value of the principal quantum number n constitute a *shell*, labeled with capital letters:

n Value	1	2	3	4	5	6	7
Letter	K	L	M	N	O	P	Q

A state with a given value of the orbital quantum number ℓ is labeled with a lower-case letter:

ℓ Value	0	1	2	3	4	5
Letter	s	p	d	f	g	h

The ground state wave function for hydrogen has no dependence on θ or ϕ. It is spherically symmetric, as are all $\ell = 0$ states (s states). The radial function is

$$\psi_{1s}(r) = \frac{1}{\sqrt{\pi a_0^3}}e^{-r/a_0} \qquad (36.22)$$

Here a_0 is the Bohr radius and the subscript $1s$ means $n = 1$, $\ell = 0$. The probability of finding the electron in a volume dV is $\psi^2\, dV$. Thus the probability $P(r)\, dr$ of finding the electron in a thin spherical shell between r and $r + dr$ is $\psi^2(4\pi r^2\, dr)$. Thus the **radial probability density** is

$$P_{1s}(r) = 4\pi r^2\psi^2 = \frac{4r^2}{a_0^3}e^{-2r/a_0} \qquad (36.23)$$

The wave function and the probability density for the first excited state, $n = 1$, $\ell = 0$, are

$$\psi_{2s}(r) = \frac{1}{4\sqrt{2\pi a_0^3}}\left(2 - \frac{r}{a_0}\right)e^{-r/2a_0} \quad \text{and} \quad P_{2s}(r) = \frac{r^2}{8a_0^3}\left(2 - \frac{r}{a_0}\right)^2 \qquad (36.24)$$

$P_{1s}(r)$ and $P_{2s}(r)$ are shown in Fig. 36-8. We see that the quantum theory of hydrogen differs markedly from the Bohr picture. The electron is not envisioned to move in an orbit, but rather it is best described by a sort of cloud of probability. The peak of the probability for the $1s$ state is at the Bohr radius a_0 (as in Bohr's theory), but there is appreciable overlap between the $1s$ state and the $2s$ state. The most probable value of r for the $2s$ state is near $5a_0$, but there is a smaller peak also near $r = a_0$. States with higher ℓ values have angular dependence, and the number of lobes of the wave function increases with increasing ℓ. This gives a directionality to where the electron is likely to be found, and this has profound consequences for chemistry.

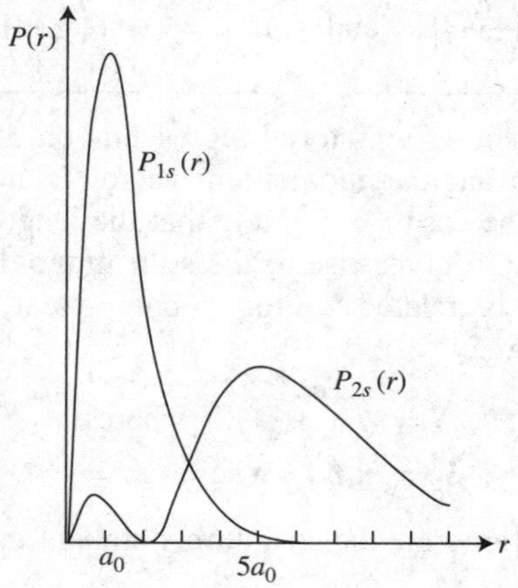

The possible values of the quantum numbers n, ℓ, m_ℓ, and m_s for the 10 lowest states of hydrogen are shown in Table 36-1.

Calculation of the energy levels of hydrogen using quantum mechanics reveals additional fine structure in the spectrum that does not appear in the Bohr model. In particular, the spin magnetic moment of the electron can interact with the effective magnetic field due to the orbital magnetic moment, resulting in different energies depending on whether the spin is "up" or "down." Further, if the probability of a transition from one level to another is calculated,

TABLE 36-1

n	ℓ	m_ℓ	m_s
1	0	0	$+\frac{1}{2}$
1	0	0	$-\frac{1}{2}$
2	0	0	$+\frac{1}{2}$
2	0	0	$-\frac{1}{2}$
2	1	0	$+\frac{1}{2}$
2	1	0	$-\frac{1}{2}$
2	1	+1	$+\frac{1}{2}$
2	1	+1	$-\frac{1}{2}$
2	1	-1	$+\frac{1}{2}$
2	1	-1	$-\frac{1}{2}$

Fig. 36-8

the intensities of the various spectral lines can be determined. It is found that some transitions have essentially zero probability of occurring. They are said to be "forbidden." One can show that the selection rules for allowed transitions are $\Delta\ell = \pm 1$ and $\Delta m_\ell = 0$ or ± 1. The reason $\Delta\ell = -1$ when a transition to a lower level occurs is that the emitted photon carries away with it angular momentum $\hbar$, and angular momentum is conserved. For the same reason $\Delta\ell = +1$ when a photon is absorbed. The experimental confirmation of these rules helps establish the idea that the photon is a "spin 1" particle; that is, it has spin angular momentum $\hbar$.

In summary, it is fair to say that the quantum theory of hydrogen is very successful, provided relativistic effects are included.

PROBLEM 36.9.   What is the probability of finding an electron at a distance of less than 1 Bohr radius from the nucleus for an electron in the ground state of hydrogen?

Solution
$$P = \int_0^{a_0} P_{1s}(r)\, dr = \int_0^{a_0} \frac{4r^2}{a_0^3} e^{-2r/a_0}\, dr$$

Let

$$v = \frac{2r}{a_0} \quad \text{and} \quad P = \frac{1}{2}\int_0^2 v^2 e^{-v}\, dv$$

From integral tables,

$$P = \tfrac{1}{2}\left[-e^{-v}(v^2 + 2v + 2)\right]_0^2 = 1 - 5\, e^{-2} = 0.323$$

PROBLEM 36.10.   How many states with $n = 3$ are possible for hydrogen?

Solution   For $n = 3$, $\ell = 0$, 1, or 2. There are $2\ell + 1$ possible values for m_ℓ, so this amounts to $[2(0) + 1] + [2(1) + 1] + [2(2) + 1] = 9$ states. For each of these states there are two possible spin states ($m_s = \pm\frac{1}{2}$), so the total number of possible states is 18.

PROBLEM 36.11. What is the *average value of r* for an electron in the ground state of hydrogen? Note that this is not the same as the most probable value of *r*.

Solution

$$r_{av} = \int_0^\infty rP(r)\,dr = \int_0^\infty \frac{4r^3}{a_0^3} e^{-2r/a_0}\,dr$$

Let

$$u = \frac{2\,r}{a_0} \quad \text{and} \quad r_{av} = \frac{a_0}{4} \int_0^\infty u^3 e^{-u}\,du$$

From integral tables,

$$r_{av} = -\frac{a_0}{4}\left[e^{-u}(u^3 + 3u^2 + 6u + 6)\right]_0^\infty = 1.5a_0 \qquad r_{av} = 1.5a_0$$

36.9 The Pauli Exclusion Principle

In 1925 Wolfgang Pauli made a remarkable discovery, now called the **Pauli exclusion principle:**

> **No two electrons in an atom can have the same four quantum numbers n, ℓ, m_ℓ, and m_s.**

To see how this comes about, first note that all of the particles ever discovered in the universe can be grouped into two classes, **bosons** and **fermions**. Fermions have spin angular momentum $\frac{1}{2}\hbar$, $\frac{3}{2}\hbar$, $\frac{5}{2}\hbar$, ..., and so on. Bosons have spin angular momentum 0, $\hbar$, $2\hbar$, $3\hbar$, ..., and so on. Electrons, protons, and neutrons are fermions, and photons and alpha particles (helium nuclei) are bosons. If a system consists of identical particles, either fermions or bosons, its physical properties cannot change if you interchange two particles. This means that on interchange of two particles the wave function can change only by a factor of $+1$ or -1. The latter is allowed because probabilities depend on ψ^2. If the wave function changes only by a factor of $+1$ on interchange of two particles (that is, it doesn't change), the wave function is said to be symmetric. If it changes by a factor of -1, it is antisymmetric. For reasons that are not clear, it turns out that the wave function for a fermion is always antisymmetric, whereas that for a boson is symmetric.

To see what this means, consider a system of two particles that can move in one dimension. Suppose $u(x)$ and $v(x)$ are two possible wave functions for the individual particles. Then a possible symmetric state for the systems could be $\psi_S = u(x_1)v(x_2) + u(x_2)v(x_1)$. Here x_1 is the position of particle 1 and x_2 is the position of particle 2. If x_1 and x_2 are interchanged, ψ is unchanged. An antisymmetric state could be $\psi_A = u(x_1)v(x_2) - u(x_2)v(x_1)$. Now if x_1 and x_2 are interchanged, ψ changes sign; that is, it is indeed antisymmetric. ψ_A is of the form a wave function must have for two fermions, for example, electrons. We see that if both electrons are in the same state; that is, $u(x) = v(x)$, then $\psi_A = 0$. Thus there is no state function for which more than one electron is in a given single particle state, that is, has the same quantum numbers. This principle has important ramifications, as illustrated in the next section.

36.10 The Periodic Table

An atom consists of a heavy nucleus containing Z protons, with total charge $+Ze$, surrounded by a cloud of Z electrons, with total charge $-Ze$. Z is the **atomic number**. Imagine you start with hydrogen, a single proton and a single electron. Now add another proton and another electron to form a helium atom. Both of these electrons can have $n = 1$, $\ell = 0$, since one can have $m_s = +\frac{1}{2}$ and the other can have $m_s = -\frac{1}{2}$. Continue in this way, adding electron after electron. Each additional electron will go into the accessible state with the lowest energy. For the most part, energy increases with increasing n and ℓ, but in some cases there are deviations from this. For example, the $4s$ state has lower energy than the $3d$ state, and the $5s$ state has

TABLE 36-2

N	ℓ	m_ℓ	m_s	SHELL	SUBSHELL	NO. IN SUBSHELL	NO. IN SHELL
1	0	0	$\pm 1/2$	K	1s	2	2
2	0	0	$\pm 1/2$	L	2s	2	
	1	$0, \pm 1$	$\pm 1/2$		2p	6	8
3	0	0	$\pm 1/2$	M	3s	2	
	1	$0, \pm 1$	$\pm 1/2$		3p	6	8
	2	$0, \pm 1, \pm 2$	$\pm 1/2$		3d	10	
4	0	0	$\pm 1/2$	N	4s	2	
	1	$0, \pm 1$	$\pm 1/2$		4p	6	18

TABLE 36-3

ELEMENT	SYMBOL	ATOMIC NUMBER, Z	ELECTRONIC CONFIGURATION
Hydrogen	H	1	$1s$
Helium	He	2	$1s^2 \cdot$
Lithium	Li	3	$1\,s^2 \cdot 2s$
Beryllium	Be	4	$1s^2 \cdot 2s^2$
Boron	B	5	$1s^2 \cdot 2s^2 2p$
Carbon	C	6	$1s^2 \cdot 2s^2 2p^2$
Nitrogen	N	7	$1s^2 \cdot 2s^2 2p^3$
Oxygen	O	8	$1s^2 \cdot 2s^2 2p^4$
Fluorine	F	9	$1s^2 \cdot 2s^2 2p^5$
Neon	Ne	10	$1s^2 \cdot 2s^2 2p^6$
Sodium	Na	11	$1s^2 \cdot 2s^2 2p^6 \cdot 3s$
Magnesium	Mg	12	$1s^2 \cdot 2s^2 2p^6 \cdot 3s^2$
Aluminum	Al	13	$1s^2 \cdot 2s^2 2p^6 \cdot 3s^2 3p$
Silicon	Si	14	$1s^2 \cdot 2s^2 2p^6 \cdot 3s^2 3p^2$

lower energy than the $4d$ state. Table 36-2 shows the possible states grouped according to *shells* (n values) and **subshells** (ℓ values) up to the $4p$ state. From Table 36-2 I see that two electrons can occupy the $1s$ subshell, two can occupy the $2s$ subshell, six can occupy the $2p$ subshell, and so on. Thus I write the ground state **electronic configuration** of magnesium ($Z = 12$) as $1s^2 2s^2 2p^6 3s^2$. The superscript indicates the number of electrons in that particular orbital. The letter shows the value of ℓ (s is $\ell = 0$, p is $\ell = 1, \dots$). The configuration of elements through silicon is shown in Table 36-3.

The letter shows the value of ℓ (s is $\ell = 0$, p is $\ell = 1, \dots$). The configuration of elements through silicon is shown in Table 36-3.

For a filled shell, the total spin angular momentum and the total orbital angular momentum are both zero ($\mathbf{S} = 0$ and $\mathbf{L} = 0$). The chemical properties of an atom are determined primarily by the electrons outside closed shells. Thus elements with closed shells plus one electron, the *alkali metals* (lithium, sodium, potassium, rubidium, cesium, and francium), have similar chemical behavior. These elements are all highly reactive, readily giving up their outermost electron and thereby becoming a positive ion. Similarly, the *halogens* (fluorine, chlorine, bromine, iodine, and astatine) are one electron short of having a closed shell. They readily grab electrons from neighboring atoms and are thus also very reactive. Elements with only closed shells (the *inert gases* helium, argon, krypton, xenon, and radon) do not tend to gain or lose an electron, and hence they do not form many molecules.

PROBLEM 36.12. What is the atomic number of the element with the electronic configuration $1s^2 2s^2 2p^5$? Is the outer subshell of this atom nearly full or nearly empty?

Solution The number of electrons is $2 + 2 + 5 = 9$, so $Z = 9$ (fluorine). The $2p$ subshell can hold six electrons, so this atom is missing one electron from a full subshell. It will readily grab an electron from a neighboring atom if it gets the chance. Fluorine is a very reactive gas. The compounds used in spray cans, fluorocarbohydrides, react with ozone in the upper atmosphere and deplete the ozone layer, perhaps exposing us to dangerous UV radiation.

PROBLEM 36.13. Show that the number of quantum states in the nth shell is $2n^2$.

Solution For a given value of n there are $2\ell + 1$ possible values of m_ℓ. ℓ ranges from $n - 1$ down to 0, so the number of states with different m_ℓ is

$$N = [2(n - 1) + 1] + [2(n - 2) + 1] + [2(n - 2) + 1] + \cdots + [2(n - n) + 1]$$
$$= (2n - 1) + (2n - 2) + \cdots + 1$$

The sum of this arithmetic series of n terms is

$$N = n\left(\frac{\text{first term} + \text{last term}}{2}\right) = n\left[\frac{(2n - 1) + 1}{2}\right] = n^2$$

But for every m_ℓ state there are two spin states, so the total number of states for given n is $2n^2$.

36.11 Summary of Key Equations

de Broglie wavelength:
$$\lambda = \frac{h}{p}$$

Time-independent Schrödinger equation:
$$\frac{d^2\psi}{dx^2} + \frac{2m}{\hbar^2}[E - U(x)]\psi(x) = 0$$

Particle in a box:
$$\psi(x) = \sqrt{\frac{2}{L}}\sin\left(\frac{n\pi x}{L}\right), \quad n = 1, 2, 3, \ldots$$
$$E_n = \frac{h^2}{8mL^2}n^2$$

Heisenberg uncertainty principle:
$$\Delta x \Delta p_x > \hbar, \qquad \Delta E \Delta t > \hbar$$

Spin magnetic moment of the electron:
$$\mu_s = -\frac{e}{m}\mathbf{S}, \quad (\mu_s)_z = \pm\mu_B$$

Bohr magneton:
$$\mu_B = \frac{e\hbar}{2m_e} = 9.27 \times 10^{-27} \text{ J/T}$$

Orbital angular momentum:
$$L = \sqrt{\ell(\ell + 1)}\hbar, \quad \ell = 0, 1, 2, \ldots (n - 1)$$
$$L_z = m_\ell\hbar$$
$$m_\ell = 0, \pm 1, \pm 2, \ldots, \pm\ell$$

Spin angular momentum:
$$S = \sqrt{s(s + 1)}\hbar, \quad s = \tfrac{1}{2} \text{ for electrons}$$
$$S_z = m_s\hbar, \quad m_s = \pm\tfrac{1}{2}$$

Selection rules:
$$\Delta\ell = \pm 1, \qquad \Delta m_\ell = 0, \pm 1$$

SUPPLEMENTARY PROBLEMS

36.14. A photon has the same momentum as an electron moving with a speed of 3×10^5 m/s. What is the wavelength of the photon?

36.15. The average kinetic energy of a gas molecule is $3/2k_B T$, where $k_B = 1.38 \times 10^{-23}$ J/K. What is the de Broglie wavelength of a helium atom (mass 6.65×10^{-27} kg) with this energy in a gas at room temperature (293 K)?

36.16. Determine the de Broglie wavelength for (a) a baseball ($m = 0.15$ kg) moving 12 m/s, (b) an electron moving 9×10^6 m/s, and (c) an electron moving at $0.9c$.

36.17. Show that Heisenberg's uncertainty principle is consistent with the diffraction effects seen in optics. To do this, imagine that you try to look at the electrons with a microscope to determine the position of one. The angular resolution of the microscope limits how accurately you can locate the electron. Suppose you bounce a photon off the electron, and it enters your microscope. In bouncing off the electron, the photon will change the electron's momentum by an indeterminate amount Δp due to its recoil. If the position of the electron is determined with an accuracy Δx, show that $\Delta x \, \Delta p \approx h$. A sketch of a possible experimental setup is shown here.

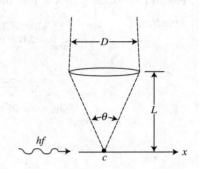

36.18. Suppose that a particle is contained in an infinite potential well extending from $x = -a$ to $x = a$. It has a wave function

$$\psi(x) = C\left(1 - \frac{|x|}{a}\right) \quad \text{for } -a \leq x \leq a$$

Determine the normalization constant C and the probability of finding the particle in the region $1/2a \leq x \leq a$.

36.19. A particle has a wave function $\psi(x) = Ce^{-x/a}$ for $x \geq 0$ and $\psi(x) = 0$ for $x \leq 0$. Determine the normalization constant C and the average value of the position x. A useful definite integral is

$$\int_0^\infty x^n e^{-bx} \, dx = \frac{n!}{b^{n+1}}$$

where is the probability density $P(x)$ a maximum?

36.20. An alpha particle (a helium nucleus composed of two protons and two neutrons) is a stable particle of mass 6.64×10^{-27} kg that can be considered a constituent of a larger nucleus. If the alpha particle is modeled as being confined in a "box" (the large nucleus) of width 1×10^{-14} m (a typical nuclear size), what is its energy and momentum in the ground state?

36.21. A pulse of light of duration Δt is emitted from a source. Show that the uncertainty relations $\Delta x \, \Delta p > \hbar$ and $\Delta E \, \Delta t > \hbar$ are equivalent in this case.

36.22. A particle has a finite probability of tunneling through a potential barrier. By applying boundary conditions to match ψ and $d\psi/dx$ outside and just inside the barrier at both faces of the barrier, we can obtain the wave function shown in Fig. 36-5. The probability of tunneling through the barrier (the transmission coefficient T) is proportional to ψ^2 on the right side of the barrier. For small probabilities, $T \simeq e^{-2Kx}$, where x is the barrier width in Fig. 36-5 and

$$K = \frac{\sqrt{2m(U-E)}}{\hbar}$$

and U is the barrier height. Tunneling is important in many electronic applications. In a certain semiconductor an electron has to tunnel through a barrier of height $\Delta E = U - E$ above the electron energy E. If the width of the barrier is 2.0 nm, what is the tunneling coefficient if $\Delta E = 1.5$ eV? If $\Delta E = 0.5$ eV? What is T if the barrier width is reduced to 0.2 nm while $\Delta E = 1.5\,\text{eV}$? Sometimes barrier widths and heights can be changed by applied voltages, and this provides a mechanism for controlling current flow.

36.23. Suppose you determine the position of an object to within a distance of $\Delta x = 0.01$ nm. What would be the uncertainty in the velocity if the object is an electron? A ping pong ball ($m = 0.0022$ kg)?

36.24. An experimenter drops a steel ball from a height y, trying to hit a spot target on the ground. The ultimate limit on his accuracy will be that determined by the Heisenberg uncertainty principle. What is this limit, Δx_2? Note that there is a possible inaccuracy Δx_1 in the position from which the ball is dropped, and even if the ball goes straight down, this would result in an error in hitting the target.

36.25. What is the minimum angle between the angular momentum vector **L** and the z-axis for the case $\ell = 3$?

36.26. The yellow light given off by the sodium vapor lamps used in many street lights results from the $3p \rightarrow 3s$ transition in sodium. If the wavelength of the light is 590 nm, what is the difference in energy between these two levels?

36.27. What is the atomic number of the element with the ground state electronic configuration $1s^2 2s^2 2p^6 3s^2$ $3p^6 4s^1 = [\ ]3p^6 4s^1$? In view of the selection rules for allowed transitions ($\Delta \ell = \pm 1$, $m_\ell = 0, \pm 1$), from which of the following excited states are transitions to the ground state allowed? (a) $[\ ]3p^5 4p^2$, (b) $[\ ]3p^5 4s^2$, (c) $[\ ]3p^6 4p^1$, (d) $[\ ]3p^6 4d^1$, (e) $[\ ]3p^6 5s^1$.

36.28. What is the electronic configuration of arsenic ($Z = 33$)?

36.29. What is the electronic configuration of iron ($Z = 26$)?

36.30. Cosmic rays are high-energy particles emitted in stellar processes. These rays rain down on us all of the time. They are typically detected for short time durations of the order of 12×10^{-15} s. This implies an uncertainty in how accurately we can determine the energy of such particles. What is this energy uncertainty?

36.31. A particle in a square potential well is described by the wave function $\psi(x) = \psi_o(x + a)$ for $-a \le x \le 0$ and $\psi(x) = \psi_o(a - x)$ for $0 \le x \le a$. What is the probability that the particle will be found in the region $0 \le x \le a/2$?

SOLUTIONS TO SUPPLEMENTARY PROBLEMS

36.14. For the electron,

$$\lambda = \frac{h}{mv} = \frac{6.63 \times 10^{-34}\,\text{J} \cdot \text{s}}{(9.11 \times 10^{-31}\,\text{kg})(3 \times 10^5\,\text{m/s})} = 2.43\,\text{nm}$$

36.15. $\frac{3}{2}k_B T = \frac{1}{2}mv^2, \quad mv = \sqrt{3\,mk_B T}$

$$\lambda = \frac{h}{mv} = \frac{6.63 \times 10^{-34}\,\text{J} \cdot \text{s}}{\sqrt{3(6.65 \times 10^{-27}\,\text{kg})(1.38 \times 10^{-23}\,\text{J/K})(293\,\text{K})}} = 7.38 \times 10^{-11}\,\text{m}$$

36.16. (a) $\lambda = \dfrac{h}{mv} = \dfrac{6.63 \times 10^{-34}\,\text{J}\cdot\text{s}}{(0.15\,\text{kg})(12\,\text{m/s})} = 3.7 \times 10^{-34}\,\text{m}$

(b) $\lambda = \dfrac{h}{mv} = \dfrac{6.63 \times 10^{-34}\,\text{J}\cdot\text{s}}{(9.11 \times 10^{-32}\,\text{kg})(9 \times 10^{6}\,\text{m/s})} = 8.1 \times 10^{-11}\,\text{m}$

(c) $p = \gamma mv,\quad \gamma = \dfrac{1}{\sqrt{1 - y^2/c^2}} = \dfrac{1}{\sqrt{1 - (0.9)^2}} = 2.3$

$\lambda = \dfrac{h}{p} = \dfrac{6.63 \times 10^{-34}\,\text{J}\cdot\text{s}}{2.3(9.11 \times 10^{-31}\,\text{kg})(0.9)(3 \times 10^{8}\,\text{m/s})} = 1.2 \times 10^{-12}\,\text{m}$

36.17. The resolution of a circular aperture of diameter D is $\theta_{\min} = 1.22(\lambda/D) \approx \lambda/D$. If L is the distance from the electron to the microscope lens, $\Delta x \approx L\theta_{\min} \approx L(\lambda/D)$. When the photon is scattered off the electron, it gives the electron some amount of momentum Δp in the x direction, but the exact amount cannot be known. However, since the photon came into the microscope with momentum h/λ, it may have transferred to the electron x momentum of the order of $\Delta p \approx h/\lambda \sin\theta$. $\sin\theta \approx D/L$, so $\Delta p \approx hD/\lambda L$. Thus $\Delta x \Delta p \approx (L\lambda/D)(hD/\lambda L) \approx h$.

36.18. $\int_{-a}^{a} \psi^2(x)\,dx = 1$, and $\psi(x)$ is an even function, so $2\int_0^a C^2[1 - (x/a)]^2 dx = 1$. Let

$$u = 1 - \frac{x}{a}, \qquad du = -\frac{1}{a}dx$$

$$2C^2 \int_1^0 u^2(-a\,du) = 2C^2 a \int_0^1 u^2\,du = 2C^2 a \frac{1}{3}u^3\Big|_0^1 = 2C^2\frac{a}{3} = 1, \quad \text{so } C = \sqrt{\frac{3}{2a}}$$

The probability of finding the particle between $1/2a$ and a is

$$P = \int_{a/2}^{a} \psi^2\,dx = \frac{3}{2a}\int_{a/2}^{a}\left(1 - \frac{x}{a}\right)^2, \quad dx = \left(\frac{3}{2a}\right)a\int_0^{1/2} u^2\,du = \left(\frac{3}{2}\right)\frac{u^3}{3}\Big|_0^{1/2} = \frac{1}{16} = 0.063$$

36.19. $\int_0^{\infty} \psi^2(x)\,dx = 1 \qquad C^2\int_0^{\infty} e^{-2x/a}\,dx = 1 \qquad -C^2\frac{a}{2}e^{-2x/a}\Big|_0^{\infty} = \frac{1}{2}aC^2 = 1 \qquad C = \sqrt{\frac{2}{a}}$

$$x_{\text{av}} = \int_0^{\infty} x\psi^2\,dx = \frac{2}{a}\int_0^{\infty} xe^{-2x/a}\,dx = \frac{2}{a}\left[\frac{1!}{(2/a)^{1+1}}\right] = \frac{1}{2}a \qquad P(x) = \psi^2(x)$$

$\psi(x)$ decreases with increasing x, so the probability density is a maximum at $x = 0$.

36.20. From Eq. 36.7,

$$E_n = \frac{h^2}{8mL^2}n^2 = \frac{(6.63 \times 10^{-34}\,\text{J}\cdot\text{s})^2}{8(6.64 \times 10^{-27}\,\text{kg})(1 \times 10^{-14}\,\text{m})^2} = 8.28 \times 10^{-14}\,\text{J} = 0.517\,\text{MeV}$$

$$E = \frac{1}{2}mv^2 = \frac{p^2}{2m}, \quad p = \frac{h}{2L} = \frac{6.63 \times 10^{-34}\,\text{J}\cdot\text{s}}{2(1 \times 10^{-14}\,\text{m})} = 3.32 \times 10^{-20}\,\text{kg}\cdot\text{m/s}$$

36.21. For a photon $p = E/c$, so $\Delta p = (1/c)\Delta E$. The uncertainty in position is of the order of the length of the wave train, that is, the distance the light travels in time Δt, so $\Delta x = c\Delta t$. Multiply:

$$\Delta x\,\Delta p = (c\Delta t)\left(\frac{1}{c}\Delta E\right) = \Delta t\,\Delta E, \quad \text{so } \Delta p\,\Delta x > \hbar \quad \text{implies } \Delta t\,\Delta E > \hbar$$

36.22. $T \simeq e^{-2Kx} = \exp\left[-\frac{2}{\hbar}\sqrt{2m(E-U)x}\right]$

$$T = \exp\left(-\frac{2}{1.05 \times 10^{-34}\,\text{J} \cdot \text{s}}\right)\left[\sqrt{2(9.11 \times 10^{-31}\,\text{kg})(1.5\,\text{eV})(1.6 \times 10^{-19}\,\text{J/eV})}\right](2 \times 10^{-9}\,\text{m}) = e^{-25.2}$$

$$T = 1.15 \times 10^{-11} \quad \text{for } U - E = 1.5\,\text{eV}$$

For

$$U - E = 0.5\,\text{eV} \qquad T = 4.8 \times 10^{-7}$$

Reducing the barrier height by a factor of 3 increases the tunneling probability by a factor of 48,000. If the barrier width is reduced to 0.20 nm, $T \simeq e^{-2.52} \simeq 0.08$. Thus increasing the width by a factor of 10 increased T by about 10^{10}.

36.23. $\Delta p\,\Delta x > \hbar, \quad \Delta p = m\Delta v$

For an electron,

$$\Delta v > \frac{\hbar}{m\Delta x} = \frac{1.05 \times 10^{-34}\,\text{J} \cdot \text{s}}{(9.11 \times 10^{-31}\,\text{kg})(1 \times 10^{-11}\,\text{m})} = 1.15 \times 10^7\,\text{m/s}$$

For a ping pong ball,

$$\Delta v > \frac{1.05 \times 10^{-34}\,\text{J} \cdot \text{s}}{(0.0022\,\text{kg})(1 \times 10^{-11}\,\text{m})} = 4.8 \times 10^{-21}\,\text{m/s}$$

Thus the uncertainty in velocity for the electron is appreciable, but it is negligible for a ping pong ball.

36.24. To fall a distance y from rest requires time t, where $y = \frac{1}{2}gt^2$. Suppose the uncertainty in the drop position is Δx_1. Then the uncertainty in the x velocity when the object is dropped is $m\Delta v > \hbar/\Delta x_1$. This velocity will cause the mass to move horizontally a distance

$$\Delta x_2 = \Delta v, \qquad t = \frac{\hbar t}{m\Delta x_1} = \frac{\hbar}{m\Delta x_1}\sqrt{\frac{2y}{g}}$$

Thus

$$\Delta x_1 \Delta x_2 > \frac{\hbar}{m}\sqrt{\frac{2y}{g}}$$

The smallest miss distance occurs for $\Delta x_1 = \Delta x_2 = \Delta x$, so

$$\Delta x = \sqrt{\frac{\hbar}{m}}\left(\frac{2y}{g}\right)^{1/4}$$

36.25. The magnitude of L is $L = \sqrt{\ell(\ell+1)}\hbar = \sqrt{3(3+1)}\hbar = \sqrt{12}\hbar$. The maximum value of $L_z = m_\ell\hbar$ with $m_\ell = 3$ is $3\hbar$. Thus the angle between L and L_z is given by

$$\cos\theta = \frac{L_z}{L} = \frac{3\hbar}{\sqrt{12}\hbar} = \frac{\sqrt{3}}{2}, \quad \theta = 30°$$

36.26. $\Delta E = hf = \frac{hc}{\lambda}$

$$= \frac{hc}{\lambda} = \frac{(6.63 \times 10^{-34}\,\text{J} \cdot \text{s})(3 \times 10^8\,\text{m/s})}{(590 \times 10^{-9}\,\text{m})(1.6 \times 10^{-19}\,\text{J/eV})} = 2.11\,\text{eV}$$

36.27. (a) One $4p$ electron goes to $3p$ ($\Delta\ell = 0$), and one goes from $4p$ to $4s$ ($\Delta\ell = -1$), so net $\Delta\ell = -1$, and this transition is allowed. (b) One electron goes from $4s$ to $3p$ ($\Delta\ell = +1$), so this means a photon is absorbed. This is not possible in going from an excited state (high energy) to lower energy (the ground state), since this requires emission of a photon. This transition would, however, be allowed for absorption of a photon. (c) One electron goes from $4p$ to $4s$ ($\Delta\ell = -1$), so this transition is allowed. (d) One electron goes from $4d$ to $4s$ ($\Delta\ell = -2$), so this transition is forbidden. (e) One electron goes from $5s$ to $4s$ ($\Delta\ell = 0$), so this transition is forbidden.

36.28. From Fig. 36-7. I see the order in which the states are filled. For each value of ℓ there are $2\ell + 1$ orbital states and 2 spin states. Thus for 33 electrons the configuration is $1s^2 2s^2 2p^6 3s^2 3p^6 3d^{10} 4s^2 4p^3$. This is sometimes written $[Ar]3d^{10}4s^2 4p^3$, where $[Ar]$ represents the configuration of argon, a closed shell element.

36.29. From Fig. 36-7 I see the order in which the states are filled. Thus for iron ($Z = 26$), the configuration is $1s^2 2s^2 2p^6 3s^2 3p^6 3d^6 4s^2 = [Ar]3d^6 4s^2$. Note that the $3d$ shell will hold 10 electrons, and instead of $3d^6 4s^2$, you might have expected $3d^8$. Instead, the last two electrons go into the $4s$ shell rather than the $3d$ shell because it is a lower energy state. This has important consequences, since the unfilled $3d$ shell now has a net angular momentum and a net orbital magnetic moment. Thus iron compounds can be magnetic.

36.30. $\Delta E \Delta t \geq h$

$$\Delta E \geq \frac{1.05 \times 10^{-34}\,\text{J} \cdot \text{s}}{12 \times 10^{-15}\text{s}} = 8.75 \times 10^{-21}\,\text{J}$$

36.31. First determine Ψ_0.

$$1 = \int\limits_{-a}^{a} \Psi(x)^2\, dx = \int\limits_{-a}^{0} \Psi_0{}^2 (x + a)^2\, dx + \int\limits_{0}^{a} \Psi_0{}^2 (a - x)^2\, dx$$

$$= \Psi_0{}^2 \frac{(x + a)^3}{3}\bigg|_{-a}^{0} + \Psi_0{}^2 \frac{(x - a)^3}{3}(-1)\bigg|_{0}^{a}$$

$$= \Psi_0{}^2 \frac{a^3}{3} + \Psi_0{}^2 \frac{a^3}{3} = \frac{2}{3}a^3 \Psi_0{}^2$$

$$\Psi_0 = \sqrt{\frac{3}{2}}\frac{1}{a^{3/2}}$$

The probability that the particle is in the region $0 \leq x \leq a/2$ is

$$P_1 = \int\limits_{0}^{a/2} \Psi(x)^2\, dx = \frac{2}{3a^3} \int\limits_{0}^{a/2} (a - x)^2\, dx$$

$$= \frac{2}{3a^3} \frac{(a - x)^3}{3}(-1)\bigg|_{0}^{a/2}$$

$$= \frac{2}{3a^3} \frac{(a/2)^3 - a^3}{3}(-1)$$

$$= \frac{7}{16}$$

CHAPTER 37

Nuclear Physics

The history of science is the story of humans' efforts to make sense out of their surroundings so as to survive and to satisfy that most basic of human traits, intellectual curiosity. The path has led them to look more and more closely at things, observing finer and finer detail. They discovered the existence of atoms, and then the constituents of atoms, namely, electrons and the nucleus. The search led to understanding the composition of nuclei in terms of protons and neutrons, and even these particles seem to be composites of tiny entities called *quarks*. Where it will all end, no one knows for sure. Maybe it is like the doggerel that goes,

> Big fleas have little fleas, that bother 'em and bite 'em.
> And little fleas have littler fleas, and so on, ad infinitum.

The journey goes on. One thing is certain. The ramifications of science have tremendous importance for the life of each of us.

37.1   Properties of the Nucleus

In 1910 Ernest Rutherford carried out experiments that did much to elucidate the structure of the atom. He scattered alpha particles from a thin foil of gold and observed what happened. Alpha particles (charge $+2e$, mass about 4 proton masses) are helium nuclei, and they are given off with energies of several megelectron-volts by naturally radioactive polonium. By using very thin foils, Rutherford reasoned that an alpha would be deflected by only a single atom. Consequently, it was expected that only deflections of a fraction of a degree would be observed. Rutherford was astounded to find that many alphas came straight back! As he said, "It was as if a 15-in shell were fired at a piece of tissue paper and it bounced straight back at you." Rutherford proposed that almost all of the mass of an atom is concentrated in the nucleus, a small, very dense particle with positive charge. Surrounding the nucleus is a thin cloud of negatively charged electrons. The scale is comparable to placing a golf ball (the nucleus) in the center of a football stadium (the atom). We now know that nuclei have the following properties:

- The nucleus is composed of nucleons, of which there are two kinds, neutrons and protons. The proton has positive charge, $e = +1.6 \times 10^{-19}$ C, and the neutron is electrically neutral. Their masses are about 1840 times the electron mass.

  Z, the *atomic number*, is equal to the number of protons in the nucleus.
  N, the *neutron number*, is equal to the number of neutrons in the nucleus.
  A, the *mass number*, is equal to the number of nucleons in the nucleus:

  $$A = N + Z$$

- The atomic number Z identifies a particular element, of which 103 are known. Some of these are unstable and do not occur in nature. A nucleus is represented by a symbol like this: $^{64}_{30}$Zn. This indicates $Z = 30$ and $A = 64$. The symbol Zn is redundant since if $Z = 30$, we know the element must be zinc. A given element can exist in different forms, called **isotopes**. Isotopes of an element all have the same Z but different values of A (and hence N). For example, isotopes of carbon are $^{11}_6C$, $^{12}_6C$, $^{13}_6C$, and $^{14}_6C$. The three isotopes of hydrogen are given names: ^1_1H is hydrogen, ^2_1H is deuterium, and ^3_1H is tritium.

- It is useful to measure nuclear masses in terms of the **unified mass unit u** (also called the *atomic mass unit*). The mass of the $^{12}_6C$ atom is defined as 12 u, so $1\,u = 1.660559 \times 10^{-27}$ kg. Since mass is related to rest energy, $E_0 = mc^2$, rest masses are also listed in units of MeV/c^2. For the proton, $E_0 = (1.67 \times 10^{-27}\,\text{kg})(3 \times 10^8\,\text{m/s})^2$, or $E_0 = 1.50 \times 10^{-10}\,\text{J} = 938\,\text{MeV}$. Useful values are given in Tables 37-1 and 37-2. Note that unified mass units for the elements listed are masses of the atoms (nucleus plus electrons), not just the nucleus.

TABLE 37-1 **Rest Masses**

PARTICLE	MASS, kg	MASS, u	MASS, MeV/c^2
(Mass unit)	1.66057×10^{-27}	1.000	931.494
Proton	1.67262×10^{-27}	1.007276	938.28
Neutron	1.67493×10^{-27}	1.008665	939.57
Electron	9.10939×10^{-31}	5.48579×10^{-4}	0.510999
^1_1H atom	1.67353×10^{-27}	1.007825	938.783
^4_2He	6.64466×10^{-27}	4.001506	3727.38
$^{12}_6C$ atom	1.99265×10^{-26}	12.000000	11,1779

TABLE 37-2 **Atomic Masses**

ELEMENT	ATOMIC MASS, u	ELEMENT	ATOMIC MASS, u	ELEMENT	ATOMIC MASS, u
^1_1H	1.007825	$^{15}_7N$	15.000109	$^{60}_{27}Co$	59.933820
^2_1H	2.014102	$^{16}_8O$	15.994915	$^{92}_{36}Kr$	91.92630
^3_2H	3.016029	$^{23}_{11}Na$	22.989768	$^{141}_{54}Ba$	140.91440
^4_2He	4.002603	$^{24}_{12}Mg$	23.985042	$^{181}_{73}Ta$	180.947992
^6_3Li	6.015121	$^{27}_{13}Al$	26.981539	$^{206}_{82}Pb$	205.974440
^7_3Li	7.016003	$^{35}_{17}Cl$	34.968853	$^{209}_{82}Pb$	208.981065
$^{10}_5B$	10.012937	$^{40}_{20}Ca$	39.962591	$^{210}_{84}Po$	209.982848
$^{12}_6C$	12.000000	$^{55}_{25}Mn$	54.938047	$^{232}_{90}Th$	232.038051
$^{14}_6C$	14.003242	$^{56}_{26}Fe$	55.934939	$^{235}_{92}U$	235.043924
$^{14}_7N$	14.003074	$^{60}_{28}Ni$	59.930788	$^{238}_{92}U$	238.050785

All nuclei are approximately spherical in shape and have about the same density. The radius of the nucleus in terms of r_0, where $r_0 = 1.2 \times 10^{-15}$ m $= 1.2$ fm and 1 fm = 1 fermi $= 10^{-15}$ m, is

$$r = r_0 A^{1/3} \tag{37.1}$$

PROBLEM 37.1.   What is the mass density of the $^{12}_6C$ nucleus?

Solution   Nucleus volume $V = 4/3\pi r^3 = (1.33\pi)(1.2 \times 10^{-15}\,\text{m})^3(12) = 8.66 \times 10^{-44}\,\text{m}^3$. The mass of the $^{12}_6C$ atom is 12.00 u, including electrons. However, the electron mass is very small, so $m \simeq 12\,u = 12 \times 1.66 \times 10^{-27}$ kg. The density is

$$\rho = \frac{m}{V} = \frac{(12)(1.66 \times 10^{-27}\,\text{kg})}{8.66 \times 10^{-44}\,\text{m}^3} = 2.3 \times 10^{17}\,\text{kg/m}^3$$

This is more than 10^{14} times as dense as water! Neutron stars have a density of this magnitude.

37.2   Nuclear Stability and Binding Energy

Protons in the nucleus repel each other because of their like electric charge. The **strong nuclear force** causes protons and neutrons to attract each other with a force sufficient to overcome the electric repulsion. This is a short-range force (it acts over about 2 fm) that is one of the three basic forces in nature. The other basic forces are gravity and the electroweak force. The latter includes both electromagnetic force and another kind of subnuclear force. The strong nuclear force is charge independent.

The **binding energy BE** of a nucleus is the energy required to completely separate the protons and neutrons of which it is composed. The mass of a nuclide is less than the mass of its constituent nucleons by an amount Δm, the **mass defect**, corresponding to a binding energy $\Delta E = \Delta mc^2$. Thus the binding energy of a nuclide $^Z_A X$ is

$$\boxed{\text{Binding energy BE} = \Delta mc^2 = (Zm_H + Nm_N - m_x)c^2} \tag{37.2}$$

Here m_H is the mass of a hydrogen atom, m_N is the neutron mass, and m_x is the mass of an atom of element X. The atomic masses include electron masses, but these cancel out since the first term includes Z electron masses, as does the last term.

PROBLEM 37.2.   What are the binding energies and the binding energy per nucleon for the following: (a) deuterium, $^2_1$H, (b) $^4_2$He, (c) $^{56}_{26}$Fe, and (d) $^{238}_{92}$U? Use Tables 37-1 and 37-2.

Solution

(a) $\Delta m = (1)(1.007825\,u) + (1)(1.008665\,u) - 2.014102\,u = 0.002388\,u$

$\Delta E = (0.002388\,u)(931.5\,\text{MeV/u}) = 2.22\,\text{MeV}$

$\text{BE/nucleon} = 1.11\,\text{MeV}$

(b) $\text{BE} = [2(1.007825\,u) + 2(1.008665\,u) - 4.002603\,u](931.5\,\text{MeV/u})$

$\text{BE} = 28.3\,\text{MeV}$ and $\text{BE/nucleon} = 7.07\,\text{MeV}$

(c) $\text{BE} = [26(1.007825\,u) + (56 - 26)(1.008665\,u) - 55.934939\,u](931.5\,\text{MeV/u})$

$\text{BE} = 492\,\text{MeV}$ and $\text{BE/nucleon} = 8.79\,\text{MeV}$

(d) $\text{BE} = [92(1.007\,825\,u) + (238 - 92)(1.008665\,u) - 238.050784\,u](931.5\,\text{MeV/u})$

$\text{BE} = 802\,\text{MeV}$ and $\text{BE/nucleon} = 7.57\,\text{MeV}$

Note that the binding energy per nucleon is greatest for iron in the above example, and less for the other nuclides. This is typical of a general behavior, as shown in Fig. 37-1. The maximum binding energy per nucleon is about 8.7 MeV/nucleon and occurs near $A \simeq 63$. This "curve of binding energy" has had profound consequences. It suggests how nuclear bombs are made, and it also shows the way to unlimited sources of energy, as we shall see later. Both of these will shape the course of civilization. Since the binding energy per nucleon is greater for $A \approx 50 - 70$ than for the heaviest nuclides, a heavy element like uranium can split apart (undergo **fission**) and yield energy. This is what happens in power reactors and in an atom bomb. Also, light elements with a low binding energy per nucleon can join together (**fusion**) and release energy. This is what happens in a hydrogen bomb and in the Sun.

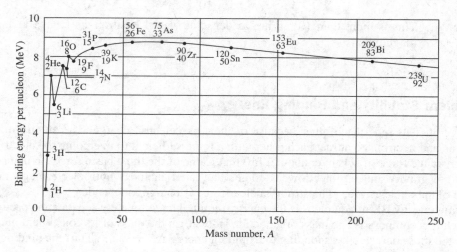

Fig. 37-1

Note that the nuclides $^{4}_{2}$He, $^{8}_{4}$Be, $^{12}_{6}$C, $^{16}_{8}$O, and $^{20}_{10}$Ne all have binding energies per nucleon appreciably higher than the smooth curve through the points for other nuclides. This is a consequence of the fact that a combination of two protons and two neutrons is particularly stable. This combination, which is a $^{4}_{2}$He nucleus, is an alpha particle. When an unstable nucleus spontaneously breaks apart, it often emits alpha particles (alpha decay).

Of the 1500 or so known nuclides, only about 260 are stable. The others spontaneously break apart; that is, they are **radioactive**. For mass numbers up to about $A \approx 40$, $N \approx Z$. For heavier nuclides, the long-range electric repulsion of the protons overcomes the short-range attractive force unless N exceeds Z. In bismuth ($Z = 83$, $A = 209$) the neutron excess is $N - Z = 43$. There are no stable nuclei with $Z > 83$, although some, for example uranium, decay very slowly and so are still present on Earth.

37.3 Radioactivity

When an unstable nuclide breaks apart, it emits radiation. Historically the principal radiations were classified as alpha, beta, and gamma, according to their penetrating power. Alpha particles are helium nuclei, $^{4}_{2}$He, with charge $+2e$. They are not highly penetrating and can be stopped by a sheet of paper or a few centimeters of air. Beta particles are energetic electrons with charge $-e$. They can penetrate several meters of air or a few millimeters of aluminum. Gamma rays are the most penetrating radiation. It takes several centimeters of lead or a meter of concrete to reduce their intensity appreciably. Gamma rays are photons of wavelength shorter than X rays. They have no electric charge. With a magnetic field we can separate the three kinds of radiation, since α's and β's are deflected oppositely, and γ's are not deflected at all.

Radioactive decay occurs randomly. Each decay event is independent of others. The rate of decay is proportional to the number N of nuclei present. Thus

$$\boxed{\frac{dN}{dt} = -\lambda N}$$

(37.3)

If N_0 nuclei are present at $t = 0$, then

$$\int_{N_0}^{N} dN = -\int_{0}^{t} \lambda \, dt \qquad \ln \frac{N}{N_0} = -\lambda t$$

The number of nuclei present after time t is thus

$$N = N_0 e^{-\lambda t} \qquad (37.4)$$

λ is the **decay constant**. It is the probability of a decay per nucleus per second. $R = dN/dt$ is the **decay rate** or **activity**. It also decreases exponentially.

$$R = \left| \frac{dN}{dt} \right| = \lambda N_0 e^{-\lambda t} = R_0 e^{-\lambda t} \qquad \text{activity} \qquad (37.5)$$

The behavior of $N(t)$ or $R(t)$ is shown in Fig. 37-2. The time for the number of nuclei or the activity to decrease by a factor of $\frac{1}{2}$ is the **half-life**, $T_{1/2}$. Thus

$$N = 0.5 \quad N_0 = N N_0 e^{-\lambda T_{1/2}} \quad \text{and} \quad \lambda T_{1/2} = \ln 2 = 0.693$$

$$T_{1/2} = \frac{0.693}{\lambda} \qquad (37.6)$$

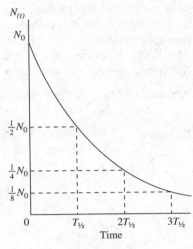

Fig. 37-2

Note that in Fig. 37-2 it does not matter where $t = 0$ is set. In a time $T_{1/2}$ the activity always decreases by $\frac{1}{2}$. After two half-lives, $N = \frac{1}{4}N_0$. After three half-lives, $N = \frac{1}{8}N_0$, and so on. Activity is measured in curies (Ci) or bequerels (Bq). 1 Ci is very large, so mCi or μCi are more frequently used. 1 Ci $= 3.7 \times 10^{10}$ decay/s and 1 Bq $= 1$ decay/s.

PROBLEM 37.3. What is the activity of 1 mg of $^{226}_{88}$Ra (radium)? Its half-life is 1620 y, and its atomic mass is 226 g/mole.

Solution One milligram contains N_0 atoms:

$$N_0 = \frac{(10^{-3}\,\text{g})(6.02 \times 10^{23}\,\text{atom/mol})}{226\,\text{g/mol}} = 2.66 \times 10^{18}\,\text{atoms}$$

$$R = \lambda N_0 = \frac{0.693}{T_{1/2}} N_0 = \frac{(0.693)(2.66 \times 10^{18})}{(1620\,\text{y})(3.17 \times 10^7\,\text{s})} = 3.6 \times 10^7\,\text{Bq} = 9.7 \times 10^{-3}\,\text{Ci}$$

PROBLEM 37.4. Radon $^{222}_{86}$Rn is a radioactive gas that can be trapped in the concrete in the basement of a house. It emits α and γ radiation with a half-life of 3.8 days. If present in high concentration, it is a significant health hazard. Suppose that a health inspector finds an activity level of 2×10^5 Bq in your basement. (a) What will the activity be after 7 days? (b) After 14 days? (c) How long must you wait until the activity has decreased to 1 Bq?

Solution $$\lambda = \frac{0.693}{T_{1/2}} = 0.181/\text{d}$$

(a) $R = R_0 e^{-\lambda t} = (2 \times 10^5\,\text{Bq})e^{-(0.181/\text{d})(7\,\text{d})} = 5.6 \times 10^4\,\text{Bq}$
(b) If $t = 14$ d, then $R = 1.56 \times 10^4$ Bq
(c) $R = 1\,\text{Bq} = (2 \times 10^5\,\text{Bq})e^{-(0.181/\text{d})t}$

$$\ln 1 = \ln(2 \times 10^5) - (0.181t)\ln e, \quad 0 = 12.2 - 0.181t, \quad t = 67.4\,\text{d}$$

Radioactive dating has been invaluable in determining the age of archeological specimens and rocks. The age of a plant or animal that died up to about 40,000 y ago can be determined with radiocarbon techniques. $^{14}_{6}$C undergoes β decay with a half-life of 5730 y. This isotope is present in air with a relative concentration of $^{14}_{6}$C to the abundant $^{12}_{6}$C of about 1.2×10^{-12}. Living organisms constantly exchange CO_2

with their surroundings, but when they die, the exchange stops and the ^{14}C begins to decay. By measuring the present activity, it is possible to estimate how long the carbon has been decaying. This assumes the relative concentration of $^{14}_6$C in the atmosphere stays constant due to nuclear reactions caused by cosmic rays. This has been corroborated by growth rings on old trees and from ice cores in Antarctica. Geological dating for more ancient times uses other isotopes also, including uranium $^{238}_{92}$U, potassium $^{40}_{19}$K, and lead $^{210}_{82}$Pb.

PROBLEM 37.5. In 1991 the remains of a stone age human were found in a glacier in the Italian Alps. He was well preserved and had a $^{14}_6$C activity of 0.12 Bq/g. Estimate the age of the Iceman.

Solution First determine the number of $^{14}_6$C atoms in 1 g of living tissue. The atomic mass of carbon is 12 g/mol, so

$$N_0 = \left(\frac{1\,g}{12\,g/mol}\right)(6.02 \times 10^{23}\,atoms/mol)(1.2 \times 10^{-12})N_0 = 6.0 \times 10^{10}\,atoms$$

$$\lambda = \frac{0.693}{T_{1/2}} = \frac{0.693}{5730\,y} = 1.2 \times 10^{-4}/y = 3.83 \times 10^{-12}/s$$

The initial activity was $R_0 = \lambda N_0 = (3.83 \times 10^{-12}/s)(6.0 \times 10^{10}) = 0.23$ Bq. After time t the activity had decreased to 0.12 Bq, so $0.12B = 0.23e^{-(1.2\times10^{-4}y^{-1})t}$. Take logarithms, and find $t = 5420$ y.

PROBLEM 37.6. $^{60}_{27}$Co is a radioactive isotope commonly used in hospitals for radiation therapy. It has a half-life of 5.24 y. What is the activity of 1 g of cobalt 60?

Solution The number of atoms in 1 g is

$$N = \frac{1\,g}{60\,g}(6.02 \times 10^{23}) = 1.00 \times 10^{22}$$

$$\lambda = \frac{0.693}{T_{1/2}} = \frac{0.693}{(5.24\,y)(3.16 \times 10^7\,s/y)} = 4.2 \times 10^{-9}/s$$

$$R = \lambda N = 4.2 \times 10^{13}\,Bq = 1140\,Ci$$

PROBLEM 37.7. A radioactive isotope has an initial activity of 6 mCi, and 24 h later the activity is 4 mCi. What is the half-life of the isotope?

Solution $$R = R_0 e^{-\lambda t}$$

Take logarithms:

$$\ln\left(\frac{R}{R_0}\right) = -\lambda t, \quad \lambda = -\frac{1}{t}\ln\left(\frac{R}{R_0}\right) = \frac{1}{t}\ln\left(\frac{R_0}{R}\right)$$

$$\lambda = \frac{1}{(24\,h)(3600\,s/h)}\ln\left(\frac{6\,mCi}{4\,mCi}\right) = 4.69 \times 10^{-6}/s$$

$$T_{1/2} = \frac{0.693}{\lambda} = 1.48 \times 10^5\,s = 41.0\,h = 1.71\,d$$

37.4 Radioactive Decay Processes

There are three principal processes of radioactive decay: alpha decay, beta decay, and gamma decay. In order for a decay to occur, the mass of the reaction products must be less than the mass of the original isotope. The energy released in a decay is the **disintegration energy** Q. Thus $Q = $ (original mass)$c^2 - $ (product masses)c^2.

When an **alpha particle** (a helium nucleus, ^4_2He) is emitted, the charge of the resulting nucleus is reduced by $2e$ and its mass number is lower by 4. Examples of such alpha decay are

$$^{238}_{92}\text{U} \rightarrow {}^{234}_{90}\text{Th} + {}^4_2\text{He} \quad \text{and} \quad {}^{226}_{86}\text{Ra} \rightarrow {}^{222}_{84}\text{Rn} + {}^4_2\text{He} \tag{37.7}$$

PROBLEM 37.8. Write the reaction for the alpha decay of radium, $^{210}_{82}\text{Po}$. Determine the energy released. See Table 37-2 for masses.

Solution The atomic number Z is reduced from 84 to 82 in alpha decay. The daughter element is thus lead, Pb. The decay reaction is $^{210}_{84}\text{Po} \rightarrow {}^{206}_{82}\text{Pb} + {}^4_2\text{He}$.

$$
\begin{aligned}
Q &= (m_{\text{Pb}} + m_{\text{He}} - m_{\text{Po}})c^2 \\
&= (205.974440\,\text{u} + 4.002603\,\text{u} - 209.982848\,\text{u})(931.5\,\text{MeV/u}) \\
&= 5.41\,\text{MeV}
\end{aligned}
$$

An interesting quantum mechanical effect occurs in alpha decay. The short-range nuclear force provides a square well that tends to trap the alpha particles in the nucleus, whereas the long-range electric repulsion tends to push the alpha particles out of the nucleus. The resulting potential energy curve $U(r)$ is shown in Fig. 37-3. The peak of the barrier is at about 30 MeV. Classically we would not expect an alpha particle with energy 4–10 MeV to be able to escape the nucleus, but quantum mechanical tunneling provides a finite chance of the alpha particle being emitted. The probability of this happening, and hence the decay rate, can be calculated. This was accomplished in 1928, and it was one of the first major achievements of quantum mechanics.

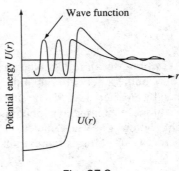

Fig. 37-3

Beta decay is the emission of an electron β^- or a positron (a positive electron β^+). A positron is the **antiparticle** of the electron. Electrons and positrons do not exist in the nucleus and are created at the time of emission. When an electron is emitted, Z increases by $+1$ for the decaying isotope (it loses charge $-e$, so its net charge increases by $+e$). The mass number does not change, since the mass number of an electron is zero. The daughter nucleus will have charge $(Z+1)e$. In order to satisfy the requirements of conservation of energy and angular momentum in beta decay, Fermi found it was necessary to postulate the existence of a new particle, the **neutrino** (the "little neutral one"). This funny character has no electric charge and seems to have no rest mass, so it travels at the speed of light. There are zillions of these little guys buzzing about, and if they have even a tiny mass, maybe they can help provide enough gravitational force to keep the universe from expanding forever, but as of now, this seems unlikely. A neutrino has spin angular momentum $1/2\hbar$. It interacts via a new force, the **weak force**. This force has been shown to be one facet of a more general force, the **electroweak force**, that describes both the weak force and the electromagnetic force. Neutrinos hardly interact with anything, and one can whiz through the Earth without being detected. Their experimental observation was a major accomplishment. Examples of beta decay are the following:

$$n \rightarrow p + {}^{\ 0}_{-1}e + \bar{\nu} \quad \text{and} \quad {}^{13}_{7}\text{N} \rightarrow {}^{13}_{6}\text{C} + {}^{\ 0}_{+1}e + \nu \tag{37.8}$$

$\bar{\nu}$ is an **antineutrino**, the antiparticle to the neutrino. In general, a neutrino is emitted in positron decay and an antineutrino is emitted in electron decay. Unstable heavy elements can undergo a series of radioactive decays. The decay of thorium $^{232}_{90}\text{Th}$ is shown in Fig. 37-4. The half-lives for various steps are shown.

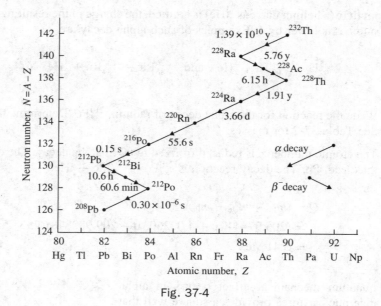

Fig. 37-4

Gamma decay results when a nucleus falls from an excited state to a lower energy state. This usually occurs immediately after an alpha decay or a beta decay that has left the nucleus in an excited state (indicated by an asterisk, X^*). The subsequent gamma emission occurs quickly, usually within 10^{-10} s. The energies of these gamma rays (photons) are typically 1 MeV to 1 GeV, much higher than the energies of visible light photons (2 eV). A nucleus doesn't change when it emits a gamma ray, other than to lower its energy. The decay of boron is representative.

$$^{12}_{5}B \rightarrow {}^{12}_{6}C^* + {}^{0}_{-1}e + \bar{\nu} \quad \text{followed by} \quad {}^{12}_{6}C^* \rightarrow {}^{12}_{6}C + \gamma \tag{37.9}$$

37.5   Nuclear Reactions

If a nucleus is bombarded by a particle such as a proton, neutron, electron, or alpha particle, it may transmutate into another element. The ancient dream of the alchemists has at last come true. Thus if particle a strikes nucleus X, producing nucleus Y and particle b, then

$$a + X \rightarrow Y + b \tag{37.10}$$

This is sometimes written $X(a, b) Y$. The **reaction energy Q** is

$$Q = \Delta mc^2 = (m_a + m_x - m_y - m_b)c^2 \tag{37.11}$$

If $Q > 0$, the reaction is *exothermic* and energy is given off, usually as kinetic energy of the products or gamma rays due to transitions between excited states of Y. If $Q < 0$, the reaction is *endothermic* and the incoming particle must have a certain minimum energy in order for the reaction to occur. If $Q = 0$, the scattering is elastic and the total kinetic energy of the system does not change. Sometimes the isotope produced in the reaction is radioactive and undergoes decay. For example,

$$^{27}_{13}Al + {}^{4}_{2}He \rightarrow {}^{30}_{15}P + n \quad \text{followed by} \quad {}^{30}_{15}P \rightarrow {}^{30}_{14}Si + {}^{0}_{+1}e + \nu \tag{37.12}$$

Such **artificially induced radioactivity** has proven of great practical importance. For example, with an accelerator such as a cyclotron, a short half-life isotope can be produced in a hospital basement and then shot up to an operating room in a pneumatic tube, where it can be used for diagnosis or therapy. It is important to use short-lived isotopes so that the patient does not receive an unacceptably large dose of radiation. Radioactive "tracers" are used in biology and chemistry processes, such as gene mapping. This artificially induced radioactivity has been of great practical importance.

PROBLEM 37.9. Complete the following nuclear reactions.

(a) $^{43}_{20}\text{Ca} \, (\alpha, \, ?) \, ^{46}_{21}\text{Sc}$

(b) $^{55}_{25}\text{Mn} \, (n, \, \gamma)?$

(c) $?(\alpha, \, p)^{17}_{8}\text{O}$

(d) $^{9}_{4}\text{Be} \, (?, \, n)^{12}_{6}\text{C}$

Solution

(a) $? = \, ^{1}_{1}\text{H}$ (proton)

(b) $? = \, ^{56}_{25}\text{Mn}$

(c) $? = \, ^{14}_{7}\text{N}$

(d) $? = \, ^{4}_{2}\text{He}$ (alpha)

37.6 Fission

When some heavy nuclei absorb a neutron, they form an unstable compound nucleus in an excited state. This compound nucleus then splits apart into two roughly equal sized pieces, a process called **fission**. Numerous possible fission paths exist for a given nucleus. For example, for uranium235, two possibilities are

$$^{1}_{0}n + {}^{235}_{92}\text{U} \rightarrow {}^{236}_{92}\text{U}^* \rightarrow {}^{141}_{56}\text{Ba} + {}^{92}_{36}\text{Kr} + 3{}^{1}_{0}n \tag{37.13}$$

$$^{1}_{0}n + {}^{235}_{92}\text{U} \rightarrow {}^{236}_{92}\text{U}^* \rightarrow {}^{140}_{54}\text{Xe} + {}^{94}_{38}\text{Sr} + 2{}^{1}_{0}n \tag{37.14}$$

The energy given off per fission event is about 200 MeV, primarily as kinetic energy of the fission fragments. This is about 10^8 times as great as the energy released per molecule in an ordinary chemical reaction such as burning coal.

PROBLEM 37.10. Determine the energy released in the reaction of Eq. 37.13. See Table 37-2 for masses.

Solution
$$Q = (m_n + m_u - m_{\text{Ba}} - m_{\text{Kr}} - 3m_n)c^2 = (m_u - m_{\text{Ba}} - m_{\text{Kr}} - 2m_n)c^2$$
$$= [235.043924\,\text{u} - 140.91440\,\text{u} - 91.92630\,\text{u} - 2(1.008665)]931.5\,\text{MeV/u}$$
$$= 173\,\text{MeV}$$

The neutrons released in one fission event can be used to induce fission in other nuclei, thereby creating the possibility of a **chain reaction**. A measure of whether a self-sustained chain reaction can occur is the **reproduction constant K** defined *as the average number of neutrons from each fission event that will cause another event*. In an atomic bomb the fission energy is released in an uncontrolled fashion. Controlled release of energy by fission of ^{235}U is used in **nuclear power reactors** to generate electricity. When $K = 1$, the reactor is said to "go critical," and the reaction can continue at a steady rate. If $K < 1$, the reaction dies out. If $K > 1$, the reactor is supercritical and a runaway reaction occurs. Power reactors maintain K close to unity.

There are numerous problems to be overcome to create a chain reaction. First, only about 0.7 percent of naturally occurring uranium is ^{235}U, and 99.3 percent is ^{238}U. It is difficult to utilize ^{238}U, so the reactor fuel has to be enriched up to a few percent of ^{235}U. Some of the neutrons released in fission are absorbed by ^{238}U, and some leak out of the reactor. In order to be most effective for inducing fission, the neutrons must be slowed down ("thermalized"). This is done by letting them collide with a moderator. In Fermi's first chain reaction, carbon was used, but modern reactors usually use heavy water (D_2O). The reaction is prevented from running away and melting down by inserting control rods, for example, cadmium, that strongly absorb neutrons. There are serious problems with using fission reactors for power generation over the long run. The supply of uranium is limited to perhaps 200 y, and the fission products are radioactive and difficult to dispose of. Solar energy and thermonuclear fusion are more promising.

PROBLEM 37.11.   A typical power plant generates about 1000 MW of electricity. If 40 percent of the energy generated by fission is turned into electricity, how much uranium235 is used in 1 y? Assume the energy released per fission event is 200 MeV.

Solution   If the energy generated from fission per year is Q, then $0.4Q = Pt$, and $P = 10^9$ W, $t = 1\,\text{y} = 3.16 \times 10^7$ s. The number of fission events is thus

$$N = \frac{(10^9\,\text{W})(3.16 \times 10^7\,\text{s})(1.6 \times 10^{13}\,\text{MeV/J})}{200\,\text{MeV}} = 2.53 \times 10^{27}$$

$$m = \frac{2.53 \times 10^{27}}{6.02 \times 10^{23}}(235\,\text{g/mol}) = 986\,\text{kg/y}$$

37.7   Nuclear Fusion

In Fig. 37-1 we see that the binding energy per nucleon of light nuclei increases with increasing Z. Thus when two nuclei join together to form a heavier nucleus, energy is released. The energy released per unit mass in this **fusion** reaction is greater than in a fission reaction. However, it is difficult to get two light nuclei close enough together so that the short-range strong nuclear force will grab them and fuse them together. It is like trying to roll a marble up the side of a very steep volcano. If the proton can make it to the top, it will fall a long way down in the center, but first it has to be able to overcome the repulsive electric coulomb barrier. For example, for a proton to approach to within a distance comparable to its diameter ($\sim 2 \times 10^{-15}$ m) to another proton, it must have enough kinetic energy to overcome a potential barrier on the order of

$$U = k\frac{e^2}{r} = \frac{(9 \times 10^9\,\text{N} \cdot \text{m})(1.6 \times 10^{-19}\,\text{C})^2}{(2 \times 10^{-15}\,\text{m})(1.6 \times 10^{-19}\,\text{J/eV})} = 7.2 \times 10^5\,\text{eV} = 0.72\,\text{MeV}$$

One way to achieve the high energies needed is to have a gas at a temperature of about 10^7 K, as is the case in the Sun. In order for a fusion reaction to be maintained in a star, there must be a sufficiently high temperature and a sufficiently high density of nuclei. Under these conditions the gravity of the star holds the reactants together, and radiation pressure and thermal effects keep it from collapsing. In our Sun a three-part proton–proton cycle produces the energy that reaches us.

$$\begin{array}{lll} {}^1_1\text{H} + {}^1_1\text{H} \rightarrow {}^2_1\text{H} + {}^0_1e + \nu & Q = 3.27\,\text{MeV} & (37.15) \end{array}$$

$$\begin{array}{lll} {}^1_1\text{H} + {}^2_1\text{H} \rightarrow {}^3_2\text{He} + \gamma & Q = 5.5\,\text{MeV} & (37.16) \end{array}$$

$$\begin{array}{lll} {}^3_2\text{He} + {}^3_2\text{He} \rightarrow {}^4_2\text{He} + {}^1_1\text{H} + {}^1_1\text{H} & Q = 12.9\,\text{MeV} & (37.17) \end{array}$$

A hydrogen bomb uses uncontrolled fusion ignited with a fission bomb. Major efforts are underway to develop controlled thermonuclear fusion. The most promising reactions for power generation are those utilizing deuterium and tritium.

$$\begin{array}{llll} \text{(D-D)} & {}^2_1\text{H} + {}^2_1\text{H} \rightarrow {}^3_1\text{H} + {}^1_1\text{H} & Q = 4.03\,\text{MeV} & (37.18) \end{array}$$

$$\begin{array}{llll} \text{(D-D)} & {}^2_1\text{H} + {}^2_1\text{H} \rightarrow {}^3_2\text{He} + n & Q = 3.27\,\text{MeV} & (37.19) \end{array}$$

$$\begin{array}{llll} \text{(D-T)} & {}^2_1\text{H} + {}^3_1\text{H} \rightarrow {}^4_2\text{He} + n & Q = 17.6\,\text{MeV} & (37.20) \end{array}$$

The engineering problems involved in producing usable thermonuclear energy are formidable. They fall in three categories:

(1) *High temperature*. Temperatures of about 10^8 K are required. At this temperature a gas is completely ionized (it is a plasma). It is not too hard to heat a rarefied gas to this temperature, but it must not be allowed to come in contact with the walls of the container, or it will immediately cool off.

(2) *Long confinement time*. The particles must be kept in contact long enough for the reaction to occur. It is very difficult to contain an electrically charged gas, and elaborate magnetic field configurations are used. The most successful is a design called a Tokomat. Huge magnetic fields are required, and it is hoped that superconducting magnets may alleviate the problems here.

(3) *High particle density*. This is necessary to increase the collision rate and hence the reaction rate. The Lawson criterion for a self-sustaining fusion reaction is that the ion density n and the confinement time τ should satisfy the condition $n\tau \geq 3 \times 10^{20}$ s/m^3.

So far we have almost reached the break-even point. The reaction generates about as much energy as is required to operate the magnets. This approach to the energy generation problem is appealing for several reasons. There is enough deuterium available in the oceans to last for a very, very long time, and it is relatively cheap. Also, problems with radioactive waste and radioactive contamination are not likely to be as serious for fusion reactors as for fission reactors.

Other fusion schemes are being investigated. **Inertial confinement** zaps a small pellet with an intense laser beam. The idea is to implode the pellet so violently that the constituents pushed inward will react in fusion. So far this hasn't worked, but the struggle goes on. Maybe you can think of something. If so, you would be the savior of all humanity.

PROBLEM 37.12. Deuterium is a stable isotope of hydrogen present with an abundance of about 0.015 percent. Calculate the energy that could be generated from 1 gal (3.8 kg) of seawater using the reaction of Eq. 37.19.

Solution The molecular mass of H_2O is 18 g/mol, so 1 gal contains N molecules.

$$N = \frac{3800\,\text{g}}{18\,\text{g}}(6.02 \times 10^{23}) = 1.27 \times 10^{26} \text{ molecules}$$

Each molecule has two hydrogen atoms, and 0.015 percent of the hydrogen is deuterium. Two deuterium nuclei are needed in each reaction, so the total energy available is

$$Q = (1.27 \times 10^{27})(1.5 \times 10^{-4})(3.27\,\text{MeV})(1.6 \times 10^{-13}\,\text{J/MeV}) = 9.97 \times 10^9 \text{ J}$$

By comparison, 1 gal of gasoline yields about 1.3×10^8 J, so the deuterium in 1 gal of seawater yields as much energy as about 80 gal of gasoline!

37.8 Summary of Key Equations

Reaction energy: $Q = [(\text{mass of reactants}) - (\text{mass of products})]c^2$

Nucleus size: $r = r_0 A^{1/3}$, $\quad r_0 = 1.2 \times 10^{-15}$ m

Radioactive decay: $N = N_0 e^{-\lambda t}$

Activity: $R = \dfrac{dN}{dt} = R_0 e^{-\lambda t}$

Radioactive half-life: $T_{1/2} = \dfrac{0.693}{\lambda}$

SUPPLEMENTARY PROBLEMS

37.13. How many electrons, protons, and neutrons are there in an atom of $^{74}_{32}$Ge?

37.14. A sample from a toxic waste dump is dissolved in $5 \, cm^3$ of water and found to have an activity of $3.0 \, mCi/cm^3$. This sample is diluted to $100 \, cm^3$, and a $10 \, cm^3$ sample of the diluted material is monitored for radioactivity. It is suspected that the radioisotope is ^{64}Cu, with a half-life of 12.7 h. If this is the case, what activity would be expected after 72 h?

37.15. Iodine tends to accumulate in the thyroid gland. One treatment for cancer of the thyroid is to inject the patient with radioactive iodine 131 that kills the cancer cells. Of course, the radiation affects other cells, too, so excessive doses must be avoided. Through biological processes the body excretes iodine exponentially with a half-life of about 7 d, so this process plus the radioactive decay reduces the activity over a period of time. The half-life of I-131 is about 8 d. How long would be required to reduce the activity in the body to 0.1 percent of its initial level?

37.16. Gamma rays are very difficult to shield against. You cannot stop a beam of gamma rays completely. You can only reduce its intensity. The intensity of the beam decreases according to $I = I_0 e^{-x/a}$. The parameter a depends on the absorbing material and is analogous to $1/\lambda$ in radioactive decay. Here I_0 is the intensity at $x = 0$, and I is the intensity after the beam has traveled a distance x. The half-value layer $X = 0.693a$ is analogous to the half-life for radioactive decay. Calculate the thickness of lead to reduce a gamma ray beam intensity to 10 percent and to 1 percent of its incident value for (a) 0.01-MeV gamma rays ($X = 0.00076 \, cm$) and (b) 5-MeV gamma rays ($X = 1.52 \, cm$).

37.17. An amateur detective investigated an old mine site where he suspected some terrorists had attempted to make some kind of nuclear bomb. In occasional visits over a period of months he recorded the following counts per hour on his homemade radiation detector. By looking at the data can you estimate the half-life of the radioisotope? The detective hoped to identify it in this way. (Not fair to use a calculator.)

Time, months	0	2	4	5.5	6	9	10	14	18
Counts per hour	1260	794	500	454	315	158	125	50	20

37.18. In studying the metabolism of a plant, a scientist adds each day $2 \, \mu Ci$ of radioactive $^{32}_{15}$P (half-life 14.3 d) to a solution in which the roots of a seedling are immersed. Assuming the plant removed a negligible amount of the radioisotope, what would be the activity of the solution immediately after the dose was added on the thirtieth day? *Hint:*

$$1 + x + x^2 + \cdots + x^{n-1} = \frac{1 - x^n}{1 - x}$$

37.19. Determine the binding energy per nucleon of $^{12}_{6}$C.

37.20. Complete the following nuclear reactions:

(a) $^{14}_{7}N(\alpha, p)$? (b) $^{2}_{1}H + ^{12}_{6}C \rightarrow ? + ^{10}_{5}B$ (c) $^{1}_{1}H(n, \gamma)$? (d) $^{2}_{1}H + ^{196}_{78}Pt \rightarrow ^{197}_{79}Au + ?$

37.21. Using the conservation laws for electric charge and mass number, determine if each of the following reactions is allowed or forbidden:

(a) $^{39}_{19}K(p, \alpha)^{36}_{17}Cl$ (b) $^{52}_{24}Cr(p, n)^{52}_{25}Mn$ (c) $^{60}_{28}Ni(\alpha, p)^{62}_{29}Cu$

(d) $^{2}_{1}H + ^{63}_{29}Cu \rightarrow ^{63}_{30}Zn + ^{1}_{1}H + ^{1}_{0}n$ (e) $^{7}_{3}Li + ^{1}_{1}H \rightarrow 2^{4}_{2}He$

37.22. $^{235}_{92}$U is unstable and decays in a series of steps, ending as the stable isotope $^{207}_{82}$Pb. How many electrons and how many alphas are emitted in this process?

37.23. A neutron is more effective in causing fission in ^{235}U if its energy is about 0.040 eV, that of a neutron at room temperature. A neutron emitted in fission typically has energy of about 1 MeV. If such a neutron loses half of its kinetic energy in each collision with a moderator nucleus, about how many collisions must it make in order to become thermalized?

37.24. Suppose that in a nuclear fission reactor the reproduction factor K is 1.0005. If the average time between successive fissions in a chain reaction is 1 ms, by what factor will the reaction rate increase in 1 s? By what factor will it increase if K = 2?

37.25. In stars hotter than the Sun, energy is produced mainly by the carbon cycle, a series of six reactions:

$$^{12}_{6}C + ^{1}_{1}H \rightarrow ^{13}_{7}N + \gamma, \quad ^{13}_{7}N \rightarrow ^{13}_{6}C + ^{0}_{1}e + \nu$$

$$^{13}_{6}C + ^{1}_{1}H \rightarrow ^{14}_{7}N + \gamma, \quad ^{14}_{7}N + ^{1}_{1}H \rightarrow ^{15}_{8}O + \gamma$$

$$^{15}_{8}O \rightarrow ^{15}_{7}N + ^{0}_{1}e + \nu, \quad ^{15}_{7}N + ^{1}_{1}N \rightarrow ^{12}_{6}C + ^{4}_{2}He$$

(a) Why is a higher temperature necessary to initiate this sequence than for the proton–proton cycle? (b) Is carbon consumed in this cycle? (c) What is the energy released in the last reaction?

37.26. Two protons, each with energy 10 keV, collide head-on. How close do they approach each other (their speeds are nonrelativistic here)?

37.27. Identify the unknown isotope X in each of the following reactions. You will need access to a periodic table to look up atomic numbers.

(a) $X \rightarrow ^{224}Ra + \alpha$
(b) $^{211}Pb \rightarrow ^{211}Bi + X$
(c) $^{7}Be + e^{-} \rightarrow X + \nu$
(d) $^{210}Po \rightarrow ^{206}Pb + X$

37.28. How many half-lives must pass until (a) 90% of a radioactive sample has decayed? (b) 99% of a radioactive sample has decayed?

SOLUTIONS TO SUPPLEMENTARY PROBLEMS

37.13. The atomic number is 32, so there are 32 protons and 32 electrons in the atom. The mass number $A = 74$, so the neutron number is $N = A - Z = 74 - 32 = 42$.

37.14. Because of the initial dilution and subsequently using 0.1 cm^3 of the solution, the initial intensity is effectively reduced by a factor of $(1/20)(1/10) = 0.005$. The initial activity of 5 cm^3 of solution was

$$(5 \text{ cm}^3)(3 \text{ mCi/cm}^3) = 15 \text{ mCi}$$

If

$$\lambda = \frac{0.693}{12.7 \text{ h}} = 0.055/\text{h}$$

then

$$R = (0.005)(15 \text{ mCi})e^{-(0.055/\text{h})(72 \text{ h})} = 1.5 \times 10^{-5} \text{ mCi}$$

37.15. $R = (R_0 e^{-\lambda_1 t})(e^{-\lambda_2 t}) = R_0 e^{-\lambda t}$, so the effective decay constant is $\lambda = \lambda_1 + \lambda_2$. The effective half-life is thus given by

$$\frac{0.693}{T} = \frac{0.693}{T_1} + \frac{0.693}{T_2}$$

$$T = \frac{T_1 T_2}{T_1 + T_2} = \frac{(8 \text{ d})(7 \text{ d})}{8 \text{ d} + 7 \text{ d}} = 3.73 \text{ d}$$

Thus

$$R = 0.001R_0 = R_0 e^{-\lambda t}$$

$$\lambda t = \left(\frac{0.693}{T}\right)t = \ln(1000), \quad t = \frac{(3.73\,\mathrm{d})(\ln 1000)}{0.693} = 37\,\mathrm{d}$$

37.16. $I = I_0 e^{-x/a}, \quad \dfrac{x}{a} = \ln\left(\dfrac{I_0}{I}\right)$

(a) 0.0025 cm, 5.05 cm,   (b) 0.005 cm, 10.1 cm

37.17. Observe that on month 4 the activity is 500 cph, and on month 10 it is 125 cph, a decrease by a factor of $\frac{1}{4}$. This required two half-lives, so $T_{1/2} = 3$ months.

37.18. The first dose has decayed for 30 d, or 15 half-lives. Its activity has decayed to $(1/2)^{15}$. The second dose has decayed for 14 half-lives and its activity has decayed to $(1/2)^{14}$. Continuing in this way we find the activity after 30 d to be

$$A = A_0\left[\left(\tfrac{1}{2}\right)^{15} + \left(\tfrac{1}{2}\right)^{14} + \cdots + \left(\tfrac{1}{2}\right)^{0}\right]$$

$$A = (2\,\mu\mathrm{Ci})\left[\frac{1 - (1/2)^{16}}{1 - 1/2}\right] = 4\,\mu\mathrm{Ci}$$

37.19. $\mathrm{BE} = (6\,m_\mathrm{H} + 6\,m_p - m_c)c^2$

$= [6(1.007825\,\mathrm{u}) + 6(1.008665\,\mathrm{u}) - 12.000000\,\mathrm{u}](931.5\,\mathrm{MeV/u})$

$= 96.163\,\mathrm{MeV}, \quad \mathrm{BE\ per\ nucleon} = \frac{1}{12}\mathrm{BE} = 7.68\,\mathrm{MeV\ per\ nucleon}$

37.20. (a) $^{17}_{8}\mathrm{O}$ (b) $^{4}_{2}\mathrm{He}$ (c) $^{2}_{1}\mathrm{H}$ (d) $^{1}_{0}n$ (e) $^{4}_{2}\mathrm{He}$

37.21. (a) Not allowed. Charge before is $19 + 1 = 20$; charge after is $17 + 2 = 19$. (b) Allowed. (c) Not allowed. Mass number before $60 + 4 = 64$; after $62 + 1 = 63$. (d) Not allowed. Charge before $29 + 1 = 30$; charge after $30 + 1 = 31$. (e) Allowed.

37.22. The mass number decreases by $235 - 207 = 28$, so 7 alphas (mass number 4) must have been emitted. The electron has zero mass number. The atomic number decreased by $92 - 82 = 10$. Seven alphas have charge $+14e$, so four electrons, each with charge $-e$, were emitted.

37.23. The neutron loses half of its energy in each collision, so it must make n collisions. Thus $(10^6\,\mathrm{eV})(1/2)^n = 0.04\,\mathrm{eV}$, and $(1/2)^n = 4 \times 10^{-8}$. Take logarithms: $n\ln(0.5) = \ln(4 \times 10^{-8})$, and $n = 24.6 \simeq 25$.

37.24. In 1 s there will be 1000 fissions, so the reaction rate will increase by a factor $f = K^{1000}$. $f = (1.0005)^{1000} = 1.65$. If $K = 2$, then $f = 2^{1000} > 10^{300}$! It is important not to let a reactor run away.

37.25. (a) In the proton–proton cycle, two protons must be near each other, each with charge $+e$. In the carbon cycle, a proton must be near a carbon nucleus with charge $+6e$, so it needs more kinetic energy.

(b) The net effect of the sequence is to convert two protons into an alpha particle. No carbon is used up.

(c) $Q = (m_\mathrm{N} + m_\mathrm{H} - m_c - m_\mathrm{He})c^2$

$= (15.000109\,\mathrm{u} + 1.007825\,\mathrm{u} - 12.000000\,\mathrm{u} - 4.002603)(931.5\,\mathrm{MeV/u})$

$= 4.97\,\mathrm{MeV}$

37.26. Energy is conserved, so $\mathrm{KE}_1 + \mathrm{KE}_2 = \mathrm{PE} = k(e^2/r)$:

$$r = \frac{ke^2}{\mathrm{KE}} = \frac{(9 \times 10^9\,\mathrm{J\text{-}m/C^2})(1.6 \times 10^{-19}\,\mathrm{C})^2}{(2 \times 10^4\,\mathrm{eV})(1.6 \times 10^{-19}\,\mathrm{J/eV})} = 72 \times 10^{-15}\,\mathrm{m}$$

37.27. (a) $^{224}_{88}\text{Ra} + ^{4}_{2}\text{He} \rightarrow ^{228}_{90}\text{Th}$

(b) $^{211}_{82}\text{Pb} \rightarrow ^{211}_{83}\text{Th} + e^{-}$

(c) $^{7}_{4}\text{Be} + e^{-} \rightarrow ^{7}_{3}\text{Li} + v$

(d) $^{210}_{84}Po \rightarrow ^{206}_{82}Pb + ^{4}_{2}He$

37.28. $N = N_0 e^{-\lambda t}$

$\lambda = \dfrac{0.693}{T_{1/2}}$

$\ln N = \ln N_0 - \lambda t$

$t = \dfrac{\ln N_0 - \ln N}{\lambda} = \dfrac{\ln(N_0/N)}{0.693} T_{1/2}$

(a) if $\frac{N}{N_0} = 0$, $t = 1.44 T_{1/2}$

(b) if $\frac{N}{N_0} = 0.01$, $t = 6.65 T_{1/2}$

APPENDIX

Physical Constants

QUANTITY	SYMBOL	VALUE
Universal gravitational constant	G	6.67×10^{-11} N · m/kg^2
Boltzmann's constant	k_B	1.38×10^{-23} J/K
Avogadro's number	A	6.02×10^{23} entities/mol
Universal gas constant	R	8.31×10^{3} J/(kmol · K)
Permittivity of free space	ϵ_0	8.85×10^{-12} O^2/(M · m^2)
Coulomb's law constant	$k = \frac{1}{4}\pi\epsilon_0$	8.99×10^{9} N · m^2/O^2
Permeability of free space	μ_0	$4\pi \times 10^{-7}$ T · m/A
Speed of light in vacuum	c	3.00×10^{8} m/s
Elementary charge	e	1.60×10^{-19} O
Electron rest mass	m_e	9.11×10^{-31} kg
Proton rest mass	m_p	1.67×10^{-27} kg
Planck's constant	h	6.63×10^{-34} J · s
	$\hbar = \frac{h}{2\pi}$	1.05×10^{-34} J · s
Atomic mass unit (unified)	u	1.66×10^{-27} kg
Bohr magneton	μ_B	9.27×10^{-24} J/T

Astronomical Data

BODY	MASS, kg	RADIUS, m	ORBITAL RADIUS, m	ORBIT PERIOD
Sun	1.99×10^{30}	6.96×10^{8}	—	—
Moon	7.35×10^{22}	1.74×10^{6}	3.84×10^{8}	27.3 d
Mercury	3.30×10^{23}	2.44×10^{6}	5.79×10^{10}	88.0 d
Venus	4.87×10^{24}	6.05×10^{6}	1.08×10^{11}	224.7 d
Earth	5.97×10^{24}	6.38×10^{6}	1.50×10^{11}	365.3 d
Mars	6.42×10^{23}	3.40×10^{6}	2.28×10^{11}	687.0 d
Jupiter	1.90×10^{27}	6.91×10^{7}	7.78×10^{11}	11.86 y
Saturn	5.69×10^{26}	6.03×10^{7}	1.43×10^{12}	29.42 y
Uranus	8.66×10^{25}	2.56×10^{7}	2.88×10^{12}	83.75 y
Neptune	1.03×10^{26}	2.48×10^{7}	4.50×10^{12}	248.0 y
Pluto	1.50×10^{22}	1.15×10^{6}	5.92×10^{12}	163.7 y

Index

Absolute, 190
Absolute temperature, 202
AC. *See* alternating current.
Acceleration, 7, 20, 21
    average, 21
    constant, 21–26
    instantaneous, 21
    negative, 21
    radial, 40
    second derivative of displacement, 21
    uniform circular, 40
Adiabatic path, 211
Air wedge, 346–347
Algebra, 3–4
    factoring, 4
    logarithms, 4
    quadratic equations, 4
Alkali metals, 394
Alpha particle, 407
Alternating current (AC) circuits, 307–314
    equations, 314
    power, 312
    resonance, 313–314
    single elements, 308–310
    transformers, 307–308
Amperes, 262
Ampere's law, 289–291
Amplitude, 150
    angular, 119
Angle, 15
Angle of incidence, 321
Angle of reflection, 321
Angular acceleration, 119
Angular frequency, 72, 150, 153, 179
Angular momentum, 131–134
    law of conservation, 132
    precession, 133
    torque, 131–132
Angular variables, 1, 118–120
    angular acceleration, 119
    angular velocity, 119
    tangential velocity, 119
Angular variables, relation to linear
    variables, 119
Angular velocity, 72, 119, 153
Antineutrino, 407
Antinode, 182
Antiparticle, 407

Approximations, 10
    bionomial expansion, 10
Archery bow, 96
Archimedes' principle, 171
Astronomical data, 416
Atomic mass unit, 402
Atomic masses, 402
Atomic number, 393
Atomic number Z, 401, 402
Atomic spectra, 376–378
Atoms, 373
    atomic spectra, 376–378
    Bohr's model, 376
    Compton effect, 375–376
    equations, 378
    photoelectric effect, 374–375
Autotransformer, 308
Average acceleration, 21
Average velocity, 20
Avogadro's number, 200
Axis of mirror, 330

Ball mill, 77
Balmer series, 377
Base of logarithms, 4
Beats, 185–186
Bernoulli's equation, 173
Binding energy BE, 403
    fission, 403
    fusion, 404
    mass defect, 403
    radioactive nuclides, 404
Bionomial expansion, 10
Block and tackle, 100
Bohr magneton, 389
Bohr theory, 390
Bohr's correspondence principle, 387
Bohr's model of the atom, 376
    Balmer series, 377
    energy level, 377
    excited state, 377
    ground state, 377
    Lyman series, 377
Boltzmann's constant, 202
Bosons, 393
British thermal units (Btus), 84
Btus. *See* British thermal units.
Bulk modulus, 144

Bungee cord, 96
Buoyancy, 171–172
Buoyancy force, 171
    Archimedes' principle, 171

Calculus, 11–12
    derivative, 11
Capacitance, 253–258
    calculation of, 253–255
    dielectrics, 256–257
    equations, 257
    farads, 253
Capacitive reactance, 309
Capacitors
    combinations of, 255
    energy storage, 256
    parallel connection, 255
    series, 255
Carnot engine, 214
Center of mass (CM), 109–111
Centrifuge, 75, 77
    centripetal, 41
Centripetal acceleration, 40,
        41, 71
    conical pendulum, 72
Centripetal force, 71
Chain reaction, 409
    heavy water, 410
Charge distributions, 236
Charged conductor, 249–250
Chinese windlass, 100
Circular motion, 71–76, 152–153
    angular frequency, 153
    angular velocity, 153
    centripetal acceleration, 71
    equations, 76
    frequency, 71
    radial acceleration, 71
    uniform, 40
Clausius form, 213
CM. *See* center of mass.
Coherent wavelengths, 343
Collisions, 107–109
    elastic, 107, 108
    equations, 112
    inelastic, 107, 108
Common logarithms, 4
Components of vector, 7
Compound microscope, 337
Compressibility, 144
Compton effect, 375–376
Concave mirror, 330
Conduction, 195–196
Conductivity, 263
Conductors, 223
Conical pendulum, 72
Conservation of energy,
        equations, 98

Conservative force, 91
    constant, 21, 36
Constant acceleration, 21–26, 36
Constructive interference, 181, 343
Continuous charge distribution, 227, 247–248
Convection, 196
Converging lenses, 333
Conversion factors, table of, 15
Convex mirror, 330
Correspondence principle, 387
Cosines, 6
Coulomb's law, 235
Cross product. *See* vector product.
Current, 262
    amperes, 262
    equations, 265
    magnetic fields and, 287–289
Current-carrying wire, 280–281
Current density, 263
Cyclic process, 213
Cyclotron frequency, 279
Cyclotron radius, 279
Cyclotron, 409

Damped oscillations, 155
De Broglie waves, 382–383
Decay constant, 405
Decay rate, 405
Degree of freedom, 202
Derivative, 11
    slope, 11
Destructive interference, 181, 343
Dielectric constant, 256–257
Dielectrics, 256–257
    polarization, 256
Diffraction, 351–356
    equations, 356
    grating, 355–356
    resolution, 354–355
    single slit, 351–353
Dimensions, 13
Dirac, P.A.M., 390
Direct current circuits, 268–274
    equations, 274
    multiloop, 271
    open circuit voltage, 268
    RC circuits, 272–274
    resistors, 269–271
Direction, 7
Disintegration energy, 406
Displacement, 20
Displacement current, 309, 318
Displacement vector, 35
Displacement, 7, 20
Diverging lenses, 333
Doppler effect for light, 367–368
    red shift effect, 367

Doppler effect, 186
Double slit interference, 343–344
Drag force, 63
Drag force in air, 87
    due to gravity, 26
Duality of light, 375

Einstein, 360
Elastic collisions, 107
Elastic limit, 143
Elastic modulus, 143
Elasticity, 143–144
    bulk modulus, 144
    compressibility, 144
    elastic modulus, 143
    equations, 145
    shear modulus, 144
    strain, 143
    stress, 143
    tensile stress, 143
Electric charge Q, 244
Electric charge, 223
    permittivity of free space, 223
    principle of superposition, 223
Electric current, 262
Electric dipole, 226
Electric fields, 223–229, 225–226
    conductors, 223
    continuous charge distribution, 227
    dipole, 226
    electric charge, 223
    equations, 229
    insulators, 223
    lines, 225
    motion of charged particle, 227
    negative charge, 225
    positive charge, 225
    quarks, 223
    semi-conductors, 223
    sinks, 225
    uniform, 227
    vector, 225
Electric flux, 234–235
    Coulomb's law, 235
Electric potential E, 244
Electric potential energy U, 244
Electric potential V, 244
Electric potential, 244–250
    charged conductor, 249–250
    continuous charge distributions, 247–248
    equations, 250
    equipotential line, 245
    equipotential surface, 245
    equipotential volume, 245
    finding the field, 246
    point charge, 246
    potential difference, 244
    potential energy, 244–246

Electric power, 264–265
Electromagnetic induction, 295–299
    energy storage, 300–301
    equations, 302
    Faraday's law, 295–298
    homopolar generator, 299
    inductance, 299–300
    magnetic materials, 301
    motional EMF, 298–299
    resonant frequency, 302
    RLC circuits, 301–302
Electromagnetic waves, 318–325
    displacement current, 318
    energy pressure, 319–320
    equations, 324–325
    Maxwell's equations, 318–319
    polarization, 320–321
    radiation pressure, 319–320
    reflection, 321–323
    total internal reflection, 323–324
    wave equation, 319–320
Electromotive force (EMF), 268
Electron diffraction, 383
Electronic configuration, 394
Electrostatic electric field, 236
Electroweak force, 407
EMF. *See* electromotive forces.
Endothermic reactions, 408
Energy, 83
Energy conservation with friction present, 94
Energy level, 377
Energy pressure, 319–320
Energy storage, 256, 300–301
    magnetic fields and, 300–301
Energy transfer, string waves and, 180
Energy, 15, 152
    definition of, 83
    equations, 86
Engine
    Carnot, 214
    gasoline, 215
Entropy, 215–217
    free expansion, 216
Equation of continuity, 172
Equation of motion, 179
Equations, 64, 76, 86, 98, 112, 126, 145, 156, 164,
    174, 187, 206, 218, 229, 240, 250, 257, 274, 283,
    302, 314, 324, 337, 356, 368, 395, 411
Equilateral triangle, 5
Equilibrium, 53, 54–57
    friction forces, 55
    mechanical advantage, 56
    normal force, 54
Equilibrium of forces, 54
Equipotential line, 245
Equipotential surface, 245
Equipotential volume, 245
Excited state, 377

Exothermic reactions, 408
External forces, 54

Factoring, 4
Falling objects, 26
Faraday's law, 295–298
    Lenz'a law, 296
Farads, 253
Fermat's principle, 323
Fermions, 393
Ferromagnetic materials, 301
Finite well, 387
First harmonic, 182
First Law of Motion, 52–53
First law of thermodynamics, 210–213
    adiabatic path, 211
    isobaric path, 211
    isochoric path, 211
    isothermal path, 211
    path independence, 211
Fission, 403, 409–410
    chain reaction, 409
    reproduction constant, 409
Fluids flow, 172
    equation of continuity, 172
    stream lines, 172
Fluids, 169–174
    Bernoulli's equation, 173
    buoyancy, 171–172
    equations, 174
    mass density, 169
    pressure, 169
Focal length, 330
Focal point, 330
Force, 7, 15
Force diagram, 54
Force, friction, 55
Forced oscillations, 155
Frame of reference, 41
Free-body diagram, 54
Free expansion, 216
Freely falling bodies, 26–28
Frequency, 71, 119
    angular velocity, 72
    angular, 72
    hertz, 71
    radians, 72
Frequency of revolutions, 119
Friction forces, 55
    reactive, 55
Friction, 91, 94
    nonconservative force, 91
Fundamental frequency, 182
Fusion, 404

Gamma decay, 408
Gases
    adiabatic process, 204, 205
    degree of freedom, 202

    equations, 206
    kinetic theory of, 203
    Maxwell-Boltzmann distribution, 203–204
    molar specific heat, 204–206
    pressure, 201–203
    root mean square speed, 203
    temperature, 201–203
    theorem of equipartition of energy, 202
Gasoline combustion, 84
Gasoline engine, 215
Gauss' Law, 234–240
    applications of, 236–240
    charge distributions, 236
    electric flux, 234–235
    electrostatic electric field, 236
    equations, 240
    induced electric field, 236
    principle of superposition, 238
    surfaces, 236
Geometry, 5–6
    equilateral triangle, 5
    gravitational, 161
    isosceles triangle, 5
    pythagorean theorem, 5–6
    right triangle, 5
    similar triangles, 5
Grating, 355–356
Gravitational potential energy, 92,
    161–162
Gravity, 160–164
    equations, 164
    Isaac Newton, 160
    motion of planets, 162–164
    potential energy, 161–162
    universal law of, 160
Greek letters, 1
    angular variables, 1
Ground state, 377

Half-life, 405
Hall coefficient, 283
Hall effect, 282–283
Halogens, 394
Heat capacity, 193
    specific, 193
    vapor, 194
Heat engine, 213
    cyclic process, 213
    working substance, 213
Heat pumps, 216
Heat transfer, 195–197
    conduction, 195–196
    convection, 196
    radiation, 196
    Stefan–Boltzmann law, 196
Heat, 193
    latent, 193
    temperature vs., 190
Heavy water, 410

Heisenberg uncertainty principle, 387–389
    wave packet, 388
Henries, 299
Hertz, 71, 150
Homopolar generator, 299
Hooke's law of force, 81
Horizontal motion, 37
Human fly ride, 73

Ideal gas law, 200–201
    Avogadro's number, 200
    mole, 200
    standard temperature and pressure, 201
    universal gasconstant, 200
Impulse, 106
    in simple harmonic motion, 152
Impulse, 106
In phase, 343
Induced electric field, 236
Inductance, 299–300
    equations, 302
    henries, 299
    inductors, 299–300
    self-inductance, 299
Inductive resistance, 309
Inductors, 299–300
Inelastic collisions, 107
Inert gases, 394
Inertial confinement, 411
Inertial frames, 52
Instantaneous acceleration, 21
Insulators, 223
Integrals, table of, 11
Interference, 181, 343–348
    coherent wavelengths, 343
    constructive, 181
        interference, 343
    destructive, 181
        interference, 343
    double slit, 343–344
    in phase, 343
    Michelson interferometer, 347
    multiple slit, 345
    out of phase, 343
    phasors, 345
    thinfilms, 345–347
Interferometer, 347
International System (SI), 13
    kilogram, 13
    length, 13
    mass, 13
    meter, 13
    second, 13
    time, 13
Irreversible processes, 213
Isobaric path, 211
Isochoric path, 211
Isosceles triangle, 5

Isothermal path, 211
Isotopes, 402

Joule heating, 264–265
Joules, 80, 84
Junction rule, 271

Kelvin form, 213
Kelvin scale, 190
Kepler's laws, 162–163
Kilogram, 13
Kinematics, 20
Kinetic energy, 82–83, 104
    law of conservation, 105
    work-energy theorem, 83
Kinetic theory, gases and, 203
Kirchhoff rules, 271

Laminar flow, 172
Latent heat, 193
    fusion, 194
    vaporization, 194
Law of conservation
    angular momentum, 132
    energy, 92
    linear momentum, 105
Law of cosines, 6
Law of sines, 6
Laws of Motion, Newton's, 52–64
Laws of nature, 13
Length contraction, 362–363
Length, 13, 15
Lenses, 332–334, 337
    optical instruments, 335–337
    thin, 332–335
Lenz's law, 296
Lever arm, 123
Linear momentum, 7, 104–105
    equations, 112
    law of conservation, 105
Lines, electric field, 225
Logarithm base, 4
Logarithms, 4
    common, 4
    natural, 4
Longitudinal pressure wave, 182–183
Loop rule, 271
Lorentz transformation equations, 361–362
Loudness, 183
Lyman series, 377

Machines, 97–98
    mechanical advantage, 97
Magnetic field dipole, 282
Magnetic field mass spectrometer, 279–280
Magnetic fields, 278
    current-carrying wire, 280–281
    cyclotron frequency, 279
    cyclotron radius, 279

Magnetic fields (*Continued*)
   energy storage and, 300–301
   equations, 283
   Hall coefficient, 283
   Hall effect, 282–283
   magnetic field dipole, 282
   motion of a charged particle, 279
   sources of, 287–291
      Ampere's law, 289–291
      currents, 287–289
   torque on a current loop, 281
Magnetic materials, 301
   ferromagnetic, 301
   paramagnetic, 301
   relative permeability, 301
Magnification, 331, 334
Magnitude, 7
Mass defect, 403
Mass density, 169
Mass number A, 401, 402
Mass spectrometer, 279–280
Mass, 13, 15
Mathematics review, 1–12
   algebra, 3–4
   approximations, 10
   calculus, 11–12
   geometry, 5–6
   scales, 10
   symbols, 1–2
   trigonometry, 5–6
   vectors, 7–10
Maxima condition, 351
Maxwell's equations, 318–319
Maxwell-Boltzmann distribution,
   203–204
Measurements, 13–17
   dimensions,13
   unit conversion, 14–16
   units, 13–14
Mechanical advantage, 56, 97
Meter, 13
Metric units, 14
Michelson interferometer, 347
Microscope, 337
Mirrors, 329–332
   equations, 337
   plane, 329–330
   spherical, 330–332
Molar specific heat, 204–206
Mole, 200
Moment of inertia, 120–121
   calculations, 121
      parallel-axis theorem, 122
      perpendicular-axis
         theorem, 122
Momentum, 104
   momentum and force,
   104
Motion equations, 28
Motion in a plane, 35

Motion of a charged particle
   electric field and, 227
   magnetic field and, 279
Motion of planets, 162–164
   Kepler's laws, 162–163
Motion, 20–28
   circular, 71–76
   constant acceleration, 36
   displacement vector, 35
   displacement, 20
   equation of, 179
   freely falling bodies, 26–28
   instantaneous velocity, 20
   key equations, 43
   Laws of, 52–64
   multi-dimensional, 35–43
   position vector, 35
   projectiles, 36–40
   relative, 41
   uniform circular, 40
   velocity, 20
Motional EMF, 298–299
Multi-dimensional
   motion, 35–43
Multiloop circuits, 271
   junction rule, 271
   Kirchhoff's rules, 271
   loop rule, 271
Multiple slit interference, 345

Natural logarithms, 4
Negative acceleration, 21
Negative charge, 225
Neutrino, 407
   antineutrino, 407
   electroweak force, 407
   weak force, 407
Neutron number N, 401, 402
Newton's Laws of Motion, 52–64
   equations, 64
   equilibrium, 54–57
   external forces, 54
   First, 52–53
   force diagram, 54
   inertial frames, 52
   nonequilibrium, 57–64
   Second, 53
   Third, 52
Newton's rings, 347
Newton, Isaac, 160
Node, 182
Nonconservative force, 91
   of a spring, 82
Nonequilibrium issues, 57–64
   terminal velocity, 63
Normal force, 54
Normalized wave function, 384
Nuclear fusion, 410–411
   inertial confinement, 411
   thermonuclear fusion, 410–411

Nuclear physics, 401–411
    binding energy BE, 403
    equations, 411
    fusion, 410–411
    nuclear reactions, 408
    nuclear stability, 403
    nucleus properties, 401–403
    radioactive decay processes,
        406–408
    radioactivity, 404–406
Nuclear reactions, 408–409
    artificially induced radioactivity,
        409
    endothermic, 408
    exothermic, 408
    fission, 409–410
    reaction energy, 408
Nuclear stability, 403
Nucleus properties, 401–403
    atomic mass unit, 402
    atomic masses, 402
    atomic number Z, 401, 402
    Ernest Rutherford, 401
    isotopes, 402
    mass number A, 401, 402
    neutron number N, 401, 402
    rest masses, 402
    unified mass unit u, 402

Objective lens, 336
Ohm's law, 263
Open circuit voltage, 268
Optical instruments, 335–337
    compound microscope, 337
    refracting telescope, 336
Order of magnitude estimates,
    16–17
Oscillations, 150–155
    damped, 155
    equations, 156
    forced, 155
    parallel axis theorem, 122
    pendulum, 153–155
    resonance, 155
    simple harmonic motion, 150
Out of phase, 343

Parallel connection, 255
Parallel resistors, 269–271
Parallel-axis theorem, 122
Paramagnetic materials, 301
Particle in a box, 385–387
    Bohr's correspondence
        principle, 387
    zero-point energy, 386
Particle in a finite well, 387
Pascal's principle, 170
Path independence, 211
Pauli exclusive principle, 393
    bosons, 393

fermions, 393
    periodic table, 393–395
Pendulum, 153–155
Perfect conductor, 263
Period, 151
    perpendicular axis theorem, 122
Periodic table, 393–395
    alkali metals, 394
    atomic number, 393
    electronic configuration, 394
    halogens, 394
    inert gases, 394
    shells, 394
    subshells, 394
Permittivity of free space, 223
Perpendicular lever arm, 123
Perpendicular-axis theorem, 122
Phase constant, 150
Phase of the wave, 179
Phase velocity, 179
Phase, 150
Phasors, 311, 345
Photoelectric effect, duality of light,
    374–375
Photons, 373
    equations, 378
    Planck's constant, 373
    quantized energy levels, 373
Physical constants, 416
Physics, 13–17
Pitch, 183
Planck's constant, 373
Plane mirror, 329–330
Plane of polarization, 320
Point charge, 246
Polarization, 256, 320–321
    plane of, 320
    unpolarized wave, 320–321
Position vector, 35
Positive charge, 225
Postulates of relativity, 360
Potential difference, 244
    voltage difference, 244
Potential energy, 91–94, 244–246
    conservative force, 91
    equations, 98
    friction, 91
    law of conservation of energy, 92
    total mechanical energy, 92
Potential energy of a spring, 95–97
Power, 15, 83–86
    alternating current (AC) circuits and,
        312
    joules, 84
Precession, 133
    spin angular momentum, 134
Pressure, 15, 169
    molecular basis, 201–203
    Pascal's principle, 170
Principle of superposition, 223, 238

Projectiles, 36–40
    trajectory, 36
Pythagorean theorem, 5–6
    law of cosines, 6
    law of sines, 6
    radians, 6

Quadratic equations, 4
Quantized energy levels, 373
Quantum mechanics, 382–395
    de Broglie waves, 382–383
    electron diffraction, 383
    equations, 395
    Heisenberg uncertainty
        principle, 387–389
    particle in a box, 385–387
    particle in a finite well, 387
    Pauli exclusive principle, 393
    quantum theory of hydrogen, 390–393
    Schrodinger equation, 383–384
    spin angular momentum, 389–390
    tunneling, 387
Quantum theory of hydrogen, 390–393
    Bohr theory, 390
    radial probability density, 391
    Schrodinger wave equation, 390
    spin magnetic quantum number, 391
Quarks, 223

Radial acceleration, 40, 71
Radial probability density, 391
Radians, 6, 72, 118
Radiation pressure, 319–320
Radiation, 196
Radioactive dating, 405–406
Radioactive decay processes, 406–408
    alpha particle, 407
    antiparticle, 407
    disintegration energy, 406
    gamma decay, 408
    neutrino, 407
Radioactive nuclides, 404
Radioactivity artificially
    induced, 409
    cyclotron, 409
Radioactivity, 404–406
    decay constant, 405
    decay rate, 405
    half-life, 405
Rayleigh criterion, 354
RC circuits, 272–274
Reaction energy, 408
Reactive forces, 55
Red shift effect, 367
Reference, frame of, 41
Reflection, 321–323
    angle of incidence, 321
    angle of reflection, 321

    index of refraction, 321
    refraction, 321, 322
    total internal, 323–324
Refracting telescope, 336
    objective lens, 336
Refraction, 321, 322
    Fermat's principle, 323
    Snell's law, 322
Refrigerators, 216
Relative motion, 41–43
    frame of reference, 41
Relative permeability, 301
Relativistic energy, 366–367
Relativistic momentum and force, 364–365
Relativistic velocity transformation, 363–364
Reproduction constant, 409
Resistance, 263
    conductivity, 263
    current density, 263
    equations, 265
    Ohm's law, 263
    perfect conductor, 263
Resistors in parallel, 269–271
Resistors in series, 269–271
Resistors, 269–271, 354–355
Resolution, Rayleigh criterion, 354
Resonance, 155
    alternating current (AC) circuits
        and, 313–314
Resonant frequencies, 182, 302
Resonant modes, 182
Rest masses, 402
Restoring force, 81
Resultant vector, 7
Reversible processes, 213
Revolutions, frequency of, 119
Right triangle, 5
Right-hand rule, 9
RLC circuits, 301–302, 310–312
Rockets, 111–112
    rotational, 120
    thrust, 112
Rolling, 125
Root mean square speed, 203
Rotational and linear motion compared, 119
Rotational equilibrium, 140–143
Rotational kinetic energy, 120–121
Rotational motion, 118–127
    angular variables, 118–120
    equations, 126
Rotational power, 125
Rotational work and power, 125–126
Rotational work, 125
Rutherford, Ernest, 401

Scalar product, 8
Scalar quantity, 7
Scalars, 7

Scales, 10
Schrodinger equation, 383–384
    normalized wave function, 384
    stationary state, 384
    wave function, 384
Schrodinger wave equation, 390
Scientific notation, 2
Second derivative of displacement, 21
Second Law of Motion, 53
    equilibrium, 53
    weight, 53
Second law of thermodynamics, 213–214
    Carnot engine, 214
    Clausius form, 213
    Kelvin form, 213
Sedimentation, 76
Sedimentation velocity, 66
Self-inductance, 299
Semi-conductors, 223
Series connection, 255
Series resistors, 269–271
Series RLC circuits, 310–312
Shear modulus, 144
Shells, 395
SHM. *See* simple harmonic motion.
SI system. *See* International System.
SI unit system, 13
Significant figures, 2
Similar triangles, 5
Simple harmonic motion
        (SHM), 150
    circular motion, 152–153
    energy, 152
    period, 151
    phase constant, 150
Simultaneity, 360–361
Sines, 6
Single element alternating circuits, 308–310
    capacitive reactance, 309
    displacement current, 309
    inductive resistance, 309
    phasors, 311
    RLC circuits, 310–312
Single slit diffraction, 351–353
    maxima, 351
Sinks, electric field, 225
Slope, 11
Snell's law, 322
Sound frequency, 183
Sound intensity, 183
Sound waves
    longitudinal pressure wave, 182–183
    loudness, 183
    pitch, 183
    sound frequency, 183
    sound intensity, 183
    standing, 184–185
Space habitat, 75

Special relativity, 360–368
    Lorentz transformation equations, 361
    Doppler effect for light, 367–368
    Einstein, 360
    equations, 368
    length contraction, 362–363
    postulates, 360
    relativistic
        energy, 366–367
        momentum and force, 364–365
        velocity transformation, 363–364
    simultaneity, 360–361
    time dilation, 362
Specific heat, 193
Speed, 15, 20
    string waves and, 180
Spherical mirror, 330–332
    axis, 330
    concave, 330
    convex, 330
    focal length, 330
    focal point, 330
    magnification, 331
    virtual images, 330
Spin angular momentum, 134, 389–390
    Bohr magneton, 389
Spin magnetic momentum, P.A.M. Dirac, 390
Spin magnetic quantum number, 389, 391
Spring Constant, 81
Spring, potential energy of, 95–97
Standard temperature and pressure, 201
Standing modes
    first harmonic, 182
    fundamental frequency, 182
Standing sound waves, 184–185
Standing waves, 181–182
    antinode, 182
    node, 182
    resonant frequencies, 182
    resonant modes, 182
Statics, 140–143
    equations, 145
    rotational equilibrium, 140–143
Stationary state, 348
Stefan–Boltzmann, 196
Strain, 143
Stream lines, 172
    laminar, 172
    turbulent, 172
Stress, 143, 145
String waves, 180–181
    energy transfer, 180
    speed, 180
Subscripts, 1
Subshells, 394
Superposition of waves, 181
    interference, 181
    superposition, 181

Symbols, 1–2
  Greek letters, 1
  scientific notation, 2
  significant figures, 2
  subscripts, 1
Systeme International unit
    system, 13
  table, 121

Tangential acceleration, 36
Tangential velocity, 119
Telescope, 336
Temperature, 190–193
  absolute, 190, 202
  heat vs., 190
  Kelvin scale, 190
  molecular basis, 201–203
  thermal equilibrium, 190
  thermal expansion, 191–192
  triple point of water, 190
  zeroth law of thermodynamics, 190
Tensile stress, 143
  elastic limit, 143
Tension, 52
Terminal velocity, 63
  to stretch a spring, 82
Theorem of equipartition of energy, 202
Thermal energy, 193
Thermal equilibrium, 190
Thermal expansion, 191–192
  volume expansion coefficient, 192
Thermodynamic processes
  entropy, 215–217
  equations, 218
  gasoline engine, 215
  heat engine, 213
  heat pumps, 216
  irreversible, 213
  laws of, 210–218
  refrigerators, 216
  reversible, 213
Thermonuclear fusion, 410–411
  issues with, 411
Thin films, interference in, 345–347
  air wedge, 346–347
Thin lenses, 332–335
  converging, 333
  diverging, 333
  magnification, 334
Third Law of Motion, 52
  tension, 52
Thrust, 112
Time dilation, 362
Time, 13, 15
Torque, 123–124, 131–132
  perpendicular lever arm, 123
Torque on a current loop, 281
Total internal reflection, 323–324
Total mechanical energy, 92

Trajectory, 36
  horizontal motion, 37
  vertical motion, 37
Transformers, 307–308
  autotransformer, 308
Transverse mechanical waves, 178–180
  angular frequency, 179
  equation of motion, 179
  phase of the wave, 179
  phase velocity, 179
  velocity, 178
Transverse sinusoidal traveling wave, 178
Triangle
  equilateral, 5
  isosceles, 5
  right, 5
  similar, 5
Trigonometry, 5–6
Triple point of water, 190
Tunneling, 387
Turbulent flow, 172

Unified mass unit u, 402
Uniform circular motion, 40
  centripetal acceleration, 40, 41
  radial acceleration, 40
Uniform electric field, 227
Unit conversion factors, 14–16
  angle, 15
  energy, 15
  force, 15
  length, 15
  mass, 15
  power, 150
  pressure, 15
  speed, 15
  time, 15
  volume, 15
  work, 15
Units of measurement, 13–14
  common prefixes, 14
  International System, 13
Universal gas constant, 200
Universal hoist (pulley), 97
Universal law of gravity, 160
Unpolarized wave, 320–321

Vapor, 194
Variables, 1
Vector field, 225
Vector multiplication
  scalar product, 8
  vector product, 9
Vector multiplication, 8
Vector product, 9
  right-hand rule, 9
Vector quantity, 7
  direction, 7
  magnitude, 7

Vectors, 7–10
  acceleration, 7
  components, 7
  displacement, 7, 35
  force, 7
  linear momentum, 7
  position, 35
  quantity, 7
  resultant, 7
  scalar quantity, 7
  tangential acceleration, 36
  velocity, 7
Velocity, 7, 20
  acceleration, 21
  average, 20
  speed, 20
Vertical motion, 37
Virtual images, 330
Voltage difference, 244
Volume expansion coefficient, 192
Volume, 15

Watt, 84
Wave equation, 319–320
Wave function, 384
Wave packet, 388
Waves, 178–187
  beats, 185–186

Doppler effect, 186
  equations, 187
  sound, 182–185
  standing, 181
  string, 180–181
  superposition of, 181
  transverse
    mechanical, 178–180
    sinusoidal traveling wave, 178
  velocity of, 178
Weak force, 407
Weight, 53
Weightlessness, 59
Windlass, 98
Work, 15, 80–82
  equations, 86
  Hooke's law force, 81
  joule, 80
  restoring force, 81
  spring constant, 81
Work-energy theorem, 82–83
Working substance, 213

Young's modulus, 143
Young, Thomas, 343–344

Zero-point energy, 386
Zeroth law of thermodynamics, 190